Stereochemistry of Organometallic and Inorganic Compounds

Volume 1

Edited by

IVAN BERNAL

Department of Chemistry, University of Houston, Houston TX 77004, U.S.A.

Elsevier

Amsterdam – Oxford – New York – Tokyo 1986

ELSEVIER SCIENCE PUBLISHERS B.V.
Sara Burgerhartstraat 25
P.O. Box 211, 1000 AE Amsterdam, The Netherlands

ELSEVIER SCIENCE PUBLISHING COMPANY INC.
52 Vanderbilt Avenue
New York, NY 10017, U.S.A.

Library of Congress Cataloging-in-Publication Data
Main entry under title:

Stereochemistry of organometallic and ingorganic
and inorganic compounds.

Includes index.
1. Stereochemistry. 2. Organometallic compounds.
3. Chemistry, Inorganic. I. Bernal, Ivan.
QD481.S763 1986 541.2'23 86-1987
ISBN 0-444-42604-3 (U.S. : set)
ISBN 0-444-42605-1 (U.S. : v. 1)

ISBN 0-444-42605-1 (Vol. 1)
ISBN 0-444-42604-3 (Series)

Printed in The Netherlands

INTRODUCTION TO THE SERIES

Early in the nineteenth century, Mitscherlich, an important chemist and pioneer crystallographer, proposed his Law of Isomorphism which stated that compounds having the same number of related atoms will be isomorphous. When the existence of normal and racemic tartaric acids was confirmed by Berzelius, who could not explain this anomaly, Mitscherlich was asked to carry out a crystallographic study of the two since his Law demanded that they be isomorphous. A number of tartrate salts were studied and the results neither solved the riddle of the two tartaric acids nor verified the Law of Isomorphism, In 1844, however, Mitscherlich startled Biot by asking him to publish a memoir in the French Academy Proceedings in which he reported that the salts of both acids having composition $Na(NH_4)C_4O_6.4H_2O$ were isomorphous even if one rotated polarized light and the other did not. In 1848 Pasteur repeated Mitscherlich's experiment and launched stereochemistry on its modern path by providing an explanation for both the phenomenon now referred to as conglomerate crystallization and the riddle of the existence of the not two, but three tartaric acids. Other researchers, such as Van't Hoff and Le Bel, soon followed despite such caustic derision as Kolbe's famous appraisal of Van't Hoff's proposals: "a Dr. van't Hoff, of the Veterinary College, Utrecht, appears to have no taste for exact chemical research. He finds it less arduous to mount his Pegasus (evidently borrowed from the Veterinary College) and to soar to his Chemical Parnassus, there to reveal in his *La Chimie dans l'Espace* how he finds atoms situated in the world's space."

The speed at which stereochemistry grew in the century following Pasteur's experiment was limited by the rate at which methods, largely chemical, could reveal the awesome complexity and beauty of *La Chimie dans l'Espace*. The developments in the fields of organometallic and inorganic chemistry, being to a great extent a consequence of the Second World War's needs and technology, proceeded at a rate heretofore unheard of, whereby an enormous amount of physical insight was introduced into these areas of chemistry. Again, as in the case of Berzelius asking for physical measurements to come to the aid of chemistry, spectroscopists, crystallographers and others responded to similar requests with a torrent of information of ever increasing quality, made possible by the introduction of computer operated and controlled machinery.

To provide a forum for the analysis of the modern avalanche of information concerning the ways in which stereochemistry affects other properties of organometallic and inorganic compounds, Elsevier Science Publishers agreed to launch a series of review volumes whose aim is to interpret such relationships for the chemical community. We would welcome your response to the aims of the series and your evaluation of Volume 1 in order to better understand your needs and respond to them in subsequent volumes.

TABLE OF CONTENTS

CHAPTER 4

STEREOCHEMISTRY OF THE BAILAR INVERSION AND RELATED METAL ION SUBSTITUTION REACTIONS

W.G. Jackson

CHAPTER 5

STEREOCHEMISTRY OF ACETYLENES COORDINATED TO COBALT

G. Pályi, G. Váradi and L. Markó

CONTRIBUTORS TO THIS VOLUME

A. Nakamura, K. Tatsumi and H. Yasuda
Department of Macromolecular Science
Faculty of Science
Osaka University
Toyonaka
Osaka 560
Japan

A.G. Nord
Section of Mineralogy
Swedish Museum of Natural History
P.O. Box 50007
S-104 05 Stockholm 50
Sweden

C. Bianchini, C. Mealli, A. Meli and M. Sabat
Instituto per lo Studio della Stereochimica ed Energetica
dei Composti di Coordinazione
Via F.D. Guerrazzi 27
50132 Firenze
Italy

W.G. Jackson
Department of Chemistry
Faculty of Military Studies
University of New South Wales
Duntroon, A.C.T. 2600
Australia

G. Pályi, G. Váradi and L. Markó
Research Group for Petrochemistry of the Hungarian Academy of Sciences
H-8200 Veszprém
Schönherz Z.u.8.
Hungary

CHAPTER 1

STEREOCHEMISTRY OF 1,3-DIENE COMPLEXES AND THE STERIC COURSE OF THEIR REACTIONS

AKIRA NAKAMURA,* KAZUYUKI TATSUMI, and HAJIME YASUDA

Department of Macromolecular Science, Faculty of Science, Osaka University, Toyonaka, Osaka 560 (Japan)

I. INTRODUCTION

1.1. Scope

1,3-Diene complexes have been prepared with a wide range of d-transition metals, and with some f-elements as well. Group 8 transition metal complexes are ubiquitous, and good reviews and books describing them are available [1-3]. Therefore, we here summarize mainly recent advances in this field, focussing our attention on diene complexes of early transition metals.

The stereochemistry of coordinated dienes has long been limited to a classical η^4-s-cis form, until the recent discovery of Cp_2Zr(s-trans-diene) [4-8]. Also, a coordination mode of s-cis diene to early transition metals has recently been found different from the classical η^4-type. These new findings have prompted us to revive the chemistry of 1,3-diene complexes. The stereochemical disposition around the metal atom becomes of interest because many bis- or tris-diene complexes can be prepared of the early transition metals with which many coordination sites are available.

In addition to the structural interest, the reactivity of some of the diene complexes of early transition metals have been investigated to find stereochemically selective reactions with many unsaturated organic compounds [7]. The stereochemistry of diene coordination is important in their own right and also because of their ability to control the stereochemistry of organic products resulting from stoichiometric and catalytic reactions. For example, s-cis zirconium complex, Cp_2Zr(diene), gives oxametallacycles with cis-C-C double bonds on reaction with ketones or aldehydes [6]. The s-trans diene complex may well be an intermediate of the catalytic diene polymerization leading to 1,4-trans polymers.

The nature of diene-metal bonding has been found to be dictated by a delicate balance of steric and electronic factors of metal atom, as well as the diene. An MO analysis of the origin of stereochemical differences among various metals revealed the importance of d-electron configuration and the energy levels [8]. In particular, donation/back donation relationships play crucial roles in the determination of the stereochemistry of the diene-metal bonding and also for the steric course of the reactions. These recent advances in the field of early transition metal chemistry have allowed us to gain a general view on the bonding and stereochemistry of the whole range of d-block transition metals.

Cyclic diene complexes have been investigated for many different cyclic systems and often exhibit intriguing stereochemical problems, e.g. remarkable non-rigidity for tricarbonyl(η^4-cyclooctatetraene)iron. These are not treated in this article since the open-chain analogues are now more important due to the variety of coordination modes and of the stereoselectivity of reactions. Due to the ease of preparation and air-stability, 1,3-diene complexes of iron(0) are most prevalent than others, but this area has been comprehensively reviewed by R. B. King in 1978 [1].

1.2. Structural Types

Homoleptic diene complexes with two or three diene ligands are rare. Considering the 18 electron rule, one might expect that a bis(diene)metal(0) with d^{10} metals, e.g. $Ni(C_4H_6)_2$ could be isolable, but finds none at present. Only bis(2,3-dimethylbutadiene)nickel(0) has been characterized by ^1H NMR [9]. Tris(diene)metal(0) has been prepared for Group 6 elements [10]. Here, the stereochemistry around the metal is an idealized trigonal prismatic structure

(C_{3h}) for Mo and W, though the refinement of their X-ray structures is unsatisfactory [10].

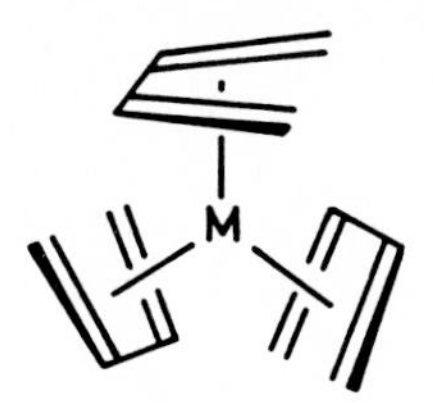

Mixed ligand bis-diene complexes are many. Auxiliary ligands such as CO, PR_3, Cp, Cl, diphosphine or η^3-allyl, are typically found in complexes such as $Fe(CO)(diene)_2$ [11], $Mn(PR_3)(diene)_2$ [12], $RhCl(diene)_2$ [13], $Ir(CO)(diene)_2$ [14] etc. The stereochemical disposition in these complexes deserves

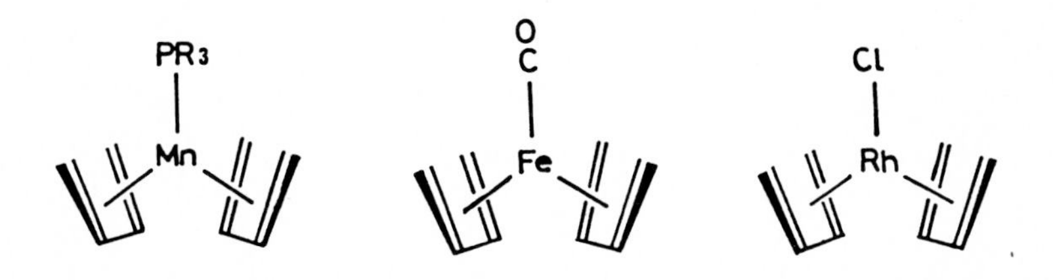

mentioning. For example, $Zr(C_4H_6)_2(dmpe)$ has a distorted octahedral structure in which the terminal carbons of two diene ligands point away from the dmpe [15]. On addition of a neutral ligand, the Zr complex becomes a 7-coordinate molecule with a pentagonal bipyramidal structure in which two dienes are in the equatorial plane [16]. $Cr(CO)_2(diene)_2$ has two diene ligands with unknown orientation [17].

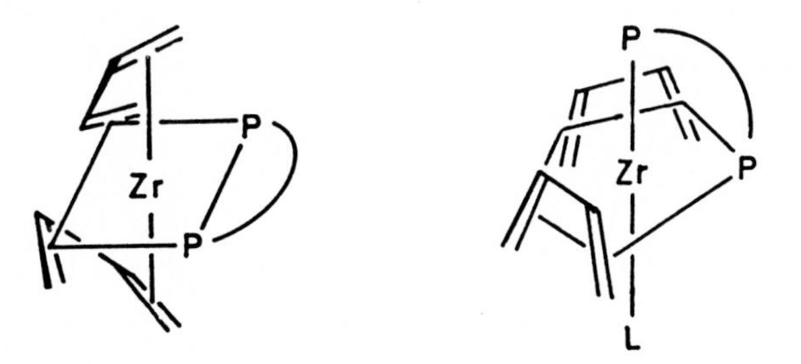

Bis-diene complex formation is facilitated by connecting the two diene parts by a suitable carbon chain. Thus, Rh(I) and Ru(0) complexes shown below have been prepared [18,19] and their structures are in accord with the stereochemical preference of these metals for coordination of two diene ligands.

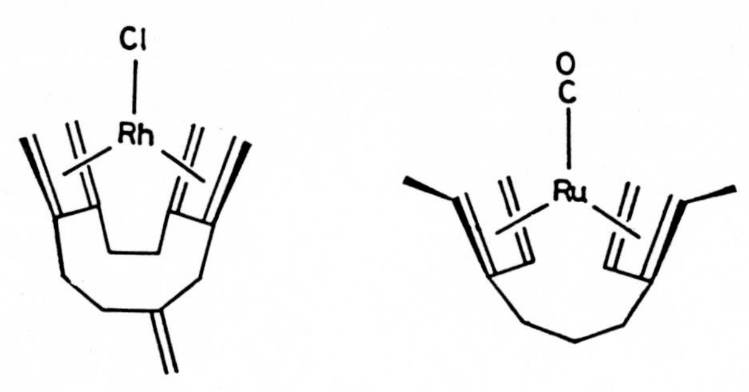

We see a variety of mono-diene complexes among the whole d-transition metal series. Illustrative examples are shown in Fig. 1. Most of the complexes obey the 18 electron rule.

Apart from these 18 electron complexes there are labile 16-electron diene complexes, e.g. $Ni(C_4H_6)(PR_3)$; $[RhCl(diene)]_2$, these are important intermediates in catalytic diene oligomerizations. Similar 16 electron complexes have recently been found for early transition metals, e.g. $CpMCl_2(C_4H_6)$, (M=Nb,Ta) [28,29].

The structural diversity shown in Figure 1 for the mononuclear complexes obscures the stereochemical features of the diene complexes, but the recent characterization of many new diene complexes of Zr and Ta has allowed us to obtain a clearer picture.

In addition to mononuclear complexes, some diene complexes of di- or tri-nuclear structures are known and they add a further variation to the family of 1,3-diene complexes (Figure 2). A μ-type s-trans diene coordination has been found in $Mn_2(CO)_8$(diene) [31], $Os_3(CO)_{10}$(diene) [32a], and $Fe_2(CO)_8$(diene) [32b]. Preference of equatorial coordination is noted for the Os cluster, but an axial-eq. chelation occurs in its s-cis isomer. The s-trans conformation is also favored in the 1,2-η^2-3,4-η^2 type complexes.

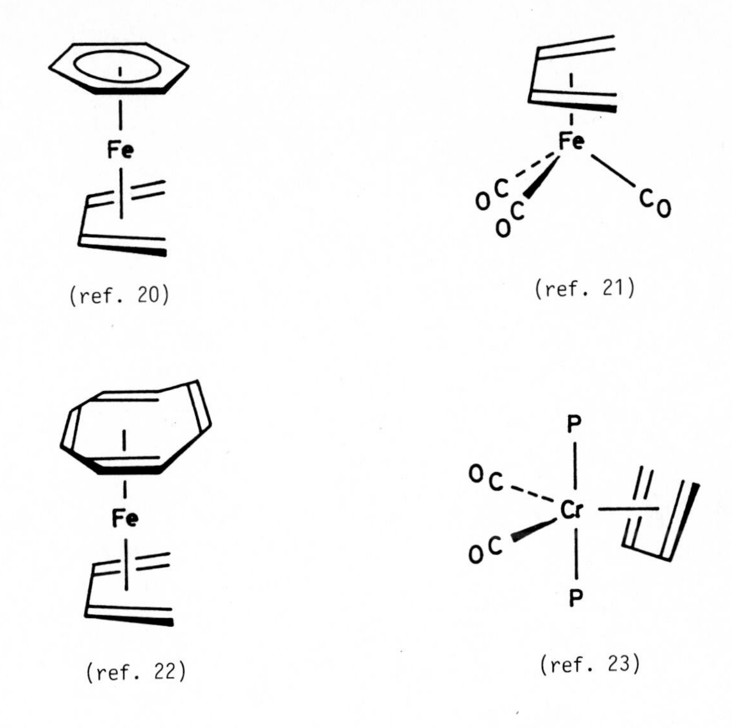

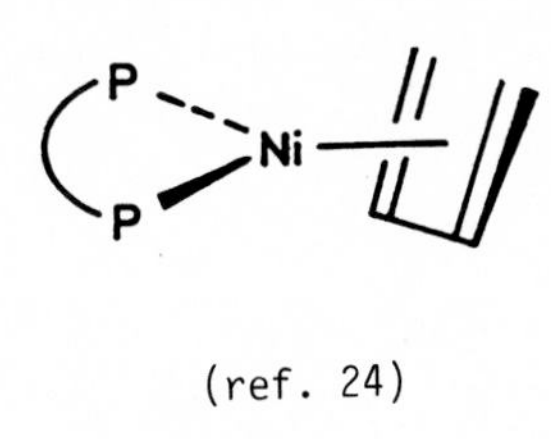

(ref. 24)

Figure 1. Structures of mono-nuclear diene complexes.

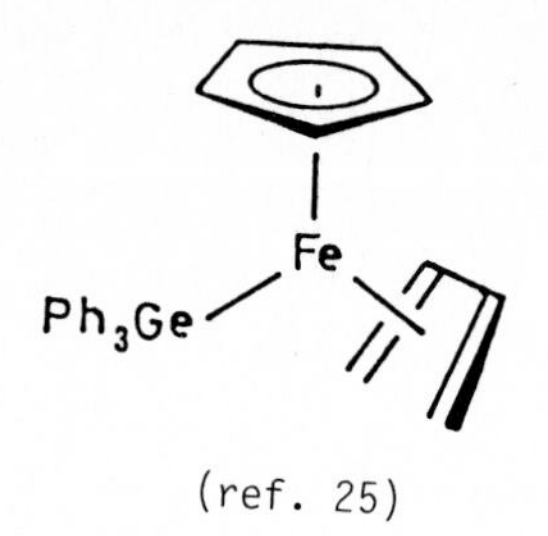

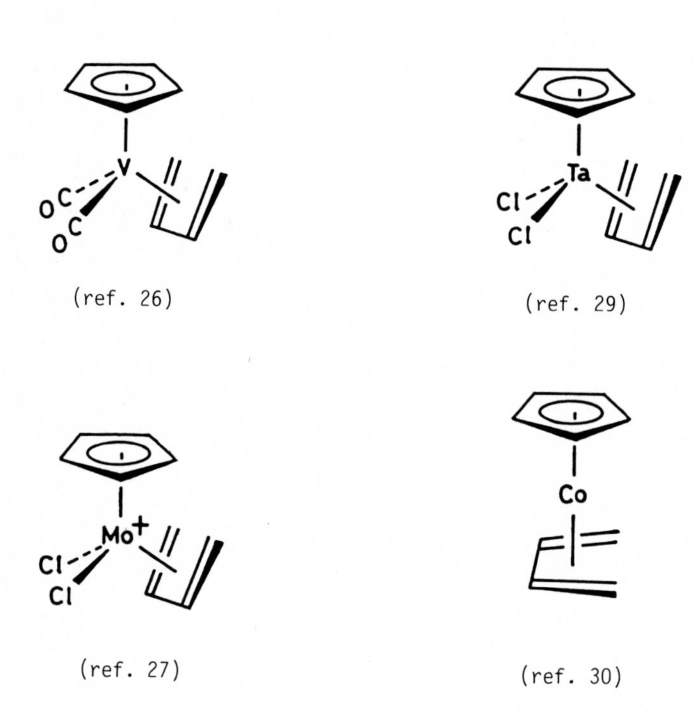

Figure 1. (Continued)

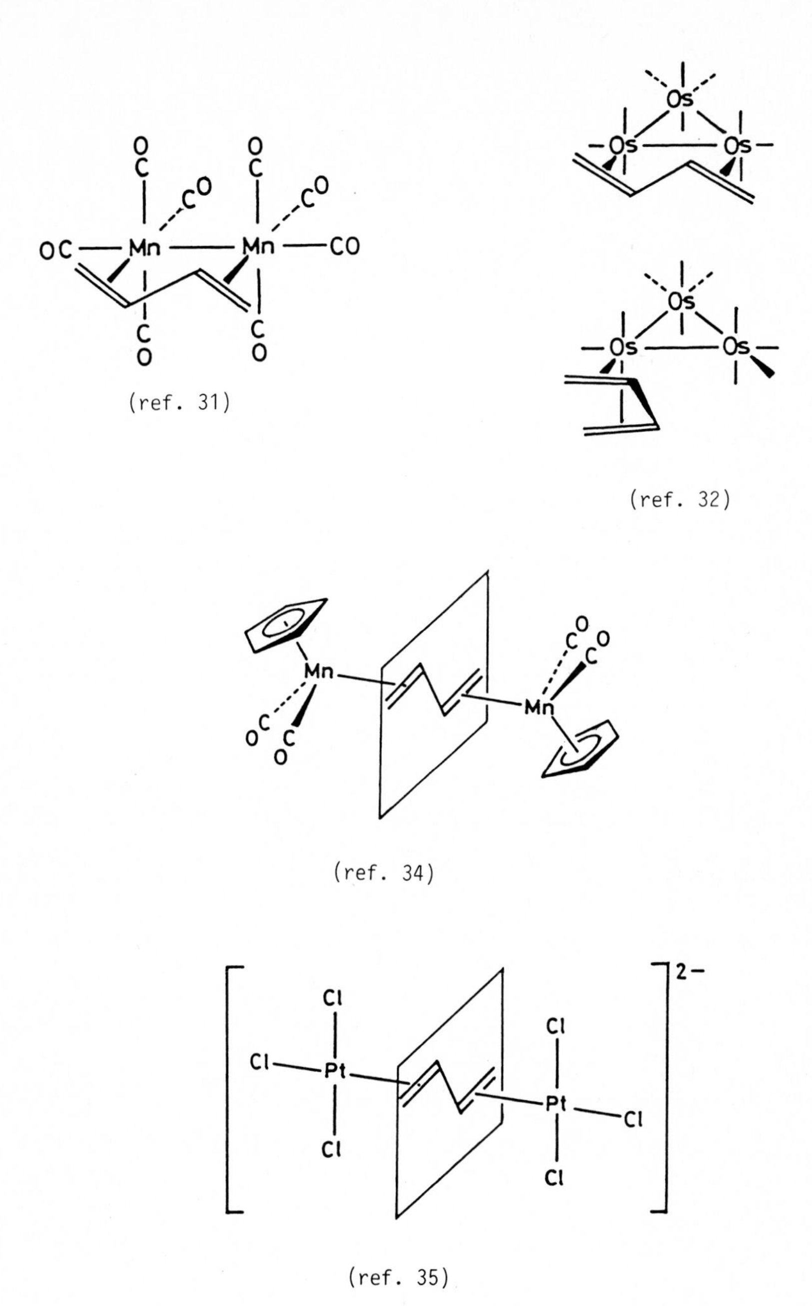

Figure 2. Structures of di- and tri-nuclear diene complexes.

II. STEREOCHEMISTRY OF COORDINATION

2.1. Mode of Coordination

1,2-η^2-Coordination

Occasionally only half of the 1,3-diene participates in bonding with a metal, leaving the other half uncoordinated. This mode of binding has been found for $Fe(CO)_4(\eta^2\text{-}C_4H_6)$ [32b], the structure of which is analogous to the corresponding olefin or alkyne complexes.

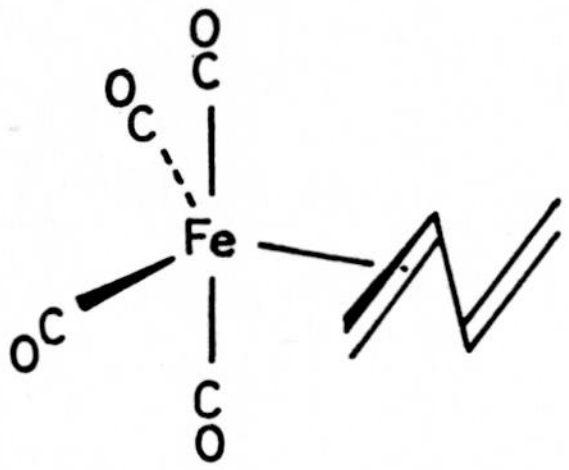

As will be described later, the 1,2-η^2-coordination geometry has often been thought to be an important intermediate in diene catalyses. For example, bis(1,2-η^2-diene)metal species, which are formed by an attack of diene on η^4-diene complexes, readily cyclizes to metallacyclopentane [35,36].

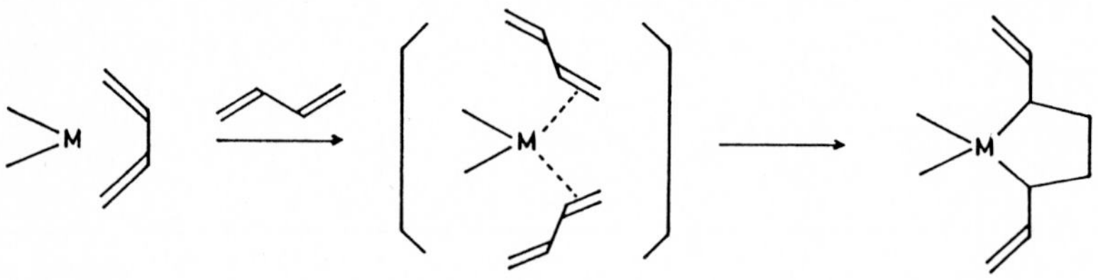

η^4-Coordination

The s-cis form had been exclusive for this type, but s-trans form is now found to be important for early transition metals. An X-ray structure of a typical s-trans diene complex is shown in Fig. 3, together with a s-cis diene complex (Fig. 4-6).

The factors determining the diene stereochemistry have been analyzed and the difference between s-cis and s-trans will be discussed later in section 2.3. Among many examples of η^4-coordination, early transition metal diene complexes are different from the majority in their bending angle and C-C bond lengths. To illustrate this point, a comparison of these parameters is shown in Fig. 7, where the difference in bonding nature is apparent. When the metal-to-central carbon length is lengthened, the coordination of diene may be better described as 1,4-η^2- or $\sigma^2\pi$ structure [4] ; i.e., C_1-C_2 bond > C_2-C_3 bond, M-C_1 bond <

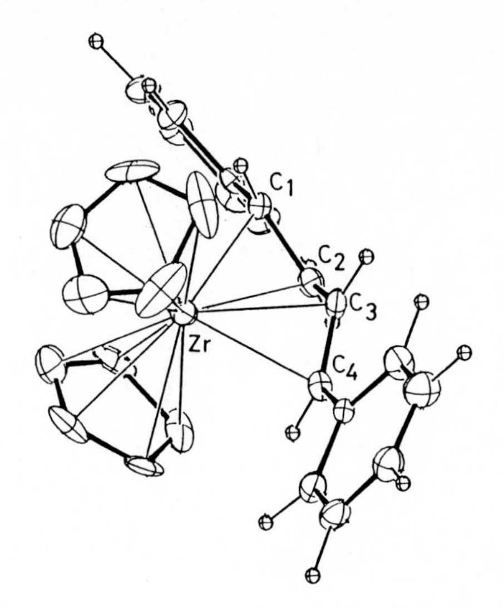

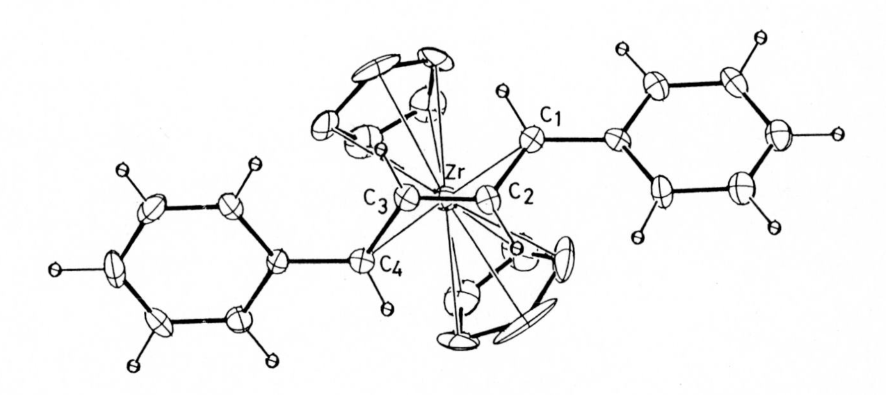

Figure 3. Structure of $Zr(\eta^5-C_5H_5)_2$(s-trans-1,4-diphenylbuta-1,3-diene) (ref. 6a).

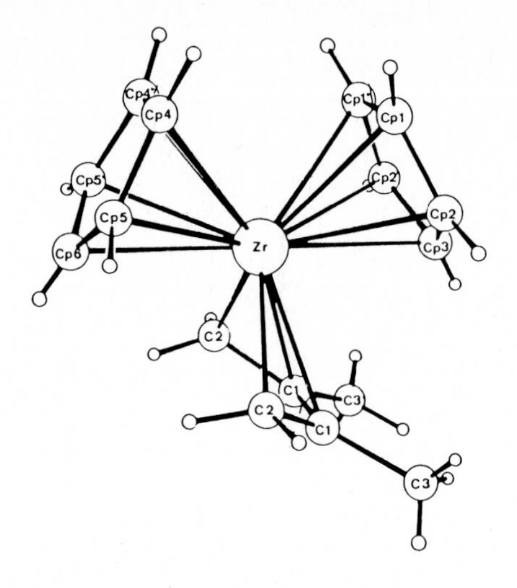

Figure 4. Structure of $Cp_2Zr(s\text{-}cis\text{-}C_6H_{10})$ (ref. 4).

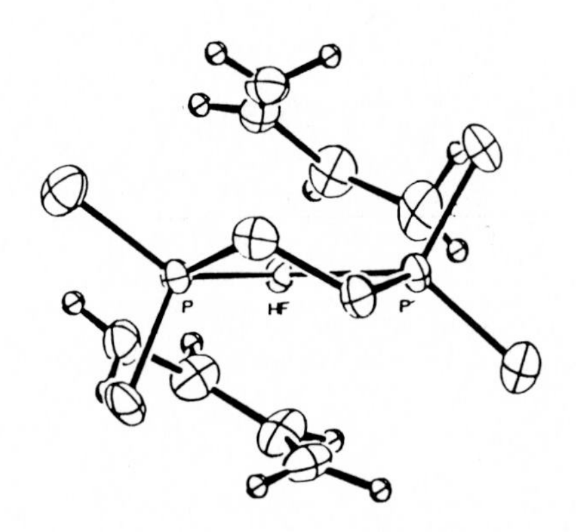

Figure 5. Structure of $Hf(\eta\text{-}C_4H_6)_2(dmpe)$ (ref. 37).

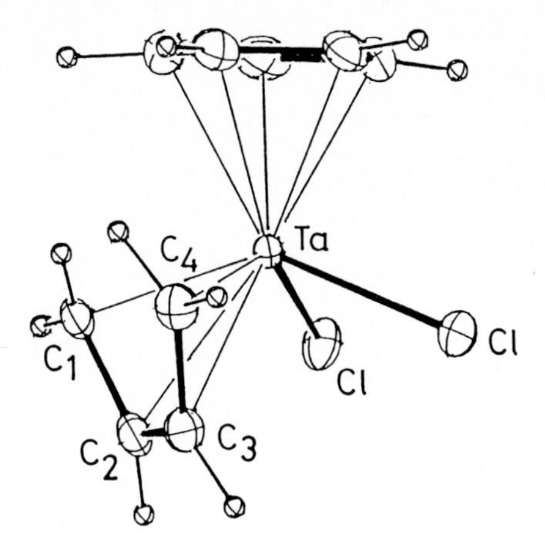

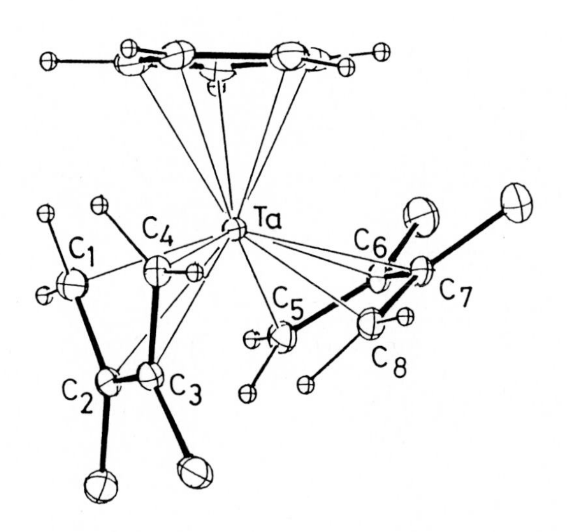

Figure 6. X-ray structures of $CpTaCl_2$(butadiene) (top) and $CpTa$(2,3-dimethyl butadiene$)_2$ (bottom) (ref. 29).

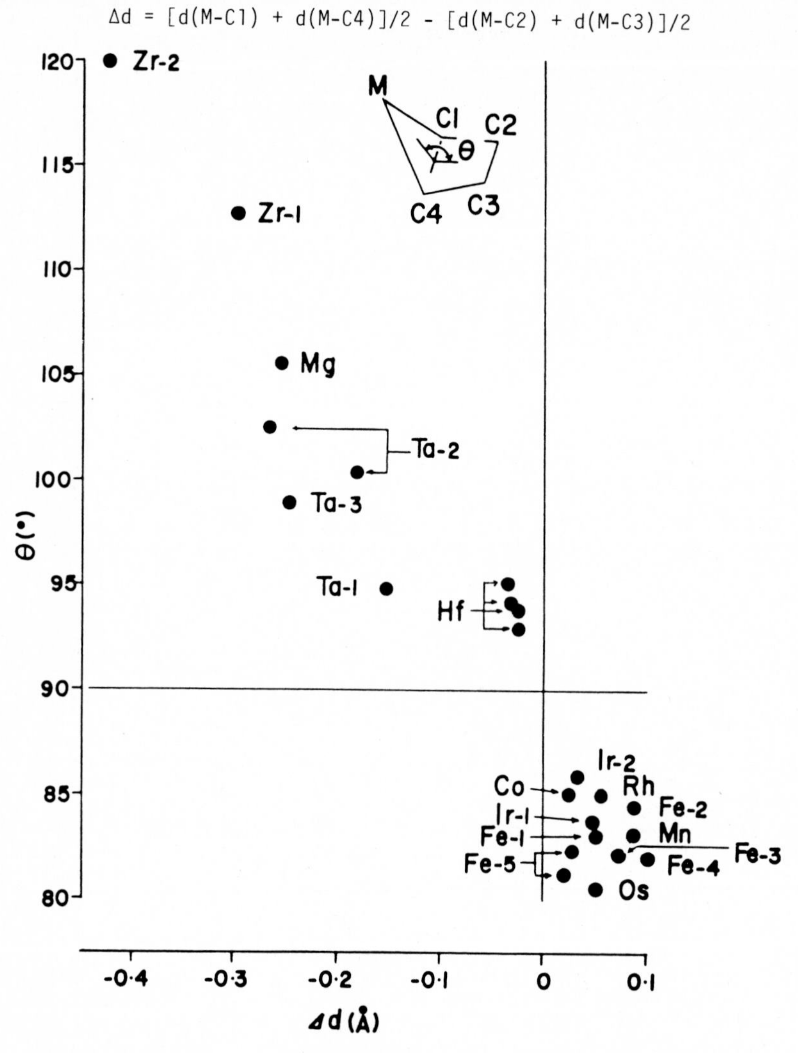

Mg; $Mg(14DPBD)(THF)_3$, Zr-1; $Cp_2Zr(23DMBD)$, Zr-2; $Cp_2Zr(23DPBD)$, Hf; $Hf(BD)_2$-(dmpe), Ta-1; $CpTaCl_2(BD)$, Ta-2; $CpTa(23DMBD)_2$, Ta-3; $Cp^*Ta(23DMBD)_2$, Mn; $Mn(CO)(BD)_2$, Fe-1; Fe(CO)(23DMBD)[glyoxal-bis(isopropyl imine)], Fe-2; Fe(CO)-(BD)(COT), Fe-3, $Fe(CO)(BD)_2$, Fe-4; $Fe(14DPBD)(CO)_3(14DPBD)$, Fe-5; Fe(14DPBD)-$(CO)_3$, Os; $OS_3(CO)_{10}(BD)$, Co; $[Co(23DMBD)(CO)_2]_2$, Rh; $Rh(BD)_2Cl$, Ir-1; Ir(BD)-$H[P(C_3H_7)_3]_2$, Ir-2; $Ir(BD)_2Cl$.

Figure 7. Correlation between the dihedral angle θ and the averaged difference in M-C bond lengths, Δd. The number following the metal indicates serial number of examples.

$M\text{-}C_2$ bond, bent angle between the $C_1\text{-}C_2\text{-}C_3\text{-}C_4$ and $C_1\text{-}M\text{-}C_4$ planes <90°.

Interconversion between the s-cis and the s-trans butadiene complexes of zirconocene occurs thermally or photochemically as shown below [4]. The activation barrier to the s-trans → s-cis isomerization was found to be

$$Cp_2Zr(\text{s-cis-butadiene}) \underset{\Delta}{\overset{h\nu}{\rightleftarrows}} Cp_2Zr(\text{s-trans-butadiene})$$

22.7 Kcal/mol ($\Delta G^{\ddagger}_{10.5°C}$). At 25°C, the two isomers are in equilibrium with the s-cis/s-trans ratio of 55/45. For the Hf analogue, Cp_2Hf(butadiene), the s-trans → s-cis barrier increases to $\Delta G^{\ddagger}_{60°C}$ = 24.7 Kcal/mol [5].

1,4-η^2 Coordination

1,3-Diene complexes of group 2,3, and 4 metals tend to form this coordination mode. In most of the cases, however, it is not so clear if the metallacyclopentane rings are planar or bent. The X-ray structure of $Mg(thf)_3$-(PhCHCHCHCHPh) is now available which shows that the MgC_4 ring puckers substantially [39].

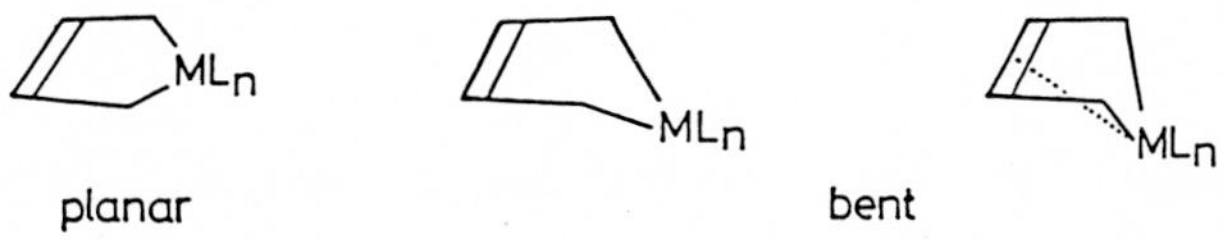

ML_n ML_n ML_n

planar bent

In the early transition metal complexes which favor bent 1,4-η^2 or $\sigma^2\pi$ bonding, their coordination geometries are sometimes not rigid. For example, Cp_2Hf(s-cis-butadiene) is fluxional at -80°C in solution through the planar metallacyclopentane structure [6b]. As described in later section the 1,4-η^2 mode is very important in the reaction chemistry of diene complexes of early transition metals. The remarkable dehydrogenation shown below, which occurs during the formation of a metallacyclopentadiene complex from $Ru(CO)_3$-(C_4H_6), is caused by the reaction of the metallacyclopentene intermediate [40].

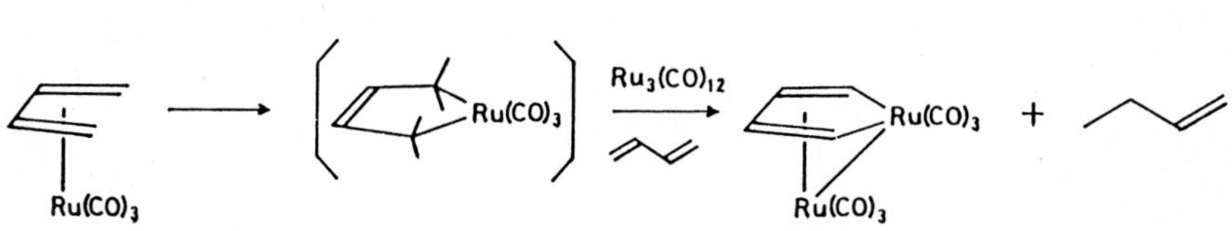

$Ru(CO)_3$ → $Ru(CO)_3$ $\xrightarrow{Ru_3(CO)_{12}}$ $Ru(CO)_3$ $Ru(CO)_3$ +

2.2 Ligand Arrangements around the Central Metal

When a metal complex contains more than one diene, there arises a wide choice for its coordination geometries. Here, we briefly discuss some limiting structures of molecules with a stoichiometry, $M(diene)_3$, $M(diene)_2L$, or $M(diene)_2L_2$. One way of viewing the ligand arrangement of these diene complexes is to consider that a diene is equivalent to two monodentate ligands. Then, $M(diene)_3$ becomes formally a hexa-coordinate complex, the two limiting configurations of which are the familiar trigonal-prism and the octahedron. Considering the orientational freedom of a diene ligand, there are two typical isomers in each of the configurations; one with a propeller-type diene conformation and the other with one diene orientated opposite to the other two. At the top portion of Figure 8, the four idealized geometries of $M(diene)_3$ are sketched, and their symmetry is given at the bottom of each structure.

There is only a single example of X-ray structures for the type $M(diene)_3$, namely $Mo(butadiene)_3$. Its geometry appears to fall in-between the idealized octahedron (C_{3h}) and trigonal-prism(D_{3h}) structures, with a propeller-type diene orientation. However the X-ray data are, unfortunately, not very accurate and further study is necessary to obtain a general idea of the favored structure of this interesting molecule.

Bis-diene complexes with two ancillary ligands, $M(diene)_2L_2$, are stereochemically analogous to $M(diene)_3$. Either of the idealized trigonal-prismatic or octahedral configuration allow for three conformational isomers, which are shown in Figure 8. The observed structure for $Hf(diene)_2(dmpe)$ [37a] is close to orientation I of the octahedral configuration. Reaction of $W(C_4H_6)_2(PMe_3)_2$ has briefly been reported, but no structural information is available [38].

The diene complex with one ligand less than $M(diene)_2L_2$ is formally penta-coordinate, thus, two idealized configurations, square-pyramid and trigonal-bipyramid, are available. Depending on the diene orientation relative to L, three conformational isomers are again possible for square-pyramid and two for trigonal-bipyramid. Among the $M(diene)_2L$ complexes with later transition metals there is a clear trend that two dienes orient in a parallel or a supine-supine fashion [11-14]. However, the situation is quite different for the early transition metal complexes, $CpM(diene)_2$ (M=Nb, Ta), where the supine-prone fashion is favored [29]. There is a good electronic reason behind this contrast, which is a subject of the following theoretical section.

It should be noted here that the electronic structure of a 1,3-diene is somewhat similar to that of the η^3-allyl anion, and that syntheses and structure characterization of the mixed ligand complexes of the type $M(\eta^3\text{-allyl})_x(diene)_y$-$L_z$ would be interesting. So far $MCp^*(\eta^3\text{-allyl})(\eta^4\text{-butadiene})$ (M=Ti, Zr, Hf) and $W(C_6H_6)(\eta^3\text{-allyl})(\eta^4\text{-butadiene})^+$ have been prepared [41-43]. In addition,

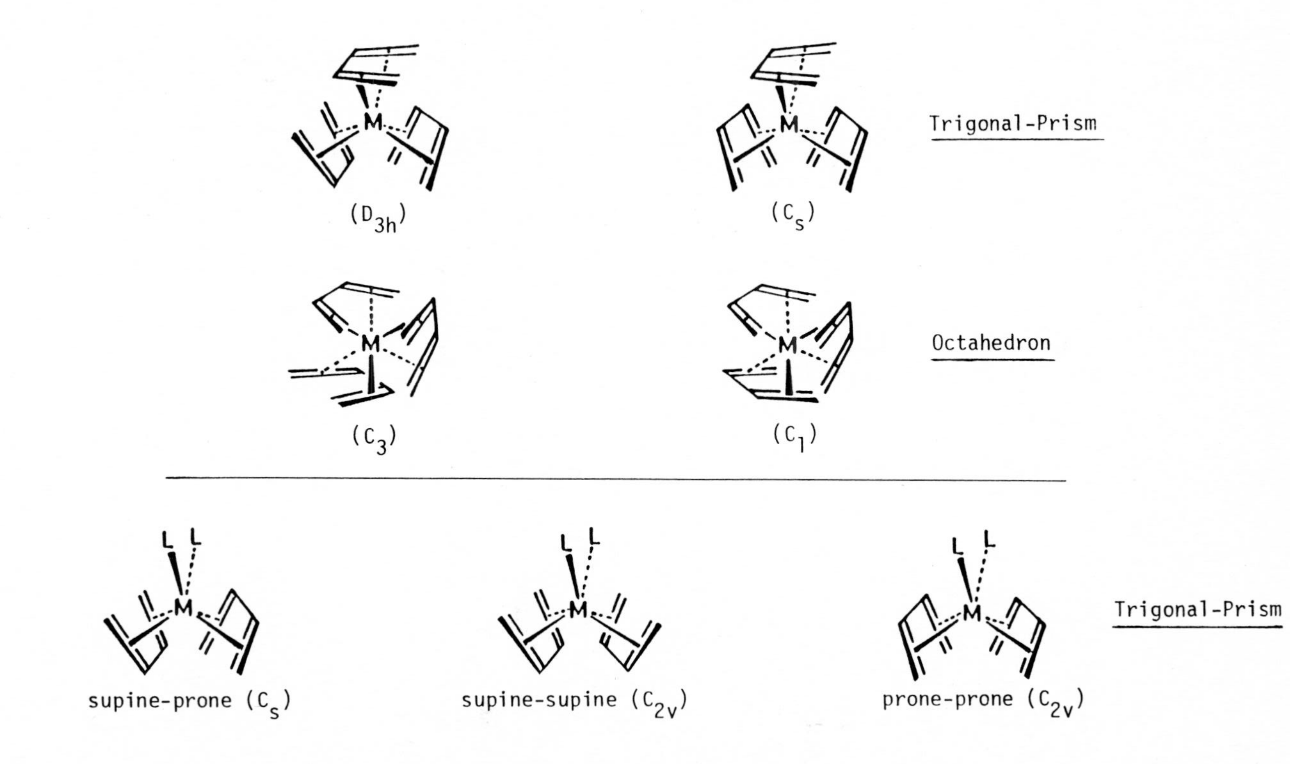

Figure 8. Limiting structures of $M(1,3\text{-diene})_3$, $M(1,3\text{-diene})_2L_2$, and $M(1,3\text{-diene})_2L$.

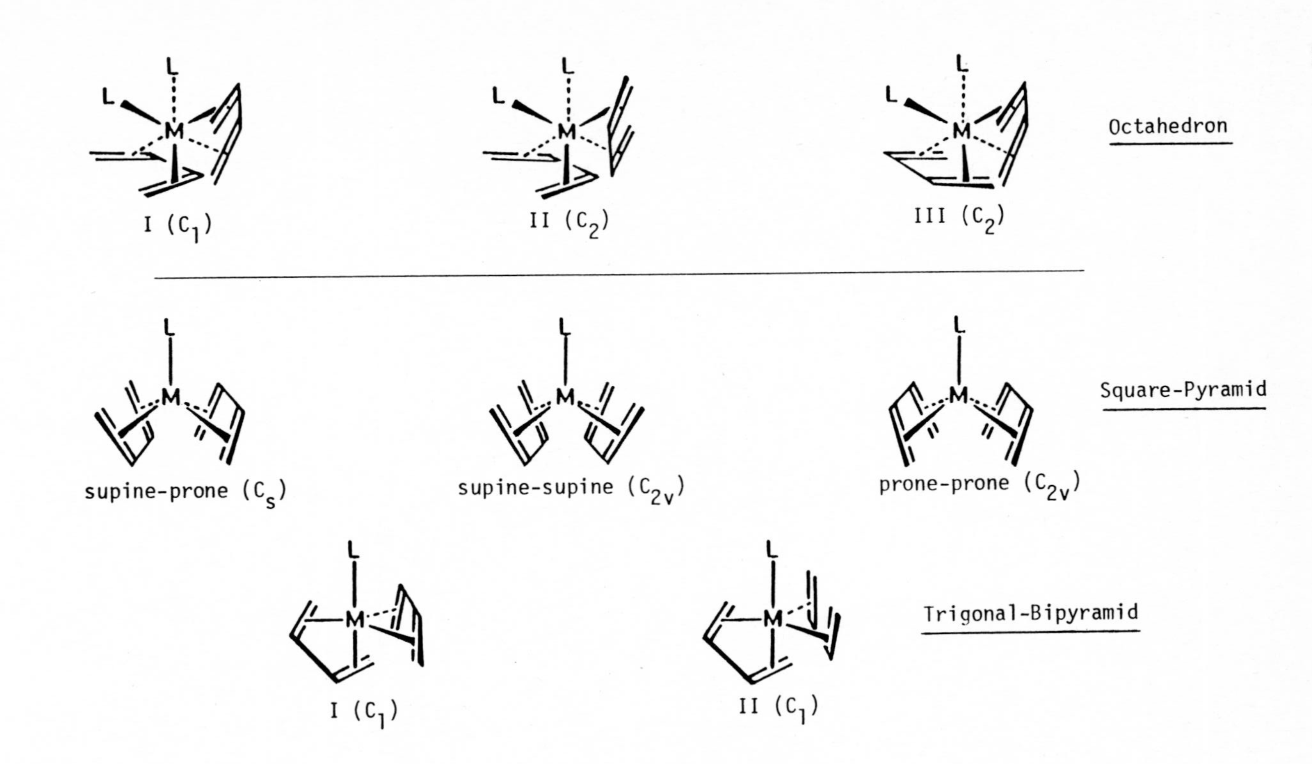

Figure 8. (Continued).

the η^4-diene ligand is readily converted to an η^3-allyl ligand by selective reaction on one of the terminal carbons. The stereochemistry of the resulting η^3-allyl complexes has a close connection with that of η^4-diene complexes, e.g., $MoCp(CO)_2(\eta^3\text{-allyl})$ vs. $MoCp(CO)_2(\eta^4\text{-diene})^+$ [44].

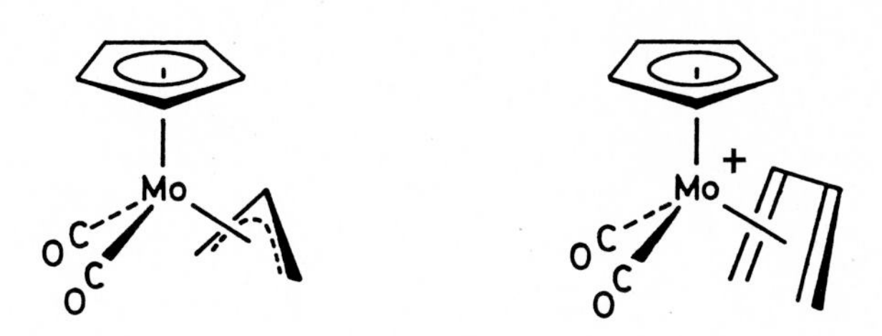

2.3. Electronic Factors Determining the Stereochemistry

The butadiene molecule has four π-type orbitals of importance. These are drawn below, numbered in the energetic order from the most stable π_1 to the highest π_4. The inner-carbon lobes are larger than the terminal-carbon lobes in the π_1 and π_4 molecular orbitals, while the trend is reversed for π_2 and π_3. The relative size of orbital lobes and nodal properties shown in 1 are common to the majority of the substituted butadiene molecules. In transition metal complexes, the 1,3-diene group carries its donor function in the two bonding π

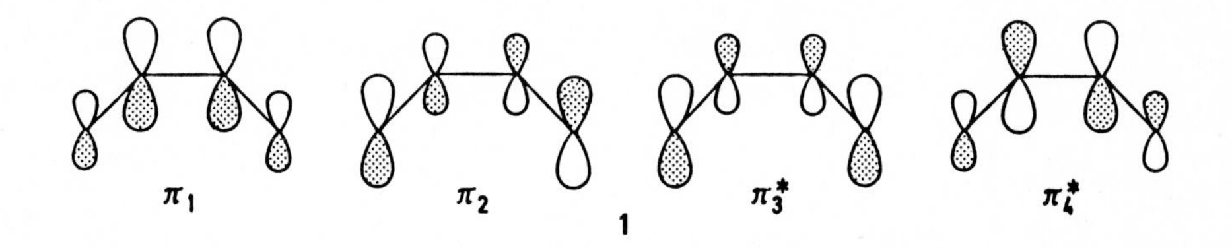

1

orbitals, and an acceptor function in the unfilled π^* orbitals. In principle, when these orbitals find suitable partners in the metal orbital set to interact with, so-called donation and back-donation types of bonding are achieved.

1,3-Dienes tend to coordinate to a metal atom in an η^4-fashion in the complexes that fulfill the above requirement for forming stable bonds. Most of the 1,3-diene complexes so far prepared with later transition metals fall in this category, a typical example is the familiar iron complex, $Fe(CO)_3(C_4H_6)$, for which extended Hückel calculations have corroborated this view. The geometrical optimization of $Fe(CO)_3(C_4H_6)$ obtained by choosing two variables, L and θ, as shown in 2 gave, in fact, an η^4-coordinated butadiene structure as the most stable form at L = 0.8 Å and θ = 15°. As will be shown later in Table 1, the calculated overlap populations tell us that all carbon atoms of

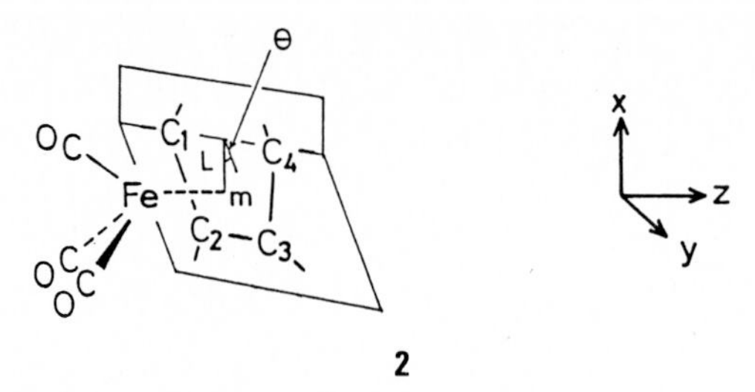

2

butadiene are almost equally involved in the Fe-butadiene bonding, and that the C_1-C_2 bond is as strong as the C_2-C_3 [8]. These theoretical results accord well with the experimentally observed structure of $Fe(CO)_3$(diene) [45].

A brief sketch of the bonding picture of $Fe(CO)_3(C_4H_6)$ is worth pursuing [46]. The frontier orbital levels of the C_{3v} $Fe(CO)_3$ fragment are shown in 3 [47]. There is a nest of three low-lying orbitals ($1a_1$ + 1e), another e set at somewhat higher energy (2e), and a high-lying spd-hybridized orbital ($2a_1$). In the d^8 electron case, 1e and $1a_1$ are fully occupied and the 2e components share the last two electrons. The butadiene π-type orbitals of 1 all find a symmetry match with these frontier orbitals of $Fe(CO)_3$. Of them, the most

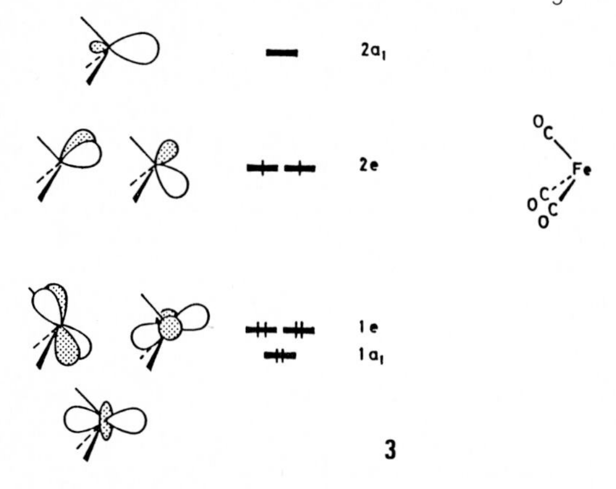

3

significant interactions occur between 2e and the butadiene π_2, π_3^* orbitals. One of the 2e, designated a' in C_s symmetry of $Fe(CO)_3(C_4H_6)$, moves down in energy due to an interaction with the unoccupied π_3^* and becomes the HOMO of the complex, 4. The occupied π_2 overlaps well with the other 2e component, a", and pushes a" up strongly. The bonding combination between π_2 and a" is shown in 5. From 4 and 5, one may easily recognize that these dominant interactions are attained when the butadiene molecule is positioned right over the central metal, viz. and η^4-coordination. Putting it in other words, the ability to form strong interactions of both 4 and 5 is the hallmark of 1,3-diene complexes with the strongest η^4-bonding.

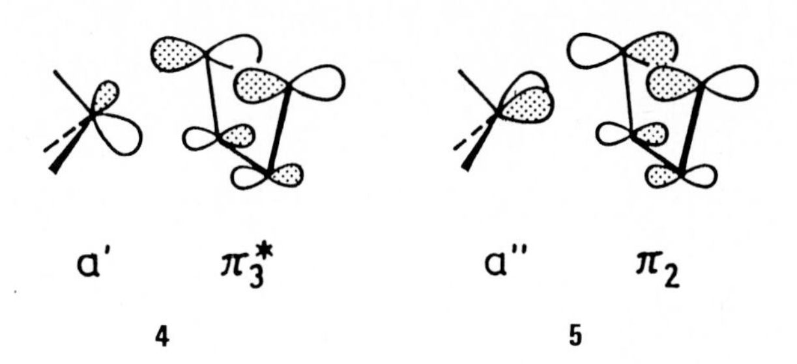

The nature of the bonding in diene complexes of early transition metals differs in a substantial way from the classical η^4-bonding in $Fe(CO)_3(C_4H_6)$. Let us examine the electronic structure of $ZrCp_2$(butadiene). When its geometry was optimized in the same manner as we did for 2, the potential minimum appeared at L = 0.3 Å and θ= 30°. The overlap populations calculated for the optimized geometry of $ZrCp_2$(butadiene) are listed in Table 1 [8]. Interestingly, the molecule contains a quite large overlap population for the Zr-$C_{1,4}$ bonds and concomitant small Zr-$C_{2,3}$ overlap populations. The central bond assumes stronger double bond nature than the terminal ones, C_1-C_2 and C_3-C_4. The larger overlap population for C_2-C_3 is a consequence. These theoretical findings are in harmony with the x-ray structure of $ZrCp_2[CH_2C(CH_3)C(CH_3)CH_2]Cl$ and suggest that a good way of describing the bonding scheme is as in 6 and not as in 7.

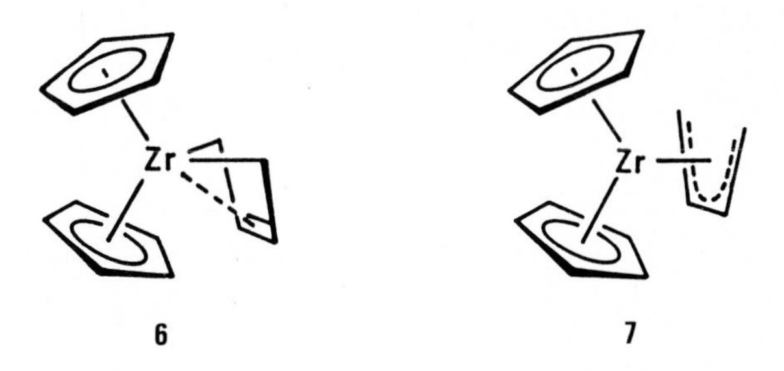

There arises an ambiguity in computing the oxidation state of Zr in structure 6. If butadiene is regarded as a neutral ligand, as we often do, then the Zr metal assumes the oxidation state of 2 with two d electrons in it. Or one may preferably count Zr tetravalent, thus a d^0 system, emphasizing the 1,4-σ character (or $\sigma^2\pi$) of the coordinated butadiene. The choice is a matter of formality, and the reality is perhaps in between these extremes. In any case, the difference in electron counting does not affect the following theoretical analysis, though we can take either view, depending on what we are looking at.

Table 1. M-C and C-C Overlap Populations (P) Calculated for the four typical Butadiene Complexes

	Fe	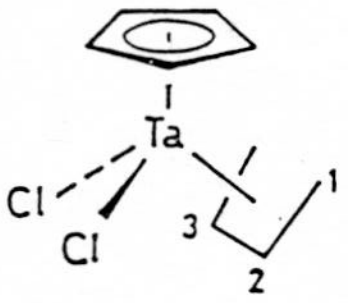 Ta	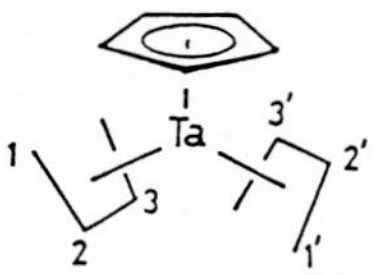 Ta		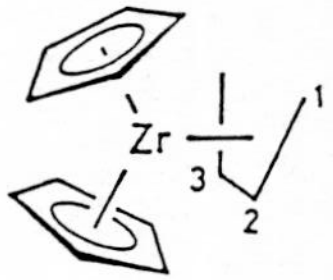 Zr
$P(M-C_1)$	0.202	0.289	0.296	0.264	0.338
$P(M-C_2)$	0.184	0.108	0.074	0.120	0.060
$P(C_1-C_2)$	0.979	0.956	0.951	0.971	0.991
$P(C_2-C_3)$	0.974	1.037	1.058	1.012	1.012

The interaction diagram for $ZrCp_2$ and butadiene is shown in Figure 9. The $ZrCp_2$ orbitals have previously been discussed in detail, and are given on the left side of Fig. 9 [48]. There are five orbitals, of which the three lower ones sit in the yz plane, while the rest are perpendicular to the plane. For the d^2 electron configuration of $ZrCp_2$, only the lowest $1a_1$ is doubly occupied. The unoccupied b_2 is responsible for the π acceptor capability of the $ZrCp_2$ fragment in the yz plane, thus interacting very well with the occupied π_2 of butadiene. This is an interaction of the donation type, corresponding to the a'-π_2 mixing, 5, for $Fe(CO)_3(C_4H_6)$. A slight back-donation interaction occurs between $1a_1$ and π_3, but it is much weaker than the a'-π_3 bonding of the Fe complex. The lack of a strong back-donation is a reason behind the 1,4-σ type structure of the Zr complex, which in turn leads to its high reactivity. Carbonyl, olefins, alkynes and even 1,3-dienes are readily inserted into the Zr-$C_{1,4}$ bond, while $Fe(CO)_3$-(C_4H_6) does not react with them (see later section).

We next consider a tantalum butadiene complex, $TaCpCl_2(C_4H_6)$ [29]. There are some similarities between the $TaCpCl_2$ fragment and $ZrCp_2$ in their frontier orbital levels as well as in the d electron count, two in number, should butadiene be regarded as neutral [49]. As shown in 8, the low-lying 1a and 1a' are analogous in shape to $1a_1$ and b_2 of $ZrCp_2$, respectively. In fact, back-donation of electrons from π_2 to 1a" occurs as effectively as the π_2-b_2 pair of the Zr case, thus as π_2-a" for $Fe(CO)_3(C_4H_6)$. Apparently, the 1a-π_3 interaction parallels the donation interaction between $1a_1$ and π_3 in $ZrCp_2(C_4H_6)$. However, the presence of 2a', which is approximately Ta $d_{y^2-x^2}$ and is able to interact with π_3 as well, makes the donation interaction somewhat different from that of $ZrCp_2(C_4H_6)$. The participation of 2a' increases the donation-type interaction to some degree, which in turn decreases its 1,4-σ bonding nature of butadiene in $TaCpCl_2(C_4H_6)$ compared with the Zr complex. Overlap populations P, summarized in Table 1, provide the above trend in a more quantitative way. The difference between P(Ta-C_1) and P(Ta-C_2) becomes smaller on going from $ZrCp_2$-(C_4H_6) to $TaCpCl_2(C_4H_6)$ to $Fe(CO)_3(C_4H_6)$. If we assume that such a difference can be an indicator of the degree of 1,4-σ bond character, the Ta complex falls in between the two extreme cases, $ZrCp_2$ and $Fe(CO)_3$. It is worthy of note that the 1,4-σ bond character decreases as the central metal moves left to right in the periodic table.

For bis(1,3-diene) complexes, one intriguing aspect is a geometrical contrast between $TaCp(1,3\text{-diene})_2$ [29] and the analogous complexes of later transition metals, e.g., $RhCl(butadiene)_2$ [13]. In an attempt to estimate relative stability of the three conformations; prone-prone, supine-supine, and supine-prone, each structure of $TaCp(butadiene)_2$ and $RhCl(butadiene)_2$ was optimized as we did for $Fe(CO)_3(butadiene)$ and $ZrCp_2(butadiene)$. Figure 10 and 11 shows the side views of these optimized geometries on the left side, and

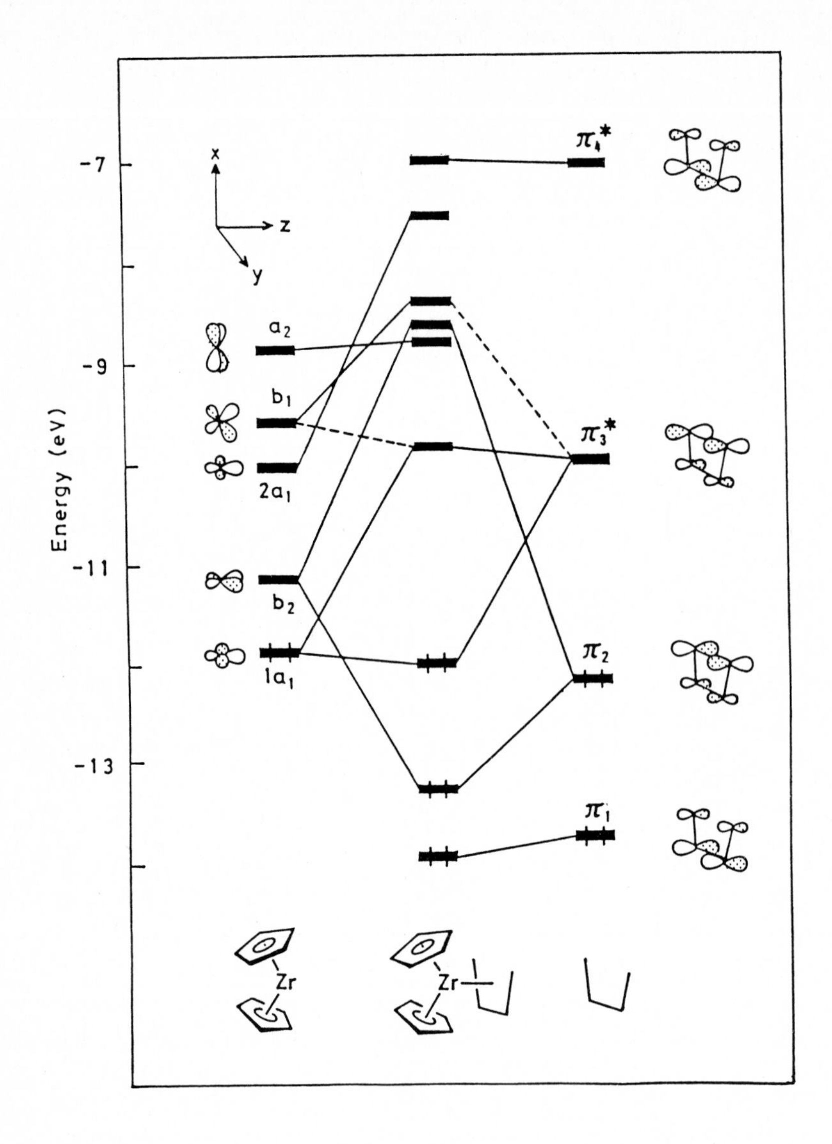

Figure 9. Orbital interaction diagram for $ZrCp_2$ and s-cis butadiene.

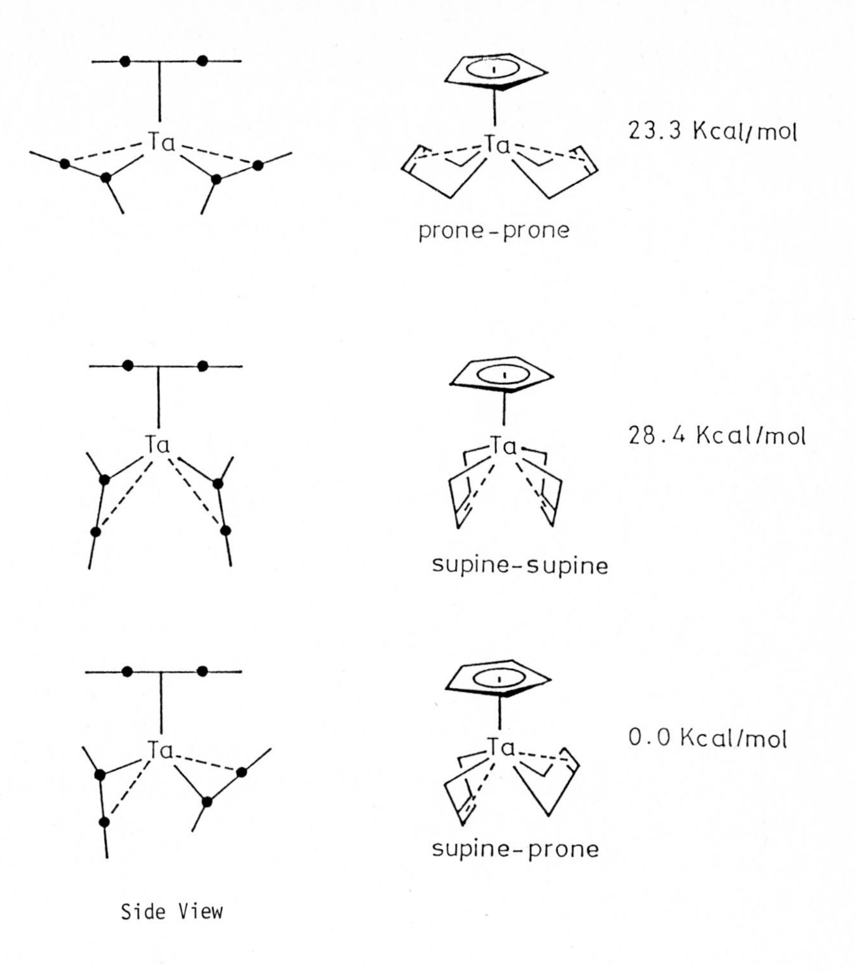

Figure 10. Geometries optimized for the three conformations of $TaCp(C_4H_6)_2$ and their relative stabilities.

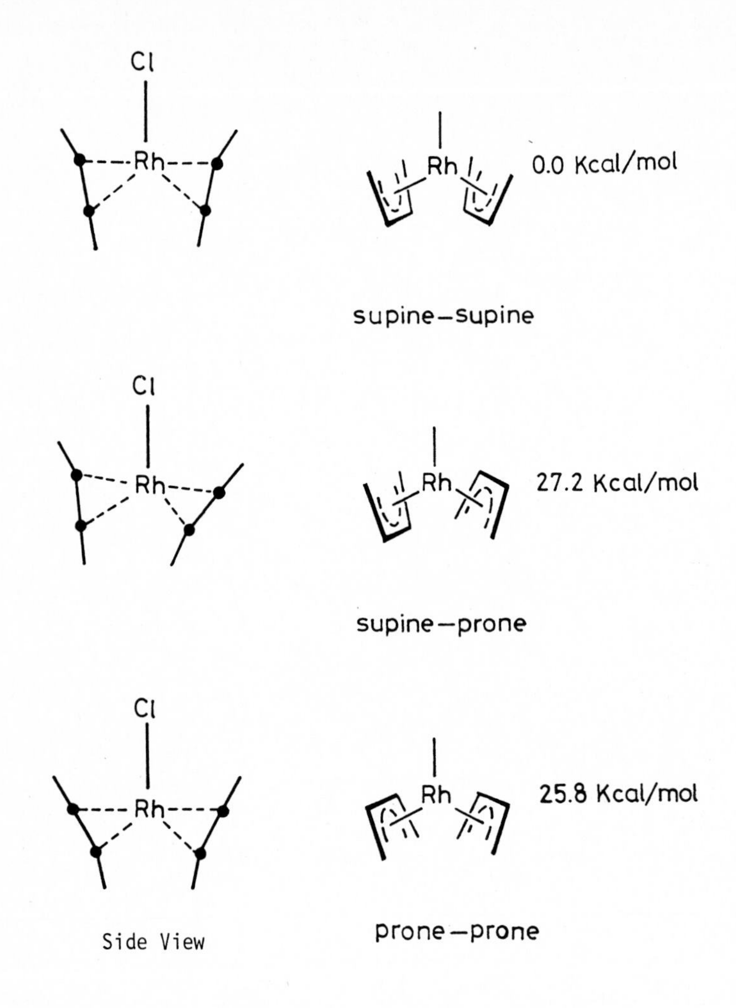

Figure 11. Geometries optimized for the three conformations of $RhCl(C_4H_6)_2$ and their relative stabilities.

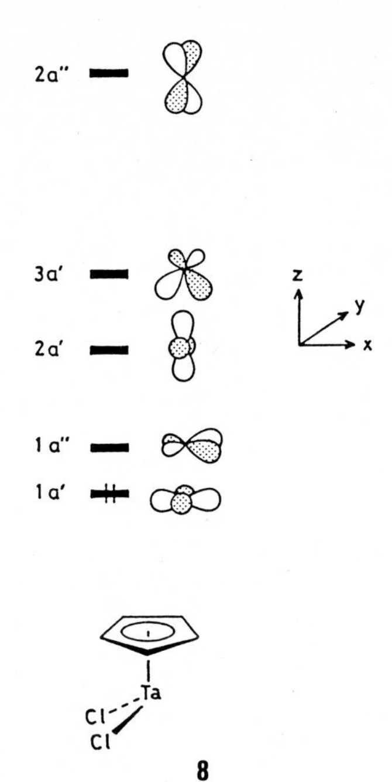

relative total energies at the right of the structures. The supine-prone conformation is clearly favored for TaCp(butadiene)$_2$, while the most stable structure of RhCl(butadiene)$_2$ is supine-supine. The alternative conformers were calculated to be much less stable than the favorable ones, in accord with the experimental observations.

The detailed analysis of orbital interactions between the TaCp fragment [50] and two butadienes lead to the conclusion that only the supine-prone conformation could allow TaCp to donate d electrons back effectively to π_3^* orbitals of both the diene ligands. Here, we regard butadiene as a neutral ligand and thus TaCp carries four d electrons. On the other hand, the strongest Rh-butadiene interactions are achieved when the complex assumes the supine-supine form.

There is another way of viewing the conformational choice of TaCp-(butadiene)$_2$. When the butadiene ligand is regarded as a dianionic ligand, as in 9, then the Ta-butadiene bond is composed of two Ta-C σ-bonds plus η^2-coordination from the inner olefinic double bond. TaCp(butadiene)$_2$ carries

$$CH_2{=}CH{-}CH{=}CH_2 \longrightarrow CH_2^{-}{-}CH{=}CH{-}CH_2^{-}$$

9

four Ta-C σ bonds, forming a so-called four-legged piano stool geometry 10. The oxidation state of Ta is now formally 5+ with no d-electrons. The d^0 piano stool molecule 10 has two low-lying vacant orbitals, x^2-y^2 and z^2, which are potential acceptors of electrons from the inner double bond of 9 [51]. The

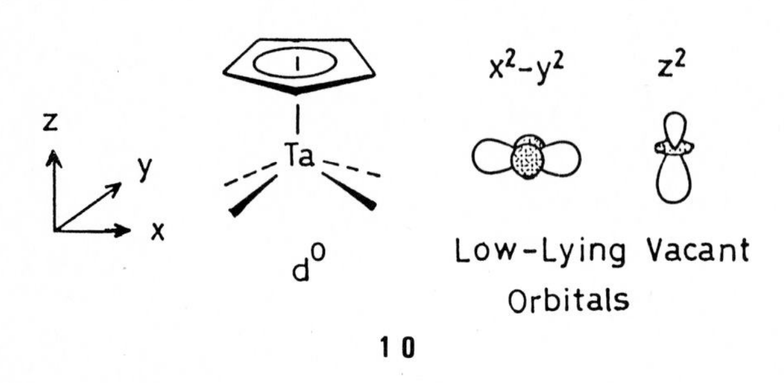

10

supine-prone conformation of $TaCp(butadiene)_2$ is best suited for this type of interactions as shown in 11.

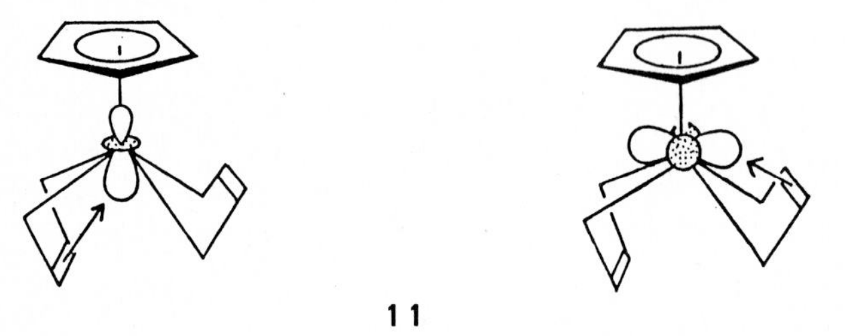

11

The Ta-C and C-C overlap populations calculated for the optimized geometry of $TaCp(butadiene)_2$ are given in Table 1, which suggest that the supine-diene has more pronounced 1,4-σ character than the prone-diene. Therefore, the nature of the bonding, and probably reactivity as well, are different between the two coordinated dienes.

While free 1,3-dienes favor the s-trans conformation slightly over the s-cis structure, coordination to a single metal atom has occurred discriminatively in the s-cis fashion. Therefore, the discovery of $ZrCp_2$(s-trans-diene) was rather surprising. In order to show the electronic origins of the novel transoid structure in the Zr complex, orbital interactions of $ZrCp_2$(s-trans-butadiene) are given in Figure 12, which can be compared with the interaction diagram for the s-cis isomer in Figure 9. The stability of the two isomers is well balanced, viz., s-cis/s-trans = 55/45 at 25°C [4]. This trend is due to electronic properties of the $ZrCp_2$ fragment, in which the presence of an occupied $1a_1$ orbital allows the metal to donate electrons back to s-trans π_3^*. In the case of later d-transition metal groups, like $Fe(CO)_3$, there is no occupied d levels suited for the back-donation interaction with s-trans π_3^*.

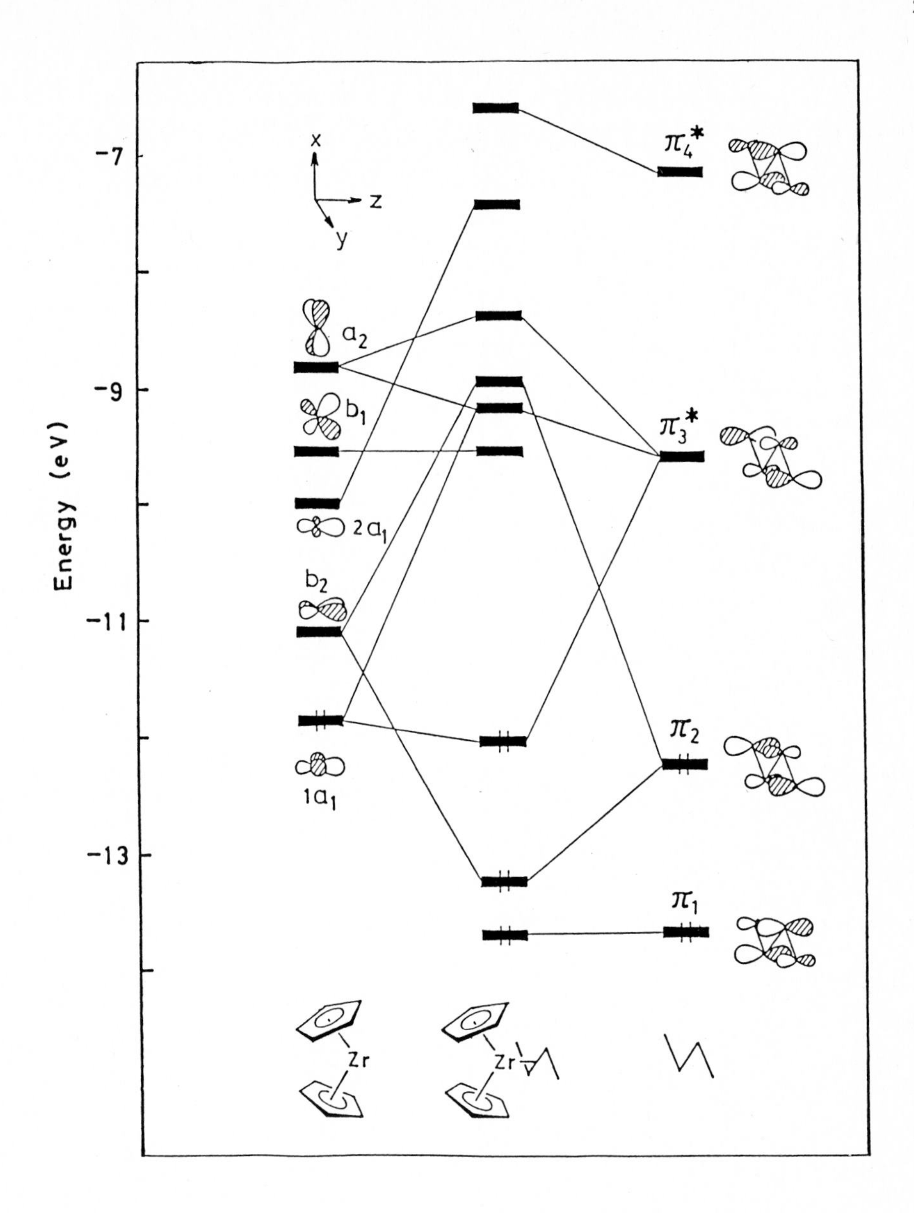

Figure 12. Orbital interaction diagram for $ZrCp_2$ and s-trans butadiene.

The $1a_1$-π_3^* back-bonding for $ZrCp_2$(s-trans-butadiene) is not very strong, but at least as strong as the corresponding interaction of $ZrCp_2$(s-cis-butadiene). As a matter of fact, the calculated energy of the optimized s-trans structure is close to that of the optimized s-cis form, favoring the s-cis only slightly (0.07 eV) [8].

The s-cis/s-trans ratio was found to be sensitive to the substitution on the diene. As shown in Figure 13, 1,4-disubstitutions induce the s-trans isomer to be favored, specially in the phenyl disubstituted case, as corroborated by calculations. The observed trend seems not to be explained by simple electronic or steric effects. We found that the deformation of the diene in the s-cis coordination disrupts the conjugation between the substituents and the diene part and thereby decreases the stability of the s-cis complex. In the s-trans complex, the conjugation is not greatly decreased, as is found by the near planarity of C=C and 1,4-substituents [8].

Substitution at the inner diene carbons enhance the preference for the s-cis complex. Actually, the isoprene of 2,3-dimethylbutadiene complexes produce solely s-cis structures. In the corresponding s-trans structure, the substituents suffer severe steric congestion with the two Cp groups [4].

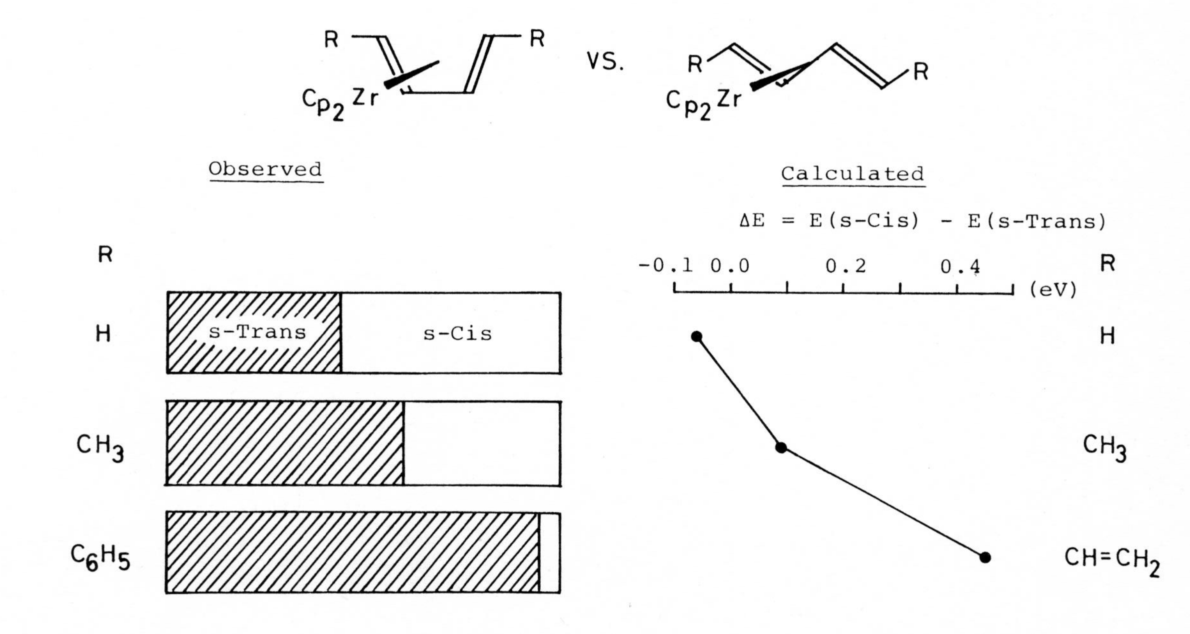

Figure 13. Observed s-cis vs. s-trans ratios of 1,4-disubstituted-butadiene complexes of zirconocene and their calculated relative energies.

III. STERIC COURSE OF THEIR REACTIONS

3.1. Addition to 1,3-Diene Ligands

Since the properties of the diene are considerably modified by coordination to a metal, the addition to the diene is critically controlled by the identity, steric and electronic properties of metal and auxiliary ligands. Protonation of $Fe(CO)_3$(diene) occurs stereospecifically at the terminal carbon, from the exo side, preserving the s-cis configuration of the diene to give a η^3-allyl complex [52a]. In the absence of coordinating anion, the protonated species is labile and a stereochemical change at the diene occurs to give a syn-methylallyl complex [53,54].

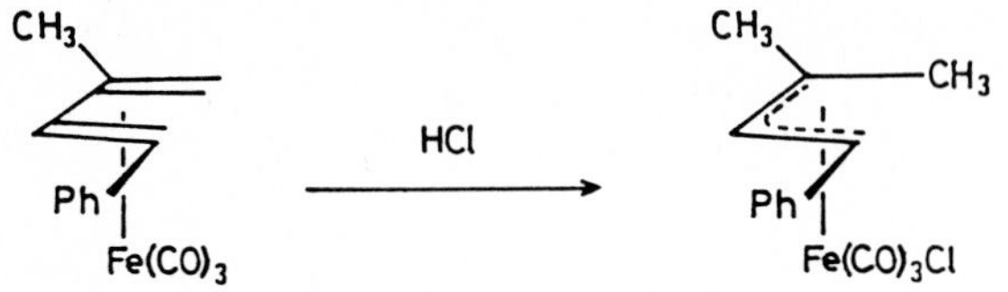

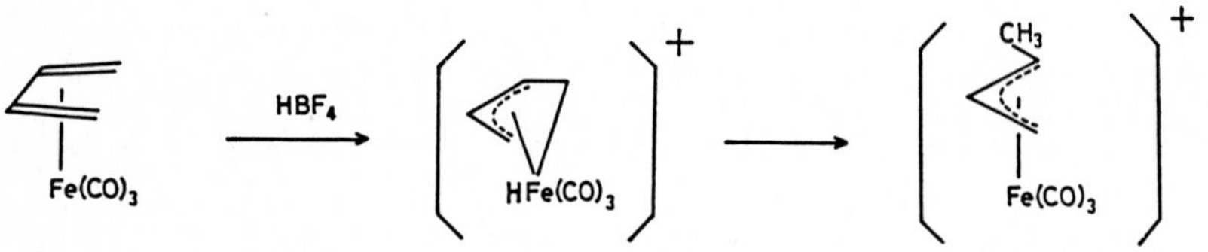

A similar change has also been observed with protonation at the CHO group. The stereochemistry of the protonation product was investigated by Lillya et al. [52b] by reaction at -120°C a mixture of S-shaped pentadienylmetal species is produced and changed to U-shaped stereoisomer by warming to -28°C. Similar stereochemical change has been observed for protonation of methylated pentadienol complexes, $Fe(CO)_3(MeCH{=}CH{-}CH{=}CH{-}CMe_2OH)$ [55].

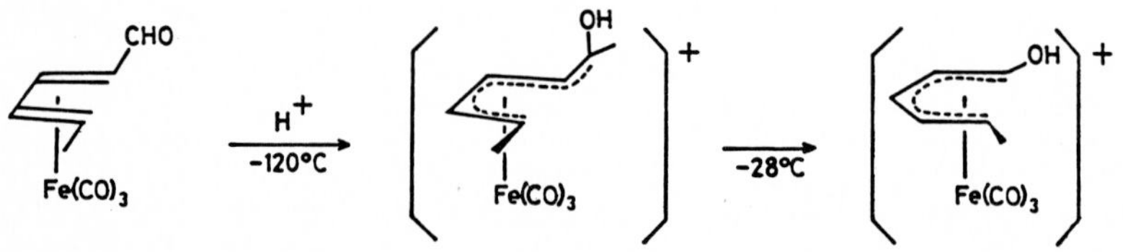

The allylmetal moiety resulting from protonation on diene-zirconium complexes is sensitive to hydrolysis. Thus, protonation of Cp_2Zr(isoprene) by D_2O in THF releases dideuterated olefin, CH_2D-$CD(CH_3)$-$CH=CH_2$, with high selectivity. The primary attack of D_2O at the Zr site, with a larger lobe of the vacant metal orbital, selectively gives a η^3-allyl-OD complex where the allyl ligand has two methyl groups at one terminal [6c]. Further protonation at this terminal gives the deuterated olefin. A stereospecific attack on the upper face of the allyl ligand will give an enantiomer in excess dictated by the chirality of the starting diene complex.

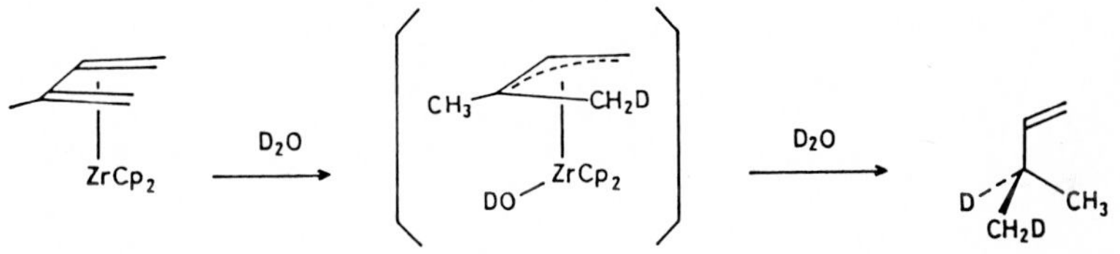

Tricarbonyl(diene)iron exhibits stereoselective addition of carbanions (e.g. $[CMe_2CN]^-$ or $[CHPh_2]^-$) on the diene [56]. Thus, tricarbonyl(butadiene) iron adds these carbanions on the unsubstituted internal carbon at -78°C to give a kinetically-favored homoallyl complex. However, an equilibration above 0°C to the more stable η-allyl product with terminal addition has been observed. The homoallyl complex is reactive to CO insertion at -78°C to give alkylated cyclobutanone after acid hydrolysis by CF_3CO_2H.

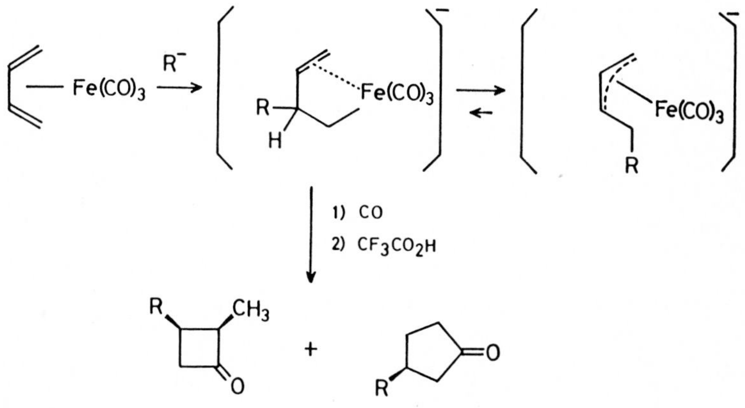

Highly labile diene-palladium(II) species, formed from Pd(II) complexes and diene, adds nucleophiles(Y) at the terminal to produce more stable η-allyl-palladium complexes. Here, the product stability determines steric course of the reaction [57]. This reaction forms a basis for catalytic oxication of butadiene to 1,4-diacetoxy-2-butene.

3.2. Insertion into 1,3-Diene Ligands

Tricarbonyl(diene)iron compounds are activated by photoirradiation for reaction with various olefins or acetylenes. For example, the following cases are reported. Note the anti-geometry of the η^3-allyl structure of the metallacycle resulting from the s-cis-butadiene ligand [58]. The difference in

$Fe(CO)_3$ + $F_2C{=}CF_2$ $\xrightarrow{h\nu}$ (metallacycle: OC, OC, CO on Fe; CF_2–CF_2)

stereochemistry from those examples observed in the corresponding reactions with Cp_2Zr(diene) may result from the stereochemical feature of the $Fe(CO)_3$ fragment which furnishes a combination of a_1 and e orbitals for interactions with diene and olefins. This combination of metal orbitals seems to favor cyclization on the metal. As an example, di-alkylation at the both terminals of a diene has been realized by the reaction of $Fe(CO)_3$(butadiene) with α,α'-dibromoacetone to give cycloheptenone in good yield [59].

$Fe(CO)_3$ $\xrightarrow{BrCH_2-CO-CH_2Br}$ (cycloheptenone)

In general, diene complexes of later transition metals are inert to addition of electrophiles such as organic carbonyl compounds. However, diene complexes of earlier transition metals add electrophiles in a selective fashion. For example, Cp_2Zr(diene) readily reacts with nitriles, Ketones, aldehydes, esters or unsaturated carbonyl compounds to give nitrile- or carbonyl-inserted complexes (eq. in the next page). The relative reactivity was found to be the following: aldehyde > ketone > ester > diene > monoene. Utilizing this order, chemoselective reactions with coordinated diene has been realized [60].

This type of reaction was thought to proceed through two different routes; one is insertion to polar metal-diene bonding and other is metallacyclization with olefin/carbonyl compounds complex, viz., [2+2] oxidative coupling of olefin and carbonyl compounds [61,62]. We have examined this reaction by starting from pure s-cis and s-trans diene complexes [61b].

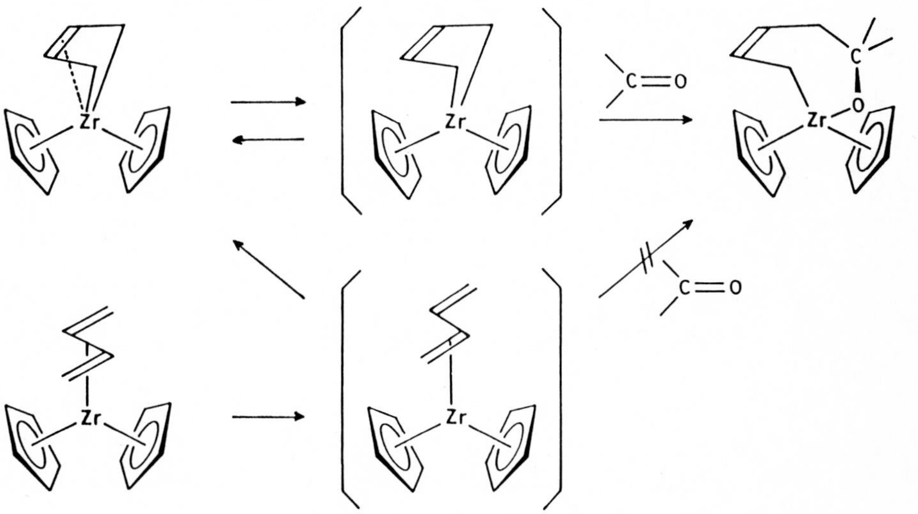

Thus, Cp'_2Zr(s-trans-butadiene) and Cp'_2Zr(s-cis-isoprene) ($Cp'=C_5Me_5$) were allowed to react with a bulky ketone, $i\text{-}Pr_2CO$, to give the corresponding addition products in high yields. The X-ray structures of these indicated very similar structures except for the methyl group at the cis-double bond. Therefore, the reaction seems to occur from the s-cis isomer and the s-trans isomer is first converted to the s-cis isomer prior to the reaction. An intermediacy of a 16-electron species with reactive zirconacyclopentene structure is thus proposed. In the case of Zr-isoprene complex, the addition was highly regioselective and occurs at the side of the methyl substituent. Since the reaction proceeds through an ionic mechanism, the electorn density at the CH_2 fragment determines the selectivity. However, in the present case, the selectivity is far more influenced by the electronic effect. An effective steric effect due to two cyclopentadienyl ligands must also be operating. The coordination site for attacking ketone seems more available by tilting the Cp rings by the steric effect of the methyl group on isoprene. Recently, opposite regioselectivity was found by Erker's group [63]. By photochemical reaction of the Zr-isoprene complex with acetone at low temperature, they found exclusive formation of a regio-isomer. The following metallacyclization mechanism was thus proposed, where dihapto coordination occurs at the vinyl group.

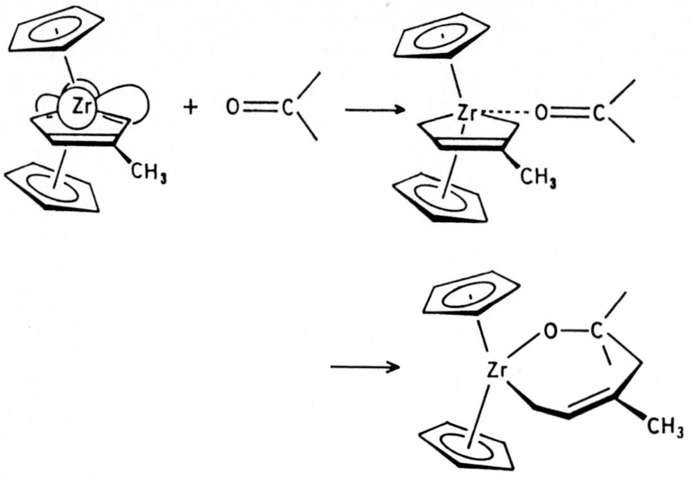

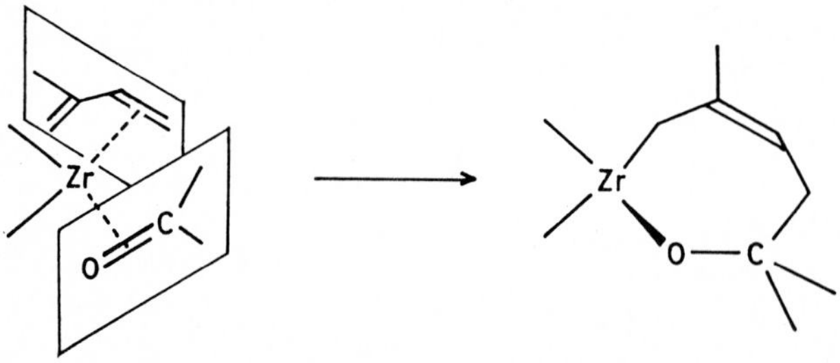

In the case of Zr-butadiene or pentadiene complex, a further insertion of carbonyl compound takes place through intermediacy of a metallacyclic allyl complex to give a dioxametallacycle with (E)-type double bond. Thus, two different carbonyl compounds are readily attached to the trans-butenylene group [64]. The stereochemistry in olefin addition is important in rationalizing the selectivity observed in catalytic co-dimerization of olefin and diene. The diene-Zr complex readily adds various olefins at 60°C to give a 7-membered metallacycle with a syn-allyl group, as illustrated below [61a].

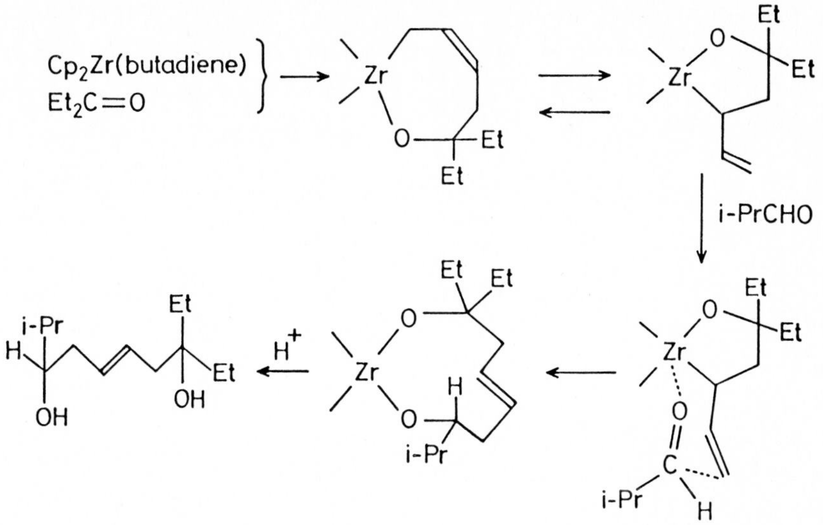

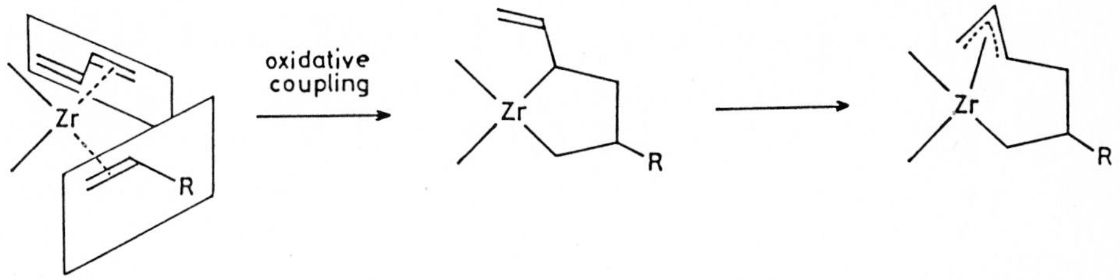

In the case of the isoprene complex, very high regioselectivity was found and was attributed to the oxidative coupling step (three-component metallacyclization) where the vinyl group of isoprene selectively enters into cyclization.

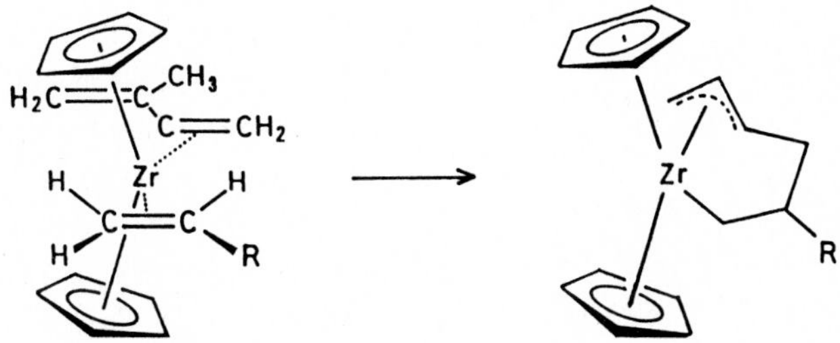

3.3. Ligand Exchange Reactions

In general, η^4-diene complexes are in equilibrium with the corresponding η^2-isomers at high temperature or upon irradiation. The ease of this change has been investigated by the ligand exchange of various Cp_2Zr(diene) complexes. Thermal equilibration at 30°C gave an order of stability as shown below [6c,60]

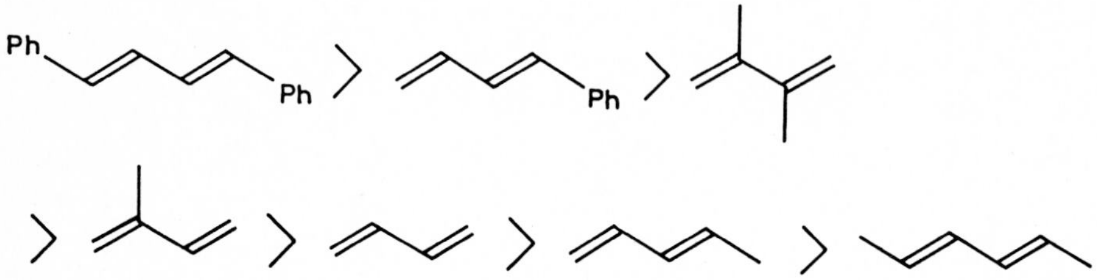

Ph ... Ph > ... Ph > ... > ... > ... > ... > ...

The order may be explained by preference for the 1,4-dihapto-structure by internally methylated dienes. The lability of 2,4-hexadiene may be due to the preference for 2,3-dihapto coordination which is readily accesible to the s-trans-tetrahapto-diene structure. Therefore, the stability order is related to the stereochemical trend for s-cis coordination.

The lability of a 2,4-hexadiene complex, $Cp_2Zr(C_6H_{10})$, has been utilized for construction of zirconacyclopentane complexes as shown below.

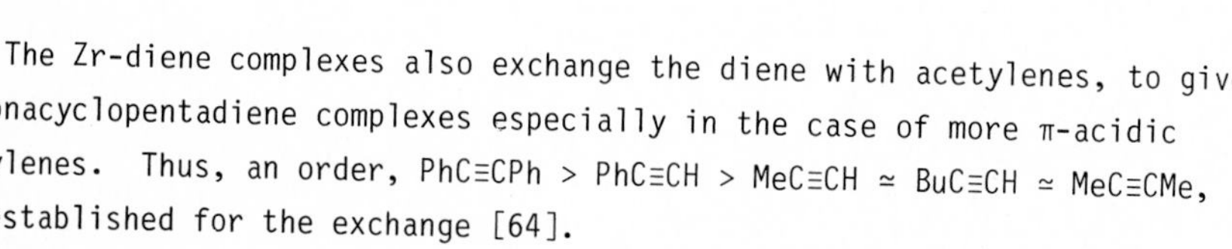

Cp_2Zr ... $\xrightarrow{2}$ Cp_2Zr ... +

The Zr-diene complexes also exchange the diene with acetylenes, to give zirconacyclopentadiene complexes especially in the case of more π-acidic acetylenes. Thus, an order, PhC≡CPh > PhC≡CH > MeC≡CH ≃ BuC≡CH ≃ MeC≡CMe, was established for the exchange [64].

The exchange reaction with olefins, acetylenes, or other dienes is in most cases accompanied by metallacyclization or oxidative coupling on Zr. General formulation of this process depicted below illustrates an important step in stoichiometric and catalytic reactions of dienes not only for Zr but also for many other metals. When metal-diene bonding is stronger in comparison with the auxiliary ligands, stereoselective exchange or substitution takes place.

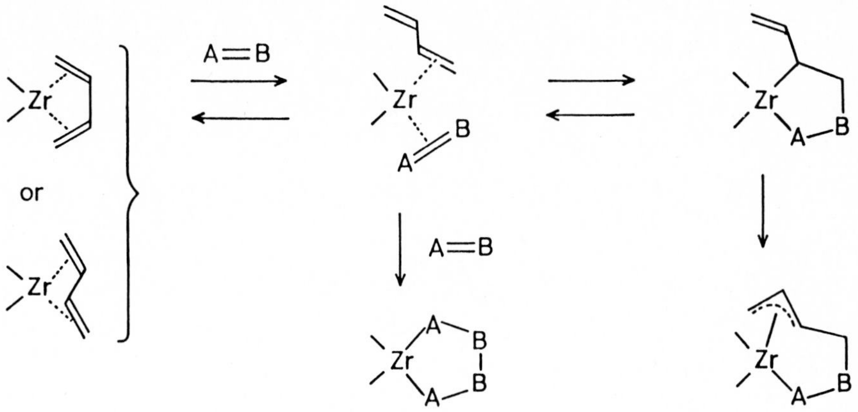

On irradiation of $Fe(CO)_3$(diene), carbonyl ligands are replaced with other coordinating molecules [65]. Thus, only equatorial CO ligand is replaced by $P(OMe)_3$ but both axial and equatorial ones are replaced by PPh_3.

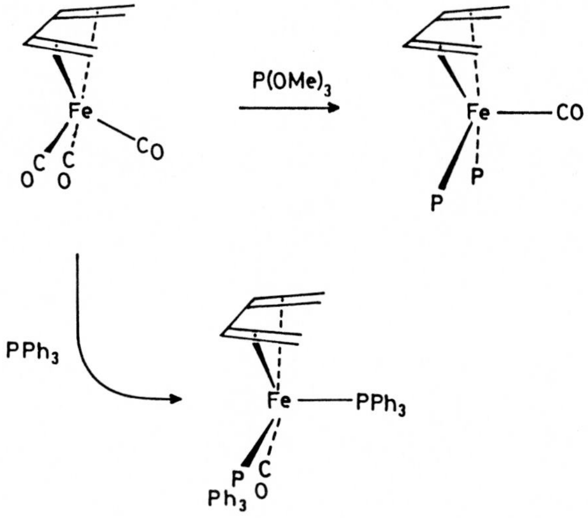

Stereochemical studies on these reactions especially for the formation of chiral metal-diene complexes will be important for further elucidation of the nature of bondings involved.

3.4. Catalytic Reactions of 1,3-Diene Complexes

Diene complexes are viable intermediates for many catalytic diene oligomerizations and polymerizations. In this context, zero-valent nickel species plays an important role. For example, an extensive research on cyclo-oligomerization of dienes has been carried out at the Max Planck Institute [66,

67]. Nickel(0)-butadiene complex is the most important catalytic species in these reactions, but has not been well-characterized due to its extreme lability. Only in a special case, i.e. $(PR_3)_2Ni($ COOR), was the structure elucidated by an X-ray analysis to involve a distorted tetrahedral species. In spite of the lack of relevant structural data [24], diene-nickel complexes, e.g. $Ni(\eta^4\text{-}C_4H_6)(L)$, are frequently postulated as the first step of the catalysis of diene reactions. The mono-diene complex adds another diene to give bis(η^2-diene)nickel species which cyclizes to nickelacyclopentane regio- and stereoselectively [68].

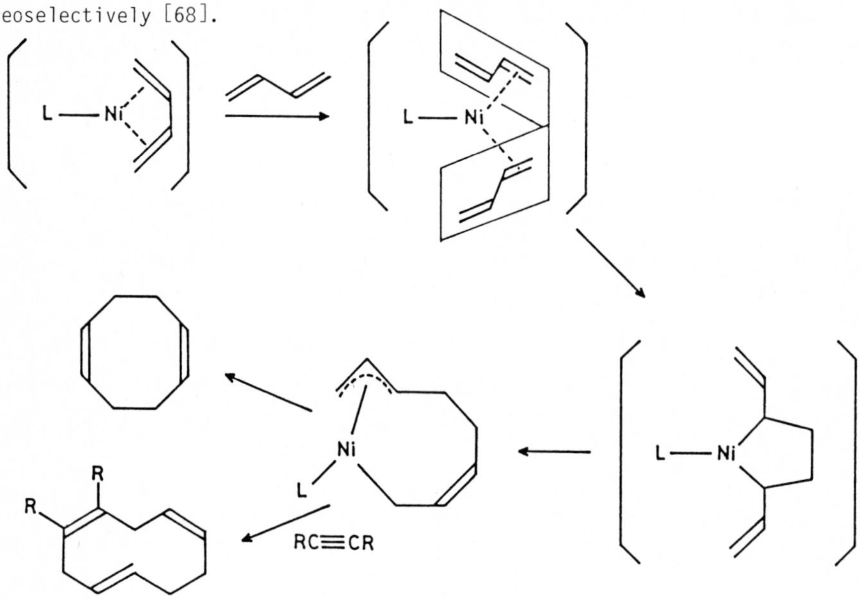

Another important diene catalysis is induced by Ziegler-type combination, e.g. organoaluminum with Ti or V compounds. Butadiene is polymerized by a system, $TiCl_3/AlEt_3$ to trans-1,4-polybutadiene, but in the presence of excess $AlEt_3$, cis,cis,-trans-cyclododecatriene was produced [69]. The identity of the halide ligand influences the stereochemistry of the polymer but not the cyclic trimer. The use of TiI_4 gives almost pure cis-1,4-polymer.

Since the intermediate catalytic species are presumably highly reactive, mechanisms involved have remained speculative. In recent years, a series of 1,3-diene complexes of Group 4 metals was prepared through various new routes and stoichiometric reactions, e.g with olefins or acetylenes, was examined. Steric and electronic effects of Cp of Cp' ligand gave been nicely utilized to stabilize the diene complexes, but sufficient reactivity still remained. The diene complex of Cp_2Ti fragment formed in situ by mixing Cp_2TiCl_2 and Mg-(butadiene) is highly reactive and thermally unstable, decomposing at -20°C precluding isolation [70]. The thermal stability was greatly improved with the corresponding Zr and Hf analogs. A butadiene complex of titanium, Cp_2Ti-(C_4H_6), prepared by mixing Cp_2TiCl_2 and Mg(butadiene) in THF, catalyzes the oligomerization of butadiene [67]. Thus, 1,3,6-octatriene was the major product, mixed with some higher oligomers. The catalytic activity of the corresponding Zr species was weaker and gives a 4,4-bonded linear dimer in high selectivity. The Hf derivative was still lower in activity. The structure of these products reflects the stereochemistry of the intermediate diene complexes. Thus, selective dimerization is thought to involve a bis(3,4-η^2-isoprene) complex which is transformed to di-isopropenylmetallacycle, and then to the dimer, as shown below. In the catalysis of isoprene, a Cp_2TiCl_2/Mg(isoprene)

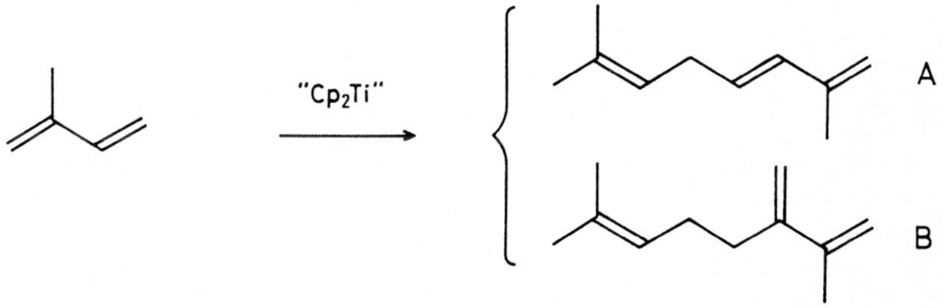

"Cp_2Ti"

A

B

system is active but less selective, giving two different linear dimers [70]. This results from intermediacy of the stereoisomeric bis(diene) intermediate. The above cyclization is followed by hydrogen migration utilizing a characteristic metal orbitals around the metal. It is the combination of metal orbitals of symmetry a_1 and b_2 which causes formation of the linear dimers in high stereoselectivity in the catalysis.

The stereochemical preference of Cp_2Zr species for A is due to stronger Zr-C bonding relative to Cp_2Ti catalysis.

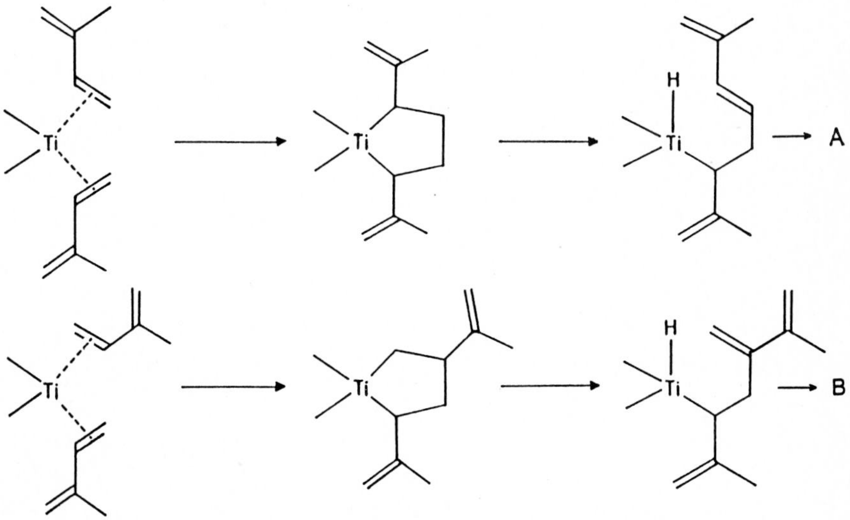

Ti → Ti → H Ti → A

Ti → Ti → H Ti → B

In all these reactions, only the vinyl part of isoprene coordinates to give the catalytic bis(diene) intermediate. This is due to the combined steric effect of two cyclopentadienyl ligands on the metal. When 2,3-dimethylbutadiene was employed under catalytic conditions, no oligomers were found. The bis-(diene) intermediate with 2,3-dimethylbutadiene is sterically too unstable to realize the coupling on the metal.

The situation is quite different in diene cyclotrimerization with mono-cyclopentadienyl titanium species [71]. Here, 2,3-dimethylbutadiene selectively enters into the final step of the catalytic cycle to give a co-oligomer shown below. With one Cp ligand on the metal, the coordination sphere is

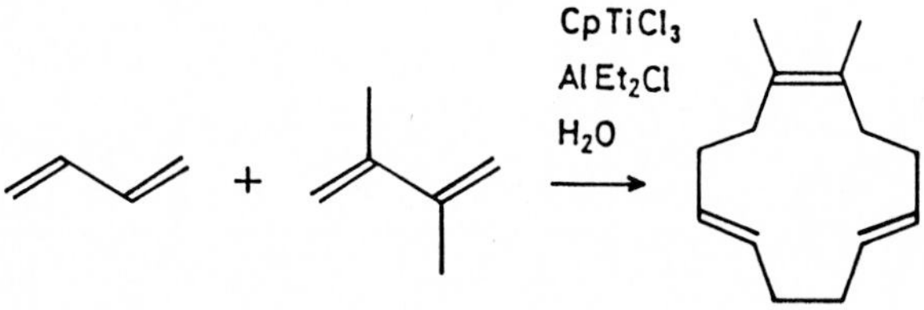

$CpTiCl_3$
$AlEt_2Cl$
H_2O

considerably widened and available metal orbitals are increased. Thus, CpTa and CpNb fragments allow stable η^4-coordination of two diene molecules and CpTi fragments coordinate(diene) and (allyl) ligands. The reactivity of these complexes critically depends on the identity of the metal and in the case of Ti, catalytic trimerization of dienes has been observed.

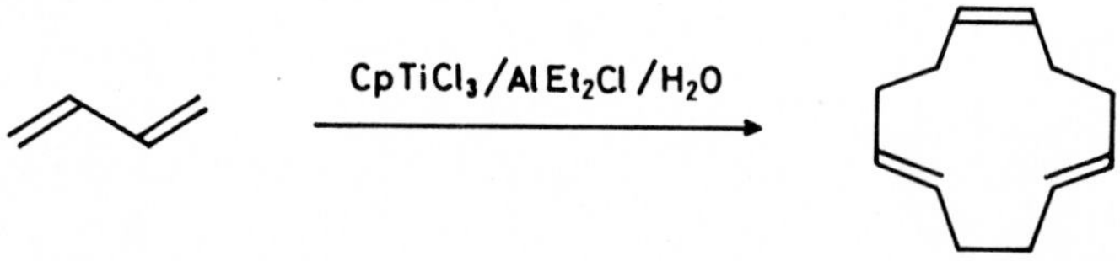

$CpTiCl_3/AlEt_2Cl/H_2O$

ctt-CDT

Here, the stereochemistry of the trimer is controlled by the coordination geometry of diene complexes and by its stereospecific reactions. The first step in the diene oligomerization is to couple two diene molecules through a labile bis(η^2-diene) complex intermediate to give a titanacycle with two trans-double bonds. The third butadiene molecule enters by s-cis-coordination to complete the catalytic cycle. Thus, selective formation of cis,trans,trans-CDT may be explained.

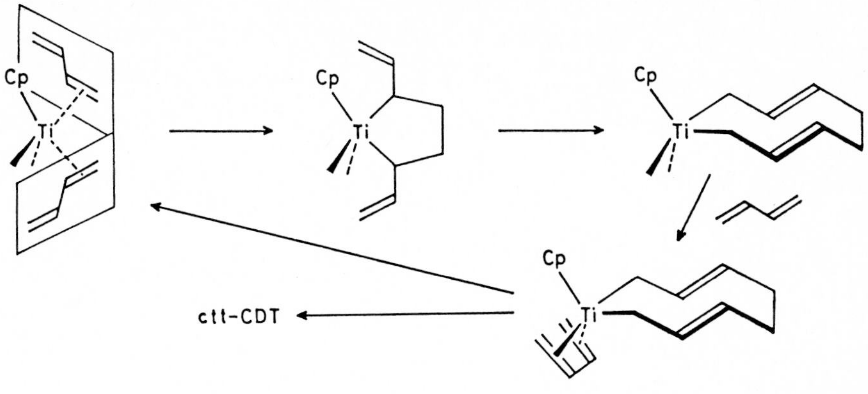

Cp Ti ctt-CDT

The cyclotrimerization is also proceed with isoprene and the following products are obtained with excellent selectivity (86%).

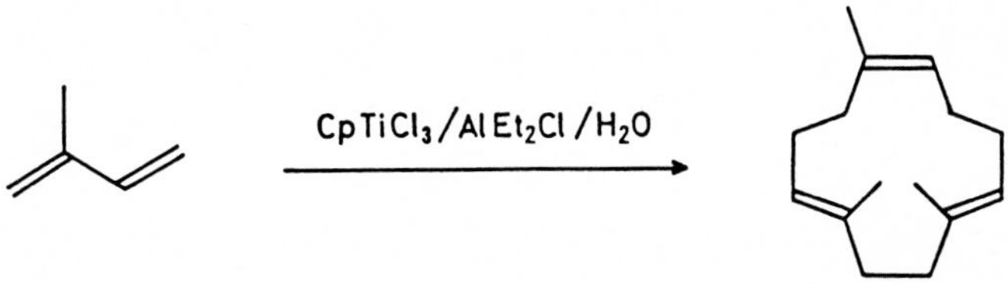

$CpTiCl_3/AlEt_2Cl/H_2O$

Carbonylbis(butadiene)iron is catalytically active in oligomerization of butadiene. A mixture of 4-vinylcyclohexane and 1,5-cyclooctadiene is produced at 70°C in the presence of Ph_3P. The phosphine ligand is added to keep the catalytic cycle by coordination to coordinatively unsaturated species. This Fe(0) species has only one vacant d-orbital for accepting a donor ligand and thus the $Fe(PPh_3)$ species has vacant sp^3 orbitals for diene catalysis. As described above, the analogous bis(diene) intermediate will be produced and further transformation gives cyclic products. By using chelating di-imine as auxiliary ligands in the catalysis, a 1:1 codimer of butadiene and acetylene was obtained selectively [72]. Thus, the intermediacy of (η^2-butadiene)(η^2-acetylene)iron(0) is implicated.

a) Dimerization

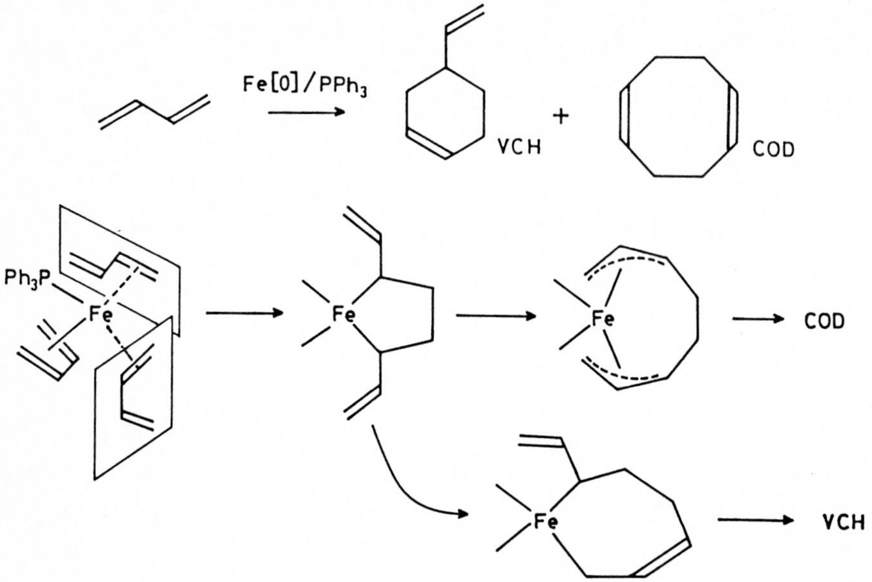

$Fe[0]/PPh_3$

VCH + COD

Ph_3P Fe → Fe → Fe → COD

Fe → VCH

b) Co-dimerization

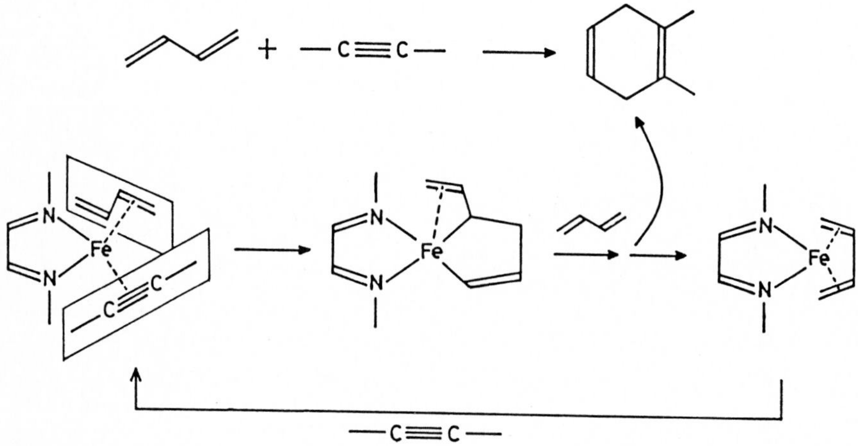

+ —C≡C— →

N Fe N C≡C → N Fe N → → N Fe N

—C≡C—

3.5. Polymerization of Dienes

Mechanisms for polymerization of 1,3-dienes have been proposed to involve η^2- or η^4-diene complexes with transition metals as intermediates [69]. In the case of selective cis-1,4-polymerization of butadiene, s-cis-η^4-coordination of butadiene has been postulated. The stereochemistry at the double bonds in polymers is thus controlled by the mode of diene coordination in the catalytic intermediate in which s-cis diene ligand is definitely prefered for low-valent later transition metals e.g. Co and Ni. As has been described earlier in this review, early transition metals such as Zr and Hf occasionally prefer the s-trans coordination. In accord with this remarkable trend, butadiene polymerizes to a trans-1,4-polymer with catalyst systems such as VCl_4-$AlEt_3$ or $TiCl_4$-$AlEt_3$. Very high catalytic activity precludes isolation and characterization of any diene complexes formed in the catalytic system. Therefore, we have calculated the stability of some plausible diene complexes as intermediates with the parameters obtained from the X-ray structures of Cp_2Zr(diene). Thus, a calculation has been performed by the EHMO method on the s-cis and s-trans butadiene complexes of $ZrCl_4$ species (C_{2v} structure). The total energies calculated for

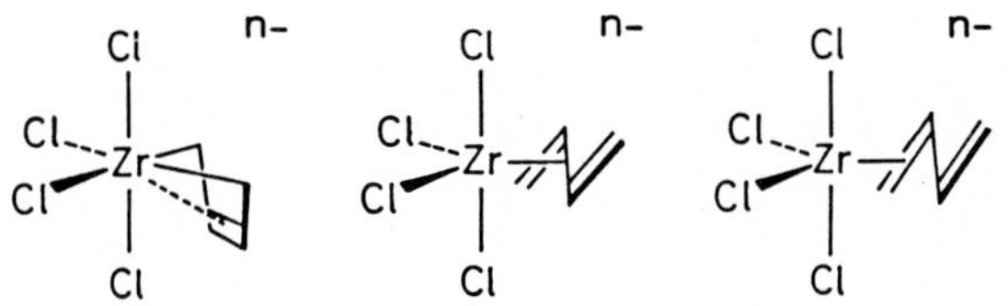

	n	s-cis ΔE	s-trans ΔE	η^2 s-trans ΔE
Zr(IV)	0	0.0	-1.0	1.8
Zr(III)	1	0.0	-0.4	9.9
Zr(II)	2	0.0	-2.0	18.4

in Kcal/mol

the s-trans complex was quite similar to the s-cis isomer just as in the case of discrete Cp_2Zr(butadiene) molecule [73]. The s-trans complex was calculated to be more stable than the corresponding η^2-isomer. Therefore, the s-trans diene coordination is proposed as an important intermediate in the "high-trans" polymerization of butadiene. The s-trans diene ligand is smoothly transformed to a syn-η^3-allyl complex by insertion to the polymer end and stabilized until the next attack of butadiene as shown below.

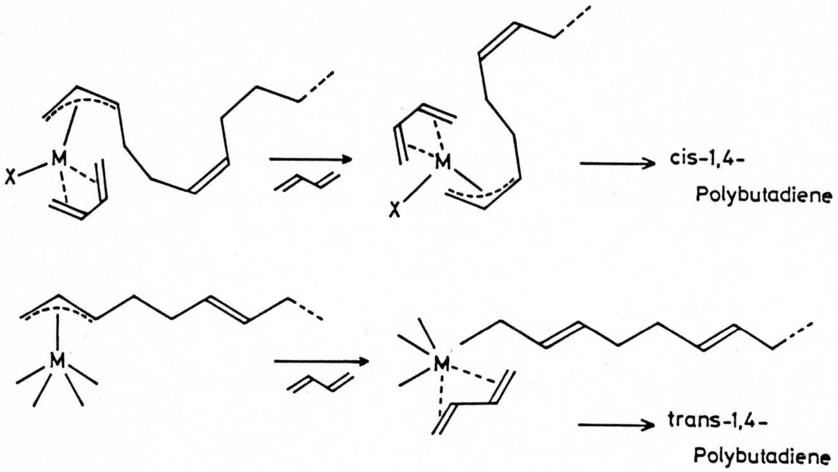

cis-1,4-Polybutadiene

trans-1,4-Polybutadiene

In some catalyst systems, e.g. $CoBr_2/PPh_3/Al(i\text{-}Bu)_3/H_2O$, 1,2-syndiotactic polybutadiene of low crystallinity [74] is produced. Recently a catalyst system, $Co(acac)_3/AlEt_2Cl/H_2O/CS_2$, was found to give the syndio polymer in very high selectivity [69d,e]. Thus, a polymer in 99.7% 1,2-syndio content was prepared and had a m.p. of 216°C. A mechanism involving s-cis coordination of butadiene was proposed, and the observed high 1,2-content was explained by selective reaction of the complexed diene with the alkylated carbon of the η^3-allyl part of the polymer chain.

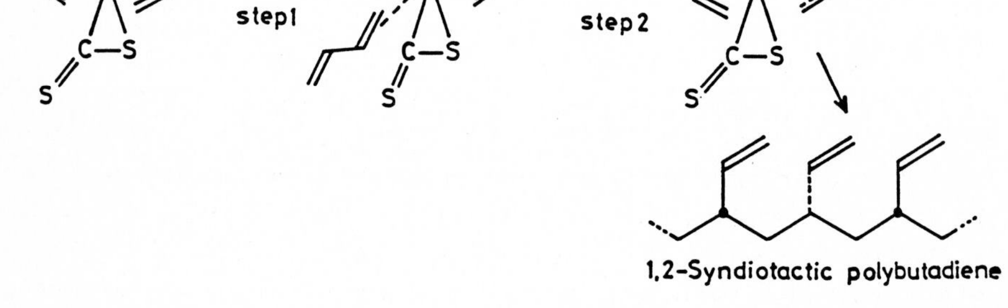

step1

step2

1,2-Syndiotactic polybutadiene

Considering the covalency of Co-C bonding in this catalytic complexes, we propose a coupling reaction between η^3-allyl and η^4-diene ligands promoted by an attack of diene. (see step 1)

Our calculations predict that the methyl or phenyl-substitution at the diene terminal enhances stability of the s-trans form. Actually, our X-ray structural analysis of the 1,4-diphenylbutadiene complex, $Cp_2Zr(PhCH=CH-CH=CHPh)$, indicated exclusive formation of s-trans form. Similarly, enhanced population of s-trans form was also observed by ^1H-NMR spectroscopy of the (E,E)-2,4-hexadiene complex, $Cp_2Zr(CH_3-CH=CH-CH=CH-CH_3)$. In the polymerization of (E)-1,3-pentadiene, a trans-1,4-polymer has been obtained by catalysis of $VCl_3/AlEt_3$ or $V(acac)_3/AlClEt_2$ on 98-99% selectivity. Here again, an s-trans pentadiene complex seems to be an intermediate [69].

A terminally disubstituted diene, e.g. (E,E)-2,4-hexadiene, has been polymerized to a stereoregular polymer with an erythro-trans-1,4-diisotactic structure by catalysis of $Ti(acac)_3/AlEt_2Cl$ or $Co(acac)_2/AlClEt_2$ at 30°C [75]. Since only s-cis coordination of the diene had been known at that time, the authors presumed the incipient s-cis coordination of the diene, followed by anti-syn isomerization of the allyl species. Now, preference of s-trans coordination is apparent, especially for early transition metals, and coordination of the diene will probably occur in s-trans fashion to give the required syn-allyl species directly which, eventually results in the trans-diisotactic polymer.

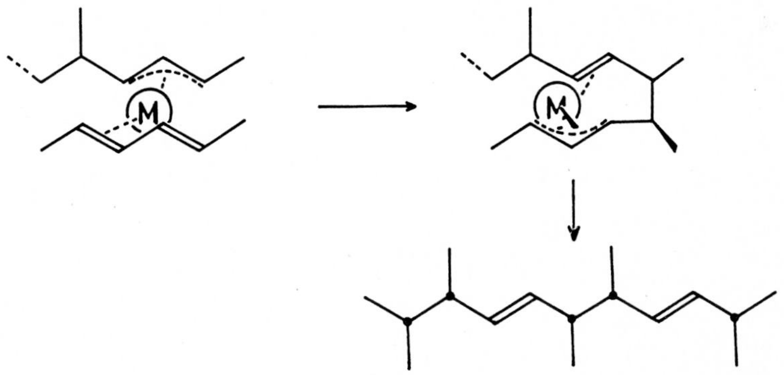

Erythro trans-1,4-diisotactic polymer

Alternating copolymers between butadiene and propylene have been prepared by catalysis of $VCl_4/AlEt_3$ at -40°C-0° [76]. The diene component is incorporated at the trans-butenylene part in the copolymer. The s-trans coordination of butadiene leading to a syn-η^3-allyl intermediate is thus proposed for this polymerization. A previous mechanism has postulated the fast anti-syn isomerization of the allyl intermediate at such a low temperature. Consideration of the inertness of the anti-allyl species for geometrical isomerization at low temperature supports the present mechanism involving an s-trans diene complex formation at the first step. The subsequent reaction of the syn-allyl species with propylene gives an alkyl-type product which now accepts butadiene to fill the coordinative unsaturation.

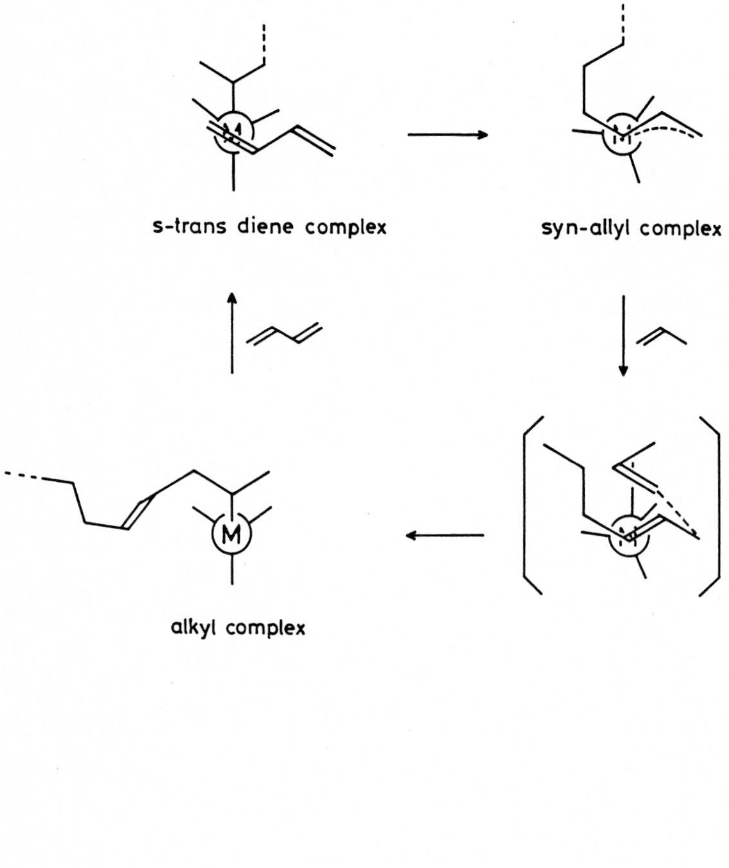

s-trans diene complex

syn-allyl complex

alkyl complex

ACKNOWLEDGEMENT

We are grateful to S. Nakayama for the illustrations and E. Kambayashi for the typing.

REFERENCES and NOTES

1. R. B. King in *The Organic Chemistry of Iron*, (Ed.), E. A. Koerner von Gustorf, F.-W. Grevels, and I. Fishler, Academic Press, New York/London, 1978, Vol.1, p.397, p.525.
2. C. Krüger, B. L. Barnett, and D. Brauer in *The Organic Chemistry of Iron*, (Ed.), E. A. Koerner von Gustorf, F.-W. Grevels, and I. Fishler, Academic Press, New York/London, 1978, Vol.1, p.1.
3. See relevant parts of *Comprehensive Organometallic Chemistry*, (Ed.), G. Wilkinson, F. G. A. Stone, and E. W. Abel, Pergamon Press, Oxford/New York, 1983, Vol.3-6.
4. (a) G. Erker, J. Wicher, K. Engel, F. Rosenfeldt, W. Dietrich, and C. Krüger, *J. Am. Chem. Soc*, 102 (1980) 6344; (b) G. Erker, J. Wicher, K. Engel, C. Krüger, and A.-P. Chiang, *Chem. Ber.*, 115 (1982) 3300, 3311; (c) U. Dorf, K. Engel, and G. Erker, *Organometallics*, 2 (1982) 462.
5. G. Erker, U. Dorf, R. Benn, R.-D. Reinhardt, and J. L. Petersen, *J. Am. Chem. Soc.*, 106 (1984) 7649.
6. (a) Y. Kai, N. Kanehisa, K. Miki, N. Kasai, K. Mashima, K. Nagasuna, H. Yasuda, and A. Nakamura, *J. Chem. Soc., Chem. Commun.*, (1982) 191; (b) H. Yasuda, Y. Kajihara, K. Mashima, K. Lee, and A. Nakamura, *Chem. Lett.*, (1981) 519; (c) H. Yasuda, Y. Kajihara, K. Mashima, K. Nagasuna, K. Lee, and A. Nakamura, *Organometallics*, 1 (1982) 388.
7. H. Yasuda, K. Tatsumi, and A. Nakamura, *Acc. Chem. Res.*, in press.
8. K. Tatsumi, H. Yasuda, and A. Nakamura, *Israel J. Chem.*, 23 (1983) 145.
9. P. W. Jolly, R. Mynott, and R. Salz, *J. Organomet. Chem.*, 184 (1980) C49.
10. P. S. Skell and M. J. McGlynchey, *Angew. Chem.*, 87 (1975) 215.
11. (a) D. A. Whiting, *Cryst. Struct. Commun.*, 1 (1972) 379; (b) A. Carbonaro and A. Greco, *J. Organomet. Chem.*, 25 (1970) 477, and see ref. 22.
12. (a) R. L. Harlow, P. J. Krusic, R. J. McKinney, and S. S. Wreford, *Organometallics*, 1 (1982) 1506; (b) M. Heberhold and A. Razavi, *Angew. Chem.*, 87 (1975) 351; (c) G. Huttner, D. Neugebauer, and A. Razavi, *Angew. Chem.*, 87 (1975) 353.
13. (a) L. Porri, A. Lionetti, G. Allegra, and A. Immirzi, *J. Chem. Soc., Chem. Commun.*, (1965) 336; (b) A. Immirzi and G. Allegra, *Acta Cryst.*, B25 (1969) 120.
14. T. C. van Soest, A. van der Ent, and E. C. Royers, *Crys. Struct. Commun.*, 3 (1973) 527.
15. S. Datta, S. S. Wreford, R. P. Beatty, and T. J. McNeese, *J. Am. Chem. Soc.*, 101 (1979) 1053.
16. R. P. Beatty, S. Datta, and S. S. Wreford, *Inorg. Chem.*, 18 (1979) 3139.
17. E. O. Fischer, H. P. Kögler, and P. Kuzel, *Chem. Ber.*, 93 (1960) 3006.
18. A. Nakamura, K. Tani, and S. Otsuka, *J. Chem. Soc. (A)*, (1969) 1404.
19. I. Noda, H. Yasuda, and A. Nakamura, *Organometallics*, 2 (1983) 1207.
20. S. D. Ittel and C. A. Tolman, *Organometallics*, 1 (1982) 1432.
21. (a) M. I. Davis and C. S. Speed, *J. Organomet. Chem.*, 21 (1970) 401; (b) O. S. Mills and G. Robinson, *Acta Cryst.* B16 (1963) 758.
22. A. Carobnaro and F. Cambisi, *J. Organomet. Chem.*, 44 (1972) 171.
23. (a) M. Kotzian, S. Özkar, and C. G. Kreiter, *J. Organomet. Chem.*, 229 (1982) 29; (b) S. Özkar and C. G. Kreiter, *J. Organomet. Chem.*, 256 (1983) 57; (c) S. Özkar and C. G. Kreiter, *Chem. Ber.*, 116 (1983) 3637.
24. H. M. Büch, P. Binger, R. Goddard, and C. Krüger, *J. Chem. Soc., Chem. Commun.*, (1983) 649.

25. V. G. Andrianov, V. P. Martynov, K. N. Anisimov, N. E. Kolobova, and V. V. Skripkin, *J. Chem. Soc., Chem. Commun.*, (1970) 1252.
26. E. O. Fischer, H. P. Kögler, and P. Kuzel, *Chem. Ber.*, 93 (1960) 3006.
27. J. L. Davidson, K. Davidson, and W. E. Lindsell, *J. Chem. Soc., Chem. Commun.*, (1983) 452.
28. (a) M. J. Bunker, M. L. H. Green, C. Couldwell, and K. Prout, *J. Organomet. Chem.*, 192 (1980) C6; (b) M. J. Bunker and M. L. H. Green, *J. Chem. Soc., Dalton Trans.*, (1981) 85.
29. H. Yasuda, K. Tatsumi, T. Okamoto, K. Mashima, K. Lee, A. Nakamura, Y. Kai, N. Kanehisa, and N. Kasai, *J. Am. Chem. Soc.*, 107 (1985) in press.
30. Y. Wakatsuki, K. Aoki, and H. Yamazaki, *J. Chem. Soc., Dalton Trans.*, (1982) 89.
31. H. E. Sasse and M. L. Ziegler, *Z. anorg. allg. Chem.*, 392 (1972) 167.
32. M. Tachikawa, J. R. Shapley, R. C. Haltiwanger, and C. G. Pierpont, *J. Am. Chem. Soc.*, 98 (1976) 4651.
33. H. D. Murdoch and E. Weiss, *Helv. Chim. Acta.*, 45 (1962) 1156.
34. M. L. Ziegler, *Z. anorg. allg. Chem.*, 355 (1967) 12.
35. V. C. Adam, J. A. J. Jarvis, B. T. Kilbourn, and P. G. Owston, *J. Chem. Soc. (D)*, (1971) 467.
36. (a) J. W. Lauher and R. Hoffmann, *J. Am. Chem. Soc.*, 98 (1976) 1729; (b) A. Stockis and R. Hoffmann, *J. Am. Chem. Soc.*, 102 (1980) 2952.
37. S. S. Wreford and J. F. Whiteney, *Inorg. Chem.*, 20 (1981) 3918.
38. M. L. H. Green, G. Parkin, C. Mingquin, and K. Prout, *J. Chem. Soc., Chem. Commun.*, (1984) 1400.
39. Y. Kai, N. Kanehisa, K. Miki, N. Kasai, K. Mashima, H. Yasuda, and A. Nakamura, *Chem. Lett.*, (1982) 1277.
40. (a) P. J. Harris, J. A. K. Howard, S. A. R. Knox, R. P. Phillips, F. G. A. Stone, and P. Woodward, *J. Chem. Soc., Dalton Trans.*, (1976) 377; (b) O. Gambino, M. Valle, S. Aime, and G. A. Vaglio, *Inorg. Chim. Acta.*, 8 (1974) 71; (c) I. Noda, H. Yasuda, and A. Nakamura, *J. Organomet. Chem.*, 250 (1983) 447.
41. A. Zwijnenburg, H. O. Oven, C. J. Gronenboom, and H. J. de Lief de Meijer, *J. Organomet. Chem.*, 94 (1975) 24.
42. (a) G. Erker, K. Berg, C. Krüger, G. Müller, K. Angermund, R. Benn, and G. Schroth, *Angew. Chem., Int. Ed. Engl.*, 23 (1984) 455; (b) K. Berg and G. Erker, *J. Organomet. Chem.*, 270 (1984) 53.
43. P. R. Brown, F. Geoffrey, N. Cloke, M. L. H. Green, and N. J. Hazel, *J. Chem. Soc., Dalton Trans.*, (1983) 1075.
44. J. W. Faller and A. M. Rosan, *J. Am. Chem. Soc.*, 99 (1977) 4858.
45. The X-ray structure analyses on $Fe(CO)_3$(1,3-diene) are numerous. See for example, (a) P. E. Riley and R. E. Davis, *Acta Crystallogr., Sect B*, 34 (1978) 3760; (b) O. S. Mills and G. Robinson, *ibid.*, 16 (1963) 758.
46. MO studies on $Fe(CO)_3$(butadiene) reported so far were concerned mostly with the assignment of photoelectron spectra. (a) J. A. Connor, L. M. R. Derrick, M. B. Hall, I. H. Hillier, M. F. Guest, B. R. Higginson and D. R. Lloyd, *Mol. Phys.*, 28 (1974) 1193; (b) K. Tatsumi and T. Fueno, *Bull. Chem. Soc. Jpn.*, 49 (1976) 929.
47. (a) L. E. Orgel, *J. Inorg. Nucl. Chem.*, 2 (1956) 315; (b) K. Tatsumi and R. Hoffmann, *Inorg. Chem.*, 20 (1981) 3771 and references therein.
48. H. H. Brintzinger and L. S. Bartell, *J. Am. Chem. Soc.*, 92 (1970) 1105; ref. 36a.
49. B. E. R. Schilling, R. Hoffmann, D. L. Lichtenberger, *J. Am. Chem. Soc.*, 101 (1979) 585, and references therein.
50. The valence orbitals of CpM are well known. See, for instance, J. W. Lauher, M. Elian, R. H. Summerville, and R. Hoffmann, *J. Am. Chem. Soc.*, 98 (1976) 3219, and references therein.
51. P. Kubáček, R. Hoffmann, and Z. Havlas, *Organometallics*, 1 (1982) 180.
52. (a) C. P. Lillya and B. R. Bonazza, *J. Am. Chem. Soc.*, 96 (1974) 2298; (b) J. W. Burrill, B. R. Bonazza, B. R. Garrett, and C. P. Lillya, *J. Organomet. Chem.*, 104 (1976) C37.

53. T. H. Whitesides and R. W. Arhart, *Inorg. Chem.*, 14 (1975) 209.
54. M. Brookhart and D. L. Harris, *Inorg. Chem.*, 13 (1974) 1540.
55. (a) T. S. Sorensen and C. R. Jablonski, *J. Organomet. Chem.*, 25 (1970) C62; (b) C. P. Lillya and R. A. Sahatjian, *J. Organomet. Chem.*, 25 (1970) C66.
56. (a) M. F. Semmelhack, J. W. Herndon, and J. Liu, *Organometallics*, 2 (1983) 1885; (b) M. F. Semmelhack and H. T. M. Le, *J. Am. Chem. Soc.*, 106 (1984) 2715, 5388.
57. J.-E. Bäckvall, *Acc. Chem. Res.*, 16 (1983) 343.
58. (a) A. Bond, B. Lewis, and M. Green, *J. Chem. Soc., Dalton Trans.*, (1975) 1109; (b) M. Green, B. Lewis, J. J. Daly, and F. Sanz, *J. Chem. Soc., Dalton Trans.*, (1975) 1118; (c) M. Bottrill, R. Goddard, M. Green, R. P. Hughes, M. K. Lloyd, B. Lewis, and P. Woodward, *J. Chem. Soc., Chem. Commun.*, (1975) 253; (d) F.-W. Grevels, U. Feldhoff, J. Leitich, and C. Krüger, *J. Organomet. Chem.*, 118 (1976) 79.
59. R. Noyori, S. Makino, and H. Takaya, *J. Am. Chem. Soc.*, 93 (1971) 1272.
60. H. Yasuda and A. Nakamura, unpublished.
61. (a) H. Yasuda, Y. Kajihara, K. Mashima, K. Nagasuna, and A. Nakamura, *Chem. Lett.*, (1981) 671, 719; (b) Y. Kai, N. Kanehisa, K. Miki, N. Kasai, M. Akita, H. Yasuda, and A. Nakamura, *Bull. Chem. Soc. Jpn.*, 56 (1983) 3735; M. Akita, H. Yasuda, and A. Nakamura, *Chem. Lett.*, (1983) 217.
62. (a) V. Skiffe and G. Erker, *J. Organomet. Chem.*, 241 (1983) 15; (b) U. Dorf, K. Engel, and G. Erker, *Angew. Chem., Int. Ed. Engl.*, 21 (1982) 914; (c) G. Erker, K. Engel, U. Dorf. J. L. Atwood, and W. E. Hunter, *Angew. Chem., Int. Ed. Engl.*, 21 (1982) 914.
63. G. Erker, K. Engel, J. L. Atwood, and W. E. Hunter, *Angew. Chem., Int. Ed. Engl.*, 22 (1983) 494.
64. H. Yasuda and A. Nakamura, unpublished results.
65. U. Feldhoff, Univ. Essen (1975), cited in ref. 1.
66. P. W. Jolly and G. Wilke, "The Organic Chemistry of Nickel", Academic Press, New York/London, 1975, Vol.2, p.213.
67. P. W. Jolly, in ref. 3, Vol.8, p.671.
68. P. W. Jolly and G. Wilke, "The Organic Chemistry of Nickel", Academic Press, New York/London, 1975, Vol.2, p.133.
69. (a) G. Natta, L. Porri, A. Carbonaro, and G. Stoppa, *Makromol. Chem.*, 77 (1964) 114; (b) G. Natta, L. Porri, and A. Carbonaro, *Makromol. Chem.*, 77 (1964) 126; (c) B. A. Dolgoplosk, E. I. Tinyakova, I. S. H. Guzman, and L.L. Afinogenova, *J. Polym. Sci., Polym. Chem. Ed.*, 22 (1984) 1535; (d) H. Ashitaka, H. Ishikawa, H. Ueno, and A. Nagasaka, *J. Polym. Sci., Polym. Chem. Ed.*, 21 (1983) 1853; (e) H. Ashitaka, K. Jinda, and H. Ueno, *J. Polym. Sci., Polym. Chem. Ed.*, 21 (1983) 1951, 1989.
70. H. Yasuda and A. Nakamura, unpublished results.
71. H. Yasuda, M. Akita, and A. Nakamura, unpublished results.
72. H. tom Dieck, *Angew. Chem., Int. Ed. Engl.*, 22 (1983) 778.
73. K. Tatsumi and A. Nakamura, to be published.
74. Y. Takeuchi, A. Sekimoto, and M. Abe, in "New Industrial Polymers", *Am. Chem. Soc. Symp. Ser. 4*, American Chemical Society, Washinton, DC, 1974, p.15, p.26.
75. M. Kamachi, N. Wakabayashi, and S. Murahashi, *Macromol.*, 7 (1974) 744.
76. (a) A. Kawasaki, I. Maruyama, M. Taniguchi, R. Hirai, J. Furukawa, *J. Polym. Sci.*, B7 (1969) 411; (b) J. Furukawa, R. Hirai, M. Nakaniwa, *J. Polym. Sci.*, B7 (1969) 671.

CHAPTER 2

STEREOCHEMISTRY OF THE PHOSPHATES OF DIVALENT METALS

ANDERS G. NORD ★

Section of Mineralogy, Swedish Museum of Natural History, P.O. Box 50007, S-104 05 Stockholm 50, Sweden

1 INTRODUCTION

Inorganic phosphates are of great interest not only to structural chemists due to their great variety in stereochemistry, but also to inorganic chemists, mineralogists, geochemists, and chemical engineerers because of their great technical importance in many different fields of chemical industry. Some examples as well as a few notes of historical interest will be given prior to the more detailed description of the stereochemistry of the phosphates of divalent metals.

The element phosphorus was discovered in 1669 by the German alchemist and merchant H. Brand during his attempts to find the "philosopher's stone" which was at that time thought to be necessary for the making of gold. The name "phosphate" was later used for the oxosalts of phosphorus acids. Orthophosphoric acid was in fact discovered as early as 1680 by Boyle, although its chemical nature was elucidated much later, 1743, by Marggraf. About 1770 Scheele recognized phosphorus as an essential element in animal bones and teeth. Gay-Lussac, Stromeyer and others explained the relationships between ortho-, di- and meta-phosphoric acids. Papers on isomorphism and solid solutions in oxosalts such as phosphates were published around 1800 by Fuchs and Mitscherlich.

The first phosphate structures to be determined were those of KH_2PO_4 (ref. 1,2), YPO_4 (ref. 3), and Ag_3PO_4 (ref. 4). The crystal structure of the well-known and technically important mineral apatite, $Ca_5(PO_4)_3(Cl,OH,F)$, was also published quite early (ref. 5). Through these studies, the tetrahedral form of the PO_4^{3-} anion was established. Today almost one thousand phosphate structures have been determined. It soon became evident that the PO_4 tetrahedra may share corners to give larger anions. These compounds are commonly called "condensed phosphates"; they may be diphosphates (formerly called pyrophosphates), tri-phosphates, metaphosphates, ultraphosphates, and so forth.

★ Present address: Swedish National Defence Research Institute, Department of Weapons Technology, P.O. Box 27322, S-102 54 Stockholm, Sweden

TABLE 1.1
Divalent-metal M^{2+} cations arranged after increasing atomic number (Z). The ionic radii (r_M, in Å units) are given for an octahedral environment of oxygen ligands after Shannon and Prewitt (6, 7). Abbreviations: CR = crystal radius, IR = effective ionic radius (cf. Ref. 7); LS = low spin, HS = high spin.

Metal "M"	Z	Cation radii (LS or no spin)		Cation radii (HS)	
		CR	IR	CR	IR
Be	4	0.59	0.45		
Mg	12	0.860	0.720		
Ca	20	1.14	1.00		
Mn	25	0.81	0.67	0.970	0.830
Fe	26	0.75	0.61	0.920	0.780
Co	27	0.79	0.65	0.885	0.745
Ni	28	0.830	0.690		
Cu	29	0.87	0.73		
Zn	30	0.880	0.740		
Sr	38	1.32	1.18		
Cd	48	1.09	0.95		
Sn	50	-	0.93		
Ba	56	1.49	1.35		
Hg	80	1.16	1.02		
Pb	82	1.33	1.19		

This chapter will summarize and discuss anhydrous and hydrous phosphate crystal structures of divalent metals and their interrelationships from different viewpoints. The contents will only discuss so-called "normal" phosphates, i.e., phosphates where the phosphorus atoms are linked solely to oxygen atoms in contrast to the situation in the "substituted" phosphates in which all or some of the phosphorus atoms are linked to atoms other than oxygen. The metal cation radii given in Table 1.1 are those of Shannon and Prewitt (ref. 6,7). So as to keep the length of the chapter within reasonable limits, hydrogen orthophosphates are only briefly discussed. Moreover, phosphates containing also other cations or anions will only be included in the discussion when considered to be of great stereochemical interest. Examples of such phosphate compounds are the minerals $Mg_2(PO_4)F$ (wagnerite), $NaFePO_4$ (trifylite), $Mg_4(PO_4)_2(OH,O,F)$ (althausite), or $Ca_{18}M_2H_2(PO_4)_{14}$ (whitlockite). Other phosphates of synthetic origin within this scope are, for instance, $Ca_4(PO_4)_2O$, $Cu_9Na_6(PO_4)_8$, $NaMgP_5O_{16}$, and $Ba_2Zn_3P_{10}O_{30}$.

The divalent-metal phosphates are here only described from chemical-structural points of view, although many of them are or have been of technical importance. This applies to the zinc orthophosphates for their potential use as components in colour television screens. Many other divalent-metal phosphates have been of great use as fertilizers, food additives, detergents, oil additives, deflocculants, or as catalysts for polymerization and other organic processes. Still more important is their use as coatings on steel to prevent rusting. Many of the hydrous and some of the anhydrous orthophosphates exist as naturally occurring minerals, which will be discussed in the following subsections. Some phosphate minerals have been known since remote antiquity. For instance, turquois, $CuAl_6(PO_4)_4(OH)_8 \cdot 4H_2O$, has been found among the remains of many ancient civilizations, e.g. in India, China, Egypt, and Mesopotamia.

As mentioned above, the crystal structures of the phosphates are interesting because of the manifold of ways to combine the PO_4 tetrahedra by corner-sharing to give more or less complicated anions. In particular the ultraphosphates display complicated and strange networks of corner-sharing phosphate tetrahedra, but interesting and unusual anionic configurations are also found in the metaphosphates. Also, the divalent M^{2+} cations may display a great variety of metal-oxygen polyhedra, with coordination numbers in the region from four to twelve. Rare coordination numbers such as five and seven occur quite frequently, primarily among the orthophosphates. The metal-oxygen polyhedra are frequently connected by the PO_4 groups by corner-sharing, sometimes by edge- and rarely by face-sharing, thus building up a three-dimensional network of the crystal structure.

While some crystal structures, such as the olivine-related structure of the rare mineral sarcopside, $(Fe,Mn,Mg)_3(PO_4)_2$, are very simple having only 21 atomic positional parameters (ref. 8), others like the α-$Ca_3(PO_4)_2$ structure may be very complicated with as many as eighteen distinct calcium atoms and a monoclinic unit cell (space group $P2_1/a$) of 4318 $Å^3$ (ref. 9).

Among the ortho-, di- and meta-phosphates isomorphism is very common and will be discussed below. Polymorphism is also common among these phosphates, and in particular the anhydrous orthophosphates display numerous interesting phase transitions. Extensive studies of $(M'M'')_3(PO_4)_2$ solid solutions having the sarcopside, graftonite or farringtonite structure (all rare minerals) have been undertaken by means of X-ray and neutron diffraction or ^{57}Fe Mössbauer spectroscopy. It has thus been shown that a crystal structure may quite easily change to another by applying high pressures or by replacing part of one divalent metal cation with another (e.g. ref. 10-12). Many systematic cation distribution studies have also been undertaken on these phases and will be summarized in the subsections to follow.

The contents have been divided into various sections (subchapters). The first and largest section (section 2) deals with orthophosphates, anhydrous as well as hydrous and hydrogen orthophosphates. The sections following will discuss diphosphates (section 3), triphosphates and related condensed phosphates (section 4), metaphosphates (section 5), and ultraphosphates (section 6). The last subchapters contain topics on various crystallographic matters such as the dimensions of phosphate tetrahedra (section 7), metal-oxygen coordination environments and distances (section 8), and close-packing in the phosphate structures here discussed (section 9). At the end of this chapter 258 references are given. The chapter contains 20 tables and 35 figures.

Notes. To save space, the greatest care has been taken to limit the large number of references. Generally, only the latest or the most relevant paper on a certain crystal structure or subject will be cited in the text. Unreferenced data like melting points which are not very important for the stereochemical discussion have been obtained from Gmelin's Handbuch der anorganischen Chemie, Auflage Acht (8). Divalent-metal phosphate compounds listed therein but for which no adequate crystallographic data seem to exist are only mentioned in very special cases. Interatomic distances, unless they are of crucial importance for the discussion, are only given with three-figure accuracy, and without standard deviations (which are usually 0.01 Å or better).

The following symbols are used throughout the text: S.G. stands for Space Group, with symbols given according to the International Tables for X-ray Crystallography (ref. 13). "V" denotes the unit cell volume in $Å^3$ units, "V_{ox}" is the unit cell volume divided by the number of oxygen atoms in the unit cell, and "Z" is the number of formula units per cell. "CN" means cation coordination number. An abbreviation such as CN = 5,5,7 means that the metal cations' coordination numbers in the compound in question are 5, 5, and 7.

"M" always refers to a divalent-metal cation. <M-O> denotes the average M-O distance within a MO_n polyhedron (CN = n), while d(X-Y) is the distance between the atoms "X" and "Y".

Some referenced papers do not contain interatomic distances or angles, or the cited values appeared to be somewhat unreliable. In such cases the present author has calculated these values from the atomic positional parameters given in the respective papers with the crystallographic DISTAN program.

2 ORTHOPHOSPHATES

This section contains six subsections on anhydrous orthophosphates, one on hydrous phosphates, and a final brief summary of hydrogen orthophosphates and miscellaneous orthophosphates, principally compounds of the type $M_2(XO_4)Y$ where X = P and Y = Cl, F, OH etc.

2.1 α-$Zn_3(PO_4)_2$, β-$Zn_3(PO_4)_2$ and related structures

In their study of the system ZnO - P_2O_5, published in 1958, Katnack and Hummel found that pure zinc orthophosphate is dimorphous (ref. 14). The low-temperature modification, called α-$Zn_3(PO_4)_2$, is stable below 942°C, where a very sluggish transition to the β form occurs. β-$Zn_3(PO_4)_2$ melts at 1060°C but is metastable at room temperature. The crystal structure of α-$Zn_3(PO_4)_2$, so far unique, was determined in 1964 by Calvo (ref. 15). This structure is built up of rather regular PO_4 tetrahedra and two kinds of ZnO_4 tetrahedra. The main structural feature is large spiral chains of the sequence -Zn(1)O_4 - PO_4 - Zn(1)O_4 - PO_4 - parallel to the two-fold screw axis (C2/c space group symmetry) and interconnected by Zn(2)O_4 tetrahedra. The spiral has a pitch equal to the length of the monoclinic _b_ axis. Both ZnO_4 tetrahedra are fairly regular. Some crystallographic data for this phase are given in Table 2.1.1.

Upon replacement of Zn^{2+} by small amounts of other divalent cations like Mn^{2+}, Fe^{2+} and Cd^{2+}, the α-$Zn_3(PO_4)_2$ structure is transformed to another phase often called "γ-$Zn_3(PO_4)_2$" although it does not exist as a pure phase but should rather be denoted γ-$(Zn,M)_3(PO_4)_2$ (see section 2.2).

TABLE 2.1.1

Crystallographic data for α-$Zn_3(PO_4)_2$ and the "β-$Zn_3(PO_4)_2$" structure types.

Compound	α-$Zn_3(PO_4)_2$	β-$Zn_3(PO_4)_2$	β'-$Mn_3(PO_4)_2$	β'-$Cd_3(PO_4)_2$
Ref.	(15)	(16)	(19)	(20)
S.G.	C2/c	$P2_1/c$	$P2_1/c$	$P2_1/c$
a (Å)	8.14	9.393	8.94	9.221
b (Å)	5.63	9.170	10.04	10.335
c (Å)	15.04	8.686	24.12	24.90
β (°)	105.13	125.73	120.8	120.7
V (Å^3)	665.4	607.2	1860	2040
V_{ox} (Å^3)	20.79	18.97	19.37	21.25
<M-O> for				
CN = 4	1.96, 1.98			
CN = 5		2.08, 2.09	2.13, 2.13, 2.14, 2.16, 2.16, 2.18	2.25, 2.25, 2.25, 2.29
CN = 6		2.16	2.22, 2.22, 2.24	2.32, 2.34, 2.34, 2.35, 2.35
Z	4	4	12	12

The crystal structure of the high-temperature modification β-$Zn_3(PO_4)_2$ was determined by Stephens and Calvo (ref. 16). There are three independent cation sites in the structure, strongly ligated by 4, 5, and 5 oxygen atoms, but with further oxygens at about 2.5 Å from the central zinc atoms. The coordination numbers (CN) for zinc might, therefore, also be given as 5,5,6. A calculation of the bond strength sums has been undertaken (ref. 17) using Brown and Shannon's formula

$$p = \Sigma s_i = \Sigma s_o(R/R_o)^{-N},$$

where s_o = 0.5, R_o = 1.947 and N = 5 for Zn-O bonds (ref. 18). The CN:s 5,5,6 gave the bond strength sums 2.030, 1.914 and 1.968 (cf. ref. 17), while CN = 4,5,5 (as preferred by Stephens and Calvo (in ref. 16) gave 1.900, 1.678, and 1.827 valence units. Since the bond strength sum "p" (in valence units) should be equal in magnitude to the electric charge of the respective central cation, the CN:s 5,5,6 are definitely favoured in this case, and will be used in the discussion to follow. The crystal structure is illustrated in Fig. 2.1.1. Some structural data are included in Table 2.1.1.

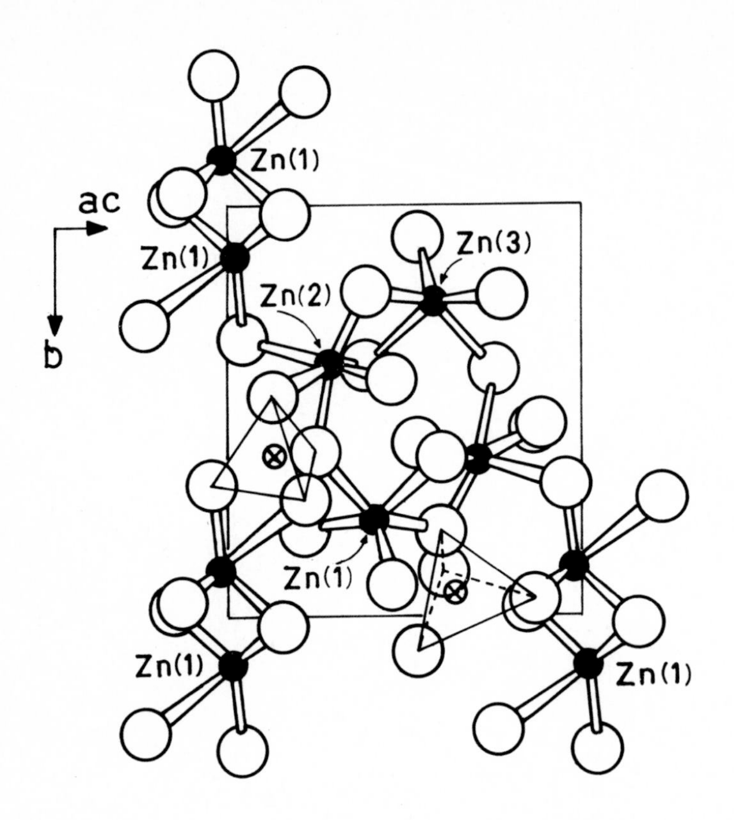

Fig. 2.1.1. The crystal structure of β-$Zn_3(PO_4)_2$. S. G. $P2_1/c$; Z = 4. ● = Zn, ⊗ = P, large open circles represent oxygen atoms. Reproduced with permission from the National Research Council of Canada from Stephens and Calvo: "Crystal structure of β-$Zn_3(PO_4)_2$" (ref. 16) in Can. J. Chem. vol. 45, pp. 2303-2312 (1967).

Two more phosphate compounds have crystal structures related to that of β-$Zn_3(PO_4)_2$. They are usually called β'-$Mn_3(PO_4)_2$ and β'-$Cd_3(PO_4)_2$, with β' instead of β to distinguish them from, but also suggest similarities to, the β-$Zn_3(PO_4)_2$ structure. The β' phases have superstructures with c' = 3c, i.e., the "c" axis is tripled with respect to the "β-$Zn_3(PO_4)_2$" subcell. They each contain nine distinct metal-oxygen polyhedra, five- or six-coordinated (cf. Table 2.1.1).

The crystal structure of β'-$Mn_3(PO_4)_2$ was also determined by Stephens and Calvo (ref. 19). According to Stephens (ref. 20), there also is a polymorph of manganese(II) orthophosphate having the graftonite structure (see section 2.3) with CN:s 6,5,5 or 7,5,5. It is quite clear that upon replacement of even small amounts of Mn^{2+} by any of the divalent cations of Mg, Fe, Co, Ni, Cu, Zn or Ca, solid solutions of the form $(Mn_{1-x}M_x)_3(PO_4)_2$ with the graftonite structure are formed (ref. 21). Such modifications exist as a natural mineral called beusite. On the other hand, attempts to obtain pure $Mn_3(PO_4)_2$-graftonite by quenching the respective sample from various temperatures or by annealing it at higher or lower temperatures failed (ref. 21). Differential thermal analyses of β'-$Mn_3(PO_4)_2$ in an inert argon atmosphere showed no distinct phase transition between room temperature and the melting point ($\sim$1370 K). The sample remained almost colourless, with a slightly pink appearance (ref. 21). On the other hand, Stephens (ref. 20) reported his Mn-graftonite to be dark-coloured, indicating that there had been some oxidation of Mn^{2+}. Since the graftonite phase is easily formed upon the substitution of small amounts of M^{2+} cations for Mn^{2+} in β'-$Mn_3(PO_4)_2$, it is reasonable to assume that also other cations like Mn^{3+} or Mn^{4+} might stabilize the graftonite structure in a similar way, and that Stephens' Mn-graftonite was somewhat impure. The reported dark colour of the sample supports this hypothesis. Moreover, when manganese(II) orthophosphate was heated with small amounts of Mn_2O_3 or MnO_2 in vacuum at 1070 K for one day, a graftonite phase with a rather dark colour was formed (ref. 21). Nevertheless, the unit cell volumes (both are monoclinic) and the structural-building metal-oxygen and phosphate polyhedra are so similar for β'-$Mn_3(PO_4)_2$ and "$Mn_3(PO_4)_2$-graftonite" that there is no obvious reason why both forms should not exist; V/Z is 155.0 $Å^3$ for β'-$Mn_3(PO_4)_2$ (ref. 19) and has been extrapolated to 155.5 $Å^3$ for the "Mn-graftonite" (ref. 21).

The solid solutions β'-$(Mn_{1-x}Cd_x)_3(PO_4)_2$ always have the β' structure for $0 < x < 1$, i.e., the same structure as the pure β' phases of manganese(II) and cadmium orthophosphate. The crystal structure of β'-$Cd_3(PO_4)_2$ has been published by Stephens (ref. 20). There are nine distinct cadmium atoms in the monoclinic unit cell (S.G. $P2_1/c$), four five-coordinated and five six-coordinated. The averaged <Cd-O> distances are around 2.25 Å for the five- and around 2.34 Å for the six-coordinated atoms, cf. Table 2.1.1. Solid solutions of $M_3(PO_4)_2$ orthophosphates in $Cd_3(PO_4)_2$ will be described in section 2.3.

2.2 The "γ-$Zn_3(PO_4)_2$" structure types

The "γ-$Zn_3(PO_4)_2$" phase, which should rather be denoted as γ-$(Zn,M)_3(PO_4)_2$ since it only exists as a solid solution, is formed upon replacement of zinc ions in α-$Zn_3(PO_4)_2$ by other divalent-metal M^{2+} cations. This was first showed by A. L. Smith (ref. 22) in 1951. Investigations performed later have shown that the "γ" phase may be stabilized at room temperature (metastable phase) with respect to α-$Zn_3(PO_4)_2$ by replacing zinc by 3-100 (atom) % Mg, 3-27 % Mn, 3-40 % Fe, 3-100 % Co, 3-33 % Ni, 3-10 % Cu, or 4-6 %

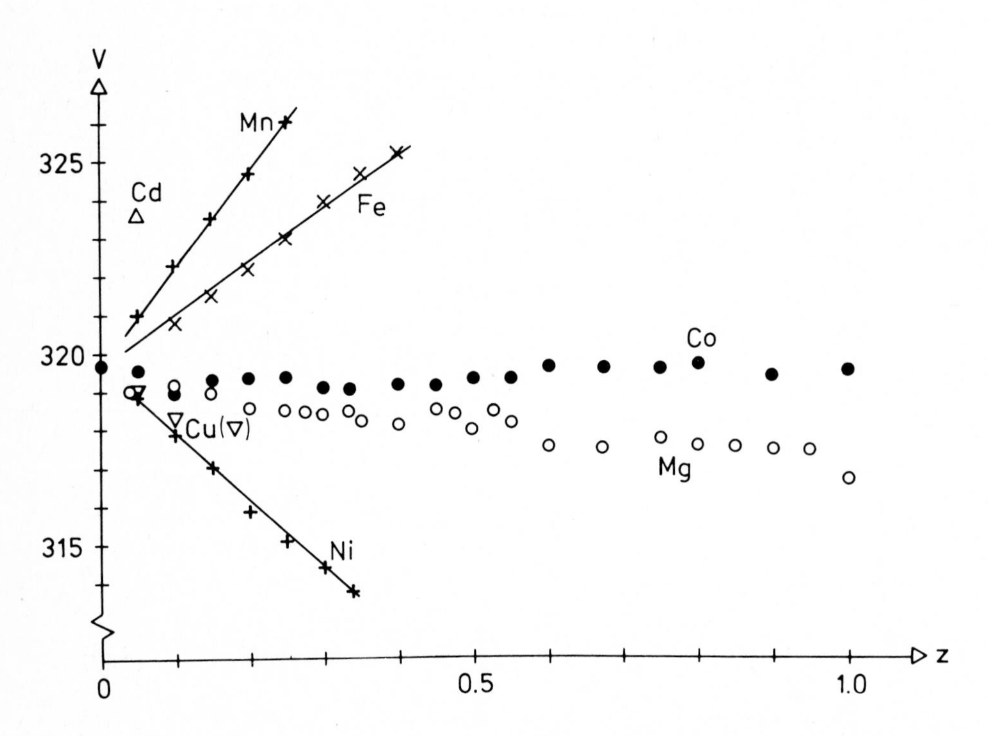

Fig. 2.2.1. Unit cell volumes (in $Å^3$ units), at 298 K, versus composition for the γ-$(Zn_{1-x}M_x)_3(PO_4)_2$ solid solutions equilibrated at 1070 K (S.G. $P2_1/n$; Z = 2). The Zn/Mg phases are represented by open circles (o), and the Zn/Co phases by filled circles (●). The figure is reproduced with permission from Pergamon Press (copyright) from Nord and Stefanidis: "Crystal chemistry of γ-$(Zn,M)_3$-$(PO_4)_2$ solid solutions" (ref. 23) in Mater. Res. Bull. vol. 16, pp. 1121-1129 (1981).

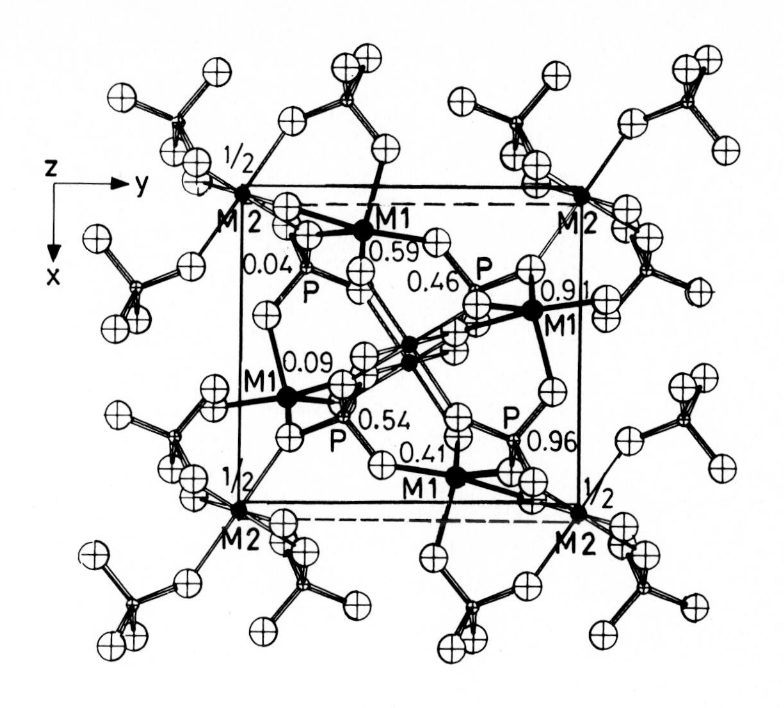

Fig. 2.2.2. An illustration of the "γ-$Zn_3(PO_4)_2$" structure, produced with the plot program ORTEP by C. K. Johnson (ref. 30). The smallest circles represent phosphorus atoms, the second largest are metal atoms (M1 or M2), and the largest circles are oxygen atoms. The figure is reproduced with permission from the Royal Swedish Academy of Science (copyright) from Nord and Kierkegaard: "Crystal chemistry of some anhydrous divalent-metal phosphates" (ref. 17) in Chemica Scripta, vol. 15, pp. 27-39 (1980).

Cd (ref. 23). Partial phase diagrams of the systems $Zn_3(PO_4)_2$ - $M_3(PO_4)_2$ have been published for M = Mg (ref. 24), Mn (ref. 25), and Cd (ref. 26). Ca may also substitute for Zn, although the $(Zn,Ca)_3(PO_4)_2$ solid solutions are unstable and cannot be quenched to room temperature (ref. 27). The variation of unit cell volume with the size and amount of incorporated M^{2+} cation is shown in Fig. 2.2.1, displaying almost linear V = f(z) dependences.

The crystal structure of "γ-$Zn_3(PO_4)_2$", stabilized by a small but unknown amount of divalent manganese, was determined in 1963 by Calvo (ref. 28), although with poor accuracy. A better structural analysis by the same author followed, based on the solid solution γ-$(Zn_{0.97}Mn_{0.03})_3(PO_4)_2$ (ref. 29). Calvo reported that one third of the cations had octahedral coordination, while two thirds had an "irregular tetrahedral configuration". However, a fifth oxygen atom is situated about 2.4 Å from the "tetrahedral" zinc atom. A bond strength calculation as described for β-$Zn_3(PO_4)_2$ shows that five-coordination (p = 2.069) is indicated rather than four-coordination (p = 1.892 valence units) (cf. ref. 17), giving the CN:s as 5 and 6 instead of 4 and 6. The five-coordinated metal atom in this structure type is henceforward always referred to as "M1", while the six-coordinated metal atom is called "M2". The structure is below described in the (monoclinic) space group $P2_1/n$ (No. 14), which is more convenient since the β angle is then around 94°. There are four five-coordinated M1 and two six-coordinated M2 sites in the unit cell (Z = 2).

The crystal structure is built up of rather regular $(M2)O_6$ octahedra, distorted trigonal bipyramids $(M1)O_5$, and PO_4 tetrahedra, as shown in Fig. 2.2.2. (Some crystallographic data for this and related structures are given in Table 2.2.1). The "octahedral" M2 atoms share oxygen atoms with six different PO_4 tetrahedra, while the M1 atoms are each coordinated by five oxygens shared by four PO_4 tetrahedra. The $(M1)O_6$ octahedron shares edges with two $(M2)O_5$ polyhedra, and the latter share an edge in pairs, thus forming infinite chains running along the monoclinic "b" axis in the crystal structure. These chains are cross-linked in the bc-plane by corner-sharing $(M2)O_6$ octahedra.

The orthophosphates $Mg_3(PO_4)_2$ (ref. 31) and $Co_3(PO_4)_2$ (ref. 32, 33) are also isotypic with γ-$Zn_3(PO_4)_2$. Relevant data are given in Table 2.2.1. The crystal structure of $Mg_3(PO_4)_2$ is far from densely packed, with a V_{ox} value of 19.79 Å^3 (ref. 31). As a matter of fact, the existence of a more closely packed polymorph has been proved, with a V_{ox} value of only 17.86 Å^3 (ref. 34, 35). The latter modification is usually denoted $Mg_3\square(PO_4)_2$ because of its structural relationship to olivine, $(Mg,Fe)_4(SiO_4)_2$, although the former structure contains ordered vacancies, "$\square$". It is isostructural with the rare mineral sarcopside. More details will be given in section 2.4. Finally, an iron-containing magnesium orthophos-

TABLE 2.2.1

Crystallographic data for "γ-$Zn_3(PO_4)_2$" and related compounds. All structures are described in space group $P2_1/n$ (Z = 2). "γ-$Zn_3(PO_4)_2$" stands for γ-$(Zn_{0.97}Mn_{0.03})_3(PO_4)_2$. All distances are given in Å units.

Compound	γ-$Zn_3(PO_4)_2$	γ-$Zn_2Mg(PO_4)_2$	$Mg_3(PO_4)_2$	$Co_3(PO_4)_2$
Ref.	(29)	(38)	(31)	(33)
a (Å)	7.545	7.569	7.596	7.552
b (Å)	8.469	8.355	8.230	8.361
c (Å)	5.074	5.059	5.078	5.063
β (°)	94.41	94.95	94.05	94.08
V (Å^3)	323.1	318.7	316.6	318.9
V_{ox} (Å^3)	20.20	19.92	19.79	19.93
CN = 5				
M-O range	1.93-2.40	1.89-2.33	1.96-2.14	1.98-2.23
<M-O>	2.05	2.05	2.03	2.06
CN = 6				
M-O range	2.01-2.23	2.03-2.22	2.03-2.18	2.06-2.17
<M-O>	2.15	2.15	2.12	2.13

TABLE 2.2.2

Cation distribution factor K_D values (compare text) and homogeneity ranges (H.R.) of $M_3(PO_4)_2$ in γ-$Zn_3(PO_4)_2$ (in mole %) for some γ-$(Zn_{1-x}M_x)_3(PO_4)_2$ solid solutions. Solubilities from Ref. (23).

Solid solution	Equilibrium temperature	K_D	H.R.(%)	Technique	Ref.
γ-$Zn_2Mg(PO_4)_2$	~1300 K	20	3-100	X-ray	(38)
γ-$(Zn_{0.50}Mg_{0.50})_3(PO_4)_2$	1070 K	35	3-100	X-ray	(42)
γ-$(Zn_{0.90}Fe_{0.10})_3(PO_4)_2$ γ-$(Zn_{0.80}Fe_{0.20})_3(PO_4)_2$ γ-$(Zn_{0.70}Fe_{0.30})_3(PO_4)_2$ γ-$(Zn_{0.65}Fe_{0.35})_3(PO_4)_2$	1070 K	30	3- 40	Mössbauer	(43)
γ-$(Zn_{0.70}Fe_{0.30})_3(PO_4)_2$	1070 K	27		Neutron	(44)
γ-$(Zn_{0.70}Ni_{0.30})_3(PO_4)_2$	1070 K	40	3- 33	Neutron	(45)
γ-$(Zn_{0.75}Mn_{0.25})_3(PO_4)_2$	1070 K	200	3- 27	Neutron	(46)
γ-$Zn_2Co(PO_4)_2$ γ-$(Zn_{0.50}Co_{0.50})_3(PO_4)_2$	1070 K	9	3-100	Neutron	(47)

phate with the approximate composition $Mg_{2.84}Fe_{0.15}Mn_{0.01}(PO_4)_2$ has been found as a rare mineral, called farringtonite (cf. ref. 36, 37).

In addition to the γ-$(Zn,M)_3(PO_4)_2$ solid solutions presented in Fig. 2.2.1, solid solutions of $M_3(PO_4)_2$ orthophosphates in either $Mg_3(PO_4)_2$ (ref. 10) or $Co_3(PO_4)_2$ (ref. 39, 40) have been prepared and equilibrated at 1070 K (1 bar) and some other temperatures. The solubilities vary quite a lot. For instance, the solubility of $Ni_3(PO_4)_2$ in magnesium orthophosphate is only 10 (mole) % (at 1070 K), while cobalt(II) and magnesium orthophosphate are soluble in each other over the whole composition range (ref. 41).

Systematic investigations of the cation distributions in these pseudo-binary systems have been undertaken with the aim to correlate cation partitionings, solubility ranges, metal cation radii, unit cell dimensions, and interatomic metal-oxygen distances. As a matter of fact, in his paper of 1963 (ref. 28), Calvo suggested that the "stabilizing" M^{2+} cations in γ-$(Zn,M)_3(PO_4)_2$ (at that time: M = Mg, Ca, Mn, or Cd) would preferentially enter the six-coordinated sites of the crystal structure, thus making the cation distribution ordered rather than random. Calvo supported his theory through emission studies of the "γ" phase under cathode ray bombardment (ref. 28).

Recent cation distribution studies of the "γ" phases based either on X-ray or neutron diffraction data or ^{57}Fe Mössbauer spectroscopy data are summarized in Table 2.2.2. The intra-crystalline Zn^{2+}/M^{2+} cation exchange in a γ-$(Zn,M)_3(PO_4)_2$ solid solution is here expressed by means of the conventional cation distribution factor K_D, defined as

$$K_D = [X_{Zn}(M1) \cdot X_M(M2)]/[X_{Zn}(M2) \cdot X_M(M1)],$$

where M1 is the five- and M2 the six-coordinated metal site (cf. Fig. 2.2.2). Assuming equal activity factors for the cations, K_D may be regarded as the equilibrium constant at the temperature in question for the cation exchange reaction

$$Zn^{2+}(M2) + M^{2+}(M1) \rightleftarrows Zn^{2+}(M1) + M^{2+}(M2).$$

A random distribution of the cations makes K_D equal to unity, while K_D increases as M^{2+} concentrates at the six-coordinated M2 sites in the crystal structure. Since K_D is greater than one (1) for all phases listed in Table 2.2.2, it is clear that Zn^{2+} has a much stronger preference for five-coordination than any of the other relevant cations. The cations (including Zn^{2+}) may thus be arranged in a sequence according to their preference for five- over six-coordination in the "γ-$Zn_3(PO_4)_2$" type structures from these K_D values. For instance, since $K_D(Zn,Mn)$ is much larger than $K_D(Zn,Mg)$, it is clear that Mn^{2+} has a much weaker tendency for five-coordination than has Mg^{2+}, and that Zn^{2+} has a much stronger preference for five-coordination than the two other cations. The sequence thus obtained from the K_D values in Table 2.2.2 is the following (at 1070 K):

$$Zn^{2+} >> Co^{2+} > Fe^{2+} > Mg^{2+} > Ni^{2+} > Mn^{2+} > Cd^{2+} > Ca^{2+} \qquad \text{(I)}.$$

The results on Cd^{2+} and Ca^{2+} have been derived from recent Mössbauer studies of ternary γ-$(Zn_{0.70}Fe_{0.25}M_{0.05})_3(PO_4)_2$ solid solutions (ref. 48), since only very small amounts of these cations, i.e. divalent cadmium and calcium, can be incorporated in zinc orthophosphate having the "gamma" type structure. The above-given sequence (I) for the cation distribution tendencies among five- and six-coordinated metal sites will be used in the discussions to follow.

There is good correlation between K_D values and homogeneity ranges, and also between cation distribution and interatomic metal-oxygen distances, preferably averaged values (cf. ref. 44). For example, in γ-$(Zn_{0.70}Ni_{0.30})_3(PO_4)_2$ the mean <M2-O> distance has decreased significantly with respect to the "γ-$Zn_3(PO_4)_2$" base structure (ref. 45), while the mean <M1-O> value is almost unchanged. This accords with the observed situation that the smaller Ni^{2+} ions preferentially occupy the octahedral M2 sites, thereby decreasing the M2-O distances, while the M1-O distances are almost unaffected since only a very limited amount of nickel enters these sites. The observed cation partitioning in this phase can be described as γ-$(Zn_{0.93}Ni_{0.07})_2^{M1}(Zn_{0.24}Ni_{0.76})^{M2}(PO_4)_2$ (ref. 45). It should finally be mentioned that good correlation also exists between unit cell dimensions and the size and amount of incorporated M^{2+} cation in this structure type (ref. 10, 23, 39).

2.3 The graftonite structure types

Graftonite is a phosphate mineral of fairly wide occurrence. Its chemical composition may be expressed as $(Fe,Mn,Ca,Mg)_3(PO_4)_2$, where the metals are listed in order of decreasing abundance. The

TABLE 2.3.1
Crystallographic data for the graftonite structure types. The space group is $P2_1/c$ (Z = 4). The graftonite mineral had the composition $(Fe_{1.8}Mn_{0.8}Ca_{0.4})(PO_4)_2$.

Compound	Graftonite	$Fe_3(PO_4)_2$	$CdZn_2(PO_4)_2$	$Cd_2Zn(PO_4)_2$
Ref.	(51)	(49)	(50)	(50)
a (Å)	8.91	8.881	9.032	9.056
b (Å)	11.58	11.169	11.417	11.86
c (Å)	6.239	6.145	5.952	6.190
β (°)	98.9	99.36	98.8	100.1
V ($Å^3$)	636.0	601.4	606.0	655.0
V_{ox} ($Å^3$)	19.87	18.79	18.94	20.47
<M-O> for				
CN = 4			1.98	
CN = 5	2.14, 2.16	2.13, 2.10	2.08	2.15, 2.23
CN = 6		2.23		
CN = 7	2.43		2.36	2.44

crystal structure of graftonite is indeed very interesting. The same structure has also been found for some synthetic compounds such as the pure end member $Fe_3(PO_4)_2$ (ref. 49) as well as the solid solutions $CdZn_2(PO_4)_2$ and $Cd_2Zn(PO_4)_2$ (ref. 50). The crystal structure was actually first solved through the two latter phases, and the structural study of the graftonite mineral was published somewhat later by the same author, Críspin Calvo (ref. 51). Some important crystallographic data are reviewed in Table 2.3.1 (above). It should also be noted that the cadmium orthoarsenate $Cd_3(AsO_4)_2$ crystallizes with the graftonite-type structure, with a unit cell volume of 723.4 $Å^3$ and with five- and six-coordinated cations (ref. 52).

The metal cations in the graftonite mineral reside in three crystallographically inequivalent positions called M1, M2, and M3, with CN:s 7, 5, and 5 (ref. 51). The MO_7 polyhedron is an irregular pentagonal bipyramid. These $(M1)O_7$ polyhedra share edges in pairs and are arranged in sheets throughout the structure. The two distinct MO_5 groups are distorted trigonal bipyramids and square pyramids, respectively. The former MO_5 polyhedra form chains almost perpendicular to the formerly-mentioned sheets, while the square pyramids connect these sheets with the chains (see Fig. 2.3.1).

The graftonite structure is very flexible so that chemical and

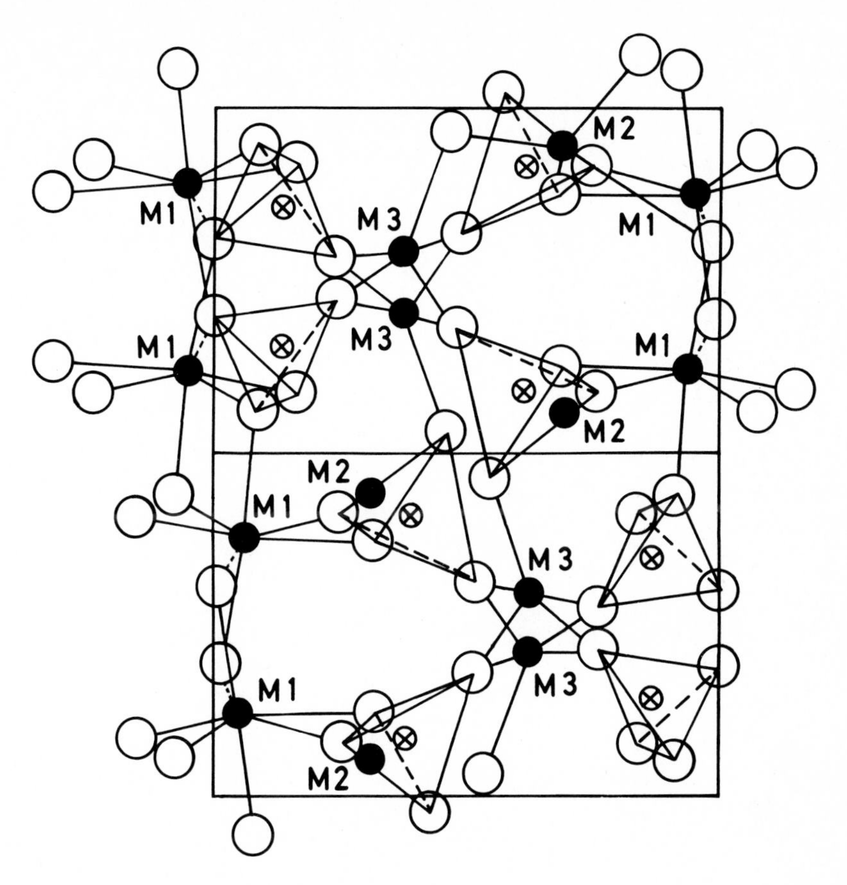

Fig. 2.3.1. The graftonite structure type (S.G. $P2_1/c$). The open circles represent oxygen atoms, ⊗ denotes phosphorus, and filled circles are metal cations at the M1, M2, or M3 sites. The M1 atoms are here shown as six-coordinated; the seventh nearest oxygen is indicated by a dotted line.

structural changes may take place although the basic framework is preserved. This can be seen from the data in Table 2.3.1, where the variation in coordination numbers is noteworthy. In the phase $CdZn_2(PO_4)_2$ for example the CN:s are 7, 4, 5, but with a "fifth" oxygen atom placed about 3 Å from the "tetrahedral" M2 cation. In $Cd_2Zn(PO_4)_2$ on the other hand these $(M2)O_4$ tetrahedra are transformed by a homogeneous change to square pyramids, so that CN = 7, 5, 5 as in the graftonite-mineral structure (ref. 50, 51). This change and the increase of the unit cell volume in the Cd,Zn-graftonites is undoubtedly a size effect.

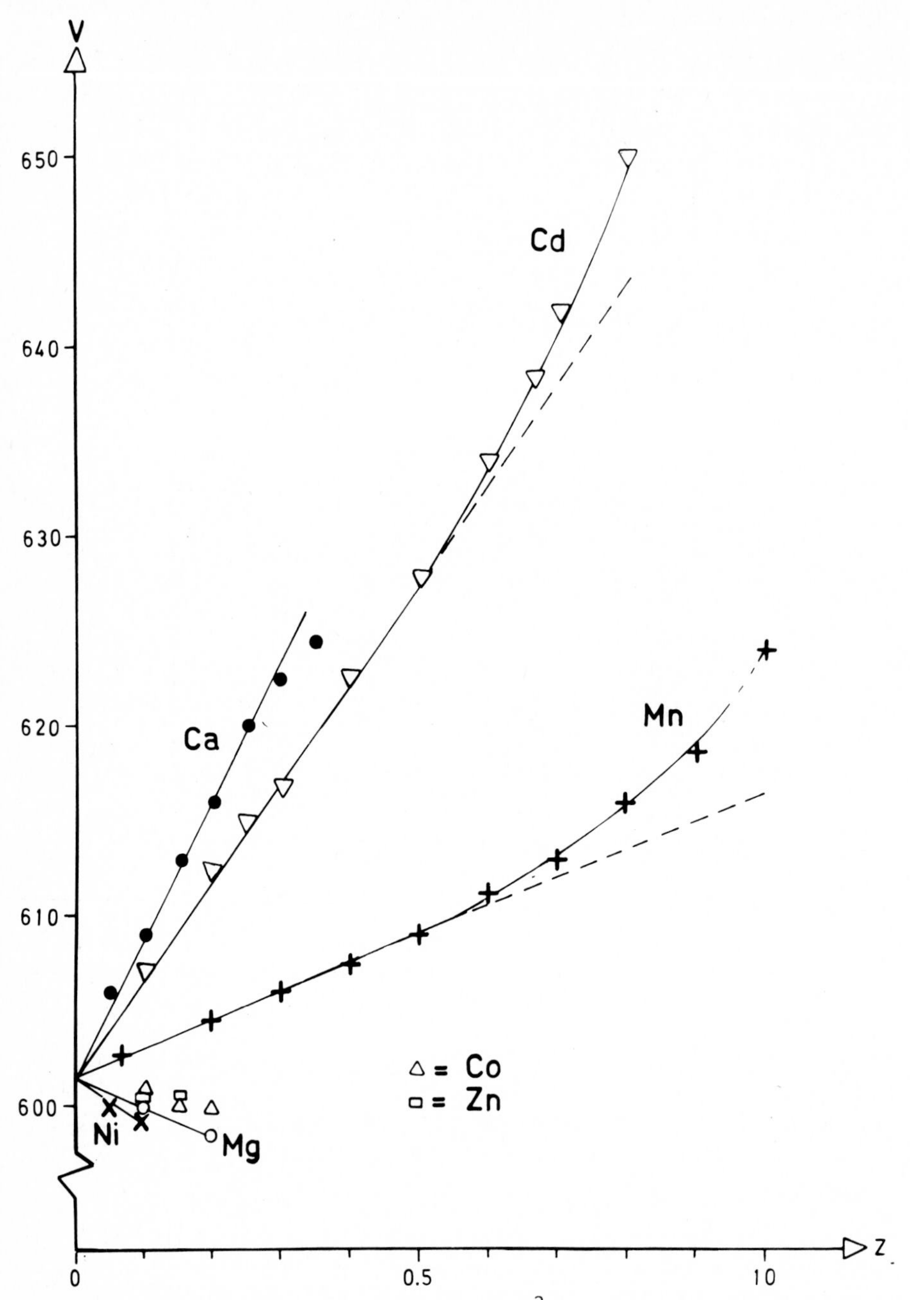

Fig. 2.3.2. Unit cell volumes V (in Å^3) for the $(Fe_{1-z}M_z)_3(PO_4)_2$ graftonite-type solid solutions. The figure is reproduced with permission from R. Oldenbourg Verlag, München (copyright) from Nord and Ericsson: "Cation distributions in $(Fe_{1-z}M_z)_3(PO_4)_2$ graftonite-type solid solutions determined by Mössbauer spectroscopy" (ref. 53) in Zeitschrift für Kristallographie vol. 161, pp. 209-224 (1982).

Pure $Fe_3(PO_4)_2$ also crystallizes with the graftonite structure at ordinary pressures, but in this compound M1 is six-coordinated so that the CN:s are 6, 5, and 5. The $(M1)O_6$ octahedron is very distorted, with Fe-O distances in the range 1.98 - 2.68 Å (ref. 49). Systematic preparations and cation distribution studies of solid solutions of $M_3(PO_4)_2$ in $Fe_3(PO_4)_2$, always equilibrated at 1070, 1 bar like the γ-$(Zn,M)_3(PO_4)_2$ types, have been carried out (ref. 53). The solubilities and variations of unit cell volume is shown in Fig. 2.3.2. The value for the hypothetical "Mn-graftonite" has been extrapolated (see section 2.1). A further extrapolation of the V = f(z) curve for the $(Fe_{1-z}Cd_z)_3(PO_4)_2$ series gives a unit cell volume of about 670 $Å^3$ for a hypothetical "Cd-graftonite", which corresponds to 168 $Å^3$ per formula unit (Z = 4). The corresponding value for β'-$Cd_3(PO_4)_2$ is very close, namely 170 $Å^3$ per formula unit. The same situation was noted for "Mn-graftonite" and the β'-$Mn_3(PO_4)_2$ modification as already mentioned in a previous section (2.1).

The cation distribution in a series of $(Fe_{1-x}Mn_x)_3(PO_4)_2$ graftonites was first determined through a combination of neutron diffraction and Mössbauer spectroscopy (ref. 12), showing that iron (Fe^{2+}) had a much higher tendency for the five-coordinated sites than manganese (Mn^{2+}). The combined results of all $(Fe,M)_3(PO_4)_2$-graftonites (ref. 53) give the following preference order for five-coordination in the graftonite-type structure:

$$Zn^{2+} > Fe^{2+}, Mg^{2+}, Co^{2+}, Ni^{2+} > Ca^{2+} > Mn^{2+} > Cd^{2+} \qquad \text{(II)}.$$

This sequence is similar to sequence (I) for the γ-$Zn_3(PO_4)_2$ type phases, although the metal-oxygen polyhedra are much more distorted in the graftonites. Ca^{2+} is the only severe misfit in a comparison between the sequences (I) and (II). Besides, the relative differences in cation preference trends are much smaller for the graftonite structure than for γ-$(Zn,M)_3(PO_4)_2$. For instance, in graftonite K_D(Zn,Mn) is only around 25 (estimated as shown in ref. 44), while it is as large as 200 in γ-$(Zn_{0.75}Mn_{0.25})_3(PO_4)_2$ (ref. 46), showing a very strong ordering of the cations with zinc at the five- and manganese at the six-coordinated sites. Sequence (II) is thus very "compressed" with respect to sequence (I), and this effect is attributed to the distorted metal-oxygen polyhedra in graftonite, which tend to decrease the differences in preference among the cations for one site over another. Of course, further studies of other structure types are necessary to verify this theory.

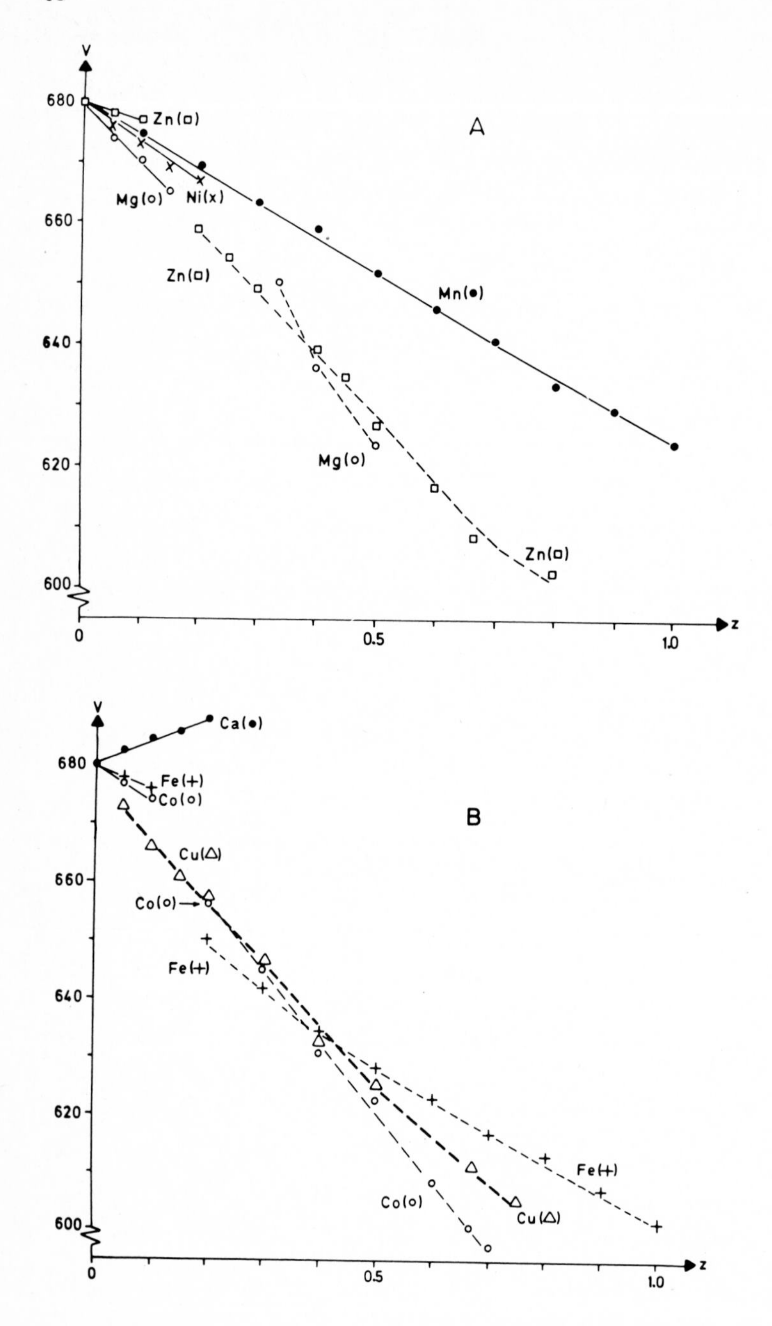

Fig. 2.3.3. Unit cell volumes V (in Å^3, always Z=4) versus z for $(Cd_{1-z}M_z)_3(PO_4)_2$ phases (1070 K). Continuous curves are β' phases and dashed curves graftonites. The figure is reproduced with permission from Pergamon Press (copyright) from Nord: "Crystal chemistry of $(Cd,M)_3(PO_4)_2$ solid solutions" (ref. 53a) in Mater. Res. Bull. vol. 18, pp. 569-579 (1983).

It was mentioned in section 2.1 that solid solutions with the graftonite structure were easily formed when Mn^{2+} in β'-$Mn_3(PO_4)_2$ was replaced by other divalent M^{2+} cations. The same is true for β'-$Cd_3(PO_4)_2$. This structure only contains five- and six-coordinated cations; 4/9 (~44 %) are five-coordinated. By substituting M^{2+} for cadmium, the graftonite structure is often formed, cf. Fig. 2.3.3. In graftonite two thirds (~67 %) of the metal cations are five-coordinated. In the $(Cd,M)_3(PO_4)_2$ solid solutions, the cations earlier shown to have the greatest tendency to five-coordination (sequences I and II), namely Zn^{2+}, Fe^{2+}, and Co^{2+}, give the largest "regions of existence" in Fig. 2.3.3 for the graftonite structure (dashed curves), while cations with a low readiness for five-coordinated sites such as Ni^{2+}, Mn^{2+} or Ca^{2+} do not give any graftonite phase at all. Thus the phase transition $\beta' \rightleftarrows$ graftonite is directly controlled by the cation site preference tendencies of the incorporated M^{2+} cations in cadmium orthophosphate, since a great preference of M^{2+} for five-coordination corresponds to a larger "region of existence" for the graftonite structure in which a larger number of available cation sites are five-coordinated (cf. ref. 53a).

2.4 The olivine-related sarcopside structure

Several non-silicate minerals crystallize with the olivine structure type, e.g., chrysoberyl (Al_2BeO_4), sinhalite ($AlMgBO_4$), and trifylite ($LiFePO_4$). In addition, Moore (ref. 8) has shown that sarcopside, $(Fe,Mn,Mg)_3(PO_4)_2$ (ref. 54), is structurally related to, although not isostructural with, the mineral olivine. Nickel(II) orthophosphate, $Ni_3(PO_4)_2$, is also isotypic with sarcopside (ref. 55), and the same is true for some other orthophosphates (see below).

The close structural relationship between the olivine $Ni_4(SiO_4)_2$ (ref. 56) and $Ni_3(PO_4)_2$-sarcopside (ref. 55) is illustrated in Fig. 2.4.1. The olivine is here for convenience denoted as $M_4(SiO_4)_2$ (M = Ni). It is orthorhombic (S.G. Pbnm) with two distinct four-fold, six-coordinated M^{2+} cation sites, here called $M1_{ol}$ and $M2_{ol}$ (point symmetries: $\bar{1}$ and m). In sarcopside, here denoted on the form $M_3\square(PO_4)_2$ (M = Ni), the symmetry has been lowered with respect to olivine to monoclinic (S.G. $P2_1/a$; $\beta \approx 91^{\circ}$).

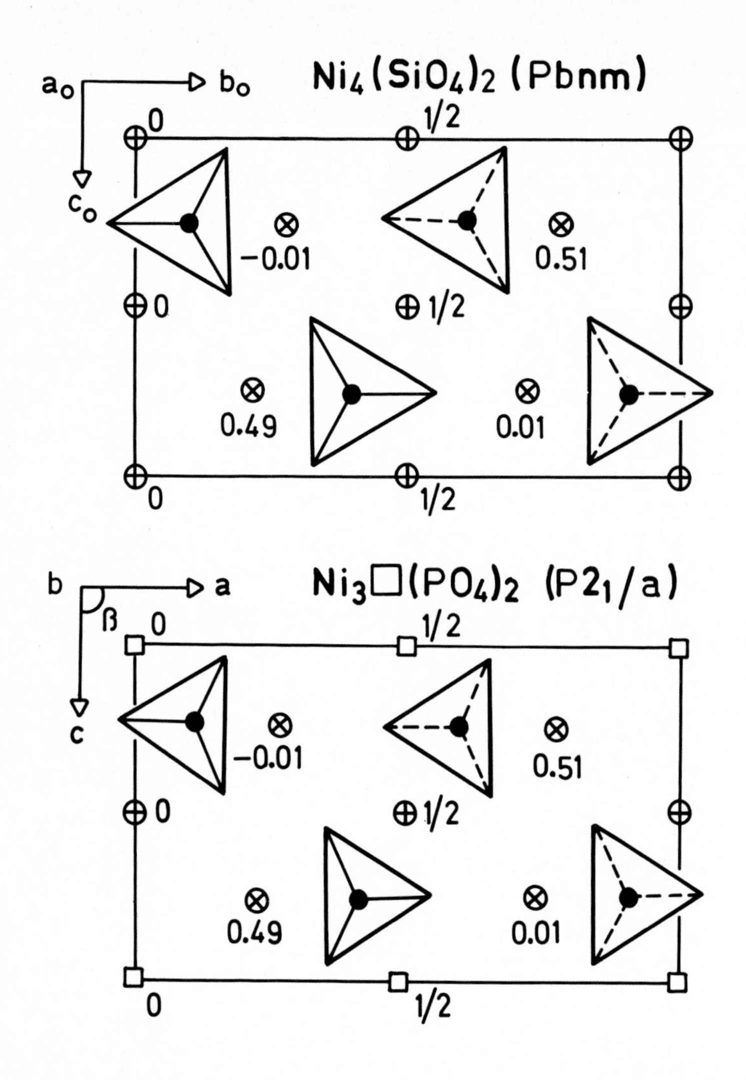

Fig. 2.4.1. Schematic drawings of the olivine and sarcopside structures, here represented by $Ni_4(SiO_4)_2$ and Ni_3□$(PO_4)_2$ (ref. 55, 56). ⊕ = M1, ⊗ = M2, □ = vacancy; filled circles are Si or P. The figure is reproduced with permission from R. Olderbourg Verlag, München (copyright) from Nord: "Crystallographic studies of olivine-related sarcopside-type solid solutions" (ref. 11) in Zeitschrift für Kristallographie vol. 166, pp. 159-176 (1984).

The four $M1_{ol}$ "olivine" sites in the unit cell have been split into a two-fold site M1 (0,0,½ and ½,½,½) and another two-fold site "□" which is vacant in Ni_3□$(PO_4)_2$ (0,0,0 and ½,½,0), while the four-fold $M2_{ol}$ sites are preserved as M2. All sites (also "□") are octahedrally coordinated. M1 and M2 have the point symmetries

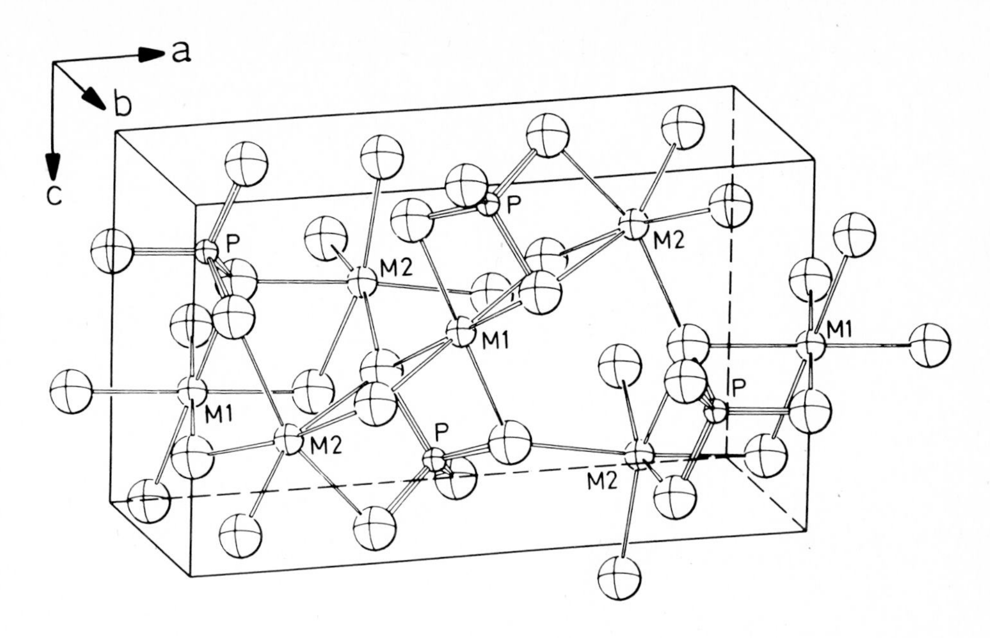

Fig. 2.4.2. An ORTEP (ref. 30) illustration of the sarcopside structure, showing an M2-M1-M2 trimer of edge-sharing MO_6 octahedra. Reproduced from ref. (11); cf. text to Fig. 2.4.1.

$\bar{1}$ and 1, respectively. In both crystal structures the "$(M2)O_6$" octahedra are slightly larger and display a somewhat larger scatter in metal-oxygen distances and O-M-O angles than "$(M1)O_6$". (The hypothetical octahedron around the "□" vacancy is still greater). All atoms, including the oxygens, occupy corresponding positions in the two structures.

Like the olivine structure, the sarcopside structure consists of hexagonally close-packed layers of oxygen atoms. In olivine, the rather regular MO_6 octahedra are connected in serrated chains; sarcopside contains trimers of edge-sharing, distorted MO_6 octahedra since the "olivine" chains have here been broken due to the vacant "□" positions; cf. ref. (8, 11) and Fig. 2.4.2.

As mentioned in section 2.1, ordinary magnesium orthophosphate (CN = 5,6) also can be transformed into a much denser, meta-stable high-pressure modification (CN = 6,6) with the sarcopside structure (ref. 34, 35). It is usually denoted $Mg_3□(PO_4)_2$ to distinguish it from the "γ-$Zn_3(PO_4)_2$" type and to indicate the struc-

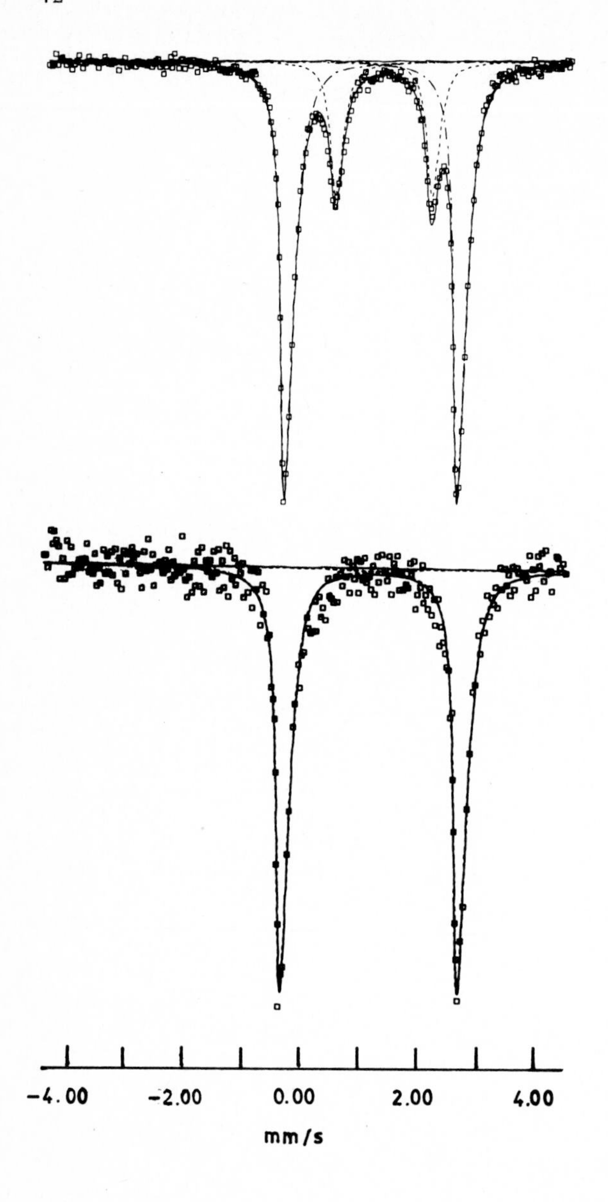

Fig. 2.4.3. ^{57}Fe Mössbauer spectrum of "γ-$Zn_3(PO_4)_2$"-type iron(II)-doped $(Mg_{0.80}Fe_{0.20})_3(PO_4)_2$ with CN = 5, 6 (above) and the spectrum for the respective sarcopside-type high-pressure modification, with CN = 6,6 (below). The figure is reproduced with permission from Acta Chemica Scandinavica (copyright) from Annersten and Nord: "A high pressure phase of magnesium orthophosphate" (ref. 35) in Acta Chem. Scand. vol. A34, pp. 389-390 (1980).

tural relationship to olivine (see also Fig. 2.4.3). Moreover, if at least 40% of the magnesium atoms in "ordinary" $Mg_3(PO_4)_2$ are substituted by divalent nickel, the crystal structure transforms to a sarcopside-type $(Mg_{1-x}Ni_x)_3\square(PO_4)_2$ (ref. 57).

A corresponding phase $Co_3\square(PO_4)_2$, isostructural with $Mg_3\square(PO_4)_2$ and sarcopside, also exists (ref. 34). Furthermore, when $Fe_3(PO_4)_2$-graftonite is hydrothermally heated to about 300°C and 800 bar, the sarcopside structure is formed (ref. 58, 59). Curiously enough,

TABLE 2.4.1

Crystallographic data for the sarcopside structure types, described in space group $P2_1/a$ (Z = 2). The sarcopside mineral had the composition $(Fe_{0.78}Mn_{0.21}Mg_{0.01})_3(PO_4)_2$.

Compound	a (Å)	b (Å)	c (Å)	β (°)	V (Å^3)	Ref.
$Ni_3(PO_4)_2$	10.107	4.700	5.830	91.22	276.9	(55)
$(Ni_{0.75}Zn_{0.25})_3(PO_4)_2$	10.150	4.707	5.870	91.11	280.4	(61)
$(Ni_{0.50}Fe_{0.50})_3(PO_4)_2$	10.306	4.717	5.959	90.92	289.6	(11)
$(Ni_{0.70}Mn_{0.30})_3(PO_4)_2$	10.267	4.726	5.961	90.82	289.2	(11)
$Mg_3\square(PO_4)_2$	10.25	4.72	5.92	90.9	287	(35)
$(Mg_{0.8}Fe_{0.2})_3\square(PO_4)_2$	10.26	4.75	5.93	90.8	289	(35)
$Co_3\square(PO_4)_2$	10.334	4.75	5.920	91.1	290.5	(34)
$Fe_3\square(PO_4)_2$	10.442	4.787	6.029	90.97	301.3	(59)
$(Fe_{0.76}Zn_{0.24})_3(PO_4)_2$	10.404	4.771	6.006	91.12	298.1	(60)
sarcopside	10.437	4.768	6.026	90.0	299.9	(8)

these two $Fe_3(PO_4)_2$ modifications are almost equally closely packed; the value of V/Z for these phases are 150.4 Å^3 for the graftonite modification with CN = 6,5,5 (ref. 49) and 150.6 Å^3 for the sarcopside-phase with CN = 6,6 (ref. 59). Thus, contrary to the situation for $Mg_3(PO_4)_2$ where the high-pressure phase is 9.3% smaller in volume (ref. 35), the sarcopside modification of iron orthophosphate cannot be regarded as a high-pressure phase. Another solid solution $(Fe_{0.76}Zn_{0.24})_3(PO_4)_2$ has also been hydrothermally prepared and crystallographically investigated by Belov et al. (ref. 60). Some crystallographic data for all these sarcopside-type phases are summarized in Table 2.4.1.

Numerous solid solutions of $M_3(PO_4)_2$ in $Ni_3(PO_4)_2$ have also been studied with the aim to determine solubility ranges and cation distributions (cf. Fig. 2.4.4; ref. 11, 57, 61). The divalent cations so far investigated in this respect may thus be arranged in the following sequence to express their preference for M1 over M2 in the olivine-related sarcopside structure (from ref. 11):

$$Ni^{2+} > Co^{2+} > Mg^{2+}, Zn^{2+} > Mn^{2+} >> Fe^{2+} \qquad (III).$$

Corresponding sequence for olivines (ref. 11, 62) is:

$$Ni^{2+} > Co^{2+} > Mg^{2+}, Fe^{2+} >> Mn^{2+} \qquad (IV).$$

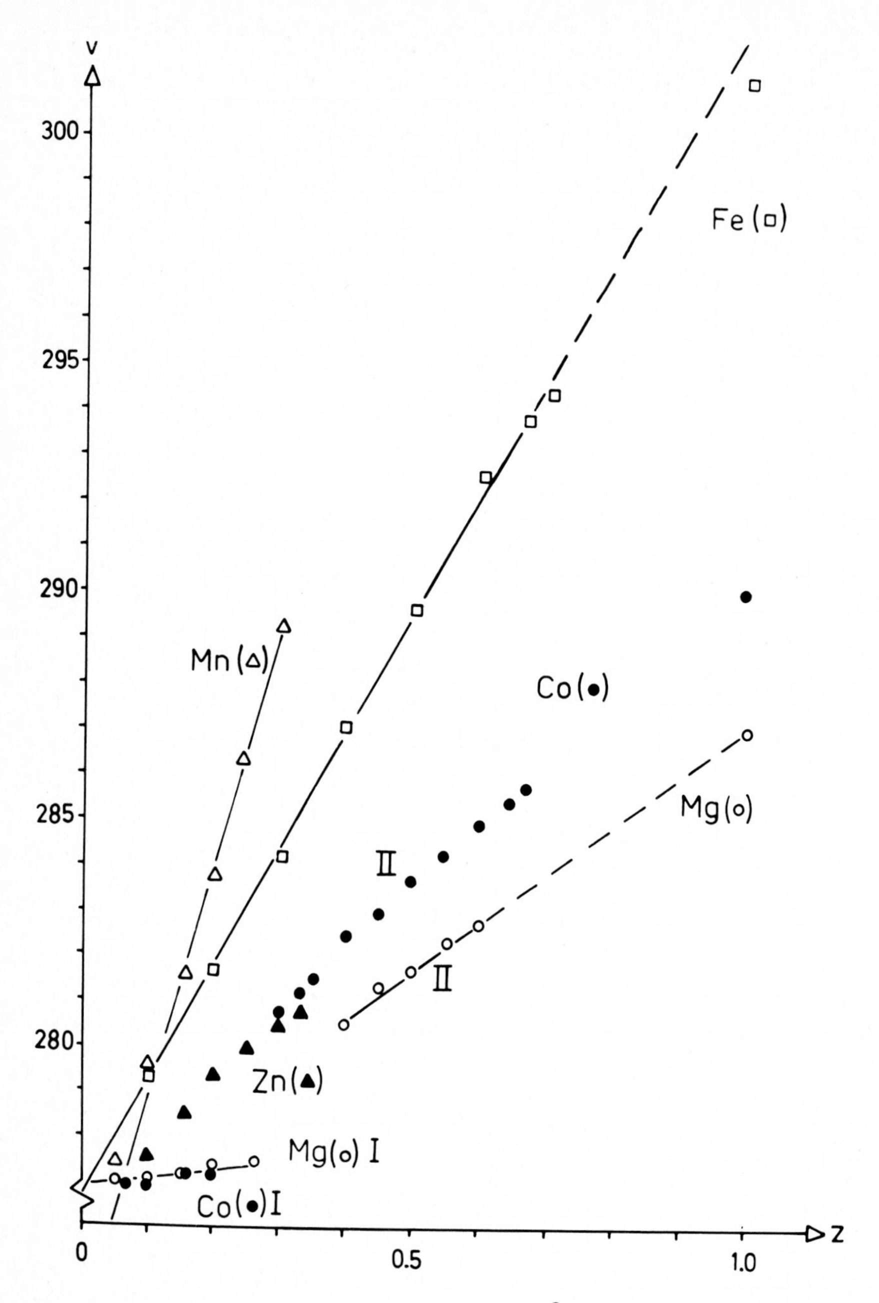

Fig. 2.4.4. Unit cell volumes V (in Å^3) versus "z" for some $(Ni_{1-z}M_z)_3(PO_4)_2$ sarcopside-type solid solutions. The values at z = 1 refer to high-pressure phases (see text); all other phases were equilibrated at 1070, 1 bar. The figure is reproduced with permission from R. Oldenbourg Verlag (copyright) from a paper by Nord (ref. 11); cf. the legend to Fig. 2.4.1.

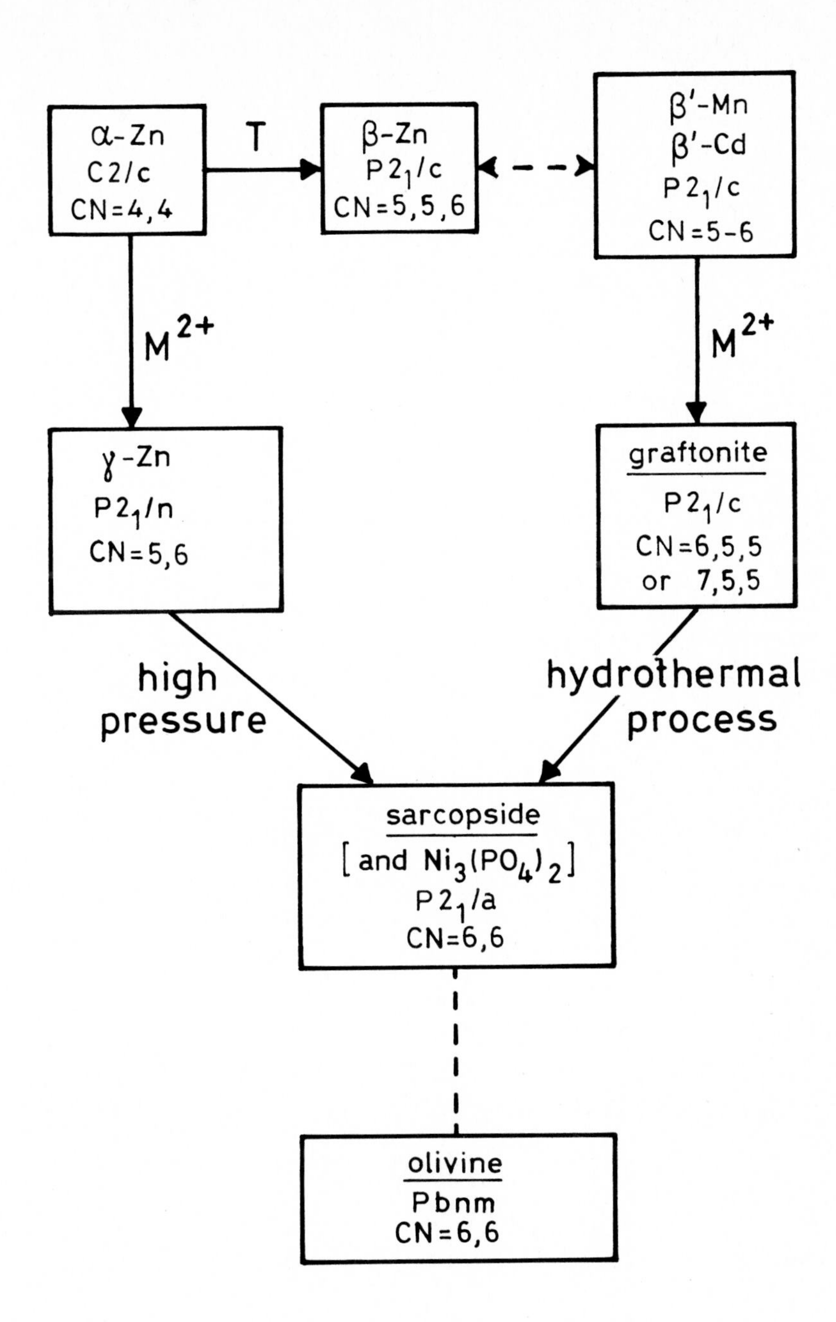

Fig. 2.4.5. Scheme illustrating the structural relationship among the divalent-metal orthophosphates discussed in sections 2.1-2.4. α-Zn stands for α-$Zn_3(PO_4)_2$, and so forth. Space group symbols and coordination numbers (CN) are also given.

The "olivine" cation preference sequence (IV) resembles the corresponding sequence (III) for the sarcopside structure, although the ordering between iron and manganese has been reversed. The fact that magnesium and iron(II) are indistinguishable in seq. (IV) is well established from studies of Mg,Fe-olivines (cf. ref. 62, 63).

Finally, a scheme illustrating the structural relationship among the divalent-metal orthophosphates discussed so far (section 2.1-2.4) is shown in Fig. 2.4.5. Space group symbols and cation coordination numbers (CN) are included.

2.5 Calcium-containing anhydrous orthophosphates

Since there are numerous calcium-containing anhydrous orthophosphates, naturally occurring minerals as well as synthetic inorganic products, a special section on these compounds was considered suitable.

For pure calcium orthophosphate, $Ca_3(PO_4)_2$, three polymorphs have been reported at ordinary pressure conditions (ref. 64). In addition, a high pressure modification has also been observed (ref. 65). One of the two high-temperature phases, called α-$Ca_3(PO_4)_2$ (S.G. $P2_1/a$; Z = 24), has been structurally investigated (ref. 9). The low ("room") temperature form, usually denoted β-$Ca_3(PO_4)_2$, is stable below 1390 K in the pure state; its crystal structure has been determined by Dickens et al. (ref. 66). Of great interest is also the mineral whitlockite, $Ca_{18}M_2H_2(PO_4)_{14}$, and the synthetic compound $Ca_4(PO_4)_2O$. Besides, many solid solutions of the type $(Ca,M)_3(PO_4)_2$ have been investigated, and there are many orthophosphate minerals containing calcium as well as other metals. Studies of these compounds are of structural-chemical interest not only in chemistry and mineralogy, but also in many other fields. For instance, calcium orthophosphates are the main inorganic constituents of bones and teeth in animals and also in abnormal human calcifications; it is of great importance as a fertilizer, and so forth.

The crystal structure of α-$Ca_3(PO_4)_2$ is extremely complicated being an inorganic compound, with as many as eighteen crystallographically inequivalent calcium atom sites. The monoclinic unit cell volume is 4318 $Å^3$ (S.G. $P2_1/a$). An approximate subcell with

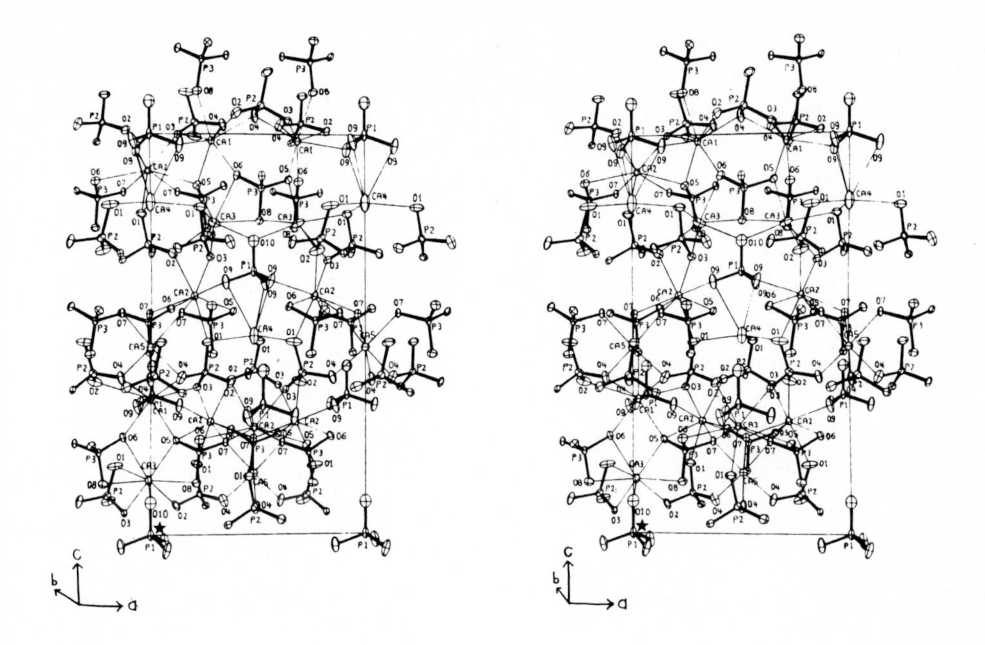

Fig. 2.5.1. Stereoscopic illustration of the crystal structure of β-$Ca_3(PO_4)_2$ (S.G. R3c). The figure is reproduced with permission from Academic Press Inc. (copyright) from Dickens, Schroeder and Brown: "The crystal structure of pure β-$Ca_3(PO_4)_2$" (ref. 66) in Journal of Solid State Chemistry, vol. 10, pp. 232-248 (1974).

b' = b/3 exists (ref. 9). The calcium atoms have coordination numbers in the range 5-9.

The structure of the modification β-$Ca_3(PO_4)_2$ is also rather complicated (S.G. R3c; Z = 21, V = 3527 $Å^3$; ref. 66). It contains five distinct Ca^{2+} ions with the coordination numbers 6, 7, 8, 8, and 9. Some interesting features in this structure are shown in Fig. 2.5.1, from ref. (66). The structure is related to that of $Ba_3(VO_4)_2$ (ref. 67).

A high-pressure phase $Ca_3(PO_4)_2$-II, prepared at 950°C and 40 kbar, has been published by Roux et al. (ref. 65). In this modification (S.G. $R\overline{3}m$, Z = 1), the two distinct calcium atoms are surrounded by ten and twelve oxygen atoms, respectively; i.e., this phase is more close-packed than the α- and β-phases of calcium orthophosphate. Solid solutions of hexagonal β-$(Ca_{1-x}M_x)_3(PO_4)_2$

forms have been prepared and studied (ref. 68), showing that the solubility of strontium and lead in β-$Ca_3(PO_4)_2$ is fairly large, while other divalent M^{2+} ions are incorporated in smaller amounts, about 10% of the metal contents. The unit cell volumes are strongly correlated to the size and amount of incorporated M^{2+} ion (ref. 68). The homogeneity ranges are principally controlled by the ionic radii: cations equi-sized with Ca^{2+} are, generally, soluble in larger amounts, while significantly smaller or larger cations, such as Ni^{2+} or Ba^{2+}, diminish the degree of solubility (the conditions are similar in apatite).

A crystallographic study of the role of Mg^{2+} as a stabilizing impurity in β-$Ca_3(PO_4)_2$ has been undertaken at the American Dental Association Health Foundation Research Unit in Washington DC (ref. 69). Up to 11.6 atom % magnesium was incorporated instead of calcium. It was evident that practically all magnesium is located at the cation site with the lowest coordination number (i.e., site M(5) with CN = 6; ref. 69). These sites may house 6/63 = 9.5 % of all divalent ions, which figure agrees well with the maximum solubility at 1620 K (11.6 atom % Mg) noted by Schroeder et al. (ref. 69). It is tempting to suggest that other smaller cations would also prefer the M(5) sites in β-$Ca_3(PO_4)_2$, which agrees well with the observed maximum solubility for nickel, copper, zinc etc as indicated in ref. (68).

Part of the pseudo-binary phase diagram $Ca_3(PO_4)_2$ - $Mg_3(PO_4)_2$ (T > 1200 K) has been published by Ando (ref. 70). This system contains an intermediate phase, $Ca_7Mg_9(Ca,Mg)_2(PO_4)_{12}$, which structure has been determined by Dickens and Brown (ref. 71). It is structurally related to glaserite, $K_3Na(SO_4)_2$ (e.g., see ref. 72, 73). The structure is complicated, with five distinct metal sites having coordination numbers between six and eight. Similar to the situation for the β-$(Ca,Mg)_3(PO_4)_2$ phases (ref. 69), magnesium is located at the site with the lowest coordination number (CN = 6).

Of great structural interest is also the mineral whitlockite. Whitlockite was first identified as a new mineral by Frondel (ref. 74, 75). X-ray powder diffraction patterns indicated great similarity between whitlockite and β-$Ca_3(PO_4)_2$, and the two phases were long considered to be equivalent. Keppler (ref. 76), though, stated that whitlockite has the approximate formula $Ca_{18}M_2H_2$-$(PO_4)_{14}$, where M (M^{2+}) is a small cation such as Mg^{2+}, Fe^{2+}, or

Mn^{2+}. This formula has now been established through crystallographic studies of natural and synthetic whitlockites (ref. 77-79). However, the whitlockite structure is closely related to that of β-$Ca_3(PO_4)_2$, with the same space group symmetry, R3c. The "M" atoms in whitlockite are six-coordinated, while the three crystallographically inequivalent calcium atoms are each surrounded by eight oxygen neighbours. The structure contains interconnected infinite chains of polyhedra parallelling the hexagonal "c" axis, with links in the chains consisting of three CaO_8 polyhedra separated by two PO_4 tetrahedra.

Although the complete extent of possible substitution of other divalent cations for calcium in whitlockite is as yet unknown, chemical analyses of natural terrestrial whitlockite minerals also indicate around 90 % Ca^{2+} and 10 % of other ("M^{2+}") divalent metals such as magnesium, iron, and manganese. This is also true for meteoritic samples, which further contain alkali and rare earth metals. A hydrothermally produced Fe-whitlockite was shown to contain 5.3 weight % iron, i.e., again around 10% of the total metal contents, in atomic percent (ref. 80), which fact seems to indicate that the smaller "M^{2+}" ions are located at the six-coordinated sites in this crystal structure.

The stereochemical conditions in apatite, $Ca_5(PO_4)_3(Cl,OH,F)$, will be briefly surveyed for comparison. There are two crystallographically non-equivalent calcium atom sites in apatite, 7- (or 8-) and 9-coordinated, respectively (ref. 5, 82). Many M^{2+} ions may substitute for calcium, e.g. smaller amounts (< 10 %) of Mg^{2+}, Fe^{2+}, or Mn^{2+}. Among the larger cations, strontium may be incorporated in large amounts as in the apatite-type called belovite with the empirical formula $(Sr,Ce,Na,Ca)_5(PO_4)_3(OH)$. Studies of partially substituted $(Ca,Sr)_5(PO_4)_3Cl$-apatites have been published by Sudarsanan and Young (ref. 82). Contrary to the conditions in calcium orthophosphate, cadmium may also replace fairly large amounts of calcium (ref. 83). Complete isomorphous miscibility between $Ca_5(PO_4)_3OH$ and $Pb_5(PO_4)_3OH$ has been demonstrated (ref. 84); cf. the existence of the mineral pyromorphite, $Pb_5(PO_4)_3Cl$. Barium, finally, is probably too large (as Ba^{2+}) as it was in the case of β-$Ca_3(PO_4)_2$, since only minor amounts of barium have been found in natural apatite minerals.

Another compound, structurally related to apatite, is $Ca_4(PO_4)_2O$, tetracalcium diphosphate oxide (ref. 85) (S.G. $P2_1$; Z = 4,

V = 797.3 Å^3). There are six seven-coordinated and two eight-coordinated calcium atoms in this structure. The dimensions of the unit cells of $Ca_4(PO_4)_2O$ and $Ca_5(PO_4)_3OH$ are simply related. Although this three-dimensional relationship in the unit-cell shapes is not carried over into details of the actual crystal structures, $Ca_4(PO_4)_2O$ does contain a layer which is similar to a layer of atoms in apatite, and an epitaxic relationship between the two compounds is conceivable (ref. 85). The compound $Ca_4(PO_4)_2O$ is also related structurally to glaserite, $K_3Na(SO_4)_2$ (ref. 72, 73). In this latter relationship, the oxide ions are counted as "extra" ions.

The relationships among the calcium-containing anhydrous orthophosphates discussed so far are displayed in Fig. 2.5.2. Note that strontium and barium orthophosphate (ref. 86) are isostructural with the high-pressure phase of calcium orthophosphate, $Ca_3(PO_4)_2$-II. There are, in addition, many calcium-bearing phosphates for which little or no adequate crystallographic or chemical information exist, e.g., the four $CaZn_2(PO_4)_2$ modifications (ref. 87), and the "fourth" (high-temperature) phase of calcium orthophosphate, usually denoted α'-$Ca_3(PO_4)_2$ (ref. 64, 88). The pseudo-binary systems $Ca_3(PO_4)_2$ - $Sr_3(PO_4)_2$ and $Ca_3(PO_4)_2$ - $Ba_3(PO_4)_2$ have in part been investigated (ref. 89, 90). The former system is dominated by solid solutions of the whitlockite-related form β-$(Ca,Sr)_3(PO_4)_2$, while the latter system is much more complicated and not known to detail. Without doubt many new intermediate phases will be found in the future in systems of this kind when given close attention, thus reflecting the manifold of compounds and crystal structures not only existing in the world of organic chemistry but also in inorganic chemistry.

Many natural minerals which contain calcium, in addition to other metals, have been found. An example has already been mentioned: whitlockite. Another example is hurlbutite, $CaBe_2(PO_4)_2$, whose crystal structure has been published by Lindbloom, Gibbs and Ribbe (ref. 91). The space group symmetry is $P2_1/a$. The structure has a framework of alternating BeO_4 and PO_4 tetrahedra with CaO_9 polyhedra (cf. Fig. 2.5.3). Other minerals which should be mentioned are herderite: $(Ca,Mn)Be(PO_4)(F,OH)$, buchwaldite: $NaCaPO_4$, hagendorfite: $(Na,Ca)(Fe,Mn)_2(PO_4)_2$, hühnerkobelite: $(Na,Ca)(Fe,Mn)_2$-$(PO_4)_2$, and nagelschmidtite: $Ca_3(PO_4)_2 \cdot 2Ca_2SiO_4$ (ref. 92). Of course there are many other minerals, cf. ref. (92).

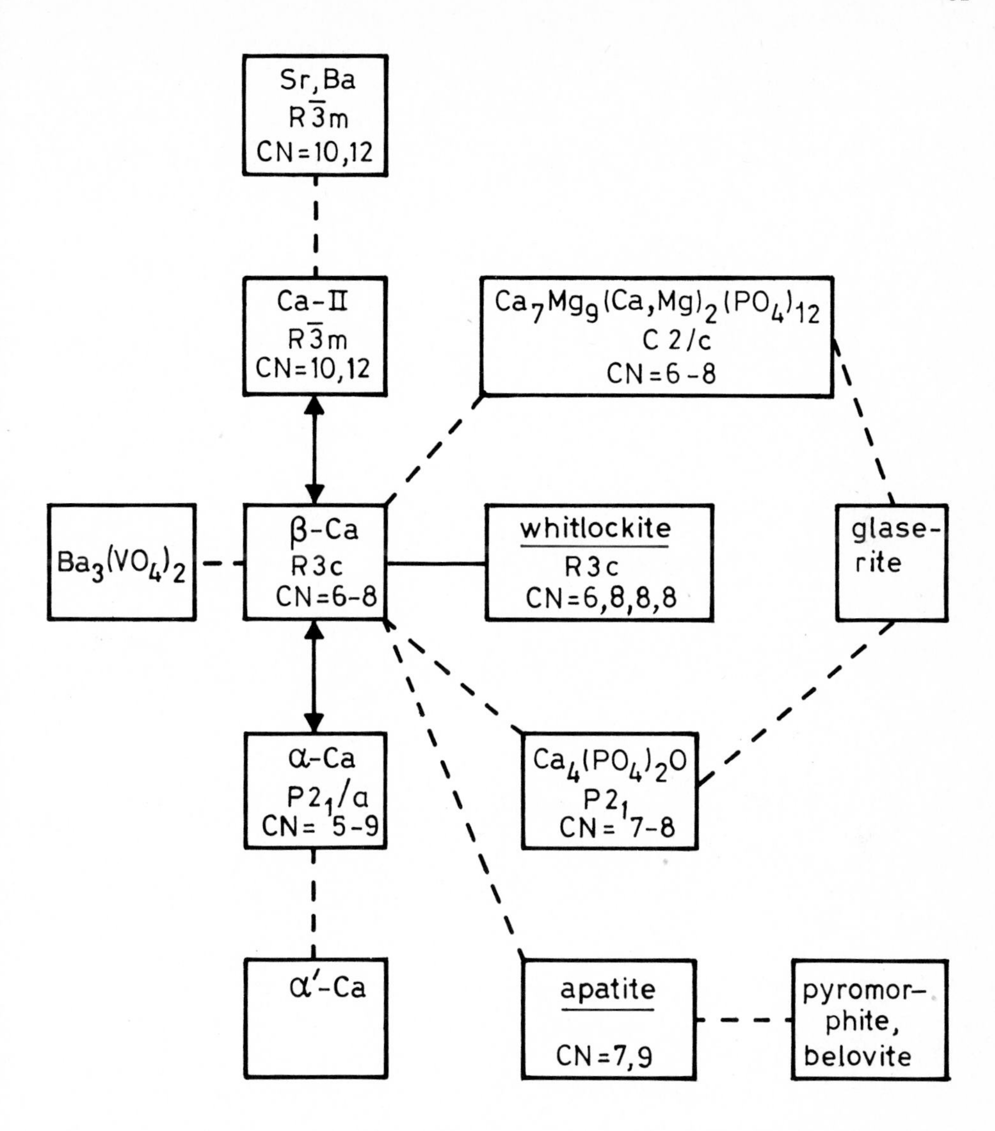

Fig. 2.5.2. Structural relationships among calcium- and calcium-containing orthophosphates and some other related compounds. α-Ca stands for α-$Ca_3(PO_4)_2$ and so forth. Space group symbols and cation coordination numbers (CN) are given.

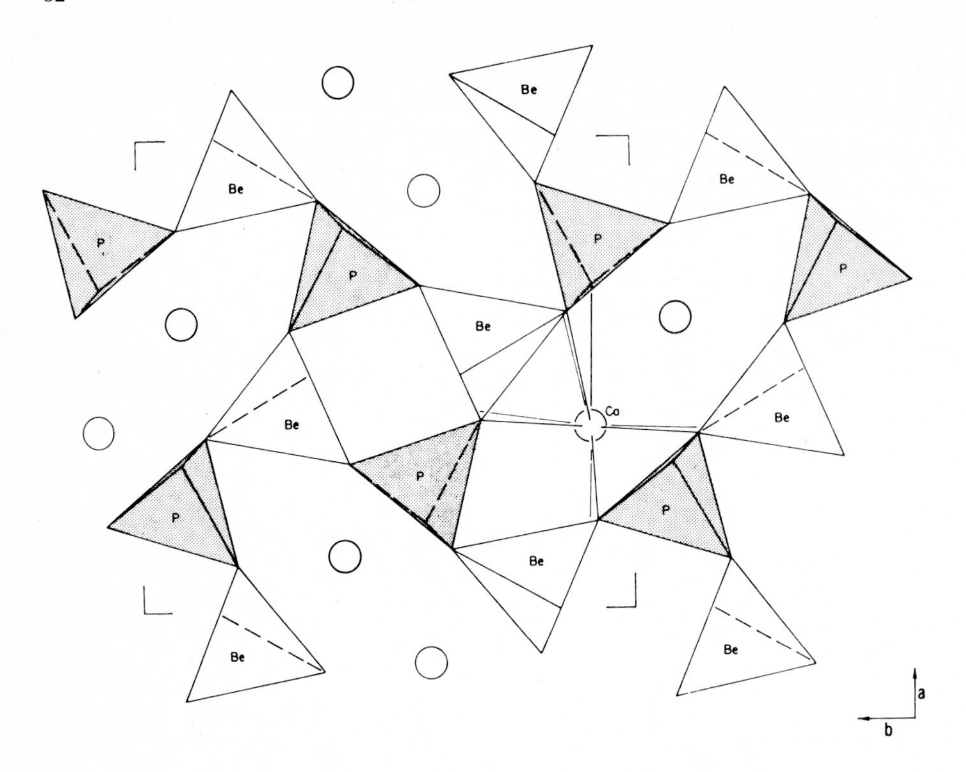

Fig. 2.5.3. The crystal structure of hurlbutite, $CaBe_2(PO_4)_2$. The figure is reproduced with permission from the Mineralogical Society of America (copyright) from Lindbloom, Gibbs and Ribbe: "The crystal structure of hurlbutite" (ref. 91) in The American Mineralogist, vol. 59, pp. 1267-1271 (1974).

2.6 Other anhydrous orthophosphates

The anhydrous orthophosphates to be briefly reviewed in this subsection contain any of the metals copper, tin, lead, or mercury as divalent cations, or mixtures of these and other metals. In addition, further interesting metal orthophosphates not fitting into any of the preceding subsections are included here, although they may not bear any similarity with the other crystal structures to be discussed.

No reliable crystallographic information exists on any orthophosphate modification on the smallest divalent cation, i.e. beryllium. However, there are a few beryllium-containing natural minerals such as hurlbutite, herderite (section 2.5), beryllonite:

$NaBePO_4$, and babephite: $BaBe(PO_4)(F,OH)$ (ref. 92), to mention some of the most important beryllium phosphate minerals.

Some crystal structures of orthophosphates containing divalent copper have been published. For instance, $Cu_3(PO_4)_2$ has been studied by Shoemaker, Anderson and Kostiner (ref. 93). The space group symmetry is $P\overline{1}$ (Z = 1), with a unit cell volume of only 140.4 $Å^3$. There are two distinct Cu^{2+} ions in the structure. One is surrounded by two oxygens at 1.924 Å and two at 1.982 Å, thus forming a somewhat distorted square-planar configuration. (The fifth nearest oxygen is at 2.95 Å; ref. 93). The other copper atom is ligated to five oxygen atoms to form an irregular CuO_5 polyhedron, so that the coordination numbers in $Cu_3(PO_4)_2$ are CN = 4, 5. The CuO_5 polyhedra occur in edge-sharing pairs in the structure.

The mineral stranskiite, $Zn_2Cu(AsO_4)_2$, is isostructural with copper(II) orthophosphate (ref. 94). In this structure there is a strong or perhaps even complete ordering of the cations among the two distinct cation sites, so that copper is located in the square-planar configuration with Cu-O distances of 2.014 and 1.893 Å, while Zn^{2+} is five-coordinated with Zn-O distances in the range 2.022 - 2.171 Å (ref. 94). Otherwise, isomorphism between a divalent-metal phosphate and the corresponding arsenate is rare because the AsO_4 tetrahedron is considerably greater than the PO_4 tetrahedron. The earlier mentioned (section 2.3) isomorphism between graftonite, $(Fe,Mn,Ca,Mg)_3(PO_4)_2$, and cadmium orthoarsenate, $Cd_3(AsO_4)_2$, is probably enabled because in the latter compound the central atom of the tetrahedra as well as the metal cations are larger than in graftonite.

Another interesting structure is that of the copper-rich phosphate $Cu_5(PO_4)_2O_2$ (ref. 95), which is also triclinic (S.G. $P\overline{1}$). It has three crystallographically non-equivalent Cu^{2+} ions, with CN = 5, 5, 6. The copper-oxygen polyhedra form a three-dimensional framework, interconnected by rather regular PO_4 tetrahedra. Two other compounds deserve to be mentioned, namely $Ca_3Cu_3(PO_4)_4$ and $Cu_9Na_6(PO_4)_8$, both containing square-planar CuO_4 groups as well as CuO_5 or CuO_6 polyhedra. The former was synthesized hydrothermally at 420°C, 3.8 kbar (ref. 96). The two calcium atoms in $Ca_3Cu_3(PO_4)_4$ are in six- and nine-coordination, and the two distinct copper-oxygen polyhedra (four- and five-coordinated Cu^{2+}) are similar to those found in $Cu_3(PO_4)_2$. The structure of Cu_9Na_6-$(PO_4)_8$, published by Senga and Kawahara (ref. 97), is triclinic

(S.G. $P\bar{1}$, Z = 1) and has no less than six kinds of copper atoms, three sodium coordination polyhedra, and four distinct PO_4 tetrahedra. Three of the Cu^{2+} ions are situated at square-planar sites, two ions occupy five-coordinated sites in distorted square pyramids, and the sixth copper is at a distorted and elongated octahedral site, so that the CN:s are 4, 4, 4, 5, 5, 6 (ref. 97) in this rather complicated structure.

The crystal structure of stannous orthophosphate, $Sn_3(PO_4)_2$, was published in 1977 by Mathew, Schroeder and Jordan (ref. 98). It crystallizes in the monoclinic space group $P2_1/c$ (Z = 4) and is built up of alternating layers of Sn^{2+} and PO_4^{3-} ions perpendicular to the monoclinic "b" axis. Each tin atom is at the apex of a trigonal pyramid with three nearest oxygen atoms, each from a different phosphate group, forming the base.

Lead orthophosphate, $Pb_3(PO_4)_2$, is polymorphous. There are the low-temperature (α) form (C2/c, CN = 9, 10) (ref. 99, 100), the high-temperature form β-$Pb_3(PO_4)_2$ (S.G. $R\bar{3}m$, CN = 10, 12) (ref. 101), and a third apatite-like form usually denoted $Pb_9(PO_4)_6$ to distinguish it from the two other modifications. This third form has a hexagonal symmetry (S.G. $P6_3/m$, CN = 6, 9) (ref. 102). In addition, quite a lot of lead-apatites have been reported in the literature (e.g. ref. 103, 104); cf. also the existence of the mineral pyromorphite, $Pb_5(PO_4)_3Cl$ (ref. 92).

The crystal structure of mercuric orthophosphate, $Hg_3(PO_4)_2$, finally, has been determined by Aurivillius and Nilsson (ref. 105). There are three distinct Hg^{2+} ions, each coordinated to two oxygen ligands with the Hg-O distances in the range 2.06-2.13 Å and the O-Hg-O angles from 163° to 170°, so that the latter group is almost linear. The structure is built up of infinite puckered nets of formula units $(Hg_3(PO_4)_2)_n$. Two such nets, with a crystallographically imposed $\bar{1}$ symmetry, run through the unit cells.

2.7 Hydrous orthophosphates

Hydrous orthophosphates of divalent metals are fairly common as minerals, much more so than the anhydrous orthophosphates. These and a few synthetic compounds are listed in Table 2.7.1. Review articles on hydrous phosphates have earlier been published by Wolffe (ref. 106, 107) and by Moore (ref. 108); recently a book on

TABLE 2.7.1
Some hydrous divalent-metal orthophosphates

Formula	Name (if mineral)	S.G.	CN	Ref.
$Mg_3(PO_4)_2 \cdot 8H_2O$	bobierrite	-	-	(109)
$Mg_3(PO_4)_2 \cdot 22H_2O$	-	$P\bar{1}$	Mg: 6,6	(110)
$Fe_3(PO_4)_2 \cdot H_2O$	-	$P2_1/a$	Fe: 5,6,6	(111)
$Fe_3(PO_4)_2 \cdot 3H_2O$	phosphopherrite	Pbna	Fe: 6,6	(112)
$Fe_3(PO_4)_2 \cdot 4H_2O$	ludlamite	$P2_1/a$	Fe: 6,6	(113)
$Fe_3(PO_4)_2 \cdot 8H_2O$	vivianite	C2/m	Fe: 6,6	(116)
$(Fe,Mn)_3(PO_4)_2 \cdot 3H_2O$	reddingite	Pmnb?	-	(118)
$(Mn,Fe)_3(PO_4)_2 \cdot 4H_2O$	switzerite	$P2_1/c$	5,6,6,6,6,6	(119)
$Co_3(PO_4)_2 \cdot H_2O$	-	$P2_1/c$	Co: 5,6,6	(120)
$Zn_3(PO_4)_2 \cdot 4H_2O$	hopeite	Pnma	Zn: 4,6	(121)
$Zn_3(PO_4)_2 \cdot 4H_2O$	parahopeite	$P\bar{1}$	Zn: 4,6	(122)
$(Zn,Co)_3(PO_4)_2 \cdot 4H_2O$	-	Pnma	-	(123)
$Zn_2(Fe,Mn)(PO_4)_2 4H_2O$	phosphophyllite	$P2_1/c$	Zn:4, Fe:6	(124)
$Ca_2Fe(PO_4)_2 \cdot 4H_2O$	anapaite	$P\bar{1}$	Ca:8, Fe:6	(126)
$Zn_2Ca(PO_4)_2 \cdot 2H_2O$	scholzite	$Pbc2_1$	Zn:4, Ca:6	(127)
$Zn_2Ca(PO_4)_2 \cdot 2H_2O$	parascholzite	monocl.		(92)
$MgCa_2(PO_4)_2 \cdot 2H_2O$	collinsite	$P\bar{1}$	Mg:6, Ca:8	(128)
$Ca_2(Mn,Fe)(PO_4)_2 2H_2O$	fairfieldite	$P\bar{1}$	Ca:8, Mn:6	(129)

phosphate minerals by Nriagu and Moore has appeared (ref. 92). It should also be mentioned that numerous hydrous orthophosphates are listed in Gmelin's Handbuch der anorganischen Chemie, although without any relevant crystallographic information for the major part of compounds. Some comments on Table 2.7.1 will be given below.

The crystal structures of $Mg_3(PO_4)_2 \cdot 8H_2O$, reported to be polymorphous with one of the phases existing as a natural mineral called bobierrite, are so far unknown (ref. 92, 109). On the other hand, the structure of $Mg_3(PO_4)_2 \cdot 22H_2O$ has been determined by Schroeder, Mathew and Brown (ref. 110), showing a complicated system of hydrogen bonds where the $Mg(H_2O)_6$ "octahedra" and PO_4 tetrahedra form a layer structure with the stacking sequence BAABAA... This latter compound contains as much as 60 weight % water, which necessitates strong hydrogen bonds to keep the structure together.

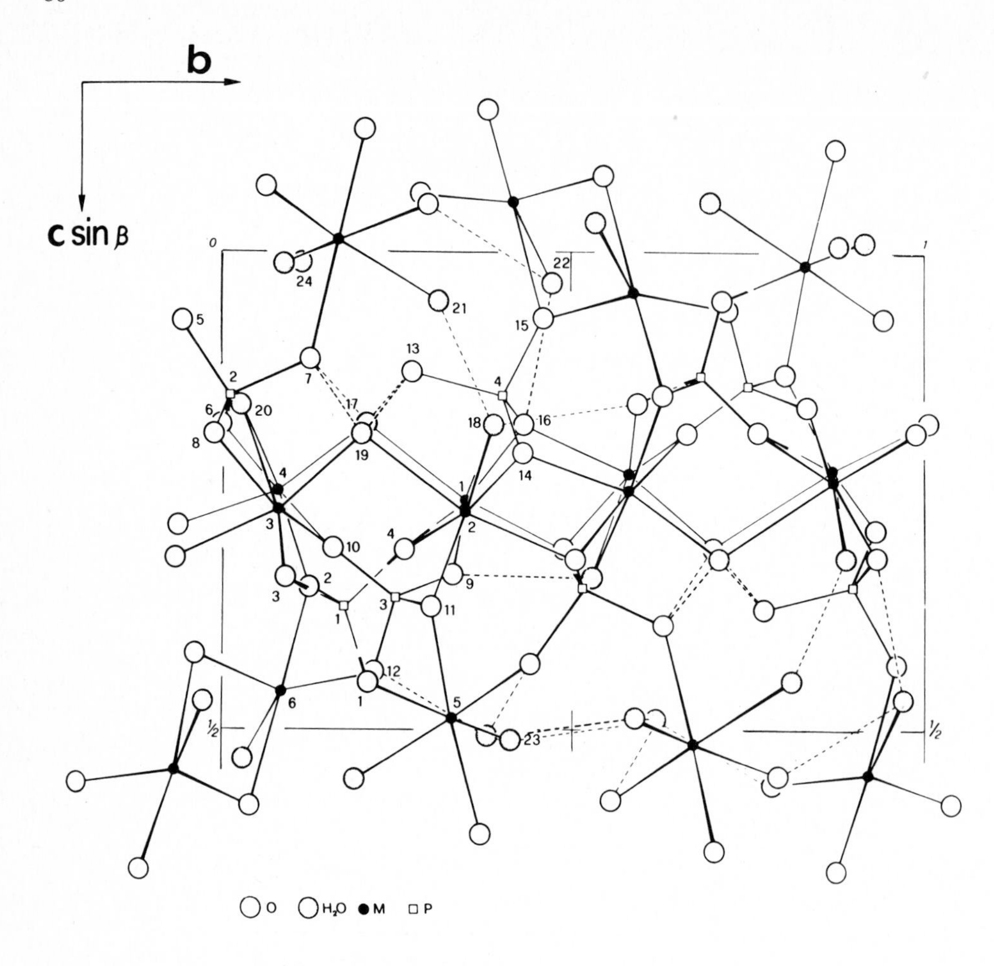

Fig. 2.7.1. The crystal structure of switzerite, $(Mn,Fe)_3(PO_4)_2\cdot 4H_2O$. The figure is reproduced with permission from Springer-Verlag, Wien (copyright) from Fanfani and Zanazzi: "Switzerite: its chemical fcrmula and structure" (ref. 119) in Tschermaks Min. Petr. Mitt. vol. 26, pp. 255-269 (1979).

The symmetry is triclinic (S.G. $P\overline{1}$). Curiously enough, yet another polymorph of $Mg_3(PO_4)_2\cdot 22H_2O$, also with $P\overline{1}$ space group symmetry, has recently been reported (ref. 110a) by Catti, Franchini-Angela and Ivaldi.

The crystal structure of $Fe_3(PO_4)_2\cdot H_2O$ was published by Moore and Araki (ref. 111). There are two six- and one five-coordinated Fe^{2+} ion, the latter with Fe-O distances in the range 1.987 - 2.243 Å, forming a complex structure which bears no obvious relationship to any other structure. However, feeble resemblances to

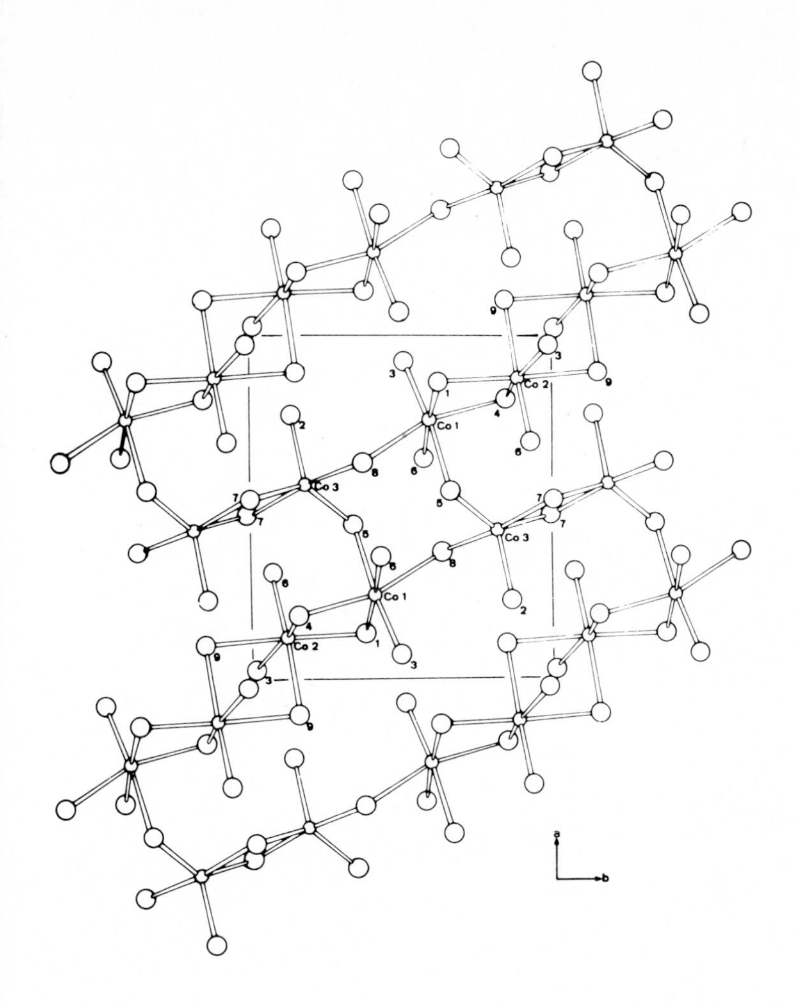

Fig. 2.7.2. The crystal structure of $Co_3(PO_4)_2 \cdot H_2O$. Projection of part of the structure onto the ab plane, centered about (½ ½ 0). The figure is reproduced with permission from The American Chemical Society (copyright) from Anderson, Kostiner and Ruszala: "Crystal structure of cobalt(II) orthophosphate monohydrate, $Co_3(PO_4)_2 \cdot H_2O$" (ref. 120) in Inorganic Chemistry vol. 15, pp. 2744-2748 (1976).

other members of the homologous series $Fe_3(PO_4)_2(H_2O)_n$ and to the graftonite structure is pointed out by the authors in question (ref. 111). The structure of phosphoferrite, $Fe_3(PO_4)_2 \cdot 3H_2O$, is based on sheets of corner- and edge-linked FeO_6 octahedra which are parallel to the a axis (S.G. Pbna). The phosphate tetrahedra are situated between these sheets resulting in a rather rigid framework of polyhedra so that the mineral crystals exhibit no good

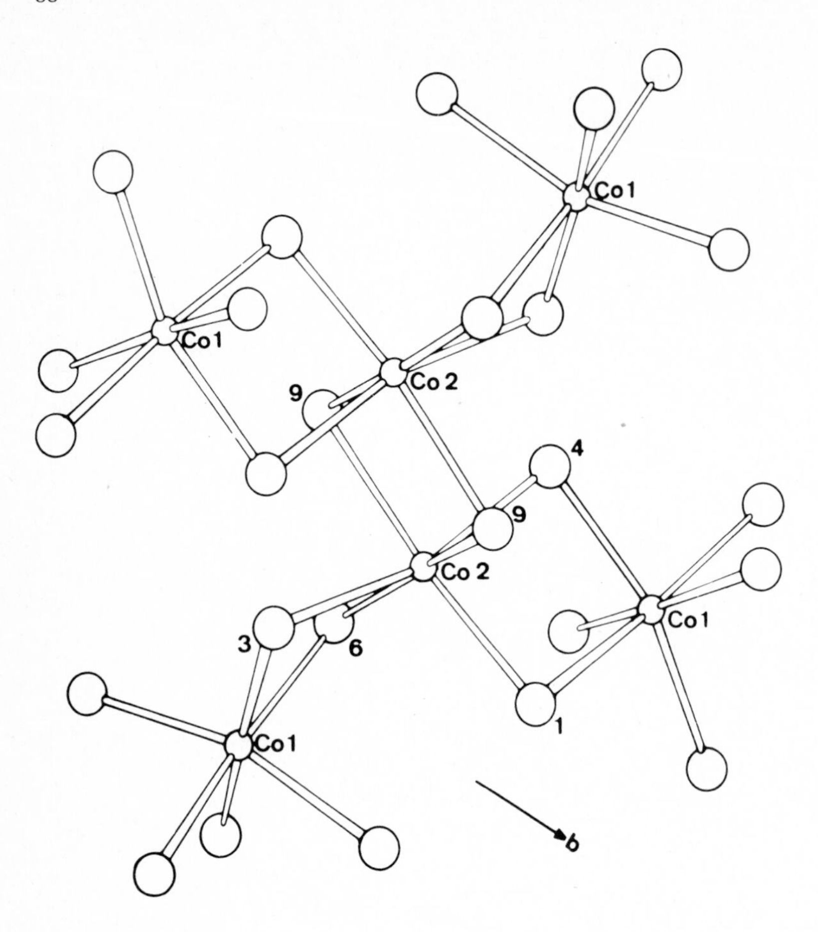

Fig. 2.7.3. Detailed part of the $Co_3(PO_4)_2 \cdot H_2O$ structure illustrating the edge-sharing linkages between the two distinct CoO_6 octahedra. Reproduced from ref. (120); cf. the text to Fig. 2.7.2.

cleavage direction (ref. 112). Vivianite, $Fe_3(PO_4)_2 \cdot 8H_2O$, is a common mineral and has been studied by many authors. The crystal structure was originally determined by Mori and Ito (ref. 115), later refined by Fejdi, Poullen and Gasperin (ref. 116). It is a comparatively simple structure, built up of two distinct FeO_6 octahedra and PO_4 tetrahedra. A magnetic structure of vivianite has been studied at very low temperatures by means of neutron diffraction by Forsyth, Johnson and Wilkinson (ref. 117).

The mineral switzerite, $(Mn,Fe)_3(PO_4)_2 \cdot 4H_2O$, has a complicated sheet structure with six crystallographically non-equivalent metal sites, of which one is five-coordinated and the other six-coordi-

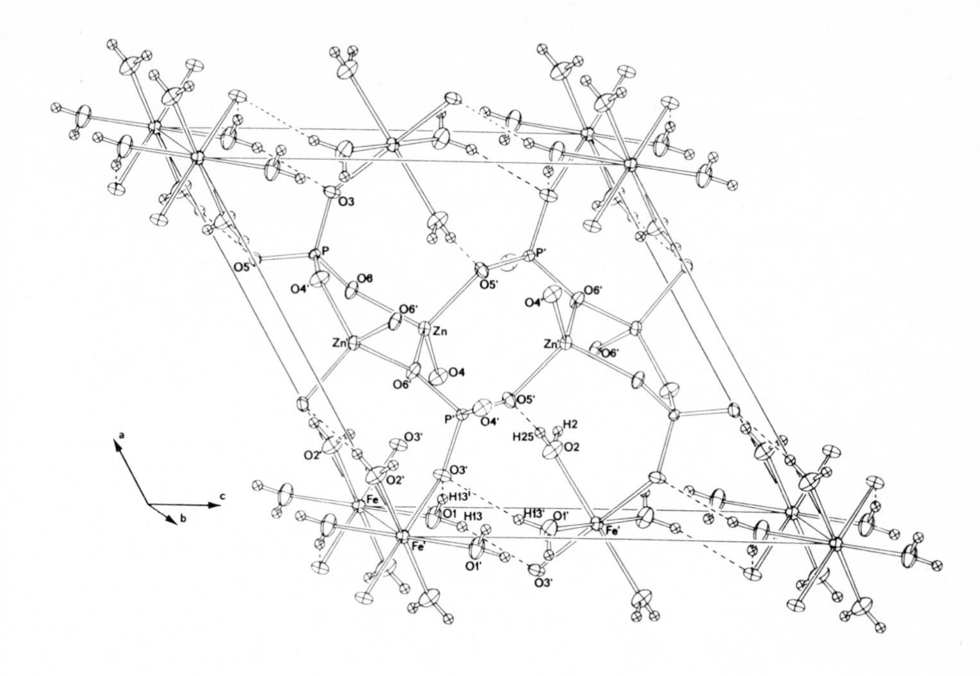

Fig. 2.7.4. Unit-cell diagram of the crystal structure of phosphophyllite, $Zn_2Fe(PO_4)_2 \cdot 4H_2O$. The figure is reproduced with permission from the Mineralogical Society of America (copyright) from Hill: "The crystal structure of phosphophyllite" (ref. 124) in The American Mineralogist, vol. 62, pp. 812-817 (1977).

nated (ref. 119). The structure is shown in Fig. 2.7.1.

Single crystals of cobalt(II) orthophosphate monohydrate, $Co_3(PO_4)_2 \cdot H_2O$, were hydrothermally grown and the corresponding structure determined by Kostiner et al. (ref. 120), cf. Figs. 2.7.2 and 2.7.3. The Co-O bond lengths around the five-coordinated cobalt atom are rather uniform (1.976-2.079 Å), while in the anhydrous $Co_3(PO_4)_2$ "1-bar" modification the five-coordinated polyhedron consists of four short bonds and one longer at 2.229 Å (ref. 33).

$Zn_3(PO_4)_2 \cdot 4H_2O$ is dimorphous; it exists in both forms as natural minerals: hopeite (ref. 121) and parahopeite (ref. 122). Phosphophyllite is a beautiful blue mineral with a perfect cleavage. It has the approximate composition $Zn_2(Fe,Mn)(PO_4)_2 \cdot 4H_2O$, where iron ($Fe^{2+}$) dominates over manganese. Its crystal structure has been determined by Hill (ref. 124); cf. Fig. 2.7.4. It is built up of ZnO_4 and PO_4 tetrahedra and unusually regular $(Fe,Mn)O_6$ octahedra. Mössbauer studies of two natural phosphophyllites have been undertaken by Ericsson and Nord (ref. 125). These results confirm Hill's structural analysis that iron is only located at the octahedral sites, so that the cation ordering is complete among the two distinct cation sites. The ^{57}Fe Mössbauer hyperfine parameters of this octahedral site, CS = 1.27 mm/s and ΔE_Q = 3.47 mm/s at 295 K, are in excellent agreement with the regular shape of the $(Fe,Mn)O_6$ octahedron, in which the metal-oxygen distances were found to be in the range 2.119-2.141 Å with an average distance of 2.130 Å (ref. 124, 125).

Among the mixed-metal hydrous orthophosphate minerals, only that of fairfieldite, $Ca_2Mn(PO_4)_2 \cdot 2H_2O$, will be mentioned in some detail. In this structure, calcium and manganese are eight- and six-coordinated, respectively (ref. 129). The structure is built up of infinite chains of MnO_6 octahedra and PO_4 tetrahedra running parallel to the _c_ axis, and the Ca^{2+} ions occupy the vacant spaces between these chains connecting them in a three-dimensional network. The structure is closely related to that of kröhnkite, $CuNa_2(SO_4)_2 \cdot 2H_2O$ (ref. 130). Finally, there are many other rather complicated orthophosphate minerals like stanfieldite: $Ca_4(Mg,Mn,Fe)_5$-$(PO_4)_6$, stewartite: $MnFe_2(PO_4)_2(OH)_2 \cdot 8H_2O$, and many other (cf. ref. 92).

2.8 Miscellaneous orthophosphate structures

A limited number of crystal structures of hydrogen orthophosphates have been published, and will be briefly dealt with in this section in connection with various other stereochemical data on phosphates of the type $M_2(PO_4)Y$ (Y = F, Cl, OH etc), and some other interesting structure types which have not been suitable to include in any of the preceding sections of subchapter 2.

Newberyite, $MgHPO_4 \cdot 3H_2O$ (S.G. Pbca) is a mineral whose structure is built up of MgO_6 octahedra sandwiched between PO_4 tetrahedra, connected by hydrogen bonds involving the water molecules. These polyhedra form layers running along [001] (ref. 131). One of the water molecules forms a bifurcated hydrogen bond. Phosphor-roesslerite, $MgHPO_4 \cdot 7H_2O$, is another magnesium hydrogen phosphate (ref. 132), chemically related to struvite, $MgNH_4PO_4 \cdot 6H_2O$ (ref. 133). Some calcium hydrogen orthophosphate structures have been determined, such as those of the minerals monetite: $CaHPO_4$ (ref. 134)

TABLE 2.8.1

Crystallographic data for the $M_2(PO_4)Y$ compounds.

Formula	Name (if mineral)	S.G.	CN	Ref.
$Mg_2(PO_4)(OH,O,F)$	althausite	Pnma	Mg:6,6	(147)
$Mg_2(PO_4)OH$	holtedahlite	hex.	-	(92)
$Mg_2(PO_4)F$	wagnerite	$P2_1/c$	5,5,5,5,6,6,6,6	(148)
$Co_2(PO_4)F$	-	$P2_1/c$	-	(149)
$Ni_2(PO_4)F$	-	$P2_1/c$	-	(149)
$Cu_2(PO_4)F$	-	$P2_1/c$	-	(149)
$Zn_2(PO_4)F$	-	$P2_1/c$	-	(149)
$(Mn,Fe)_2(PO_4)OH$	triploidite	$P2_1/a$	5,5,5,5,6,6,6,6	(152)
$(Fe,Mn)_2(PO_4)OH$	wolfeite	$P2_1/a$	-	(153)
$(Fe,Mg,Mn,Na,H)_2$-$(PO_4)OH$	satterlyite	hex.	-	(92)
$(Mn,Fe)_2(PO_4)F$	triplite	I2/a	6,6	(154)
$(Fe,Mn,Mg,Ca)_2$-$(PO_4)(F,O)$	zwieselite	monocl.		(92)
$Mn_2(PO_4)F$	-	C2/c	6,6	(155)
$Cu_2(PO_4)F$	-	C2/c	6,6	(156)
$Cd_2(PO_4)F$	-	C2/c	6,6	(157)
$Cu_2(PO_4)OH$	libethenite	Pnnm	5,6	(158)
$Zn_2(PO_4)OH$	tarbuttite	$P\bar{1}$	5,5	(159)
$Mg_2(PO_4)Cl$	-	$Pna2_1$	6,6	(160)
$Mn_2(PO_4)Cl$	-	$Pna2_1$	6,6	(161)
$Fe_2(PO_4)Cl$	-	C2/c	6,6,6	(162)
$Co_2(PO_4)Cl$	-	C2/c	6,6,6	(163)
$Ca_2(PO_4)Cl$	"Cl-spodiosite"	Pbcm	8,8	(165)
$CaMg(PO_4)F$	isokite	C2/c	Ca:7, Mg:6	(166)
$CaBe(PO_4)(OH,F)$	herderite	$P2_1/a$	Ca:8, Be:4	(168)
$Sn_2(PO_4)OH$	-	$P2_1/n$	Sn: 3,3	(169)

and brushite: $CaHPO_4 \cdot 2H_2O$ (ref. 135); and the synthetic compound $Ca(H_2PO_4)_2$, containing an unusual hydrogen bonding system extensively studied by Dickens, Prince, Schroeder and Brown (ref. 136). Several tin(II) hydrogen orthophosphate structures have been reported (ref. 137-139), and also the crystal structures of the compounds $BaHPO_4$ (ref. 140), $HgHPO_4$ (ref. 140a), and $Pb(H_2PO_4)_2$

(ref. 140b).

The important mineral apatite, $Ca_5(PO_4)_3(F,OH,Cl)$, has already been discussed in section 2.5. It is not only a mineral of great industrial value but also the main inorganic constituent of bones and teeth in animals. However, there are also many other phosphate compounds present in human tissues, such as octacalcium phosphate (ref. 141), whitlockite: $Ca_{18}M_2H_2(PO_4)_{14}$, β-$Ca_3(PO_4)_2$, $Ca_2P_2O_7 \cdot 2H_2O$, struvite: $MgNH_4PO_4 \cdot 6H_2O$, newberyite: $MgHPO_4 \cdot 3H_2O$, and many other types (ref. 92).

Stereochemically related to apatite are the minerals (or related synthetic compounds) pseudo-malachite: $Cu_5(PO_4)_2(OH)_4$ (ref. 142), pyromorphite: $Pb_5(PO_4)_3Cl$ (ref. 92), two isomorphs of arsenoclasite: $Co_5(PO_4)_2(OH)_4$ and $Mn_5(PO_4)_2(OH)_4$ (ref. 143), and many other phases.

Compounds of the type $M_2(XO_4)Y$, in which "M" is a divalent metal like Mg, Fe, Zn, Ca etc, "X" is phosphorus or arsenic, and "Y" is F, Cl or OH, are very common as minerals. The so far known $M_2(PO_4)Y$ phosphate types are summarized in Table 2.8.1. Summarizing papers on such compounds have earlier been published by Richmond (ref. 144), and by Klement and Haselbeck (ref. 145, 146). Brief comments to the more important compounds, from a stereochemical viewpoint, will be made below, with reference to Table 2.8.1. It should also be pointed out that discussions and comments on the dimensions of phosphate tetrahedra are given in section 7 for all phosphate types of this chapter.

Althausite, $Mg_2(PO_4)(OH,O,F)$, is a natural mineral with an interesting crystal structure (S.G. Pnma). There are one MgO_5F octahedron and two distinct MgO_4F polyhedra, all highly distorted (ref. 147). The five-coordinated magnesium ions are situated in distorted trigonal bipyramids. Although many of the "$M_2(XO_4)Y$" compounds and minerals are isomorphous with each other, the structure of althausite is unique but slightly related to the structure of wagnerite, $Mg_2(PO_4)F$, which is chemically close and also contains five- and six-coordinated magnesium atoms. Another mineral that is chemically close to althausite is holtedahlite, $Mg_2(PO_4)OH$. Holtedahlite might thus be regarded as a polymorph form of althausite. The former is again the magnesium analogue to satterlyite, $(Fe,Mg,Mn,Na,H)_2(PO_4)OH$, which in its turn is a hexagonal polymorph of wolfeite, $(Fe,Mn)_2(PO_4)OH$ (ref. 92).

The complicated crystal structure of wagnerite, $Mg_2(PO_4)F$, was

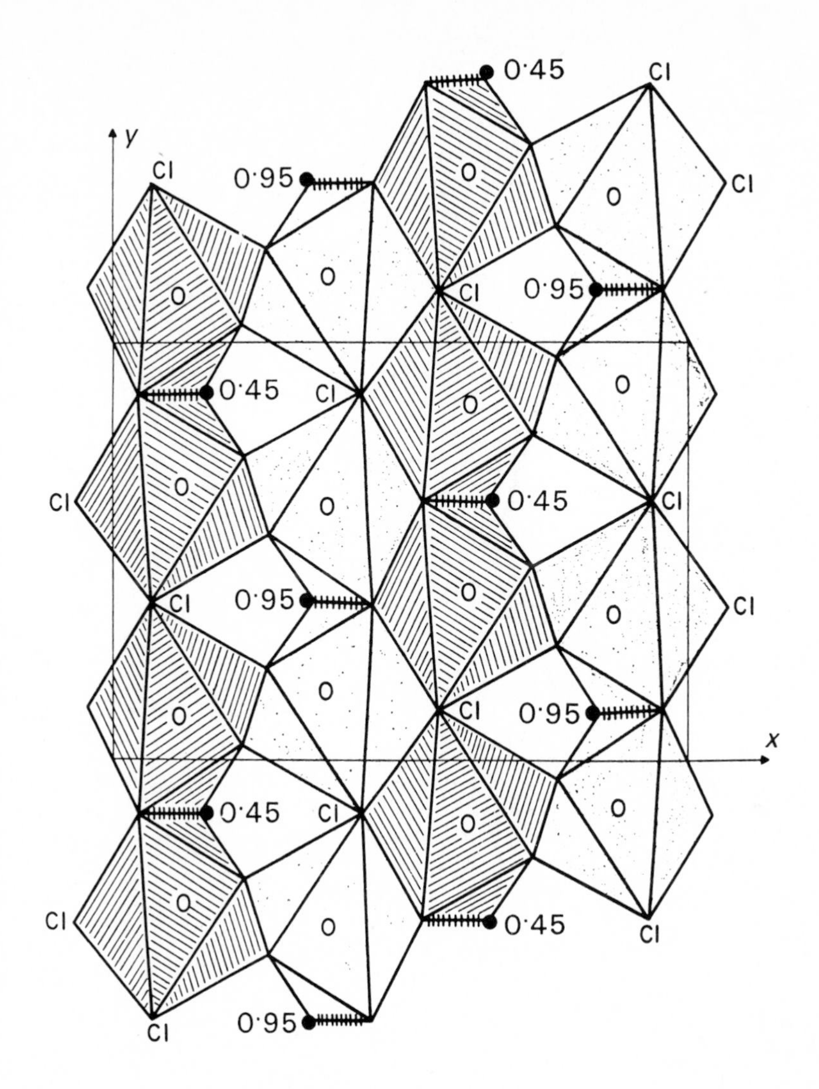

Fig. 2.8.1. Polyhedral representation of the crystal structure of $Mn_2(PO_4)Cl$, viewed down the c axis (S.G. $Pna2_1$). Manganese atoms are represented by open circles, phosphorus atoms by solid circles. All corners not labelled "Cl" are oxygen atoms.
The figure has been reproduced with permission from the International Union of Crystallography (copyright) from Rea and Kostiner: "The crystal structure of manganese chlorophosphate, $Mn_2(PO_4)Cl$" (ref. 161) in Acta Crystallographica, vol. B28, pp. 2505-2509 (1972).

determined in 1967 by Coda, Giuseppetti and Tadini (ref. 148). It contains four five- and four six-coordinated magnesium atom sites. Solid solutions of the wagnerite structure have been studied by Auh and Hummel (ref. 149), who found complete substitution for

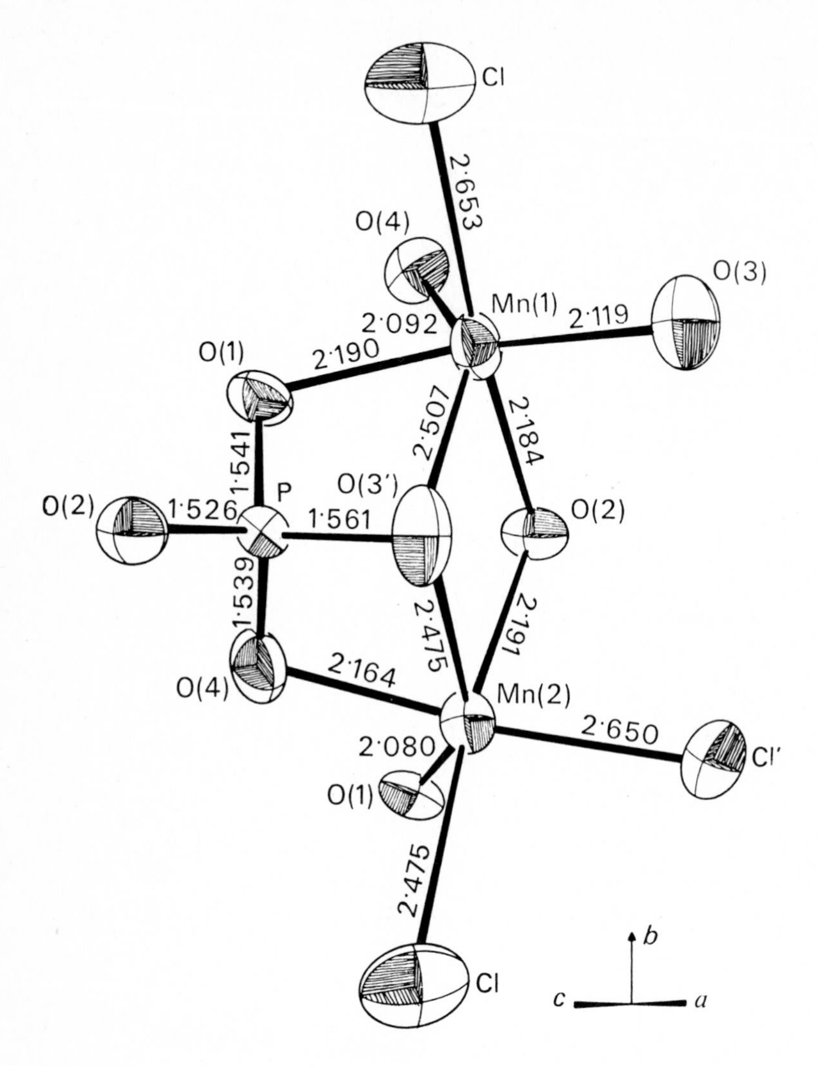

Fig. 2.8.2. The structural unit of $Mn_2(PO_4)Cl$. Reproduced from ref. (161); cf. the text to figure 2.8.1.

Co^{2+}, Ni^{2+}, Cu^{2+}, and Zn^{2+} for magnesium in this structure. It is indeed noticeable that a wagnerite-type of the form $Ni_2(PO_4)F$ should exist, since divalent nickel has an extremely low tendency for five-coordination (e.g., see ref. 150, 151), and has so far only been definitely demonstrated in α-$Ni_2P_2O_7$ (section 3).

Other important mineral groups are triploidite-wolfeite (ref. 152, 153), with the hexagonal satterlyite modification as a related phase (ref. 92), and the triplite-zwieselite group (ref. 154), to which belong the synthetic compounds $Mn_2(PO_4)F$ (ref. 155),

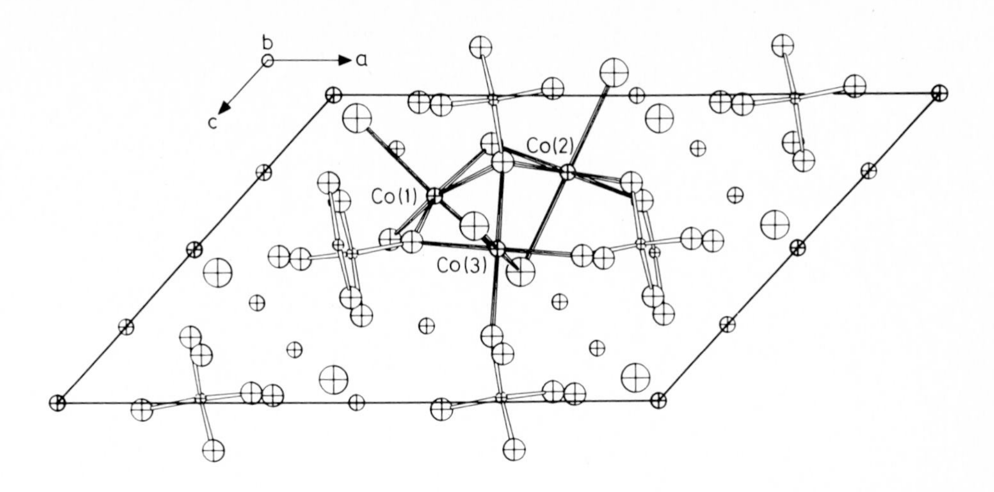

Fig. 2.8.3. The $Co_2(PO_4)Cl$ structure, viewed down the monoclinic b axis (S.G. C2/c; Z = 8). The circles, in order of increasing magnitude, are P, Co, O, and Cl atoms. Only one of the trimers of octahedra is shown for clarity (compare text).

$Cu_2(PO_4)F$ (ref. 156), and $Cd_2(PO_4)F$ (ref. 157). The triplite structure type is composed of isolated phosphate tetrahedra, joined by vertices to distorted MO_4Y_2 octahedra. These octahedra are linked into chains by shared edges. There are two sets of chains, parallel to a and b respectively in the monoclinic unit cell (ref. 154). Although the three other compounds mentioned are isotypic with triplite, there are significant differences noted in the position of the halogene atoms.

The mineral libethenite, $Cu_2(PO_4)OH$, is not isotypic with triplite but with andalusite, Al_2SiO_5, and the minerals of the olivenite group: $Cu_2(AsO_4)OH$ (ref. 158). The libethenite structure is built up of distorted $CuO_4(OH)_2$ octahedra and $CuO_4(OH)$ trigonal bipyramids. Tarbuttite, $Zn_2(PO_4)OH$, is triclinic (S.G. $P\bar{1}$), with two kinds of ZnO_5 polyhedra, forming zig-zag chains through the structure (ref. 159).

$Mg_2(PO_4)Cl$ and $Mn_2(PO_4)Cl$ are isostructural (ref. 160, 161). The structure is built up of MO_5Cl and MO_4Cl_2 octahedra, which are linked to each other and to a PO_4 tetrahedron by edge-sharing to form small structural units forming a three-dimensional network (Figs. 2.8.1 and 2.8.2). The space group is $Pna2_1$.

A very unusual structural building unit is displayed by the isostructural compounds $Fe_2(PO_4)Cl$ (ref. 162) and $Co_2(PO_4)Cl$ (ref. 163). This crystal structure type is built from PO_4 tetrahedra and three distinct MO_4Cl_2 octahedra (M = Fe or Co). The PO_4 tetrahedra are slightly distorted, and the MO_4Cl_2 octahedra are still more so, with a large spread in the distances between the metal atoms and their ligand atoms. Each $M(1)O_4Cl_2$ polyhedron shares one face with $M(2)O_4Cl_2$ and an adjacent face with $M(3)O_4Cl_2$ (see Fig. 2.8.3). These three connected polyhedra form a trimer of an unusual type with face-sharing octahedra, also found in the structure of the mineral seamanite, $Mn_3(PO_4)B(OH)_6$ (ref. 164). The trimers are further linked together, giving infinite chains of face-sharing MO_4Cl_2 polyhedra parallel to the crystallographic c axis. This structural feature is most uncommon and not at all similar to that of, e.g., $Mg_2(PO_4)Cl$ or $Mn_2(PO_4)Cl$, which have edge-sharing MO_4Cl_2 octahedra (ref. 160, 161). In spite of this fact, the volume occupied by one formula unit of $Co_2(PO_4)Cl$ (i.e., 104.2 $Å^3$) is very close to the corresponding value observed for $Mg_2(PO_4)Cl$ (104.1 $Å^3$; ref. 160), to which should be added the well-known fact that the cations Co^{2+} and Mg^{2+} are almost equally sized (ref. 6, 7). The face-sharing CoO_4Cl_2 octahedra in $Co_2(PO_4)Cl$ induce very short metal-metal distances (around 3 Å), but the structure also contains irregular voids which, on the average, makes the structure less closely packed than the face-sharing polyhedra in the trimers would otherwise suggest.

$Ca_2(PO_4)Cl$, sometimes called "chlorine-spodiosite", is isotypic with $Ca_2(CrO_4)Cl$ (ref. 165). This structure type is formed by discrete XO_4^{3-} tetrahedra which appear to be held together primarily by the calcium ions. The space group symmetry is Pbcm, with four formula units per unit cell. The XO_4 tetrahedra are much more regular in $Ca_2(PO_4)Cl$ than in $Ca_2(CrO_4)Cl$ (ref. 165).

The mineral isokite: $CaMg(PO_4)F$ (ref. 166) is structurally analogous to tilasite: $CaMg(AsO_4)F$ (ref. 167), with seven- and six-coordinated metal cations. Herderite, $CaBe(PO_4)(OH,F)$ (ref. 168) has been investigated by Lager and Gibbs in 1974. The structure consists of sheets of corner-sharing PO_4 and $BeO_3(OH)$ tetrahedra linked by sheets of edge-sharing $CaO_6(OH)_2$ polyhedra. Finally, the crystal structure of the compound $Sn_2(PO_4)OH$ has been determined by Jordan (ref. 169).

3 DIPHOSPHATES

3.1 Introductory notes

In diphosphates (formerly often called "pyrophosphates" because they were formed upon the pyrolysis of orthophosphates), the phosphate tetrahedra are connected in pairs through corner-sharing. In principle, three different conformations of the $P_2O_7^{4-}$ ion exist: the skew, staggered, and eclipsed forms (see Fig. 3.1.1). Usually the P - O - P link is angular, but linear groups have also been reported (cf. below).

The first diphosphate structure to be determined was that of ZrP_2O_7, in 1935, by Levi and Peyronel (ref. 170). This structure, however, has later turned out to be more complicated than was originally thought (ref. 171). The divalent-metal diphosphates are often dimorphous so that there is a low (usually: α) and a high (β) phase which transform reversibly into each other. Many anhydrous compounds have been reported in the literature, while at present only two crystal structures of hydrous diphosphates of divalent metals have been reported, although many other are described with no adequate crystallographic information.

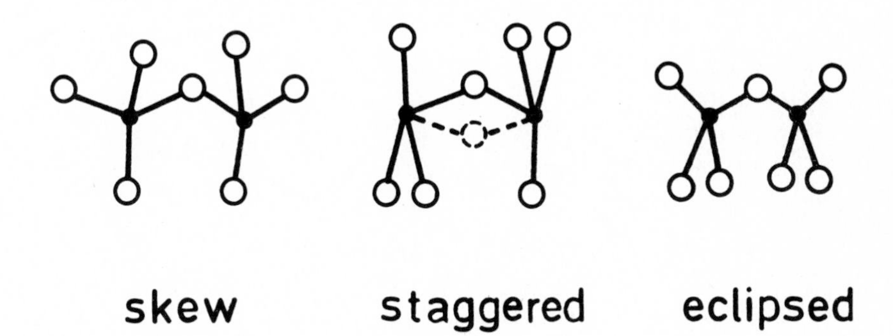

Fig. 3.1.1. Diphosphate ion configurations: skew (left), staggered (middle), and eclipsed form (right). ●= phosphorus, o = oxygen.

3.2 Anhydrous diphosphates

Many of the anhydrous $M_2P_2O_7$ diphosphates belong to one of two groups of isomorphous compounds, namely the α-$Mg_2P_2O_7$ type, or the β-$Mg_2P_2O_7$ type, often called the "thortveitite" type. A large part of this section will describe and discuss these two structure types, while other crystal structures are accorded only briefer descriptions. The diphosphate structures are summarized in Table 3.2.1, which gives some basic crystallographic information for the forms in question.

Of the two groups of isostructural diphosphates just mentioned, the first consists of α-$Mg_2P_2O_7$, α-$Co_2P_2O_7$, and α-$Ni_2P_2O_7$. The structure type was determined in 1967 by Lukaszewicz (ref. 172) and has later been studied by other authors as well (ref. 175 - 177). The crystal structure is preferably described in space group $B2_1/c$ rather than in the conventional setting $P2_1/c$ to facilitate a comparison with the β-$Mg_2P_2O_7$ ("thortveitite-type") structure (cf. Fig. 3.2.1). In the three α-$Mg_2P_2O_7$-type forms mentioned, the metal cations are five- and six-coordinated, and the P-O-P angles of the skew diphosphate anions are around 140° (ref. 175-177).

β-$Mg_2P_2O_7$, $Mn_2P_2O_7$, β-$Ni_2P_2O_7$, β-$Cu_2P_2O_7$, β-$Zn_2P_2O_7$ and probably also β-$Co_2P_2O_7$ (ref. 178-183) all belong to another large structure group isotypic with the mineral thortveitite, $(Sc,Y)_2Si_2O_7$, originally determined by Zachariasen (ref. 173). This structure type is found also in other $M_2X_2O_7$ compounds such as diarsenates and divanadates. The structure is somewhat more densely packed than that of the α-$Mg_2P_2O_7$ type, probably because all cations are six-coordinated and the diphosphate ions are of the staggered form. The space group symmetry is C2/m, indicating that the P-O-P bridge should be linear, which is unusual and has in fact been doubted by some authors, as will be exemplified below.

The structures of α-$Mg_2P_2O_7$ and β-$Mg_2P_2O_7$ are very similar, as shown in Fig. 3.2.1. In β-$Mg_2P_2O_7$ (the "thortveitite" type) there is only one distinct magnesium atom, which is octahedrally coordinated to six oxygens. MgO_6 share edges and form sheets parallel to the ab plane where the cations lie in two-thirds of the octahedral "holes" of rather close-packed oxygens. The P-O-P link is supposed to be linear (however: see below!).

In the α phase the P-O-P angle is bent to 144° (ref. 175). This implies that the cations' coordinations change. There are now two

TABLE 3.2.1

Crystallographic data of divalent-metal diphosphates. Abbreviations: ecl = eclipsed, stg = staggered diphosphate anion (cf. Fig. 3.1.1). The data are partly taken from ref. (17) by Nord and Kierkegaard.

Compound	Temp(^{o}C)	S.G.	V_{ox} (Å^3)	Metal-oxygen data CN: <M-O> /in Å/		Diphosphate anion data form	P-O(bridge) /in Å/ values		P-O-P(^{o})	Ref.
α-$Mg_2P_2O_7$	25	$B2_1/c$	17.14	5: 2.04	6: 2.10	stg	1.57	1.61	144	(175)
α-$Co_2P_2O_7$	25	$B2_1/c$	17.20	5: 2.05	6: 2.12	stg	1.57	1.60	143	(176)
α-$Ni_2P_2O_7$	25	$B2_1/c$	16.78	5: 2.01	6: 2.08	stg	1.60	1.61	137	(177)
β-$Mg_2P_2O_7$	95	C2/m	16.86	6: 2.07		stg	1.56	1.56	180	(178)
β-$Ni_2P_2O_7$	580	C2/m	16.62	6: 2.12		stg	1.53	1.53	180	(180)
β-$Cu_2P_2O_7$	100	C2/m	17.14	6: 2.17		stg	1.54	1.54	180	(181)
β-$Zn_2P_2O_7$	140	C2/m	17.02	6: 2.11		stg	1.57	1.57	180	(182)
$Mn_2P_2O_7$	25	C2/m	18.04	6: 2.20		stg	1.58	1.58	(166)*	(183)
$Fe_2P_2O_7$	25	P1	17.51	6: 2.16	6: 2.21	skew	1.58	1.62	153	(184)
δ-$Ni_2P_2O_7$	(25)**	$P2_1/a$	16.36	6: 2.10		stg	1.56	1.56	180	(185)
α-$Cu_2P_2O_7$	25	C2/c	17.20	6: 2.18		stg	1.58	1.58	157	(186)
α-$Zn_2P_2O_7$	25	I2/c	17.23	5: 2.03	5: 2.04	1 skew	1.60	1.60	139	(187)
				6: 2.09		2 stg	1.57	1.60	148	
$Cd_2P_2O_7$	25	$P\bar{1}$	19.36	6: 2.31	7: 2.46	ecl	1.60	1.64	132	(188)
α-$Ca_2P_2O_7$	25	$P2_1/n$	20.53	8: 2.50	8: 2.54	ecl	1.58	1.62	130	(189)
β-$Ca_2P_2O_7$	(25)**	$P4_1$	19.26	7: 2.43	8: 2.46	1 ecl	1.62	1.64	130	(190)
				7: 2.45	9: 2.57	2 ecl	1.59	1.62	138	
α-$Sr_2P_2O_7$	25	Pnma	22.54	9: 2.68	9: 2.72	ecl	1.59	1.61	131	(191)

*) Compare text; **) Metastable phase.

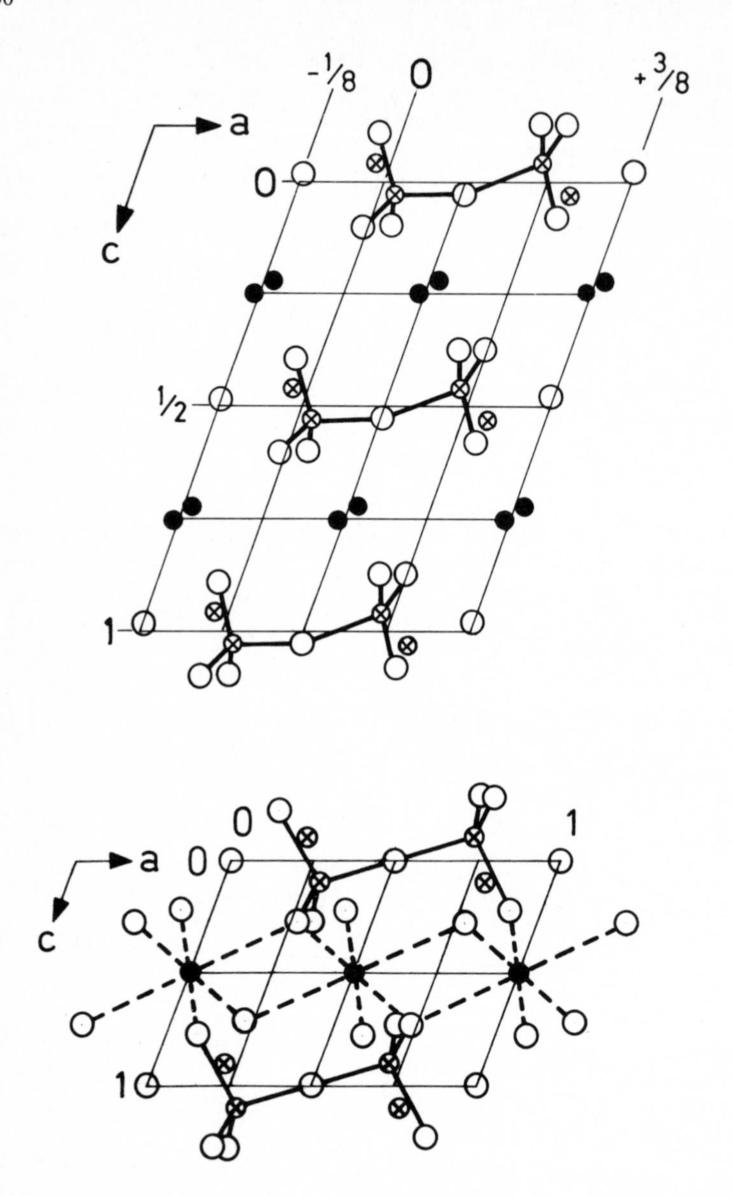

Fig. 3.2.1. Major features of the crystal structures of α-$Mg_2P_2O_7$ ($B2_1/c$; above) and β-$Mg_2P_2O_7$ (C2/m; below). The figures are drawn to the same scale. ● = magnesium, ⊗ = phosphorus, o = oxygen. The figure is reproduced with permission from the Royal Swedish Academy of Science (copyright) from Nord and Kierkegaard: "Crystal chemistry of some anhydrous divalent-metal phosphates" (ref. 17) in Chemica Scripta vol. 15, pp. 27-39 (1980).

distinct cation sites, one five- and the other six-coordinated. The structure still reminds very much of the β phase, but in detail the bonding about the cations is different. In the MgO_6 octahedron the Mg-O distances are in the range 2.059-2.142 Å (ref. 175), which might be compared with 2.02-2.15 Å in β-$Mg_2P_2O_7$ (ref. 178). The MgO_5 polyhedron in α-$Mg_2P_2O_7$ has an irregular shape, with Mg-O distances ranging from 1.985 to 2.120 Å (ref. 175). However, there is a sixth oxygen about 3.35 Å from the magnesium atom, obviously a reminescent of the "MgO_6" octahedron in β-$Mg_2P_2O_7$. The phase transition between the two $Mg_2P_2O_7$ is rather strange and has been studied by many authors, e.g. Crispin Calvo (ref. 175, 178).

The β-$Mg_2P_2O_7$ type structure refinements are as a rule somewhat inaccurate, since the diffraction data had to be recorded at elevated temperatures in all cases but one ($Mn_2P_2O_7$). However, all refinements show large thermal motions for the P-O-P bridge oxygen atoms. The same goes for thortveitite-like structures other than those listed in Table 3.2.1.

This suggests that the bridge oxygens may be statistically disordered in such a way that P-O-P appears to be linear although each individual link is bent. Some available data in Table 3.2.1 give a P-O(bridge) grand mean of 1.60 Å for diphosphates with non-linear P-O-P groups, and a grand mean value of 1.53 Å where P-O-P was reported to be linear (thortveitite structure types). Thus the latter (and shorter) distance may be an artefact of an over-simplified structural model in which P - O' - P appears to be linear (Fig. 3.2.2). If the O' bridge atom is moved from its "linear" position to the position O_{br} in the figure, the P - O - P link will be bent, and the P - O_{br} value will now increase to approximately 1.60 Å. Trigonometrical calculations with the values given in Fig. 3.2.2 show the P -O_{br} -P angle to be around 150°, in good agreement with the corresponding values reported for diphosphates with "non-linear" anions in Table 3.2.1.

A third reason for doubting the linear P-O-P conformation is that these corner-sharing PO_4 tetrahedra are more regular than they should be according to theories of Cruickshank (ref. 174; cf. also section 7). Valuable comments on a similar problem have been given in a paper by Bernal et al. (ref. 183a). For this reason, extensive studies of $Mn_2P_2O_7$, the only diphosphate with the thortveitite structure stable at room temperature, has been undertaken (ref. 183).

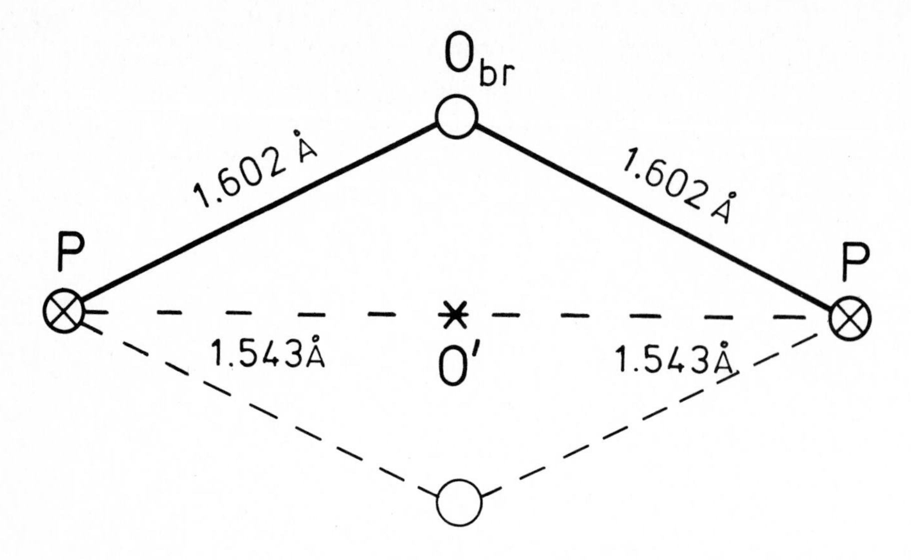

Fig. 3.2.2. Plausible statistical disorder of the $P_2O_7^{4-}$ anion in the β-$Mg_2P_2O_7$ structure type (compare text).

Manganese(II) diphosphate, $Mn_2P_2O_7$, was first structurally investigated by Lukaszewicz and Smajkiewicz (ref. 193), although with poor accuracy (R_F = 0.18). It has recently been reinvestigated on the basis of X-ray and neutron diffraction data (ref. 183), to show that a "non-linear" P-O-P model for the diphosphate anion, still in the centro-symmetrical space group C2/m, gives more realistic temperature factors and interatomic distances. In this model the bridging oxygen is "split" in a statistical way so that 50 per cent is located at (0,y,0) and 50% at (0,-y,0) instead of 100% at (0,0,0) as with the "linear" model. The same model has earlier been applied on the thortveitite-like compounds β-$Cu_2P_2O_7$ (ref. 181) and $Mn_2V_2O_7$ (ref. 194), again indicating that the "split-atom" model favours over the "linear" model, although there are still some structural ambiguities which suggest that even the "split-atom" model is likely to be a slightly oversimplified model, and the question as to the true configuration of the $P_2O_7^{4-}$ anion in the thortveitite-type structure is not yet definitely settled (cf. ref. 183).

The crystal structure of $Fe_2P_2O_7$, determined by direct methods and refined in space group P1 by Stefanidis and Nord (ref. 184),

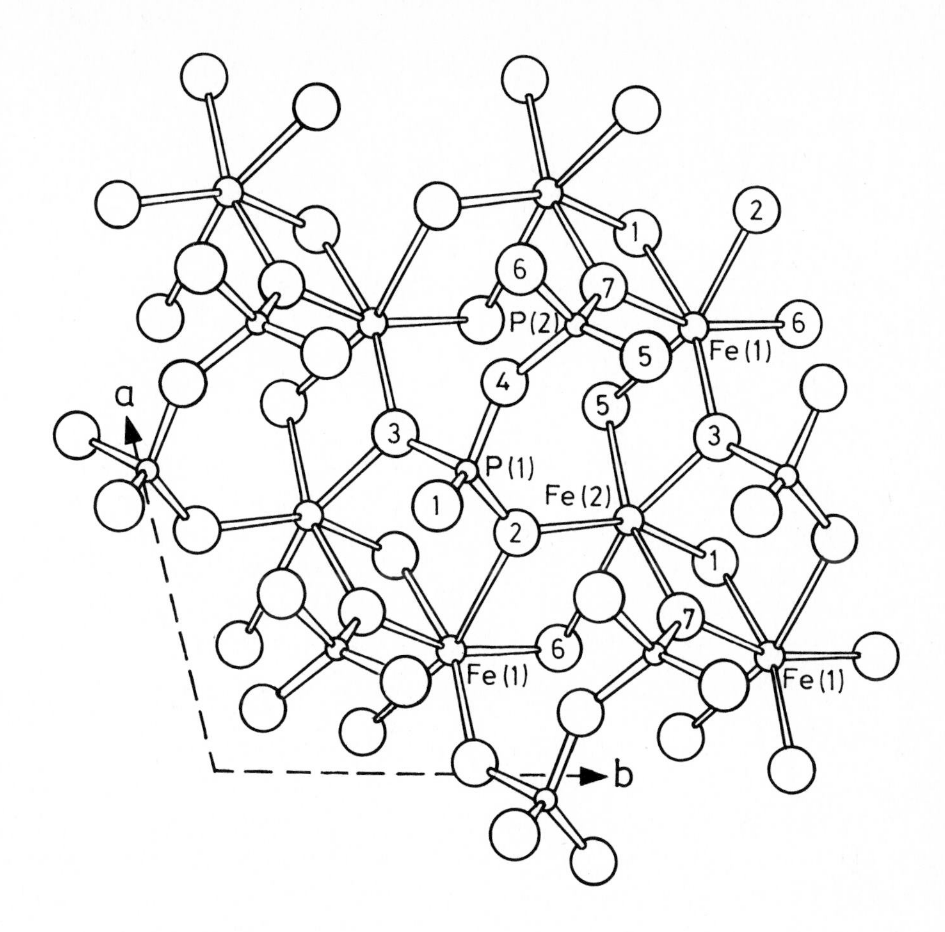

Fig. 3.2.3. An ORTEP plot of the crystal structure of $Fe_2P_2O_7$ (space group P1 assumed). Approximately four formula units are shown. For clarity spheres of the same sizes have been used for each atom type: small circles for phosphorus, medium-sized for iron, and large circles for oxygen atoms. The figure is reproduced with permission from R. Oldenbourg Verlag, München (copyright) from Stefanidis and Nord: "The crystal structure of iron(II) diphosphate, $Fe_2P_2O_7$" (ref. 184) in Zeitschrift für Kristallographie vol. 159, pp. 255-264 (1982).

independently refined in $P\bar{1}$ by Hoggins, Swinnea and Steinfink (ref. 195), is structurally closely related to the thortveitite structure. The structure is built up of distorted edge-sharing FeO_6 octahedra, in connection with corner-sharing $P_2O_7{}^{4-}$ groups. The crystal structure is illustrated in Fig. 3.2.3. The two distinct PO_4 tetrahedra are joined together in pairs through corner-sharing, thus forming diphosphate ions which are of the "skew" type, with a P-O-P angle of 153° (still space group P1 assumed). In the refinements in the centrosymmetrical space group by Hoggins et al. (ref. 195), the thermal motion of the bridging oxygen is indeed very large. It is possible that the true structure of $Fe_2P_2O_7$ is somewhere between these two structural suggestions.

For other investigated $M_2P_2O_7$ diphosphates, the P-O-P angle is in the region 130-157° except for the thortveitite types, where a linear angle (180°) is most often reported (cf. Table 3.2.1).

Some attention will also be given to the stereochemistry of the other anhydrous diphosphate compounds listed in Table 3.2.1. For instance, nickel diphosphate may crystallize in a third polymorph form called δ-$Ni_2P_2O_7$, structurally related to thortveitite and with a reported P-O-P angle of 180° (ref. 185). The low temperature forms α-$Cu_2P_2O_7$ and α-$Zn_2P_2O_7$ (ref. 186, 187) are also related to the thortveitite structure, although with different space group symmetries. The cationic environments are given in Table 3.2.1.

$Cd_2P_2O_7$ is triclinic (ref. 188), with six- and seven-coordinated cations, and shows some similarity to those $M_2X_2O_7$ compounds, where "M" has a fairly large radius and "X" a small one. On the other hand, the structure differs substantially from those found for other divalent-metal diphosphates although it bears some relationship, in terms of packing, to the β-$Ca_2P_2O_7$ structure.

Calcium diphosphate is dimorphous, with "α" as the low- and "β" as the high-temperature modification (ref. 189, 190), both with metal-oxygen polyhedra having higher "M" coordination numbers (see Table 3.2.1). α-$Sr_2P_2O_7$ (ref. 191, 192) bears a slight resemblance to α-$Ca_2P_2O_7$. A high-temperature phase β-$Sr_2P_2O_7$, presumably isotypic with β-$Ca_2P_2O_7$, as well as $Ba_2P_2O_7$ phases, although without any relevant crystallographic data, have also been reported (ref. 196, 197). Crystallographic studies of solid solutions in the pseudo-binary system $Sr_2P_2O_7$ - $Mg_2P_2O_7$ have been carried out by Calvo (ref. 198). $Pb_2P_2O_7$, a triclinic phase, has been investigated by X-ray diffraction, but the crystal structure has not yet been determined (ref. 199). The same is true for $Sn_2P_2O_7$ (ref. 200). No adequate stereochemical information exists for any $Be_2P_2O_7$ phase. Structural relationships among the anhydrous divalent-metal diphosphates are shown in Fig. 3.2.4.

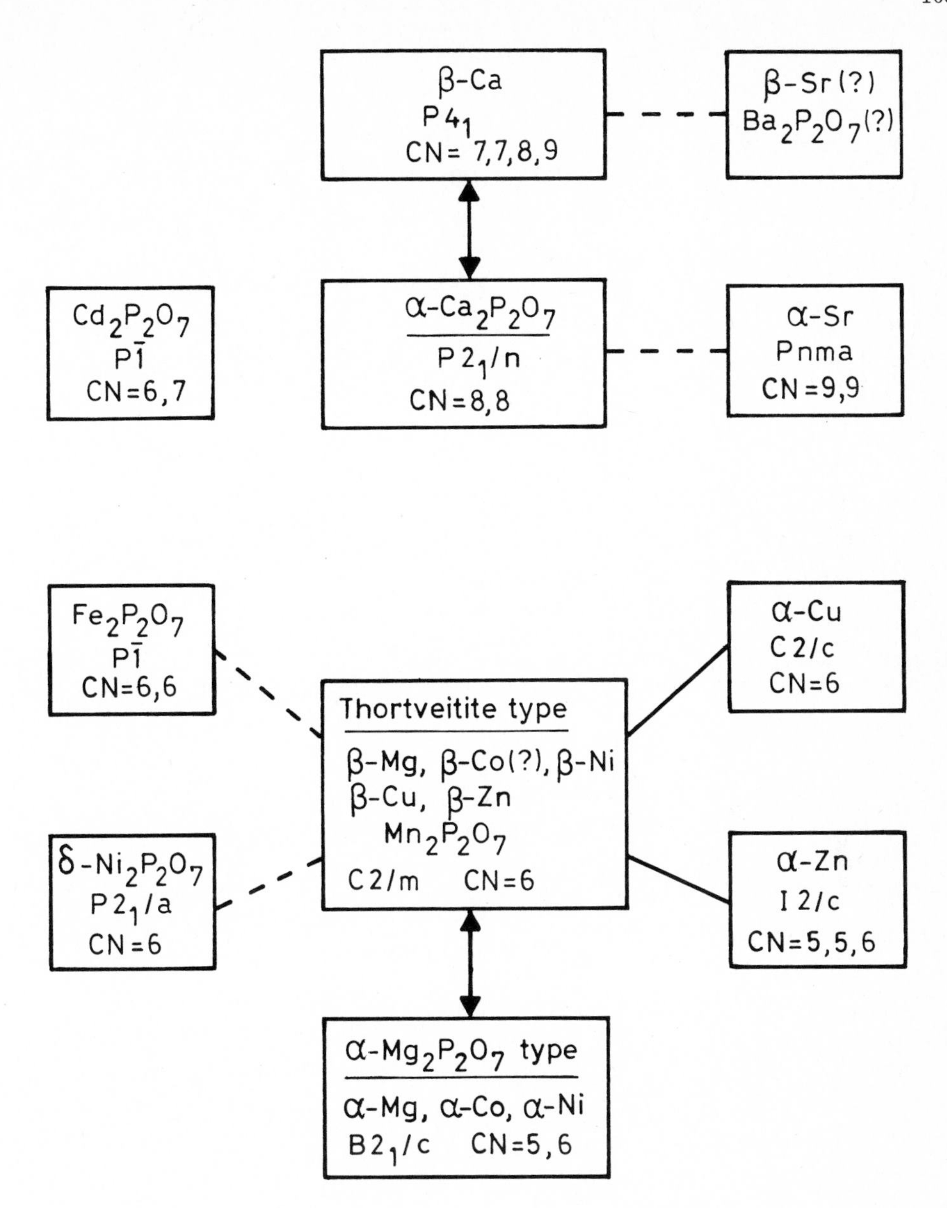

Fig. 3.2.4. Structural relationships among the anhydrous divalent-metal $M_2P_2O_7$ diphosphates. α-Mg stands for α-$Mg_2P_2O_7$, and so forth. Space group symbols and metal coordination numbers (CN) are also given.

3.3 Hydrous diphosphates

Only two crystal structures of hydrous divalent-metal diphosphates have been reported so far, viz. $Mn_2P_2O_7 \cdot 2H_2O$ (ref. 201) and $Ca_2P_2O_7 \cdot 2H_2O$ (ref. 202). The former phase ($P2_1/n$) has two distinct MnO_6 octahedra (mean Mn-O: 2.191 and 2.194 Å) and a P-O-P angle of 128°. $Ca_2P_2O_7 \cdot 2H_2O$, which exists rarely in human tissues and abnormal calcifications, crystallizes in space group $P\bar{1}$, with two CaO_7 groups (mean Ca-O: 2.44 and 2.46 Å, respectively) and a diphosphate anion having a P-O-P angle of 123°.

4 TRIPHOSPHATES AND RELATED CONDENSED PHOSPHATES

In the diphosphates, two PO_4 groups are linked together by corner-sharing to form an $P_2O_7^{4-}$ anion. In a similar way, three PO_4 tetrahedra may be linked in a (short) chain to form a $P_3O_{10}^{5-}$ anion, which is the main structural unit of triphosphates of divalent metals of the form $M_5(P_3O_{10})_2$. In this way, more than three PO_4 groups may be connected in a yet longer chain, without branches, forming anions of the general formula $P_nO_{3n+1}^{-(n+2)}$. An infinitely long un-branched chain ($n = \infty$) gives a metaphosphate anion of the formula PO_3^- (see next section).

Comparatively few stereochemical details of triphosphates or related phosphates of the "P_nO_{3n+1} family" containing divalent-metal M^{2+} cations have been published so far, but a few crystal structures of these will be listed below. The first triphosphate crystal structure to be determined in this field was that of zinc triphosphate heptadecahydrate, $Zn_5(P_3O_{10})_2 \cdot 17H_2O$, which was published in 1975 by Averbuch-Pouchot, Durif and Guitel (ref. 203). The triclinic structure (space group $P\bar{1}$) has zinc ions which are coordinated to four, six and six oxygen atoms. The two independent P-O-P angles are 128 and 131°, respectively.

Two other triphosphates which deserve to be mentioned in this section are $Zn_2HP_3O_{10} \cdot 6H_2O$ (ref. 204) and $Zn_2NaP_3O_{10} \cdot 9H_2O$ (ref. 205). The P-O-P angles in these triphosphates are 129 and 130°, respectively 130 and 129° (ref. 204, 205).

In the pseudo-binary system MgO - P_2O_5 studied by Berak (ref. 206), there is a slight indication of the existence of an intermediate phase $Mg_5(P_3O_{10})_2$ but, so far, this compound has not been prepared. Similar compounds are briefly mentioned in Gmelin's Handbuch der anorganischen Chemie.

In 1978, Smolin et al. (ref. 207) published the crystal structure of $Na_3Mg_2P_5O_{16}$, which contains a new type of phosphate radical $P_5O_{16}^{7-}$ with five inter-connected PO_4 tetrahedra forming a chain. The P - O - P angles in this extremely rare (from a stereochemical point of view) anion are in the range 127-137°, while the O-P-O angles within the individual, rather distorted, PO_4 tetrahedra are in the range 98-116° (ref. 207). The P-O bond distances are in the range 1.48 - 1.66 Å (ref. 207). The space group symmetry is P2/a. The only (non-equivalent) magnesium atom is octahedrally coordinated to six oxygens, while one sodium is six- and the other, rarely enough, five-coordinated. The pentaphosphate group P_5O_{16} has a formal charge of -7. The short chain is bent like a horseshoe and has a crystallographically imposed 2 symmetry (ref. 207).

5 METAPHOSPHATES

5.1 Brief introduction

In metaphosphate structures, the PO_4 tetrahedra are connected in such a way that, with the exception of the end groups in infinitely long un-branched chains, each tetrahedron shares corners with two other PO_4 groups so that two oxygens are bridged and two are unbridged (terminal) atoms. The metaphosphates may contain either cyclic anions, or long polyphosphate chains of the type $(PO_3^-)_n$. The first metaphosphate to be investigated stereochemically was $Al_4(P_4O_{12})_3$ studied by Pauling and Sherman in 1937 (ref. 208). The structure contains cyclic tetrametaphosphate rings of the form $P_4O_{12}^{4-}$.

Many other anionic conformations have now been found in metaphosphates by X-ray diffraction techniques, e.g., trimetaphosphate ions in $Na_3P_3O_9 \cdot 6H_2O$ (ref. 209), pentametaphosphate groups in $Na_4(NH_4)P_5O_{15} \cdot 4H_2O$ (ref. 210), hexametaphosphate anions in $Na_6P_6O_{18} \cdot 6H_2O$ (ref. 211), octametaphosphate cyclic rings in $(NH_4)_2$-$Cu_3P_8O_{24}$ (ref. 212), and even a decametaphosphate structure has been published: $Ba_2Zn_3P_{10}O_{30}$ (ref. 213). Infinitely long polymetaphosphate chains were first found in Kurrol's and Maddrell's salts,

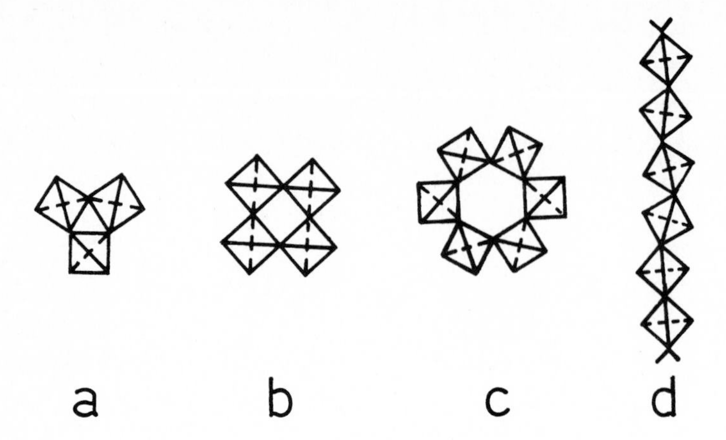

Fig. 5.1.1. Examples of different anionic configurations in metaphosphate structures: (a) trimetaphosphate $P_3O_9^{3-}$, (b) tetrametaphosphate $P_4O_{12}^{4-}$, (c) hexametaphosphate $P_6O_{18}^{6-}$, and (d) polymetaphosphate chain $(PO_3^-)_n$, where "n" is a very large number.

$(NaPO_3)_n$ (ref. 214, 215). Today, six different kinds of anionic polyphosphate chains with regard to stereochemical periodicity and conformation are known (ref. 216, 217). Examples of some anionic metaphosphate configurations are given in Fig. 5.1.1.

Among divalent-metal metaphosphates, the major part of crystal structures have turned out to be tetrametaphosphates, so far; therefore these will be discussed first in section 5.2, with the crystal structures of trimetaphosphates and other polymetaphosphates to follow.

5.2 Tetrametaphosphates

Thilo and Grunze (ref. 218-220) have done some pioneer work in the field of stereochemistry on tetrametaphosphates. They showed, from X-ray powder diffractograms and paper chromatography on phosphate solutions, that the metaphosphates of the divalent ions of magnesium, manganese, iron, cobalt, nickel, copper, zinc and cadmium are isomorphous and probably tetrametaphosphates of the form $M_2P_4O_{12}$ (M = Mg, Mn etc). Spectroscopic measurements (IR) by Steger (ref. 221) gave similar results. Later, relevant unit cell parameters of the monoclinic unit cells were published (in 1968) by

TABLE 5.2.1
Data for $M_2P_4O_{12}$ tetrametaphosphates and hydrous modifications.

Compound	S.G.	V_{ox} ($Å^3$)	Cation coordination: CN: <M-O> /in Å/		Ref.
$Mg_2P_4O_{12}$	C2/c	17.60	6: 2.06	6: 2.09	(224)
$Cu_2P_4O_{12}$	C2/c	17.80	6: 2.09	6: 2.12	(223)
$Co_2P_4O_{12}$	C2/c	17.76	6: 2.11	6: 2.08	(225)
$Ni_2P_4O_{12}$	C2/c	17.30	6: 2.06	6: 2.07	(226)
$NiCoP_4O_{12}$	C2/c	17.50	6: 2.05	6: 2.10	(227)
$NiZnP_4O_{12}$	C2/c	17.48	6: 2.09	6: 2.12	(227)
$NiMgP_4O_{12}$	C2/c	17.38	6, 6		(227)
$Mn_2P_4O_{12}$	C2/c	18.92	6, 6		(222)
$Fe_2P_4O_{12}$	C2/c	18.12	6, 6		(229)
$Zn_2P_4O_{12}$	C2/c	17.72	6, 6		(222)
$Cd_2P_4O_{12}$	C2/c	20.05	6, 6		(222)
$Ca_2P_4O_{12}4H_2O$	$P2_1/n$	21.10	7: 2.40		(230)
$Pb_2P_4O_{12}4H_2O$	$P2_1/n$	21.13	8		(231)

Beucher and Grenier (ref. 222). The complete crystal structure was finally solved independently by means of the compounds $Cu_2P_4O_{12}$ (ref. 223) and $Mg_2P_4O_{12}$ (ref. 224), cf. Table 5.2.1. The space group is C2/c, with two unique six-coordinated cation sites. These MO_6 octahedra are connected by edge-sharing in chains of the form $(M1)O_6$ - $(M2)O_6$ - $(M1)O_6$ - $(M2)O_6$ - running through the structure. The anions are certainly tetrametaphosphate rings, arranged in sheets parallel with $(10\bar{1})$ and with a crystallographically imposed $\bar{1}$ symmetry; see Figs. 5.2.1 and 5.2.2.

The crystal structures of some tetrametaphosphates isomorphous with the above mentioned two compounds have been refined on the basis of neutron powder diffraction data, utilizing the Rietveld (ref. 228) full-profile refinement technique, namely $Co_2P_4O_{12}$, $Ni_2P_4O_{12}$, $NiCoP_4O_{12}$, and $NiZnP_4O_{12}$ (ref. 225-227). The averaged metal-oxygen distances for the six tetrametaphosphates mentioned so far are well correlated with the cation radii as given by Shannon and Prewitt (ref. 6, 7); cf. Table 5.2.1. It is noticeable, though, that the CuO_6 octahedra in $Cu_2P_4O_{12}$ are much more distorted than in the other $M_2P_4O_{12}$ tetrametaphosphates, due to the Jahn-Teller effect. The P - O - P angles in all phases are usually around 130°. The studies of the two solid solutions $Ni(M)P_4O_{12}$,

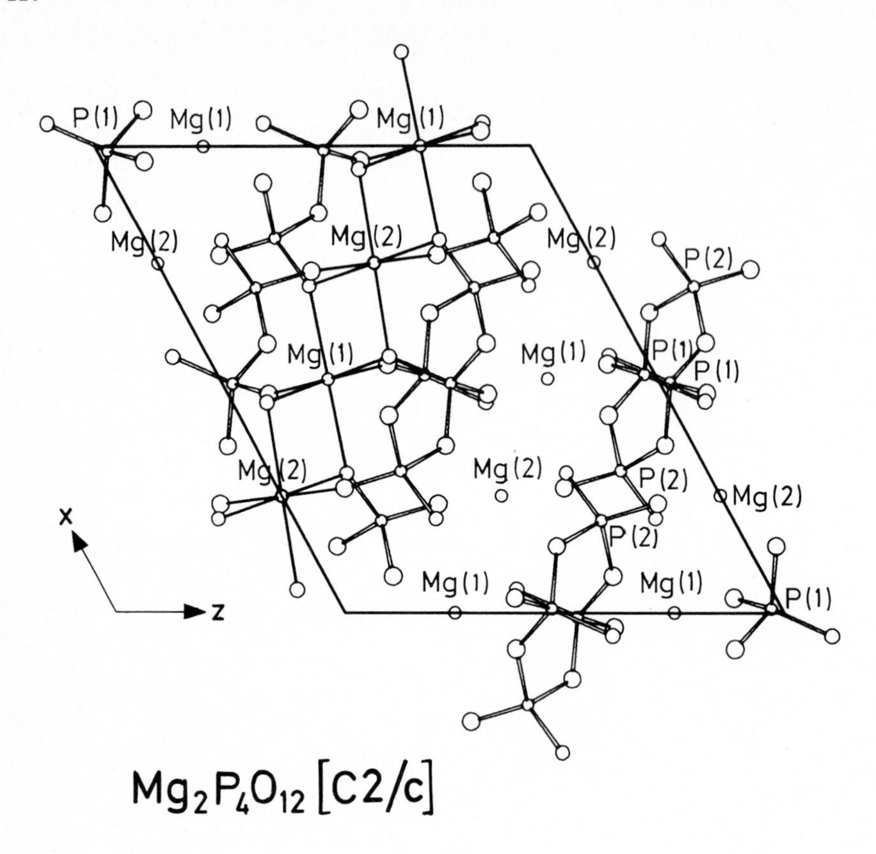

Fig. 5.2.1. The crystal structure of $Mg_2P_4O_{12}$ (S.G. C2/c; Z = 4), showing about one unit cell. The figure is reproduced with permission from the Royal Swedish Academy of Science (copyright) from Nord and Kierkegaard: "Crystal chemistry of some anhydrous divalent-metal phosphates" (ref. 17) in Chemica Scripta, vol. 15, pp. 27-39 (1980).

with M = Co or Zn (ref. 227), were undertaken with the aim to determine the cation distribution among the two octahedrally coordinated metal cation sites.

It is interesting that so many among the divalent-metal metaphosphates have the same crystal structure at ordinary conditions (room temperature, 1 bar pressure). Furthermore, almost all of these isostructural $M_2P_4O_{12}$ tetrametaphosphates may be compressed by about five volume per cent at 1000°C and 80 kbar, thus forming

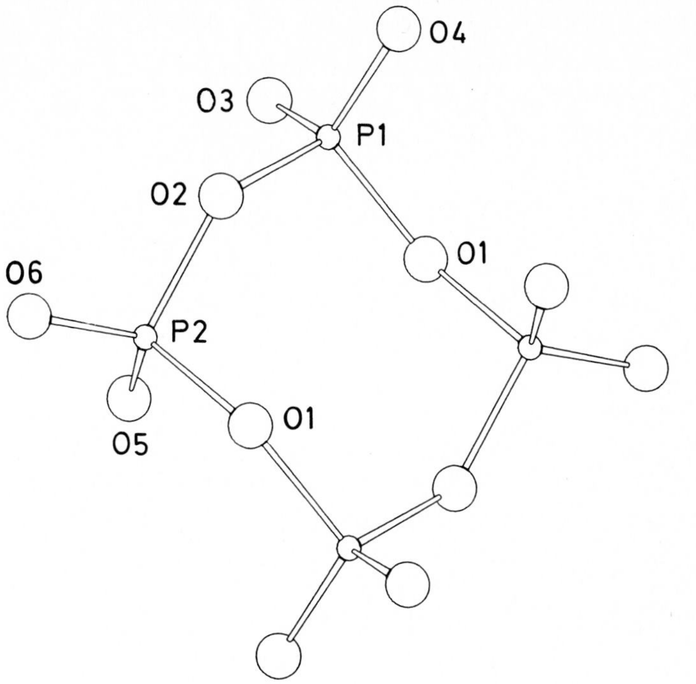

Fig. 5.2.2. The tetrametaphosphate anion in $Mg_2P_4O_{12}$.

another common monoclinic high-pressure phase with an unknown structure but presumably also with tetrametaphosphate anions, and therefore denoted $M_2P_4O_{12}$-II (ref. 229).

Two hydrous tetrametaphosphate structures have also been determined: $Ca_2P_4O_{12} \cdot 4H_2O$ (ref. 230) and $Pb_2P_4O_{12} \cdot 4H_2O$ (ref. 231). The unit cell dimensions of these two compounds are very similar, and they crystallize in the same space group, $P2_1/n$. However, beside this resemblance there are striking differences in the two structures. For instance, Ca^{2+} is seven- and Pb^{2+} is eight-coordinated. Moreover, in the former phase the cation polyhedra are connected to isolated pairs while in the latter they are connected by edge-

TABLE 5.3.1
Structural data for some hydrous divalent-metal trimetaphosphates.

Compound	S.G.	Coordination number (CN) and <M-O> values /in Å/	Ref.
$Cd_3(P_3O_9)_2\cdot 10H_2O$	$P2_1/n$	6: 2.27 6: 2.29	(232)
$Cd_3(P_3O_9)_2\cdot 14H_2O$	$P\bar{3}$	6: 2.29	(233)
$Sr_3(P_3O_9)_2\cdot 7H_2O$	Pnma	7: 2.62 8: 2.63 9: 2.66	(234)
$Ba_3(P_3O_9)_2\cdot 6H_2O$	$P\bar{1}$	9: 2.85 9: 2.91 9: 2.87	(235)
$Ba_2Zn(P_3O_9)_2\cdot 10H_2O$	C2/c	Ba (9): 2.85 Zn (6): 2.06	(236)
$Pb_3(P_3O_9)_2\cdot 3H_2O$	$P4_12_12$	6: 2.59 7: 2.64	(237)

sharing, forming infinite chains (ref. 230, 231).

5.3 Trimetaphosphates

Crystal structures of some trimetaphosphates, all hydrous, have been determined (cf. Table 5.3.1). The P - O - P angles within the ring-formed $P_3O_9^{3-}$ ions are usually around 130°, as in the tetrametaphosphates. The structure of $Sr_3(P_3O_9)_2\cdot 7H_2O$ contains three distinct metal atoms with the coordination numbers 7, 8, and 9.

5.4 Some other metaphosphate structure types

The tetrametaphosphate modification $Zn_2P_4O_{12}$ transforms to another phase, a polymetaphosphate $Zn(PO_3)_2$, at 745°C. There are three modifications of cadmium metaphosphate: the tetrametaphosphate $Cd_2P_4O_{12}$ and two forms of $Cd(PO_3)_2$. Their relationships are (ref. 239, 240):

$$Cd_2P_4O_{12} \xrightarrow{300°C} 2\ \alpha\text{-}Cd(PO_3)_2 \underset{}{\overset{835°C}{\rightleftarrows}} 2\ \beta\text{-}Cd(PO_3)_2$$

Four different calcium metaphosphate phases have been reported, but the crystallographic knowledge of these is so far rather meagre. The above mentioned $M(PO_3)_2$ forms of zinc and cadmium metaphosphate contain chain-formed PO_3^- aions (cf. Table 5.4.1). The crystal structure of the zinc polymetaphosphate $Zn(PO_3)_2$ has been published by Averbuch-Pouchot, Durif and Bagieu-Beucher (ref. 238). The space group symmetry is C2/c. The zinc atoms are six-coordi-

TABLE 5.4.1

Data for polymetaphosphates containing chain-formed PO_3^- anions.

Compound	S.G.	V_{ox} (Å^3)	Coordination numbers (CN) and <M-O> values /in Å/	Ref.
$Zn(PO_3)_2$	C2/c	16.97	6: 2.09	(238)
α-$Cd(PO_3)_2$	Pbca	19.30	6: 2.32	(239)
β-$Cd(PO_3)_2$	$P2_12_12_1$	19.48	6: 2.31	(240)
$Ba(PO_3)_2$	$P2_12_12_1$	21.11	10:2.93	(241)
$Pb(PO_3)_2$	$P2_1/c$	20.85	7: 2.58 8: 2.64	(242)
$BaCd(PO_3)_4$	$P2_1/n$	20.65	Cd (6): 2.29 Ba (9): 2.83	(243)
$Be(PO_3)_2$-II	$P2_1/n$	17.93	4: 1.59	(244)
$Be(PO_3)_2$-III	$C222_1$	18.19	4: 1.59	(244a)

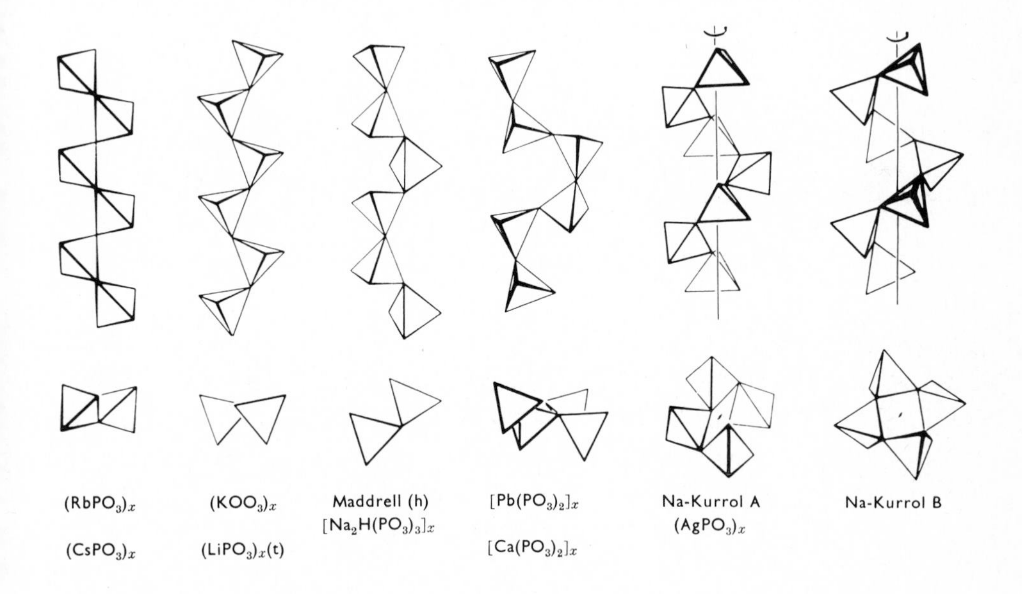

Fig. 5.4.1. Stereochemical drawings of the six hitherto known chain-formed polymetaphosphate anions. The figure is reproduced with permission from The International Union of Crystallography (copyright) from Jost: "Die Struktur des Bleipolyphosphats und allgemeiner Überblick über Polyphosphat-strukturen" (ref. 242) in Acta Crystallographica, vol. 17, pp. 1539-1544 (1964).

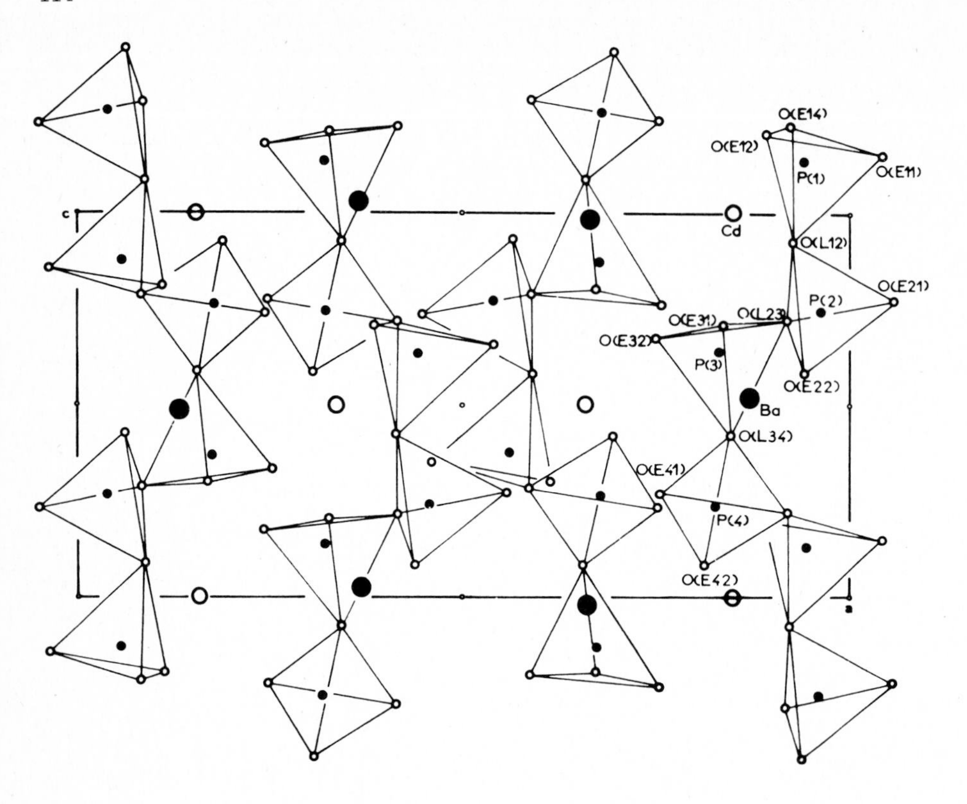

Fig. 5.4.2. The crystal structure of $BaCd(PO_3)_4$. The figure is reproduced with permission from The International Union of Crystallography (copyright) from Averbuch-Pouchot, Durif and Guitel: "Structure cristalline du polyphosphate de cadmium-baryum, CdBa-$(PO_3)_4$" in Acta Crystallographica, vol. B31, pp. 2453-2456 (1975) (ref. 243).

nated with Zn-O distances in the range 2.00-2.22 Å (ref. 238). These somewhat distorted ZnO_6 octahedra are connected by edge-sharing in chains running in the c axis direction, also parallel to the $(PO_3^-)_n$ chains. The PO_4 tetrahedra are somewhat distorted, with P-O distances in the range 1.48 - 1.60 Å and O-P-O angles from 102.5 to 118.8°; the only non-equivalent P-O-P angle is 135.5° (ref. 238).

A good summary of the stereochemistry of PO_3^- polymetaphosphate anionic chains has been given by Jost in his paper on the crystal structure of $Pb(PO_3)_2$ (ref. 242), cf. Fig. 5.4.1. The crystal structures of one of the polymetaphosphates listed in Table 5.4.1, namely that of $BaCd(PO_3)_4$ (ref. 243), is shown in Fig. 5.4.2. The

P - O - P angles of the polymetaphosphates of divalent metals vary more than was the case for the ring-shaped metaphosphate anions; they are in the range 128-154°.

Beryllium metaphosphate was described in 1957 by Thilo and Grunze (ref. 218), who concluded that their modification was probably a polymetaphosphate. To date three different polymorphs named $Be(PO_3)_2$-I, -II and -III have been reported in the literature. For the phase $Be(PO_3)_2$-I only X-ray powder diffraction data have been given, cf. ref. (243a). This modification is sometimes called α-$Be(PO_3)_2$. The crystal structure of $Be(PO_3)_2$-II has been determined by Averbuch-Pouchot et al. (ref. 244); the space group symmetry is $P2_1/n$. The structure of $Be(PO_3)_2$-III, at 119°C, has been studied by Schultz and Liebau (ref. 244a). The structure is related to that of silica K (keatite); the space group symmetry is $C222_1$, although pseudo-tetragonal. As in $Be(PO_3)_2$-II, the beryllium atoms are coordinated to four oxygens forming BeO_4 tetrahedra. The Be-O distances are in the range 1.55-1.66 Å (mean: 1.59 Å). There are three distinct PO_4 tetrahedra, connected by corner-sharing in $(PO_3^-)_n$ chains running through the structure. The P-O-P angles are 137.5 and 139.9^{o}, respectively. The polyphosphate chains are very strongly folded, having eight phosphate tetrahedra in the identity period, instead of four as in $Be(PO_3)_2$-II (ref. 244, 244a). The $Be(PO_3)_2$-III structure may thus be regarded as a three-dimensional framework of corner-shared BeO_4 and PO_4 tetrahedra. Yet another polymorph of beryllium polymetaphosphate, having $P2_1$ symmetry, has been reported (cf. ref. 244a, 244b).

Finally, a structure containing a very unusual phosphate anion will be discussed: the structure of $Ba_2Zn_3P_{10}O_{30}$ ($P2_1/n$), which has a ring-shaped decametaphosphate anion. It was published in 1982 by Bagieu-Beucher et al. (ref. 213). The P-O-P angles in this compound are in the range 123-144°. The coordination numbers of zinc are 4, 6, and 6, while barium is coordinated to nine oxygens. The BaO_9 polyhedra are connected to each other to form large channels with the ZnO_6 octahedra, in which are located the $P_{10}O_{30}$ rings and the ZnO_4 tetrahedra (ref. 213). Although this compound contains the only 10-tetrahedron ring anion synthesized and identified so far, it is worth mentioning that 12-tetrahedron ring anions have been found in silicates and germanates (ref. 245, 246). Finally, pseudo-binary and pseudo-ternary metaphosphate phase diagrams have been published, such as the system $Zn(PO_3)_2$ - $Cd(PO_3)_2$ - $Mg(PO_3)_2$ by Brown and Hummel (ref. 247).

6 ULTRAPHOSPHATES

Ultraphosphates include all condensed phosphates where at least some of the phosphate tetrahedra are linked by corner-sharing to three other PO_4 groups. (In metaphosphates, each phosphate tetrahedron is linked to two other PO_4 groups). The negative charge per phosphorus atom in the ultraphosphate anions is thus around zero, so that cationic metal contents are also low in this type of compounds. Although comparatively few ultraphosphate structures have been determined so far, almost endless series of structures are theoretically possible from a geometrical-mathematical point of view. The terminology is quite straight-forward: all ultraphosphates may be expressed as $[P_nO_{3n+p}]^{(n+2p)-}$ with $p < 0$ ($p = 0$ gives metaphosphates: $P_nO_{3n}^{n-}$). Furthermore, $p = -1$ gives "mono-ultraphosphates", $p = -2$ gives "di-ultraphosphates", and so forth. Ultraphosphates of the lantanide series have been recognized as good laser crystals, and without doubt many other important technical applications will soon be found for this interesting group of phosphate compounds.

Only a few divalent-metal ultraphosphate structures have been determined till date. The first was CaP_4O_{11}, a mono-ultraphosphate, published in 1974 by Tordjman, Bagieu-Beucher and Zilber (ref. 248). Soon afterwards, MnP_4O_{11} was structurally investigated and shown to be isostructural with CaP_4O_{11} (ref. 249). The space group symmetry for these two compounds is $P2_1/n$ ($Z = 4$). Some crystallographic data are given below; the figures quoted first refer to MnP_4O_{11} (ref. 249), while those given within parentheses refer to CaP_4O_{11} (ref. 248). In this structure type, the anions form large sheets (see Fig. 6.1). The anionic pattern may be described as consisting of P_8O_{24} rings, each one connected to four other rings. It may also be regarded as built up of infinite $(PO_3)_n$ chains, interconnected here and there by additional phosphate tetrahedra to form sheets throughout the structure. The unit cell volume is 915 (980) Å^3, with $Z = 4$, to give V_{ox} values of 20.80 (22.27) Å^3.

All cations are coordinated to six oxygens with an average metal-oxygen distance of 2.20 (2.33) Å; the metal-oxygen bond length ranges are 2.14 - 2.27 (2.30 - 2.36) Å. The crystallographically non-equivalent MO_6 octahedra connect adjacent anionic sheets. The phosphate tetrahedra building up the ultraphosphate anions are distorted, and the reported P-O distances spread in the

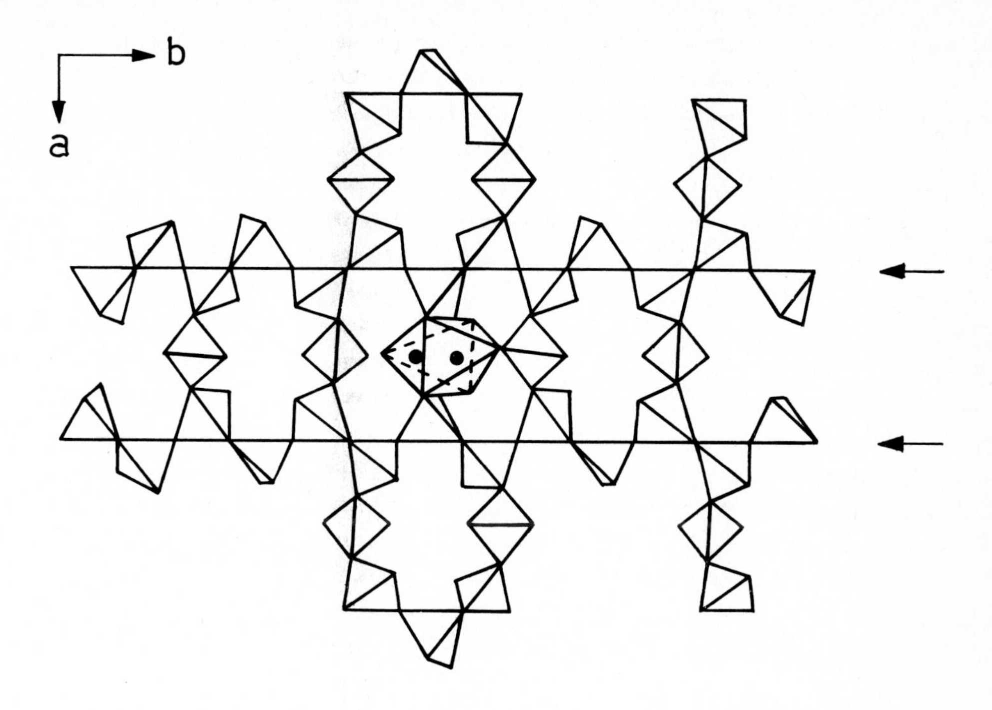

Fig. 6.1. The structural network of MP_4O_{11} (M = Mn or Ca). ● = metal cation "M". For clarity only one MO_6 octahedron is shown. The two arrows indicate the $(PO_3)_n$ chains mentioned in the text.

region 1.42 - 1.68 (1.43 - 1.65) Å, while the P-O-P angles are in the range 124 - 144° (126 - 142°) and the O-P-O angles 98 - 125° (99 - 124°) (ref. 249, 248).

Unit cell dimensions and X-ray powder diffraction data of the ultraphosphate MgP_4O_{11} (orthorhombic) have also been published (ref. 249, 250). The crystal structure of $Cd_2P_6O_{17}$, also a mono-ultraphosphate, has been determined by Antsischkina, Porai-Koschits, Minatjeva and Ivanova (ref. 251), displaying a large network of corner-sharing distorted PO_4 tetrahedra with O-P-O angles in the range 99-122° and P-O distances from 1.44 to 1.61 Å; the P-O-P angles are 124.5, 132.1 or 141.2°. The space group symmetry is $P2_1/n$ (Z = 2; V = 699.5 Å^3). The cadmium atoms are coordinated to six oxygen atoms, with a mean Cd-O distance of 2.28 Å (ref. 251). In addition, the structure of $Sr_2P_6O_{17}$ has also been published by Antsichkina et al. (ref. 252). This compound crystallizes in the monoclinic space group $P2_1$ (Z = 2; V = 644.5 Å^3) and contains two distinct strontium atoms with coordination numbers 7 and 8, respectively. The average Sr-O distances are 2.57 and 2.62 Å. The phosphate tetrahedra are strongly distorted, with P-O distances in the range 1.40 - 1.63 Å and O-P-O angles from 97 to 124°. It is clear that the packing of atoms is much more effective in $Sr_2P_6O_{17}$ than in $Cd_2P_6O_{17}$, since the unit cell volume is smaller (644.5 Å^3) for the strontium compound, although strontium as an ion (Sr^{2+}) is significantly larger than Cd^{2+} (V = 699.5 Å^3 in $Cd_2P_6O_{17}$).

7 DIMENSIONS OF THE PHOSPHATE TETRAHEDRA

Prior to the advent of X-ray crystallography, the phosphate ion was considered to constitute a regular configuration with the phosphorus atom in its center. This was also postulated for other XO_4 anions like the sulphate and orthosilicate groups. Some authors proposed a square-planar configuration, others a tetrahedral. Structural studies in the nineteen twenties of KH_2PO_4 (ref. 1, 2), YPO_4 (ref. 3), and Ag_3PO_4 (ref. 4) established the tetrahedral form for the phosphate ion.

Many hundred inorganic phosphate compounds have now been structurally analysed by means of X-ray or neutron single-crystal or powder diffraction techniques. These studies show that "free" (unlinked) phosphate tetrahedra in orthophosphates are almost regular, even if perfectly symmetric tetrahedra are indeed rare. An ideally tetrahedral PO_4^{3-} ion with point group symmetry Td has many symmetry elements: three two-fold axes, four three-fold axes, three four-fold inversion axes, and six planes of symmetry. Ions with ideal Td symmetry are only to be expected in dilute solutions or in highly symmetrical lattice environments. IR and Raman spectra have actually provided evidence for full Td symmetry in aqueous solutions, while phosphate ions in the crystalline state usually show slight deviations from this ideal configuration (ref. 253). More pronounced distortions appear when two or more phosphate tetrahedra are connected to form larger anions like diphosphates, triphosphates, metaphosphates, ultraphosphates, etc. Distortions also occur in hydrogen phosphates. This will be exemplified for the divalent-metal phosphates, but only briefly, since review articles on PO_4 dimensions have been published earlier (cf. ref. 17, 253, 254, 255).

TABLE 7.1

Dimensions of phosphate tetrahedra in anhydrous orthophosphates. All distances are given in Ångström units (Å) and all angles in degrees (°). The abbreviations used here and in Tables 7.2 - 7.6 are described in the text.

Compound/ mineral	NT	P-O total range (Å)	Grand mean P-O (and σ)	Δ(PO)	O-P-O range	Δ(OPO)	Ref.
α-$Zn_3(PO_4)_2$	1	1.50-1.58	1.53(1)	0.08	105-117	12	(15)
β-$Zn_3(PO_4)_2$	2	1.50-1.56	1.53(1)	0.04	105-115	6	(16)
β'-$Mn_3(PO_4)_2$	6	1.49-1.58	1.537(5)	0.05	104-115	8	(19)
β'-$Cd_3(PO_4)_2$	6	1.44-1.64	1.54(6)	0.15	100-117	12	(20)
"γ-$Zn_3(PO_4)_2$"	1	1.52-1.56	1.54(1)	0.04	109-112	3	(29)
$Zn_2Mg(PO_4)_2$	1	1.50-1.58	1.54(2)	0.08	106-113	7	(38)
$Mg_3(PO_4)_2$	1	1.51-1.54	1.527(8)	0.03	104-112	8	(31)
$Co_3(PO_4)_2$	1	1.52-1.55	1.538(3)	0.03	105-113	8	(33)
sarcopside *	1	1.51-1.59	1.54(1)	0.08	102-114	8	(8)
$Ni_3(PO_4)_2$	1	1.52-1.60	1.548(2)	0.08	102-112	10	(55)
graftonite **	2	1.52-1.58	1.55(2)	0.05	103-115	9	(51)
$Fe_3(PO_4)_2$	2	1.52-1.54	1.534(4)	0.02	102-115	10	(49)
$CdZn_2(PO_4)_2$	2	1.52-1.57	1.54(1)	0.04	104-112	6	(50)
$Cd_2Zn(PO_4)_2$	2	1.50-1.60	1.56(2)	0.09	102-113	9	(50)
$Cu_3(PO_4)_2$	1	1.51-1.57	1.546(3)	0.06	105-112	7	(93)
α-$Ca_3(PO_4)_2$	12	1.49-1.57	1.538(4)	0.03	104-115	6	(9)
β-$Ca_3(PO_4)_2$	3	1.50-1.55	1.532(4)	0.02	106-116	6	(66)
$Ca_3(PO_4)_2$-II	1	1.51-1.57	1.54(1)	0.06	107-111	4	(65)
$Ca_8Mg_{10}(PO_4)_{12}$	6	1.52-1.57	1.538(3)	0.03	103-114	8	(71)
$Ca_4(PO_4)_2O$	4	1.51-1.55	1.536(4)	0.03	105-114	6	(85)
$Cu_5(PO_4)_2O_2$	1	1.50-1.58	1.538(4)	0.08	105-111	6	(95)
$Sn_3(PO_4)_2$	2	1.51-1.55	1.535(7)	0.03	104-114	8	(98)
$Pb_3(PO_4)_2$ $P6_3/m$	1	1.55-1.56	1.55(2)	0.01	106-111	5	(102)
$Pb_3(PO_4)_2$ C2/c	1	1.48-1.56	1.53(1)	0.08	105-113	8	(99)
$Pb_3(PO_4)_2$ $R\bar{3}m$	1	1.49-1.57	1.53(3)	0.08	108-111	3	(101)
$Hg_3(PO_4)_2$	2	1.53-1.56	1.54(1)	0.02	104-114	10	(105)
$CaBe_2(PO_4)_2$	2	1.51-1.55	1.530(5)	0.03	104-114	8	(91)
MEAN:			1.537	0.05		7	

*$(Fe_{0.78}Mn_{0.21}Mg_{0.01})_3(PO_4)_2$; **$(Fe_{1.8}Mn_{0.8}Ca_{0.4})(PO_4)_2$

TABLE 7.2

Phosphate tetrahedra in hydrous orthophosphates

Compound/ mineral	NT	P-O total range (Å)	Grand mean P-O (and σ)	Δ(PO)	O-P-O range	Δ(OPO)	Ref.
$Mg_3(PO_4)_2 \cdot 22H_2O$	1	1.54-1.55	1.545(5)	0.01	109-111	2	(110)
$Fe_3(PO_4)_2 \cdot H_2O$	2	1.54-1.56	1.545(4)	0.03	102-113	8	(111)
$Fe_3(PO_4)_2 \cdot 3H_2O$	1	1.54-1.55	1.544(2)	0.01	107-111	4	(112)
$Fe_3(PO_4)_2 4H_2O$ 298 K	1	1.54-1.55	1.542(2)	0.01	108-110	2	(113)
$Fe_3(PO_4)_2 4H_2O$ 4.2 K	1	1.52-1.55	1.527(7)	0.03	108-112	4	(114)
$Fe_3(PO_4)_2 \cdot 8H_2O$	1	1.53-1.56	1.542(2)	0.03	107-112	5	(116)
switzerite	4	1.52-1.57	1.54(1)	0.04	107-111	3	(119)
$Co_3(PO_4)_2 \cdot H_2O$	2	1.52-1.54	1.532(4)	0.02	103-113	8	(120)
hopeite	1	1.51-1.56	1.537(5)	0.05	105-112	7	(121)
parahopeite	1	1.53-1.57	1.545(6)	0.04	100-113	13	(122)
phosphophyllite	1	1.51-1.57	1.534(2)	0.06	105-111	6	(124)
anapaite	1	1.52-1.56	1.542(2)	0.04	105-113	8	(126)
collinsite	1	1.53-1.56	1.543(5)	0.03	105-113	8	(128)
fairfieldite	1	1.51-1.56	1.546(7)	0.05	104-112	8	(129)
MEAN:			1.540	0.03		6	

Geometrical data for the PO_4 tetrahedra of most phosphates discussed in this summary are given in the Tables 7.1 - 7.6. Although many of the phosphates contain more than one crystallographically unique PO_4 tetrahedron, all this information is not given to save space, but concentrated in the following way (cf. ref. 17): "NT" indicates the number of unique PO_4 tetrahedra in the compound in question. The "P - O total range" and "Grand mean P - O" are based on all various phosphate tetrahedra in that compound. The concept Δ(PO), though, reflects the (average) distortion for individual PO_4 tetrahedra and has been obtained in the following way: the differences $[(P - O)_{max} - (P - O)_{min}]$ is calculated for each phosphate ion, and these differences are then averaged over all tetrahedra to give Δ(PO) for that compound.

Rather than citing standard deviations for the grand mean P-O values, only the average sigma value, "σ", for individual P - O bond distances is given for the respective compound. This may seem unorthodox, but since most authors have their own ways of calculating standard deviations of average values, the present approach

TABLE 7.3

Phosphate dimensions of $M_2(PO_4)Y$ compounds.

Compound	NT	P-O total range (Å)	Grand mean P-O (and σ)	Δ(PO)	O-P-O range	Δ(OPO)	Ref.
$Mg_2(PO_4)(OH,O,F)$	2	1.52-1.55	1.542(1)	0.02	107-110	2	(147)
$Mg_2(PO_4)F$	4	1.52-1.55	1.539(7)	0.02	105-112	5	(148)
$(Mn,Fe)_2(PO_4)OH$	4	1.50-1.56	1.53(1)	0.04	106-112	4	(152)
$(Mn,Fe)_2(PO_4)F$	1	1.53-1.54	1.536(6)	0.01	107-111	4	(154)
$Mn_2(PO_4)F$	1	1.53-1.54	1.537(2)	0.01	108-111	3	(155)
$Cu_2(PO_4)F$	1	1.52-1.55	1.537(3)	0.03	106-112	6	(156)
$Cd_2(PO_4)F$	1	1.53-1.55	1.537(4)	0.02	107-111	4	(157)
$Cu_2(PO_4)OH$	1	1.53-1.57	1.545(3)	0.04	106-111	5	(158)
$Zn_2(PO_4)OH$	1	1.52-1.55	1.53(2)	0.03	106-112	6	(159)
$Mg_2(PO_4)Cl$	1	1.52-1.56	1.540(3)	0.04	104-114	10	(160)
$Mn_2(PO_4)Cl$	1	1.53-1.56	1.542(4)	0.03	105-113	8	(161)
$Fe_2(PO_4)Cl$	1	1.49-1.57	1.536(3)	0.08	106-112	6	(162)
$Co_2(PO_4)Cl$	1	1.48-1.57	1.536(8)	0.09	106-113	7	(163)
$Ca_2(PO_4)Cl$	1	1.53-1.55	1.541(2)	0.02	107-114	7	(165)
$CaBe(PO_4)(OH,F)$	1	1.51-1.55	1.541(3)	0.04	107-113	6	(168)
$Sn_2(PO_4)OH$	1	1.52-1.54	1.534(6)	0.02	108-110	2	(169)
MEAN:			1.538	0.04		5	

is judged to be more straightforward.

As regards O - P - O angles, these are given as an overall range and Δ(OPO) values, the latter defined analoguously to Δ(PO). The <O-P-O> grand mean values are omitted since even in distorted tetrahedra these are very close to the "ideal" value of 109.47°. For condensed phosphates, with corner-sharing PO_4 tetrahedra, the P-O-P angles are also given.

From the Δ(PO) and Δ(OPO) values of Tables 7.1 - 7.3 it is clear that all well-defined orthophosphate structures display fairly regular PO_4 tetrahedra, with a grand mean P - O value of 1.538 Å. Corresponding grand mean values reported earlier are 1.536 Å (ref. 253), 1.537 Å (ref. 254), and 1.537 Å (ref. 255). The latter values are valid for orthophosphates containing cations of all kinds and with a variety of positive charges. For the present divalent-metal orthophosphates, Δ(PO) is usually around 0.05 Å, and Δ(OPO) around 7° (mean values).

TABLE 7.4

Phosphate ion dimensions in various diphosphates.

Compound	NT	P-O total range (Å)	Grand mean P-O (and σ)	Δ(PO)	O-P-O range	Δ(OPO)	P-O-P angle	Ref.
α-$Mg_2P_2O_7$	2	1.47-1.61	1.535(10)	0.09	102-115	11	144	(175)
α-$Co_2P_2O_7$	2	1.50-1.60	1.534(10)	0.08	104-114	10	143	(176)
α-$Ni_2P_2O_7$	2	1.52-1.61	1.55(3)	0.08	103-114	11	137	(177)
α-$Cu_2P_2O_7$	1	1.51-1.58	1.538(7)	0.08	104-112	8	157	(186)
α-$Zn_2P_2O_7$	3	1.49-1.60	1.539(7)	0.08	103-114	9	139,148	(187)
$Cd_2P_2O_7$	2	1.45-1.65	1.54(2)	0.15	103-117	12	132	(188)
α-$Ca_2P_2O_7$	2	1.49-1.62	1.534(8)	0.10	105-115	9	130	(189)
β-$Ca_2P_2O_7$	4	1.48-1.64	1.543(9)	0.11	102-117	8	130,138	(190)
α-$Sr_2P_2O_7$	2	1.48-1.61	1.52(1)	0.11	105-112	6	131	(191)
$Mn_2P_2O_7 \cdot 2H_2O$	2	1.50-1.62	1.543(4)	0.10	104-115	11	128	(201)
$Ca_2P_2O_7 \cdot 2H_2O$	2	1.50-1.62	1.544	0.10	105-115	9	123	(202)
MEAN:			1.537	0.09		10		
β-$Mg_2P_2O_7$	1	1.53-1.56	1.54(1)	0.03	103-113	10	180	(178)
β-$Ni_2P_2O_7$	1	1.51-1.53	1.51(1)	0.02	102-113	11	180	(180)
δ-$Ni_2P_2O_7$	1	1.52-1.56	1.53(1)	0.04	105-114	9	180	(185)
β-$Cu_2P_2O_7$	1	1.50-1.54	1.52(1)	0.04	105-112	7	180	(181)
β-$Zn_2P_2O_7$	1	1.55-1.57	1.56(1)	0.02	102-113	11	180	(182)
$Mn_2P_2O_7$*	1	1.52-1.57	1.531(3)	0.05	104-112	8	180	(183)
$Mn_2P_2O_7$**	1	1.52-1.58	1.536(3)	0.06	101-113	12	166	(183)
MEAN:			1.53	0.03		9		

TABLE 7.5

Phosphate dimensions in triphosphates and in the pentaphosphate $Na_3Mg_2P_5O_{16}$.

Compound	NT	P-O range	Grand mean	Δ(PO)	O-P-O range	Δ(OPO)	P-O-P angles	Ref.
$Zn_5(P_3O_{10})_2 17H_2O$	3	1.49-1.63	1.543(4)	0.11	103-117	11	128,131	(203)
$Zn_2HP_3O_{10} \cdot 6H_2O$	3	1.48-1.62	1.536(5)	0.12	103-118	11	129,130	(204)
$Zn_2NaP_3O_{10} \cdot 9H_2O$	3	1.48-1.63	1.544(4)	0.12	103-118	11	129,130	(205)
$Na_3Mg_2P_5O_{16}$	3	1.48-1.66	1.538(2)	0.12	98-118	13	127,137	(207)
MEAN:			1.540	0.12		12		

*Linear P-O-P model; **Split-atom model

TABLE 7.6

Phosphate ion dimensions in metaphosphates and ultraphosphates.

Compound	NT	P-O total range (Å)	Grand mean P-O (and σ)	Δ(PO)	O-P-O range	Δ(OPO)	P-O-P angles	Ref.
TRIMETAPHOSPHATES								
$Cd_3(P_3O_9)_2 \cdot 10H_2O$	3	1.48-1.60	1.543(4)	0.12	102-121	17	127,131, 133	(232)
$Cd_3(P_3O_9)_2 \cdot 14H_2O$	1	1.46-1.59	1.540(3)	0.13	103-118	15	132	(233)
$Sr_3(P_3O_9)_2 \cdot 7H_2O$	3	1.45-1.64	1.54(1)	0.16	99-120	21	125,127, 130	(234)
$Ba_2Zn(P_3O_9)_2 10H_2O$	3	1.44-1.65	1.55(1)	0.17	102-121	16	123,127, 130	(236)
$Ba_3(P_3O_9)_2 \cdot 6H_2O$	6	1.46-1.64	1.547(6)	0.14	99-123	17	126(x3), 127(x4), 129(x2)	(235)
$Pb_3(P_3O_9)_2 \cdot 3H_2O$	3	1.45-1.66	1.54(3)	0.14	99-117	15	122,122, 128	(237)
MEAN:			1.543	0.14		17		
TETRAMETAPHOSPHATES								
$Mg_2P_4O_{12}$	2	1.46-1.60	1.539(5)	0.13	101-119	16	134,139	(224)
$Cu_2P_4O_{12}$	2	1.47-1.60	1.538(6)	0.12	101-119	15	136,138	(223)
$Ni_2P_4O_{12}$	2	1.48-1.63	1.54(1)	0.10	100-120	18	130,134	(226)
$Co_2P_4O_{12}$	2	1.48-1.66	1.55(2)	0.18	100-122	20	131,135	(225)
$Pb_2P_4O_{12} \cdot 4H_2O$	2	1.45-1.62	1.54(2)	0.14	102-121	17	128,143	(231)
$Ca_2P_4O_{12} \cdot 4H_2O$	2	1.47-1.61	1.543(3)	0.13	101-119	17	127,135	(230)
MEAN:			1.542	0.13		17		
DECAMETAPHOSPHATE								
$Ba_2Zn_3P_{10}O_{30}$	5	1.48-1.62	1.542(3)	0.12	97-119	18	123,129,130, 134,144	(213)
POLYMETAPHOSPHATES								
$Zn(PO_3)_2$	1	1.48-1.60	1.540(2)	0.12	103-119	16	136	(238)
$Ba(PO_3)_2$	2	1.43-1.65	1.55(1)	0.17	96-122	18	116,125	(241)
$Pb(PO_3)_2$	4	1.45-1.62	1.52(1)	0.13	100-123	15	138,138, 143,147	(242)
$BaCd(PO_3)_4$	4	1.46-1.60	1.536(4)	0.11	103-119	15	132,135, 137,141	(243)
$Be(PO_3)_2$-III	2	1.45-1.60	1.52(2)	0.11	99-121	19	137,140	(244a)
MEAN:			1.533	0.13		17		
ULTRAPHOSPHATES								
MnP_4O_{11}	4	1.42-1.68	1.561(15)	0.21	98-125	18	124,129,135, 136,144	(249)
CaP_4O_{11}	4	1.43-1.65	1.539(15)	0.17	99-124	19	126,131,137, 138,142	(248)
$Cd_2P_6O_{17}$	3	1.44-1.61	1.529(5)	0.16	99-122	21	124,132,141	(251)
MEAN:			1.543	0.18		19		

Distortion occurs when the PO_4 tetrahedra bond to each other. For instance, the triphosphates display a $\Delta(PO)$ mean value of 0.12 Å and a $\Delta(OPO)$ mean of 11°. These distortion measures are still greater for the metaphosphates, in particular $\Delta(OPO)$, and the effect is, again, more pronounced among the ultraphosphates showing very distorted PO_4 tetrahedra with O-P-O angles in the region 98-125°. For the ultraphosphates, $\Delta(PO)$ is as great as 0.18 Å, and $\Delta(OPO)$ is 19° (mean values; cf. Table 7.6).

Distortion of the phosphate tetrahedra is also displayed in the hydrogen orthophosphates such as $MgHPO_4 \cdot 3H_2O$ (131), $Ca(H_2PO_4)_2$ (ref. 136), $SnHPO_4$ (ref. 137), $HgHPO_4$ (ref. 256), and $Pb(H_2PO_4)_2$ (ref. 257). In these cases the P - O····H bond distance has increased while the other P-O distances within the tetrahedron have contracted to give a "normal" average <P-O> value. Normally, the P - O····H bond is about 0.1 Å longer than the other P-O distances. As regards O - P - O angles, these are usually in the region 102-118°, with an average value within each HPO_4 group close to the ideal value of 109.47^{o}.

From numerous crystallographic studies it has long been known that whenever PO_4 groups (or other XO_4 groups as well) are connected by corner-sharing, the "bridging" $P\text{-}O_{br}$ distances are always longest while the terminal P-O distances (to "free" oxygen atoms) have contracted so as to preserve the average P-O value. This average value is practically constant within all kinds of phosphates, i.e. around 1.54 Å (Tables 7.1 - 7.6), while individual P-O distances seldom deviate more than 0.1 Å from this value. As early as 1961, Cruickshank (ref. 174) explained such distortions of XO_4 tetrahedra in terms of π-bonding. He also showed that the average <X-O> distance would be fairly equal for regular and distorted XO_4 tetrahedra. Corbridge further clarified this effect in 1971 (ref. 253), and so did Baur in 1974 (ref. 255). Baur has also developed empirical methods of predicting bond lengths and bond angles in PO_4 and other XO_4 tetrahedra, and he has obtained some empirical relationships between d(P-O), d(O-O), O-P-O angles and similar geometrical quantities (ref. 255).

The divalent-metal phosphates listed in Tables 7.1 - 7.6 show good agreement with earlier results and with theory, but there is one curious exception: the β-$Mg_2P_2O_7$ type diphosphates with reported P - O - P angles of 180° display rather regular PO_4 tetrahedra, although their PO_4 groups are connected in pairs. This is remarkable and does not accord with theory. If P-O-P is not linear but bent, as suggested earlier in the section on diphosphate structures (section 3), this would make the PO_4 tetrahedra of this structure type more distorted. As mentioned earlier, the studies of $Mn_2P_2O_7$ (ref. 183) and $Mn_2V_2O_7$ (ref. 194) by means of X-ray and neutron diffraction techniques do indicate a statistical disorder of the P_2O_7 and V_2O_7 anions, respectively, i.e., their bridging oxygen atoms are disordered to give a non-linear X-O-X link and somewhat more distorted XO_4 tetrahedra, which facts better agree with theory and with results on similar structural configurations.

8 METAL-OXYGEN COORDINATION ENVIRONMENTS AND DISTANCES

This section reviews metal-oxygen coordination environments and distances reported in various divalent-metal phosphate structures. A summary of grand mean <M-O> distances for various coordination numbers is given in Table 8.1. These values have been calculated using all M-O distances appearing in the phosphate structures presented in this chapter, with the exception of very doubtful (or inaccurate) crystal structure determinations. In addition, corresponding values from an extensive study of divalent-metal coordination environments in inorganic oxide and oxosalt structures are given, within parentheses. The latter data have been obtained from a recent compilation by Nord and Kierkegaard (ref. 258) based on about 1700 published mineral and other inorganic oxosalt crystal structures containing medium-sized divalent cations of the metals Mg, Ca, Mn, Fe, Co, Ni, Cu, Zn, and Cd.

Table 8.1 shows that the metal-oxygen distances for the phosphates of divalent metals are closely related to those of the more extensive compilation of such data (ref. 258), which is also a logical expectation. It is, therefore, the more interesting to study coordination environments for the cations involved in this summary. Table 8.2 contains the relative frequencies (in percent) of the various MO_n polyhedra for the respective M^{2+} cations. Corresponding values from the more extensive compilation (ref. 258) are again inserted within parentheses. This table should thus be read as here exemplified for magnesium (Mg^{2+}). Only five- and six-coordinated Mg^{2+} environments have been reported for all magnesium-containing phosphates with known crystal structure mentioned in the present summary, i.e., pure magnesium phosphates or phosphate compounds

TABLE 8.1.

Grand mean metal-oxygen distances <M-O> (in Å units) for various coordination numbers (CN) in divalent-metal phosphate structures. Corresponding values from the more extensive summary of 1700 oxide and oxosalt structures (258) are given within parentheses. "Z" is the atomic number, r_M is the "octahedral" metal cation radius ("IR") after Shannon and Prewitt (6,7).

M^{2+}	Z	r_M (Å)	CN=2	CN=4[a]	CN=5	CN=6	CN=7	CN=8	CN=9	CN=10	CN=12
Be	4	0.45		1.61							
Mg	12	0.720		 (1.951)	2.04 (2.035)	2.08 (2.091)	 (2.234)	 (2.273)			
Ca	20	1.00			2.35 (2.354)	2.37 (2.374)	2.45 (2.460)	2.51 (2.497)	2.56 (2.561)	2.59 (2.614)	2.74 (2.686)
Mn	25	0.83		 (2.082)	2.15 (2.142)	2.21 (2.198)	 (2.37)	 (2.378)	 (2.44)		
Fe	26	0.78		 (2.04)	2.10 (2.114)	2.16 (2.112)		 (2.52)			
Co	27	0.745		 (1.957)	2.04 (2.040)	2.11 (2.112)		 (2.50)			
Ni	28	0.690	 (1.68)	 (1.88)	2.01 (2.01)	2.08 (2.065)		 (2.43)			
Cu	29	0.73		1.95[b] (1.950[b])	2.04 (2.03)	2.14 (2.135)	 (2.23)	 (2.54)			
Zn	30	0.740		1.96 (1.971)	2.06 (2.053)	2.11 (2.113)					
Sr	38	1.18					2.59	2.62	2.67	2.70	2.86
Cd	48	0.95		 (2.195)	2.26 (2.252)	2.31 (2.283)	2.42 (2.415)	 (2.434)	 (2.555)		
Ba	56	1.35							2.87	2.94	3.02
Hg	80	1.02	2.10								
Pb	82	1.19				2.58	2.64		2.73	2.82	2.92

[a]Usually tetrahedral configuration. [b]Square-planar configuration.

TABLE 8.2

Summary of M^{2+} ion coordination frequencies. The values are given in percent for each cation. Corresponding values from the more extensive summary of 1700 oxosalt structures (258) are given within parentheses.

M^{2+}	CN=2	CN=3	CN=4[a]	CN=5	CN=6	CN=7	CN=8	CN=9	CN=10	CN=12
Be	100									
Mg				15	85					
			(2)	(6)	(90)	(1)	(1)			
Ca				4	13	46	22	11	2	2
				(1)	(25)	(23)	(38)	(8)	(2)	(3)
Mn				47	53					
			(3)	(8)	(85)	(.5)	(3)	(.5)		
Fe				19	81					
			(5)	(7)	(87)		(1)			
Co				33	67					
		(2)	(4)	(4)	(89)		(1)			
Ni				12	88					
	(1)	(1)	(5)	(3)	(89)		(1)			
Cu			12	38	50					
			(20)	(17)	(62)	(.7)	(.3)			
Zn			32	23	45					
		(2)	(46)	(12)	(40)					
Sr						22	22	34	11	11
Cd				20	65	15				
			(4)	(7)	(67)	(8)	(9)	(5)		
Ba								50	42	8
Hg	100									
Pb					22	11		34	22	11

[a] Usually tetrahedral configuration, although square-planar data are also included for Cu and Ni.

containing magnesium and some other metal(s). Among all these, 15% of the distinct magnesium ions are five- and 85% are six-coordinated. In the more extensive summary of divalent-metal oxosalts (ref. 258), coordination numbers between 4 and 8 have been found, with the relative frequencies 2, 6, 90, 1, and 1 % (Table 8.2). It should be stressed that only compounds having solely oxygen ligands around the metal cation(s) have been included in the two summaries. This means that all $M_2(PO_4)Y$ compounds have been excluded, since the metals are surrounded by oxygen and other ("Y") ligands like fluorine and chlorine.

In the present compilation of phosphate structures, the small beryllium ion has only been found with a tetrahedral environment of oxygen atoms. (Divalent cations of Be, Sr, Ba, Sn, Pb, or Hg were not included in ref. 258). Six-coordination is most common for all other small and medium-sized cations, while the larger ones like Ca^{2+}, Sr^{2+}, Ba^{2+}, Pb^{2+} etc. more often are surrounded by at least seven oxygen ligands, often 10 - 12. Mercury (Hg^{2+}), though, is unique with its tendency for a two-fold, linear or almost linear, coordination.

It is indeed noteworthy that five-coordination has been found for all phosphates of medium-sized cations, even for the comparatively large Cd^{2+} and Ca^{2+} ions. Moreover, it is striking that five-coordination around any of these cations is so much more common in the phosphate structures than in the larger group of oxides and oxosalts of various kinds (sulphates, nitrates, borates, silicates, arsenates, vanadates, chromates, etc). For instance, 15% of all distinct Mg^{2+} ions in magnesium-containing phosphates are five-coordinated, while only 6% in the general group of oxosalts are five-coordinated. For divalent nickel, this difference is still greater: 12% of the Ni^{2+} ions in Ni-containing phosphates are five-coordinated, while five-coordination in the general group of oxosalts is extremely rare (only 3 %).

Among the phosphates, five-coordinated cations are most frequently observed in the orthophosphate structures. It is perhaps logical to assume that isolated (un-linked) PO_4 tetrahedra more easily may adapt themselves to five-coordination around M^{2+} than larger anions of linked PO_4 groups or infinite sheets of ultraphosphate anions. On the other hand, the general group of oxosalts may also contain small unlinked anions such as SO_4^{2-}, NO_3^-, AsO_4^{3-} etc; therefore, it is possible that the size of the PO_4 tetrahedron, in comparison to the size of the medium-sized divalent cations (usually with a radius around 0.70-0.75 Å) is especially favourable, from a geometrical point of view, for these cations to surround themselves with a five-coordinated environment of oxygen atoms.

Often, in some of the phosphate structures containing two (or more) different metal cations, there are several kinds of cation sites with different coordination numbers such that the cations may select various coordination polyhedra. In these cases, the interesting matter of cationic preferences for specific sites can be studied.

An early study of this problem was the determination of the structure of β-$(Ca_{1-x}Mg_x)_3(PO_4)_2$ (ref. 69), where magnesium, as expected, preferentially entered the smaller, six-coordinated sites, while the significantly larger Ca^{2+} ions were located at the other sites with CN:s in the range 6-9. However, in other studies it has clearly been shown that the size effect is negligible. Systematic investigations of the cationic preference of five- over six-coordination in the farringtonite ("γ-$Zn_3(PO_4)_2$") and graftonite structure types have already been discussed (sections 2.2 and 2.3) and so have cation distributions in the olivine-related sarcopside structure (section 2.4).

9 CLOSE-PACKINGS IN THE PHOSPHATE STRUCTURES

The degree of close-packing in an oxosalt crystal structure is principally determined by the oxygen atoms. In a phosphate crystal structure, the oxygen atoms can never be arranged in a truly close-packed array (cubic or hexagonal close-packing where each oxygen has twelve neighbours) because the metal cations are too large to be placed in the small voids between adjacent oxygen atoms. However, the concept V_{ox} is a useful tool to judge how densely packed a crystal structure is, and will be used in the forthcoming discussions.

The concept V_{ox} has earlier been defined as the unit cell volume "V" divided by the number of oxygen atoms in the unit cell ($Z \cdot p$ in a $M_mP_nO_p$ phosphate). The phosphate structures described in the present chapter have V_{ox} values in the range 16-25 $Å^3$, whereas structures built up of close-packed oxygen atoms usually have V_{ox} values in the range 14-15 $Å^3$. Some V_{ox} values have been given in the crystallographic tables of the preceding subsections. Values quoted below have been taken from the papers referenced there.

A first example is illustrated by magnesium orthophosphate, $Mg_3(PO_4)_2$, with V_{ox} = 19.79 $Å^3$ for the ordinary farringtonite phase containing five- and six-coordinated cations, whereas the olivine-related sarcopside-type high-pressure modification, usually denoted $Mg_3\square(PO_4)_2$ (CN = 6, 6), has V_{ox} = 17.86 $Å^3$, i.e., the volume has been reduced by about 9.8 %. On the other hand, the two established crystallographic forms of iron(II) orthophosphate have almost identical values of V_{ox} in spite of differences in cation coordination environments: the graftonite phase has V_{ox} = 18.79 $Å^3$ (CN = 5,5,6), while the sarcopside phase has V_{ox} = 18.83 $Å^3$ (CN = 6, 6). This close relationship in packing makes the hydrothermally induced phase transition from graftonite to sarcopside difficult to explain; it seems that heat and pressure favours a change of the atoms to the sarcopside structure.

In connection with these two anhydrous phases, hydrous iron(II) orthophosphates will also be included for comparison. The degree of packing in this kind of compounds, which hold small to moderate amounts of water in the crystal structure, are fairly equal to that of the two anhydrous iron(II) orthophosphates just mentioned: the V_{ox} values are 18.78 $Å^3$ in $Fe_3(PO_4)_2 \cdot H_2O$, 18.69 $Å^3$ in phosphoferrite, $Fe_3(PO_4)_2 \cdot 3H_2O$, and 18.71 $Å^3$ in ludlamite, $Fe_3(PO_4)_2 \cdot 4H_2O$. Vivianite ($Fe_3(PO_4)_2 \cdot 8H_2O$), though, has a larger value (V_{ox} = 19.25 $Å^3$).

Calcium orthophosphate is also polymorphic. The α-$Ca_3(PO_4)_2$ phase has V_{ox} = 22.49 $Å^3$ (CN = 5 - 9), the "β" phase V_{ox} = 21.00 $Å^3$ (CN = 6 - 9), while the high-pressure modification $Ca_3(PO_4)_2$-II has V_{ox} = 18.55 $Å^3$ (CN = 10, 12), i.e. a much denser structure, which is also reflected in the higher coordination numbers. The smallest V_{ox} value among the orthophosphates is found in $Ni_3(PO_4)_2$ (CN = 6, 6; sarcopside structure), with V_{ox} = 17.31 $Å^3$. It is possible that $Be_3(PO_4)_2$, if it exists, has a still lower value. Of course, orthophosphates of metals with the largest cations (Sr^{2+}, Pb^{2+}, Ba^{2+}) have the largest V_{ox} values, often in the region 22-25 $Å^3$.

The packing tendencies displayed among the orthophosphates are also generally valid for the groups of condensed phosphates. Among the $M_2P_2O_7$ diphosphates, for instance, the "α" phases (usually with five- and six-coordinated cations) have larger V_{ox} values than the corresponding "β" phases, which very often crystallize with the thortveitite-type structure which has only one crystallographically distinct, octahedrally coordinated, metal cation. This point is exemplified by magnesium diphosphate, with V_{ox} = 17.14 Å^3 for the "α" and 16.86 Å^3 for the "β" phase. The smallest V_{ox} value among all diphosphates is observed for δ-$Ni_2P_2O_7$, with V_{ox} = 16.36 Å^3. Again, large values are noted for the diphosphates containing larger cations. For example, V_{ox} = 21.98 Å^3 for $Pb_2P_2O_7$ and 22.54 Å^3 for α-$Sr_2P_2O_7$.

The meta- and ultra-phosphates will be compared as a group since some interesting details may thus be better displayed. There is a large group of isostructural $M_2P_4O_{12}$ tetrametaphosphates (M = Mg, Mn, Fe, Co, Ni, Cu, Zn, and Cd) with V_{ox} values from 17.33 Å^3 (for nickel tetrametaphosphate) to 20.05 Å^3 (for $Cd_2P_4O_{12}$). High-pressure phases, with as yet unknown structure, have been reported for most of these tetrametaphosphates. The latter are denoted $M_2P_4O_{12}$-II and have still smaller V_{ox} values; for instance, 16.54 Å^3 for $Ni_2P_4O_{12}$-II. Furthermore, a few of these metaphosphates also exist in a crystalline state with chain-shaped polymetaphosphate anions, such as $Zn(PO_3)_2$ (V_{ox} = 19.83 Å^3), i.e., a less closely packed structure than was noted for the tetrametaphosphate modification $Zn_2P_4O_{12}$ (with V_{ox} = 17.72 Å^3).

The ultraphosphates contain comparatively small amounts of metal and may thus appear to be fairly close-packed in spite of a large metal cation. For instance, $Sr_2P_6O_{17}$ has a V_{ox} value as low as 18.96 Å^3 (CN = 7, 8) although Sr^{2+} is a rather large cation. The isomorphous phases MnP_4O_{11} and CaP_4O_{11} have V_{ox} values of 20.80 and 22.27 Å^3, respectively (CN = 6).

In a comparison between ortho-, di- and meta-phosphates, it is clear that for a certain cation the "cation-rich" orthophosphates have comparatively large V_{ox} values with their 1.5 metal cations per PO_4 tetrahedron and their isolated (un-linked) phosphate groups. Oxygen close-packing is most favourable for the diphosphates with only one cation per PO_4 group and with all phosphate tetrahedra connected in pairs through corner-sharing. Although there is only 0.5 metal cation per phosphate tetrahedron in the metaphosphates, the bulkier tetra- and poly-metaphosphate anions do not allow close-packing of the oxygens, and V_{ox} increases slightly in these structures.

The magnesium phosphates may again be used to exemplify this: $Mg_3(PO_4)_2$ has V_{ox} = 19.79 Å^3, the diphosphates (α and β) 17.14 and 16.86 Å^3 respectively, and the tetrametaphosphate $Mg_2P_4O_{12}$ has

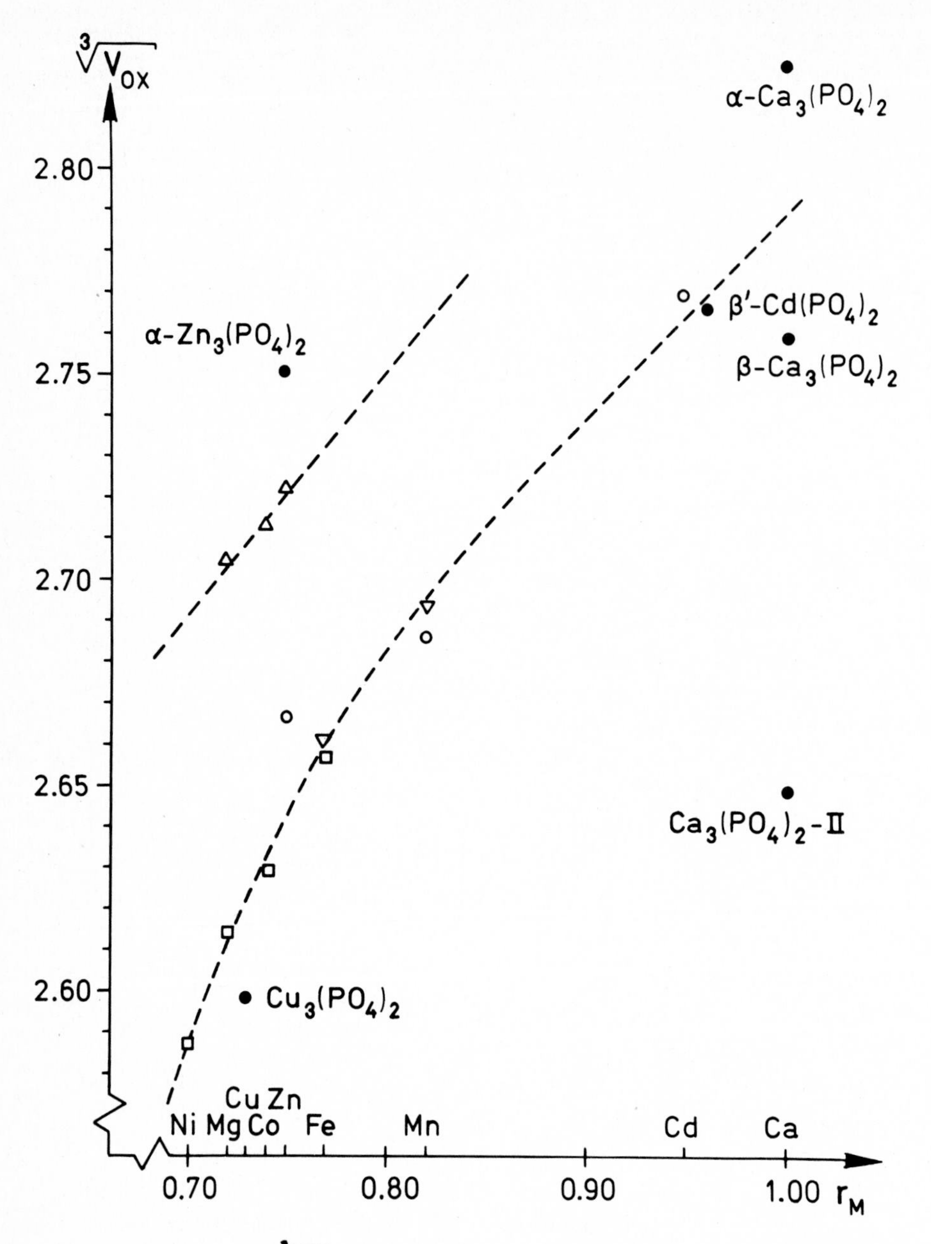

Fig. 9.1. Diagram of $\sqrt[3]{V_{OX}}$ versus r_M (Å) for orthophosphates. Symbols: ▲ = farringtonite-type structures (CN = 5,6), o = β-$Zn_3(PO_4)_2$ types (CN = 5 - 6), ◻ = sarcopside types (CN = 6, 6), ▼= graftonite types (CN = 5 - 7); filled circles are explained in the figure. The figure is in part reproduced, with permission from the Royal Swedish Academy of Science (copyright), from Nord and Kierkegaard: "Crystal chemistry of some anhydrous divalent-metal phosphates" (ref. 17) in Chemica Scripta vol. 15, pp. 27-39 (1980).

V_{ox} = 17.60 $Å^3$. (The high-pressure phase $Mg_2P_4O_{12}$-II has V_{ox} = 16.83 $Å^3$). The same tendency is observed for other series of divalent-metal phosphates like those of manganese, iron, cobalt, nickel, copper, zinc, and cadmium.

Nord and Kierkegaard (ref. 17) have earlier shown that for each structure type among the divalent-metal phosphates (and often even more generally) there is a strong correlation between V_{ox} values and cation radii, r_M. This trend is displayed in Figs. 9.1 - 9.3 using plots of $\sqrt[3]{V_{ox}}$ versus r_M (according to Shannon and Prewitt; ref. 6, 7) for ortho-, di- and meta-phosphates, respectively.

Fig. 9.1 shows the degree of close-packing among the $M_3(PO_4)_2$ orthophosphates. Obviously the orthophosphates can be sorted into two groups, with the exception of the densely packed high-pressure phase $Ca_3(PO_4)_2$-II. The first (and smallest) group consists of structures which are far from close-packed, namely α-$Zn_3(PO_4)_2$ (CN = 4, 4) and the farringtonite-type ("γ-$Zn_3(PO_4)_2$") structures (CN = 5, 6). This group is indicated in Fig. 9.1 by a dotted line.

The second group is formed by orthophosphates with slightly higher metal coordination numbers (CN usually around 6) and consequently with somewhat more closely packed structures. The compounds belonging in this group are connected by a dotted curve in Fig. 9.1, thus showing that there is a clear correlation between V_{ox} and r_M for these orthophosphates not only for each structure type (graftonite, sarcopside, β'-$Mn_3(PO_4)_2$ etc) but also for the entire class of compounds. One may say that for this class the V_{ox} value mainly depends on the cation radius, r_M.

The diphosphates $M_2P_2O_7$ are displayed in a similar way in Fig. 9.2. The metal coordination numbers are generally also 6, and the diagram clearly reveals that practically all diphosphates, excepting the two $Ca_2P_2O_7$ modifications, fall on an indicated line described by the approximate equation (cf. ref. 17)

$$\sqrt[3]{V_{ox}} = \sqrt[3]{10.8} + 0.52 \cdot r_M \qquad (1)$$

thus showing a strong correlation between V_{ox} and r_M.

The tetrametaphosphates form two large groups (see Fig. 9.3). The first, less closely-packed group, contains the "ordinary" tetrametaphosphates $M_2P_4O_{12}$ (CN = 6, 6) roughly satisfying the equation

$$\sqrt[3]{V_{ox}} = \sqrt[3]{10.9} + 0.54 \cdot r_M \qquad (2)$$

which is almost the same expression as eq. (1) for the diphosphates.

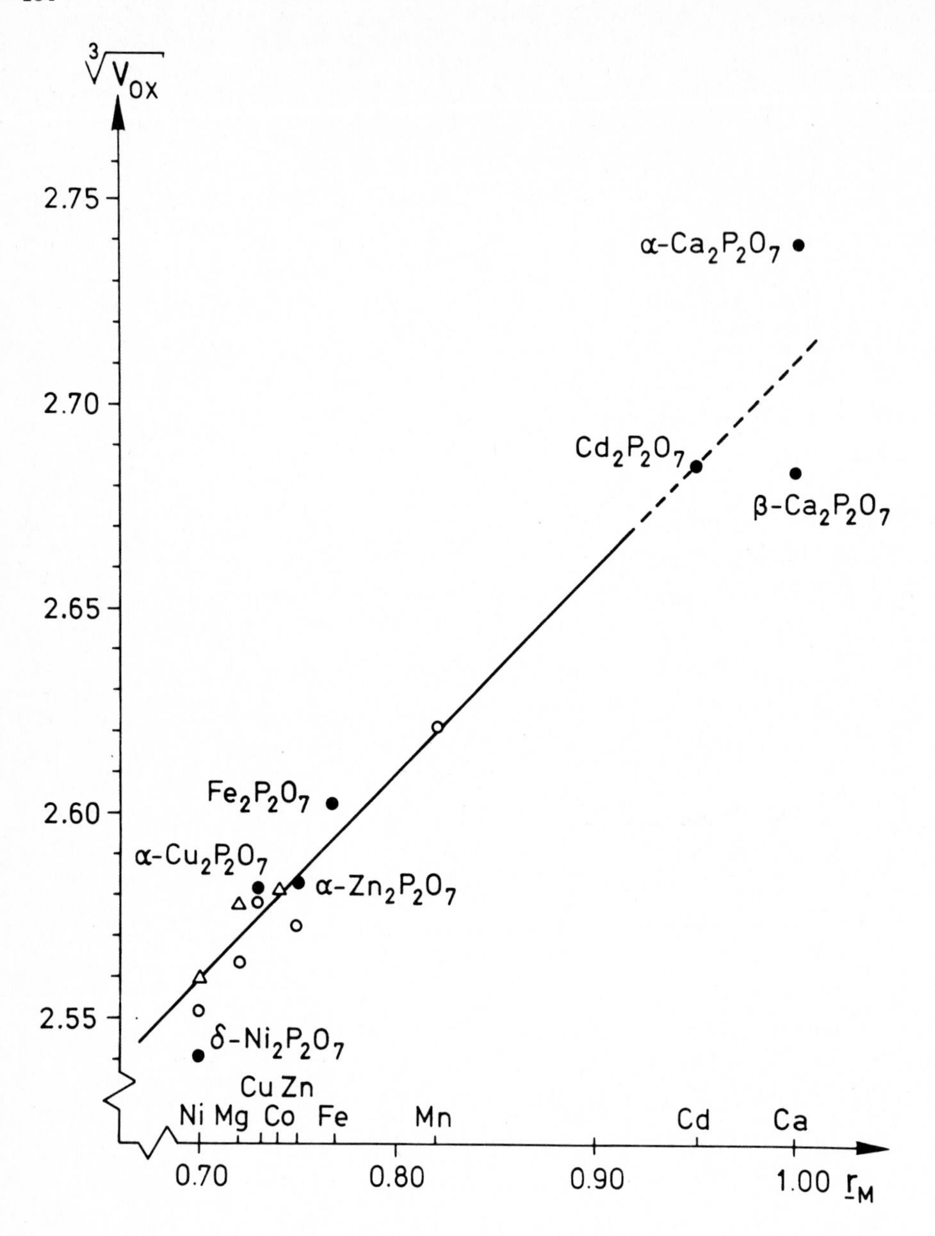

Fig. 9.2. Diagram of $\sqrt[3]{V_{ox}}$ versus r_M (Å) for diphosphates. Symbols: **Δ** = α-$Mg_2P_2O_7$ structure types (CN = 5, 6), o = β-$Mg_2P_2O_7$ types (CN = 6); filled circles are explained in the figure. The figure is partly reproduced from ref. (17); cf. the text to Figure 9.1.

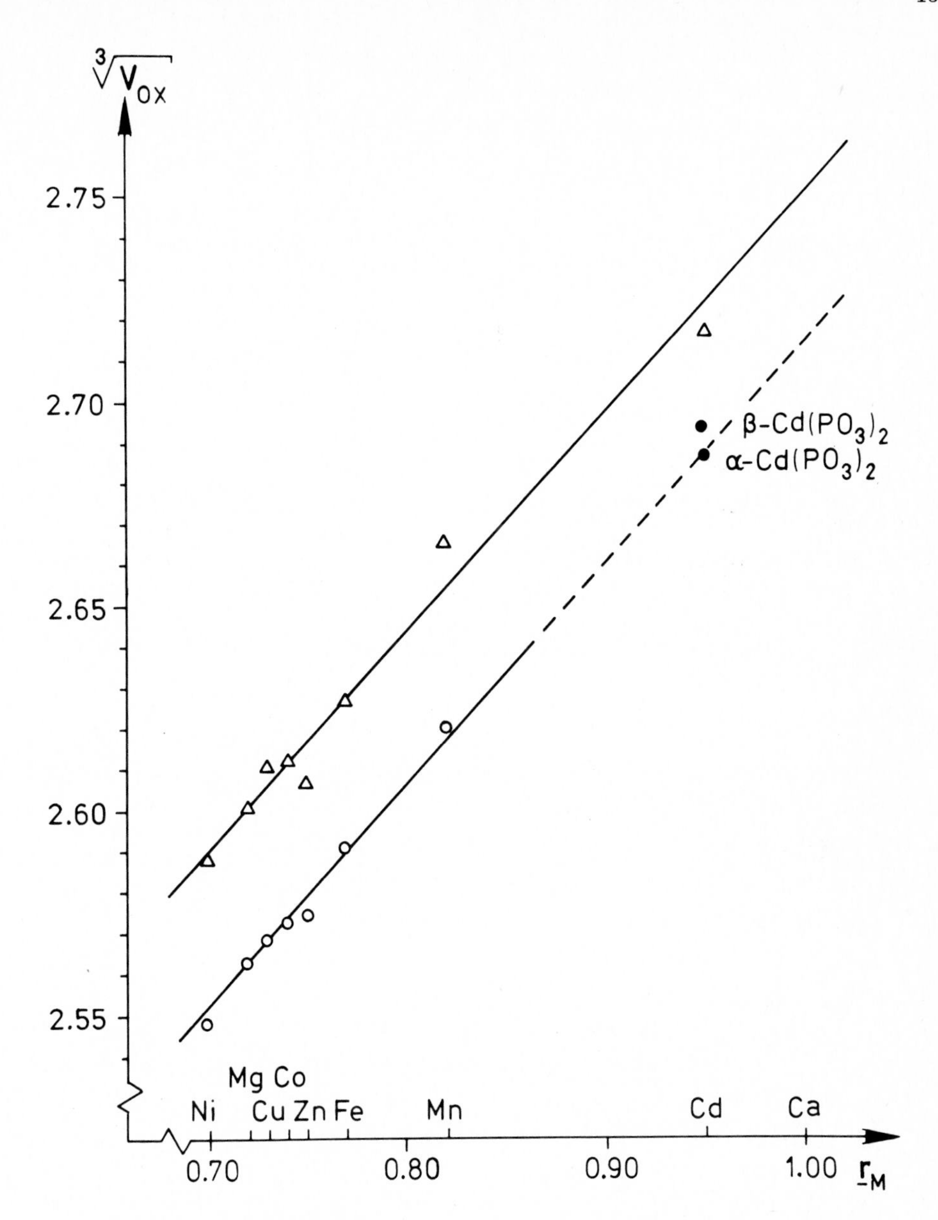

Fig. 9.3. Diagram of $\sqrt[3]{V_{ox}}$ versus r_M (Å) for metaphosphates. Symbols: ▲ = $M_2P_4O_{12}$ (CN = 6, 6), o = $M_2P_4O_{12}$-II (structure unknown); filled circles are explained in the figure. The figure is partly reproduced from ref. (17); cf. the text to Fig. 9.1.

The high-pressure phases $M_2P_4O_{12}$-II (with an unknown crystal structure) form a second group with higher densities (Fig. 9.3), i.e., smaller unit cell volumes. The correlation between V_{ox} and r_M is given by

$$\sqrt[3]{V_{ox}} = \sqrt[3]{10.3} + 0.54 \cdot r_M \qquad (3)$$

which is similar to eq. (2). From equations (2) and (3) one may easily deduce that

$$\Delta V_{ox} = V_{ox}(M_2P_4O_{12}) - V_{ox}(M_2P_4O_{12}\text{-II}) \approx 0.60 + 0.29 \cdot r_M \qquad (4),$$

i.e., ΔV_{ox} is about 0.8 Å^3 for all medium-sized divalent metal cations considered in this study (cf. also ref. 17). This corresponds to a volume contraction of about 5 % for the transition $M_2P_4O_{12}(C2/c) \longrightarrow M_2P_4O_{12}$-II.

Regarding the polymetaphosphates included in Fig. 9.3, the two $Cd(PO_3)_2$ modifications lie on, or very nearly on, the "$M_2P_4O_{12}$-II" line. Both phases have six-coordinated cadmium ions (ref. 239, 240) which is probably also the case for the $M_2P_4O_{12}$-II phases judging from their unit cell volumes. Moreover, also $Zn(PO_3)_2$ is rather close ($\sqrt[3]{V_{ox}} = 2.60$, $r_M = 0.75$; not included in Fig. 9.3) which very well agrees with the fact that the zinc ions in this crystal structure are six-coordinated (ref. 238).

In addition to the above discussion of V_{ox} values, it should be stressed that there is also a strong correlation between average <M - O> bond distances and cation radii for each coordination number, particularly when isostructural phosphates are compared. The β-$Zn_3(PO_4)_2$, "γ-$Zn_3(PO_4)_2$", α-$Mg_2P_2O_7$ and $M_2P_4O_{12}$ structural groups may serve as examples (cf. Tables 2.1.1, 2.2.1, 3.2.1 and 5.2.1 in the preceding sections). However, a more detailed investigation in this line is hampered by the lack of accurate interatomic distances data.

10 CONCLUDING REMARKS

A majority among the phosphates of divalent metals structurally investigated so far have been shown to be polymorphous, usually dimorphous (at ordinary pressures). Structural relationships as well as isomorphism between various phosphates occur frequently. The phosphate structures are built of PO_4 tetrahedra and MO_n polyhedra ($4 \lesssim n \lesssim 12$; CN = n) united to form three-dimensional networks. As pointed out earlier, irregular coordination environments with CN = 5 and CN = 7 are rather common. Regarding symmetry, it is noteworthy that a very large number of phosphates crystallize in centric, monoclinic space groups.

Isomorphism among these phosphates is also fairly common and the divalent-metal phosphates often form groups of isostructural compounds like the farringtonite, sarcopside, and graftonite orthophosphates, the α-$Mg_2P_2O_7$ and β-$Mg_2P_2O_7$ type diphosphates, and the tetrametaphosphate groups $M_2P_4O_{12}$ and $M_2P_4O_{12}$-II.

Equi-sized cations often form isotypic compounds with the same anion. This is especially true for the medium-sized cations (Mg^{2+}, Mn^{2+}, Fe^{2+}, Zn^{2+} etc), whereas the calcium, strontium, barium or lead phosphates generally crystallize with structures not related to those of the smaller cations. Moreover, equi-sized cations often substitute for each other in naturally occurring minerals.

Undoubtedly the cation sizes are important in predicting isomorphism, but equi-sized cations are by no means the only criterion. The anions or at least the anionic configurations are also important. Among the divalent phosphates there are many different orthophosphate structure groups, but only a few metaphosphate structure types. The reason may be that large and rigid anions like $P_4O_{12}{}^{4-}$ can only arrange themselves geometrically with M^{2+} cations in a few ways, while M^{2+} and isolated (un-linked) phosphate tetrahedra may easily be arranged in several different three-dimensional patterns. It is to be noted that the two ultraphosphates MnP_4O_{11} and CaP_4O_{11} are isostructural as if the large anionic sheets helped to preserve the same structure although the cations differ in size. A similar example concerned with the same effect is given by the metavanadates MgV_2O_6 (ref. 259) and CaV_2O_6 (ref. 260). Although the calcium ion is almost 50% larger than that of magnesium, the two compounds are isomorphous. The anions are VO_6 octahedra sharing corners and edges to form large anionic sheets throughout the structure.

However, some of the criteria for isomorphism presented above may just be coincidences, and it is certainly not easy to suggest any definite rules governing the absence or presence of isomorphism among the phosphates of divalent metals. It is possible that theoretical calculations of bond strengths and bond energies may shed some light upon this problem. Such studies might also be useful to postulate which crystal structure is stable at specified physical conditions and to explain phase transitions. It has been shown by many examples in this chapter that a phosphate, especially an orthophosphate, often transforms to another polymorph or another crystal structure type when pressure and/or heat are applied, or by partly replacing the M^{2+} cation with another cation. More work is needed to increae the crystallographic knowledge of the important and very interesting stereochemistry of the phosphates of divalent metals.

REFERENCES

1 O. Hassel, Z. Elektrochem., 31 (1925) 523-529.
2 S.B. Hendricks, Amer. J. Sci., 14 (1927) 269-287.
3 L. Vegard, Phil. Mag., 4 (1927) 511-525.
4 R.W.G. Wyckoff, Amer. J. Sci., 10 (1925) 107-118.
5 S. Náray-Szabó, Z. Kristallogr., 75 (1930) 387-398.
6 R.D. Shannon and C.T. Prewitt, Acta Crystallogr., B25 (1969) 925-946.
7 R.D. Shannon, Acta Crystallogr., A32 (1976) 751-767.
8 P.B. Moore, Amer. Mineral., 57 (1972) 24-35.
9 M. Mathew, L.W. Schroeder, B. Dickens and W.E. Brown, Acta Crystallogr., B33 (1977) 1325-1333.
10 A.G. Nord and T. Stefanidis, Mater. Res. Bull., 15 (1980) 1183-1191.
11 A.G. Nord, Z. Kristallogr., 166 (1984) 159-176.
12 A.G. Nord and T. Ericsson, Amer. Mineral., 67 (1982) 826-832.
13 International Tables for X-ray Crystallography, Vol. I, Kynoch Press, Birmingham, 1968.
14 F.L. Katnack and F.A. Hummel, J. Electrochem. Soc., 105 (1958) 125-133.
15 C. Calvo, Can. J. Chem., 43 (1965) 436-445.
16 J.S. Stephens and C. Calvo, Can. J. Chem., 45 (1967) 2303-2312.
17 A.G. Nord and P. Kierkegaard, Chemica Scripta, 15 (1980) 27-39.
18 I.D. Brown and R.D. Shannon, Acta Crystallogr., A29 (1973) 266-282.
19 J.S. Stephens and C. Calvo, Can. J. Chem., 47 (1969) 2215-2225.
20 J.S. Stephens, Ph.D. Thesis, McMaster University, Hamilton, Ontario (1967).
21 A.G. Nord, Mater. Res. Bull., 17 (1982) 1001-1010.
22 A.L. Smith, J. Electrochem. Soc., 98 (1951) 363-368.
23 A.G. Nord and T. Stefanidis, Mater. Res. Bull., 16 (1981) 1121-1129.
24 J.F. Sarver, F.L Katnack and F.A. Hummel, J. Electrochem. Soc. 106 (1959) 960-963.
25 F.A. Hummel and F.L. Katnack, J. Electrochem. Soc., 105 (1958) 528-533.
26 J.J. Brown and F.A. Hummel, J. Electrochem. Soc., 110 (1963) 1218-1223.
27 E.R. Kreidler and F.A. Hummel, Inorg. Chem., 6 (1967) 524-528.
28 C. Calvo, J. Phys. Chem. Solids, 24 (1963) 141-149.
29 C. Calvo, Personal communication (1973).

30 C.K. Johnson, ORTEP: a Fortran thermal-ellipsoid plot program for crystal structure illustrations. ORNL-3794, Oak Ridge, Tenn. (1965).
31 A.G. Nord and P. Kierkegaard, Acta Chem. Scand., 22 (1968) 1466-1474.
32 A.G. Nord, Acta Chem. Scand., A28 (1974) 150-152.
33 J.B. Anderson, E. Kostiner, M.C. Miller and J.R. Rea, J. Solid State Chem., 14 (1975) 372-377.
34 G. Berthet, J.C. Joubert and E.F. Bertaut, Z. Kristallogr., 136 (1972) 98-105.
35 H. Annersten and A.G. Nord, Acta Chem. Scand., A34 (1980) 389-390.
36 E.R. DuFresne and S.K. Roy, Geochim. Cosmochim. Acta, 24 (1961) 198-205.
37 L.H. Fuchs, E. Olsen and E. Gebert, Amer. Mineral., 58 (1973) 949-951.
38 A.G. Nord, Mater. Res. Bull., 12 (1977) 563-568.
39 A.G. Nord, Acta Chem. Scand., A36 (1982) 95-99.
40 A.G. Nord, H. Annersten, T. Ericsson, G. Åberg and T. Stefanidis, Chemica Scripta (1985), In press.
41 A.G. Nord and T. Stefanidis, Z. Kristallogr., 153 (1980) 141-149.
42 A.G. Nord and T. Stefanidis, Unpublished results.
43 H. Annersten, T. Ericsson and A.G. Nord, J. Phys. Chem. Solids, 41 (1980) 1235-1240.
44 A.G. Nord, J. Appl. Crystallogr., 17 (1984) 55-60.
45 A.G. Nord and T. Stefanidis, Acta Crystallogr., B37 (1981) 1509-1511.
46 A.G. Nord and T. Stefanidis, Polyhedron, 1 (1982) 349-353.
47 A.G. Nord, Acta Crystallogr., B40 (1984) 191-194.
48 A.G. Nord and T. Ericsson, Amer. Mineral. (1985), In press.
49 E. Kostiner and J.R. Rea, Inorg. Chem., 13 (1974) 2876-2880.
50 C. Calvo and J.S. Stephens, Can. J. Chem., 46 (1968) 903-915.
51 C. Calvo, Amer. Mineral., 53 (1968) 742-750.
52 G. Engel and W. Klee, Z. Kristallogr., 132 (1970) 332-339.
53 A.G. Nord and T. Ericsson, Z. Kristallogr., 161 (1982) 209-224.
53a A.G. Nord, Mater. Res. Bull., 18 (1983) 569-579.
54 M. Websky, Neues Jahrb. Mineral., (1868) 606-607.
55 C. Calvo and R. Faggiani, Can. J. Chem., 53 (1975) 1516-1520.
56 G.A. Lager and E.P. Meagher, Amer. Mineral., 63 (1978) 365-377.
57 A.G. Nord and T. Stefanidis, Phys. Chem. Minerals, 10 (1983) 10-15.
58 E. Mattievich and J. Danon, J. Inorg. Nucl. Chem., 39 (1977) 569-580.
59 T. Ericsson and A.G. Nord, Amer. Mineral., 69 (1984) 889-895.
60 J.K. Kabalov, M.A. Simonov, O.V. Jakubovich, N.A. Jamnova and N.V. Belov, Dokl. Akad. Nauk. SSSR, 210 (1973) 830-832.
61 A.G. Nord, Neues Jahrb. Mineral., (1982) 422-432.
62 G.E. Brown, in P.H. Ribbe (Ed.) Orthosilicates, Reviews in Mineralogy, Vol. 5; Mineralogical Society of America, Wash., 1980, pp. 275-381.
63 G. Nover and G. Will, Z. Kristallogr., 155 (1981) 27-45.
64 R.W. Nurse, J.H. Welch and W. Gutt, J. Chem. Soc., (1959) 1077-1083.
65 P. Roux, D. Louër and G. Bonel, Compt. Rend. Acad. Sci. Paris, 286 (1978) 549-551 (Ser. C).
66 B. Dickens, L.W. Schroeder and W.E. Brown, J. Solid State Chem., 10 (1974) 232-248.

67 P. Süsse and M.J. Buerger, Z. Kristallogr., 131 (1970) 161-174.
68 A.G. Nord, Neues Jahrb. Mineral., (1983) 489-497.
69 L.W. Schroeder, B. Dickens and W.E. Brown, J. Solid State Chem., 22 (1977) 253-262.
70 J. Ando, Bull. Chem. Soc. Japan, 31 (1958) 201-205.
71 B. Dickens and W.E. Brown, Tschermaks Miner. Petr. Mitt., 16 (1971) 79-104.
72 B. Gossner, Neues Jahrb. Mineral., Geol. Beil.-Band 57A (1928) 89-116.
73 A.G. Nord, Ph. D. Thesis, University of Stockholm, Stockholm 1974.
74 C. Frondel, Amer. Mineral., 26 (1941) 145-152.
75 C. Frondel, Amer. Mineral., 28 (1943) 215-232.
76 U. Keppler, Neues Jahrb. Mineral., (1965) 171-176.
77 R. Gopal and C. Calvo, Nature, 237 (1972) 30-32.
78 R. Gopal, C. Calvo, J. Ito and W.K. Sabine, Can. J. Chem., 52 (1974) 1155-1164.
79 C. Calvo and R. Gopal, Amer. Mineral., 60 (1975) 120-133.
80 M. Corlett and U. Keppler, Naturwissenschaften, 53 (1966) 105.
81 P. Rahmdohr and H. Struntz, Klockmann's Lehrbuch der Mineralogie, 16. Aufl., überarbeitet und erweitert. F. Enke Verlag, Stuttgart 1982.
82 K. Sudarsanan and R.A. Young, Acta Crystallogr., B36 (1980) 1525-1530.
83 O.E. Piro, M.C. Apella, E.J. Baran and B.E. Rivero, Revue Chim. Mineral., 19 (1982) 11-18.
84 M. Müller, Helvetica Chim. Acta, 30 (1947) 2069-2080.
85 B. Dickens, W.E. Brown, G.J. Kruger and J.M. Stewart, Acta Crystallogr., B29 (1973) 2046-2056.
86 W.H. Zachariasen, Acta Crystallogr., 1 (1948) 263-265.
87 E.R. Kreidler and F.A. Hummel, Inorg. Chem., 6 (1967) 524-528.
88 J.H. Welch and W. Gutt, J. Chem. Soc., (1961) 4442-4444.
89 J.F. Sarver, M.V. Hoffman and F.A. Hummel, J. Electrochem. Soc., 108 (1961) 1103-1110.
90 E.R. Kreidler, J. Electrochem. Soc., 118 (1971) 923-929.
91 J.T. Lindbloom, G.V. Gibbs and P.H. Ribbe, Amer. Mineral., 59 (1974) 1267-1271.
92 J.O. Nriagu and P.B. Moore, Phosphate minerals, Springer-Verlag, Berlin Heidelberg 1984.
93 G.L. Shoemaker, J.B. Anderson and E. Kostiner, Acta Crystallogr., B33 (1977) 2969-2972.
94 C. Calvo and K.Y. Leung, Z. Kristallogr., 130 (1969) 231-233.
95 M. Brunel-Laügt and J.C. Guitel, Acta Crystallogr., B33 (1977) 3465-3468.
96 J.B. Anderson, E. Kostiner and F.A. Ruszala, J. Solid State Chem., 39 (1981) 29-34.
97 Y. Senga and A. Kawahara, Acta Crystallogr., B36 (1980) 2555-2558.
98 M. Mathew, L.W. Schroeder and T.H. Jordan, Acta Crystallogr., B33 (1977) 1812-1816.
99 D.M.C. Guimaraes, Acta Crystallogr., A35 (1979) 108-114.
100 L.H. Brixner, P.E. Bierstedt, W.F. Jaep and J.R. Barkley, Mater. Res. Bull., 8 (1973) 497-504.
101 H.N. Ng and C. Calvo, Can. J. Phys., 53 (1975) 42-51.
102 M. Hata, F. Marumo and S.I. Iwai, Acta Crystallogr., B36 (1980) 2128-2130.
103 I. Mayer and A. Semadja, J. Solid State Chem., 46 (1983) 363-366.

104 B. Eiberger and M. Greenblatt, J. Solid State Chem., 41 (1982) 44-50.
105 K. Aurivillius and B.A. Nilsson, Z. Kristallogr., 141 (1975) 1-10.
106 C.W. Wolfe, Amer. Mineral., 25 (1940) 738-753.
107 C.W. Wolfe, Amer. Mineral., 25 (1940) 787-809.
108 P.B. Moore, Amer. Mineral., 50 (1965) 2052-2062.
109 T. Kanazawa, T. Umegaki and M. Shimizu, Bull. Chem. Soc. Japan, 52 (1979) 3713-3717.
110 L.W. Schroeder, M. Mathew and W.E. Brown, J. Phys. Chem., 82 (1978) 2335-2340.
110a M. Catti, M. Franchini-Angela and G. Ivaldi, Z. Kristallogr., 155 (1981) 53-64.
111 P.B. Moore and T. Araki, Amer. Mineral., 60 (1975) 454-459.
112 P.B. Moore and T. Araki, Inorg. Chem., 15 (1976) 316-321.
113 S.C. Abrahams and J.L. Bernstein, J. Chem. Phys., 44 (1966) 2223-2229.
114 S.C. Abrahams, J. Chem. Phys., 44 (1966) 2230-2237.
115 H. Mori and T. Ito, Acta Crystallogr., 3 (1950) 1-6.
116 P. Fejdi, J.F. Poullen and M. Gasparin, Bull. Minéral., 103 (1980) 135-138.
117 J.B. Forsyth, C.E. Johnson and C. Wilkinson, J. Phys. C, Solid State Phys., 3 (1970) 1127-1139.
118 P. Eversheim and W. Kleber, Acta Crystallogr., 6 (1953) 215-216.
119 L. Fanfani and P.F. Zanazzi, Tschermaks Mineral. Petr. Mitt., 26 (1979) 255-269.
120 J.B. Anderson, E. Kostiner and F.A. Ruszala, Inorg. Chem., 15 (1976) 2744-2748.
121 A. Whitaker, Acta Crystallogr., B31 (1975) 2026-2035.
122 G.Y. Chao, Z. Kristallogr., 130 (1969) 261-266.
123 N.M. Antraptseva and L.N. Schtjegrov, Zhurnal Neorg. Xim., 28 (1983) 2818-2823.
124 R.J. Hill, Amer. Mineral., 62 (1977) 812-817.
125 T. Ericsson and A.G. Nord, Neues Jahrb. Mineral., (1984) 193-197.
126 M. Catti, G. Ferraris and G. Ivaldi, Bull. Minéral., 102 (1979) 314-318.
127 K. Taxer, Amer. Mineral., 60 (1975) 1019-1022.
128 P.D. Brotherton, E.N. Maslen, M.W. Pryce and A.H. White, Austral. J. Chem., 27 (1974) 653-656.
129 L. Fanfani, A. Nunzi and P.F. Zanazzi, Acta Crystallogr., B26 (1970) 640-645.
130 B.R. Rao, Acta Crystallogr., 14 (1961) 738-744.
131 F. Abbona, R. Boistelle and R. Haser, Acta Crystallogr., B35 (1979) 2514-2518.
132 A. Whitaker, Z. Kristallogr., 137 (1973) 194-219.
133 A. Whitaker and J.W. Jeffery, Acta Crystallogr., B26 (1970) 1429-1440.
134 B. Dickens, J.S. Bowen and W.E. Brown, Acta Crystallogr., B28 (1971) 797-806.
135 N.A. Curry and D.W. Jones, J. Chem. Soc., (1971) 3725-3729.
136 B. Dickens, E. Prince, L.W. Schroeder and W.E. Brown, Acta Crystallogr., B29 (1973) 2057-2070.
137 A.F. Berndt and R. Lamberg, Acta Crystallogr., B27 (1971) 1092-1094.
138 R. Herak, B. Prelesnik, M. Čurić and P. Vasić, J. Chem. Soc., Dalton, (1978) 566-569.
139 R.C. McDonald and K. Eriks, Inorg. Chem., 19 (1980) 1237-1241.

140 G. Burley, J. Res. Natl. Bur. Stand., 60 (1958) 23-27.
140a E. Dubler, L. Beck, L. Linowsky and G.B. Jameson, Acta Crystallogr., B37 (1981) 2214-2217.
140b P. Vasič, B. Prelesnik, R. Herak and M. Čurič, Acta Crystallographica B37 (1981) 660-662.
141 W.E. Brown, Nature, 196 (1962) 1048-1050.
142 S. Ghose, Acta Crystallogr., 16 (1963) 124-128.
143 F.A. Ruszala, J.B. Anderson and E. Kostiner, Inorg. Chem., 16 (1977) 2417-2422.
144 W.E. Richmond, Amer. Mineral., 25 (1940) 441-479.
145 R. Klement and H. Haselbeck, Zeitschr. anorg. allg. Chemie, 334 (1964) 27-36.
146 R. Klement and H. Haselbeck, Zeitschr. anorg. allg. Chemie, 336 (1965) 113-128.
147 C. Rømming and G. Raade, Amer. Mineral., 65 (1980) 488-498.
148 A. Coda, G. Giuseppetti and C. Tadini, Accad. Nazion. dei Lincei, Rend. della Classe Scienze fis., mat., nat., 43 (1967) 212-224.
149 K. Auh and F.A. Hummel, Can. Mineral., 12 (1974) 346-351.
150 R.G. Burns, Mineralogical applications of crystal field theory. University press, Cambridge (1970).
151 A.G. Nord and T. Stefanidis, Acta Crystallogr., B37 (1981) 1509-1511.
152 L. Waldrop, Z. Kristallogr., 131 (1970) 1-20.
153 C. Frondel, Amer. Mineral., 34 (1949) 692-705.
154 L. Waldrop, Z. Kristallogr., 130 (1969) 1-14.
155 J.R. Rea and E. Kostiner, Acta Crystallogr., B28 (1972) 2525-2529.
156 J.R. Rea and E. Kostiner, Acta Crystallogr., B32 (1976) 1944-1947.
157 J.R. Rea and E. Kostiner, Acta Crystallogr., B30 (1974) 2901-2903.
158 P. Keller, H. Hess and F. Zettler, Neues Jahrb. Mineral., 134 (1979) 147-156.
159 G. Cocco, L. Fanfani and F.P. Zanazzi, Z. Kristallogr., 123 (1966) 321-329.
160 J.R. Rea and E. Kostiner, Acta Crystallogr., B28 (1972) 3461-3464.
161 J.R. Rea and E. Kostiner, Acta Crystallogr., B28 (1972) 2505-2509.
162 J.B. Anderson, J.R. Rea and E. Kostiner, Acta Crystallogr., B32 (1976) 2427-2431.
163 A.G. Nord and T. Stefanidis, Cryst. Struct. Commun., 10 (1981) 1251-1257.
164 P.B. Moore and S. Ghose, Amer. Mineral., 56 (1971) 1527-1538.
165 M. Greenblatt, E. Banks and B. Post, Acta Crystallogr., 23 (1967) 166-171.
166 T. Deans, J.D.C. McConnell and R. Pickup, Amer. Mineral., 40 (1955) 776.
167 K.W. Bladh, R.K. Corbett, W.J. McLean and R.B. Laughon, Amer. Mineral., 57 (1972) 1880-1884.
168 G.A. Lager and G.V. Gibbs, Amer. Mineral., 59 (1974) 919-925.
169 T.H. Jordan, L.W. Schroeder, B. Dickens and W.E. Brown, Inorg. Chem., 15 (1976) 1810-1814.
170 G.R. Levi and G. Peyronel, Z. Kristallogr., 92 (1935) 190-209.
171 L.O. Hagman and P. Kierkegaard, Acta Chem. Scand., 23 (1969) 327-328.
172 K. Łukaszewicz, Bull. Acad. Pol. Sci, ser. chim., 15 (1967) 53-57.

173 W.H. Zachariasen, Z. Kristallogr., 73 (1930) 1-6.
174 D.W.J. Cruickshank, J. Chem. Soc., (1961) 5486-5504.
175 C. Calvo, Acta Crystallogr., 23 (1967) 289-295.
176 N. Krishnamachari and C. Calvo, Acta Crystallogr., B28 (1972) 2883-2885.
177 K. Łukaszewicz, Bull. Acad. Pol. Sci., ser. chim., 15 (1967) 47-51.
178 C. Calvo, Can. J. Chem., 43 (1965) 1139-1146.
179 J.F. Sarver, Trans. Brit. Ceram. Soc., 65 (1966) 192-198.
180 A. Pietraszko and K. Łukaszewicz, Bull. Acad. Pol. Sci., ser. chim., 16 (1968) 183-187.
181 B.E. Robertson and C. Calvo, Can. J. Chem., 46 (1968) 605-612.
182 C. Calvo, Can. J. Chem., 43 (1965) 1147-1153.
183 T. Stefanidis and A.G. Nord, Acta Crystallogr., C40 (1984) 1995-1999.
183a K.H. Pannell, Y.S. Chen, K. Belknap, C.C. Wu, I. Bernal, M.W. Creswick and H.N. Huang, Inorg. Chem., 22 (1983) 418-427.
184 T. Stefanidis and A.G. Nord, Z. Kristallogr., 159 (1982) 255-264.
185 R. Masse, J.C. Guitel and A. Durif, Mater. Res. Bull., 14 (1979) 337-341.
186 B.E. Robertson and C. Calvo, Acta Crystallogr., 22 (1967) 665-672.
187 B.E. Robertson and C. Calvo, J. Solid State Chem., 1 (1970) 120-133.
188 C. Calvo and P.K.L Au, Can. J. Chem., 47 (1969) 3409-3416.
189 C. Calvo, Inorg. Chem., 7 (1968) 1345-1351.
190 N.C. Webb, Acta Crystallogr., 21 (1966) 942-948.
191 L.O. Hagman, I. Jansson and C. Magnéli, Acta Chem. Scand., 22 (1968) 1419-1429.
192 J.C. Grenier and R. Masse, Bull. Soc. franc. Minéral. Crist., 92 (1969) 91-92.
193 K. Łukaszewics and R. Smajkiewicz, Rocz. Chem., 35 (1961) 741-744.
194 A.G. Nord, Neues Jahrb. Mineral., (1984) 283-288.
195 J.T. Hoggins, J.S. Swinnea and H. Steinfink, J. Solid State Chem., 47 (1983) 278-283.
196 R.C. Ropp, M.A. Aia, C.W.W. Hoffman, T.J. Veleker and R.W. Mooney, Anal. Chem., 31 (1959) 1163-1166.
197 ASTM File Card 9-48.
198 C. Calvo, J. Electrochem. Soc., 115 (1968) 1095-1096.
199 L.H. Brixner, P.E. Bierstedt and C.M. Foris, J. Solid State Chem., 6 (1973) 430-432.
200 J.K. Lees and P.A. Flinn, J. Chem. Phys., 48 (1968) 882-889.
201 Schneider and R.L. Collin, Inorg. Chem., 12 (1973) 2136-2139.
202 N.S. Mandel, Acta Crystallogr., B31 (1975) 1730-1734.
203 M.T. Averbuch-Pouchot, A. Durif and J.C. Guitel, Acta Crystallogr., B31 (1975) 2482-2486.
204 M.T. Averbuch-Pouchot and J.C. Guitel, Acta Crystallogr., B32 (1976) 1670-1673.
205 M.T. Averbuch-Pouchot and J.C. Guitel, Acta Crystallogr., B33 (1977) 1427-1431.
206 J. Berak, Rocz. Chem., 32 (1958) 17-21.
207 Y.I. Smolin, Y.F. Shepelev, A.I. Domanskii and J. Majling, Sov. Phys. Crystallogr., 23 (1978) 715-717.
208 L. Pauling and J. Sherman, Z. Kristallogr., A96 (1937) 481-487.
209 V. Caglioti, G. Giacomello and E. Bianchi, R.C. Accad. Ital., 3 (1942) 761-775.

210 K.H. Jost, Acta Crystallogr., B28 (1972) 732-738.
211 K.H. Jost, Acta Crystallogr., 19 (1965) 555-560.
212 M. Laügt and J.C. Guitel, Z. Kristallogr., 141 (1975) 203-216.
213 M. Bagieu-Beucher, A. Durif and J.C. Guitel, J. Solid State Chem., 45 (1982) 159-163.
214 K. Plieth and C. Wurster, Zeitschr. anorg. Chemie, 267 (1951) 49-61.
215 K. Dornberger-Schiff, F. Liebau and E. Thilo, Naturwiss., 41 (1954) 551-553.
216 E. Thilo, Angew. Chemie, 4 (1965) 1061-1070.
217 K.H. Jost, Acta Crystallogr., 17 (1964) 1539-1544.
218 E. Thilo and I. Grunze, Zeitschr. anorg. allg. Chemie, 290 (1957) 209-222.
219 E. Thilo and I. Grunze, Zeitschr. anorg. allg. Chemie, 290 (1957) 223-237.
220 I. Grunze, Dissertation, Humboldt- Universität, Berlin (1956).
221 E. Steger, Zeitschr. anorg. allg. Chemie, 294 (1958) 146-154.
222 M. Beucher and J.C. Grenier, Mater. Res. Bull., 3 (1968) 643-648.
223 M. Laügt, J.C. Guitel, I. Tordjman and G. Bassi, Acta Crystallogr., B28 (1974) 201-208.
224 A.G. Nord and K.B. Lindberg, Acta Chem. Scand., A29 (1975) 1-6.
225 A.G. Nord, Cryst. Struct. Commun., 11 (1982) 1467-1474.
226 A.G. Nord, Acta Chem. Scand., A37 (1983) 539-543.
227 A.G. Nord, Mater. Res. Bull., 18 (1983) 765-773.
228 H.M. Rietveld, J. Appl. Crystallogr., 2 (1969) 65-71.
229 M. Bagieu-Beucher, M. Gondrand and M. Perroux, J. Solid State Chem., 19 (1976) 353-357.
230 M. Schneider, K.H. Jost and H. Fichtner, Zeitschr. anorg. allg. Chemie, 500 (1983) 117-122.
231 H. Worzala, Zeitschr. anorg. allg. Chemie, 421 (1976) 122-128.
232 M.T. Averbuch-Pouchot, A. Durif and J.C. Guitel, Acta Crystallogr., B32 (1976) 1894-1896.
233 M.T. Averbuch-Pouchot, A. Durif and J.C. Guitel, Acta Crystallogr., B32 (1976) 1533-1535.
234 I. Tordjman, A. Durif and J.C. Guitel, Acta Crystallogr., B32 (1976) 205-208.
235 R. Masse, J.C. Guitel and A. Durif, Acta Crystallogr., B32 (1976) 1892-1894.
236 A. Durif, M.T. Averbuch-Pouchot and J.C. Guitel, Acta Crystallogr., B31 (1975) 2680-2682.
237 M. Brunel-Laügt, I. Tordjman and A. Durif, Acta Crystallogr., B32 (1976) 3246-3249.
238 M.T. Averbuch-Pouchot, A. Durif and M. Bagieu-Beucher, Acta Crystallogr., C39 (1983) 25-26.
239 M. Bagieu-Beucher, J.C. Guitel, I. Tordjman and A. Durif, Bull. Soc. franc. Minéral. Cristallogr., 97 (1974) 481-484.
240 M. Bagieu-Beucher, M. Brunel-Laügt and J.C. Guitel, Acta Crystallogr., B35 (1979) 292-295.
241 J.C. Grenier, C. Martin, A. Durif, D. Tranqui and J.C. Guitel, Bull. Soc. franc. Minéral. Cristallogr., 90 (1967) 24-31.
242 K.H. Jost, Acta Crystallogr., 17 (1964) 1539-1544.
243 M.T. Averbuch-Pouchot, A. Durif and J.C. Guitel, Acta Crystallogr., B31 (1975) 2453-2456.
243a S. Jaulmes, Rev. Chim. Minéral., 1 (1964) 617-671.

244 M.T. Averbuch-Pouchot, A. Durif and I. Tordjman, Acta Crystallogr., B33 (1977) 3462-3464.
244a E. Schultz and F. Liebau, Z. Kristallogr., 154 (1981) 115-126.
244b E. Schultz, Dissertation, Universität Kiel, (1974).
245 B.A. Maksimov, Y. A. Kharitonov and N.V. Belov, Dokl. Akad. Nauk SSSR, 213 (1973) 1072-1075.
246 O. Baumgartner and H. Völlenkle, Z. Kristallogr., 146 (1977) 261-270.
247 J.J. Brown and F.A. Hummel, J. Electrochem. Soc., 111 (1964) 660-665.
248 I. Tordjman, M. Bagieu-Beucher and R. Zilber, Z. Kristallogr. 140 (1974) 145-153.
249 L.X. Minatjeva, M.A. Porai-Koschits, A.S. Antsischkina, V.G. Ivanova and A.V. Lavrov, Koord. Xim., 1 (1975) 421-428.
250 A.V. Lavrov, L.X. Minatjeva and L.S. Gyzejeva, Neorg. Mater., 9 (1973) 1466-1467.
251 A.S. Antsischkina, M.A. Porai-Koschits, L.X. Minatjeva and V.G. Ivanova, Koord. Xim., 5 (1979) 268-275.
252 A.S. Antsischkina, M.A. Porai-Koschits, L.X. Minatjeva, V.G. Ivanova and A.V. Lavrov, Koord. Xim., 4 (1978) 448-454.
253 D.E.C. Corbridge, Bull. Soc. franc. Minéral. Cristallogr., 94 (1971) 271-299.
254 R.D. Shannon and C. Calvo, J. Solid State Chem., 6 (1973) 538-549.
255 W.H. Baur, Acta Crystallogr., B30 (1974) 1195-1215.
256 E. Dubler, L. Beck, L. Linowsky and G.B. Jameson, Acta Crystallogr., B37 (1981) 2214-2217.
257 P. Vasić, B. Prelesnik, R. Herak and M. Čurić, Acta Crystallogr., B37 (1981) 660-662.
258 A.G. Nord and P. Kierkegaard, Chemica Scripta (1985), In press.
259 H. N. Ng and C. Calvo, Can. J. Chem., 50 (1972) 3619-3624.
260 J.C. Bouloux, G. Perez and J. Galy, Bull. Soc. franc. Minéral. Cristallogr., 95 (1972) 130-133.

CHAPTER 3

TRANSITION METAL COMPLEXES WITH CARBON DISULFIDE. CORRELATIONS BETWEEN STEREOCHEMISTRY AND REACTIVITY

CLAUDIO BIANCHINI, CARLO MEALLI, ANDREA MELI and MICHAL SABAT

Instituto per lo Studio della Stereochimica ed Energetica, dei Composti di Coordinazione, Via F.D. Guerrazzi 27, 50132 Firenze, Italy

ABBREVIATIONS

acac	acetylacetonate
AO	Atomic Orbital
bipy	2,2′-bipyridyl
Bu	butyl
C-D-D	Chatt-Dewar-Duncanson
CDFC	Cambridge Data File Coordinates
Cp	η^5-cyclopentadienyl
Cy	cyclohexyl
dppm	bis(diphenylphosphino)methane
diphos	1,2-diphenylphosphino ethane

EHMO	Extended Hückel Molecular Orbital
FMO	Fragment Molecular Orbital
HOMO	Highest Occupied Molecular Orbital
LMO	Localized Molecular Orbital
LUMO	Lowest Unoccupied Molecular Orbital
Me	methyl
MO	Molecular Orbital
np_3	tris(2-diphenylphosphinoethyl)amine
Ph	phenyl
Pr	propyl
R	alkyl
SOMO	Singly Occupied Molecular Orbital
triphos	1,1,1-tris(diphenylphosphinomethyl)ethane
VSEPR	Valence Shell Electron Pair Repulsion
VB	valence bond

1 INTRODUCTION

Since the synthesis of the first CS_2 complex, $Pt(\eta^2-CS_2)(PPh_3)_2$ reported by Baird and Wilkinson in 1966 (ref. 1), the coordination chemistry of carbon disulfide has continued to develop rapidly and to yield novel discoveries. It is now apparent that CS_2 complexes can function as excellent and, in some cases, unique starting materials in inorganic, organometallic and organic syntheses.

The need for a detailed knowledge of all the aspects of the coordination chemistry of carbon disulfide is obvious. In addition, the challenging task of developing a C_1 chemistry for the inert CO_2 molecule may be facilitated from observations of the behaviour of CS_2 in analogous circumstances.

At least four reviews have appeared on the subject (ref. 2-5). The present article will concentrate on three main topics.

First, the known structures of mononuclear and polynuclear transition metal complexes of CS_2 will be presented for an immediate, pictorial evaluation of the bonding capabilities of the triatomic molecule. Obviously, the geometrical features of the complexes and any systematic trends observed from structural parameters contain invaluable information, not only about the bonding, but also about eventual reactivity patterns.

Next, the reactivity of CS_2 metal complexes toward any sort of reagents will be reviewed and discussed in detail.

Finally, correlations will be drawn between the experimental, structural and chemical data and the electronic nature of the compounds. The number of papers dealing with the electronic structure of transition metal CS_2 complexes is very limited (ref. 6-8). Therefore, instead of reporting published results in detail, the final part of the article will discuss, in terms of commonly accepted qualitative MO arguments, the overall bonding situation in certain model compounds. Whenever possible, the implications for further reactivity of the compounds will be pointed out.

Before proceeding with the discussion, the reader must be warned about the selection rules adopted to limit the number of compounds under consideration. In principle, the real carbon disulfide complexes should be restricted to those which contain the triatomic in its intact form: namely, the compounds in which the carbon and sulfur atoms of CS_2 make coordinative bonds only to one or more metal atoms. On the other hand, and in conformity with a very useful interpretational point of view, certain transition metal fragments can be formally replaced by common organic residues (ref. 9). Accordingly, a number of compounds where an organic group substitutes an isolobal transition metal fragment were considered pertinent to the discussion. Indeed, the two types of compounds exhibit similar stereochemical and structural features and, in some cases, comparable reactivity. These assumptions increase the number of CS_2 complexes; on the other hand, we have excluded the compounds where the carbon atom of the CS_2 grouping is not linked to the metal or is not sp^2 hybridized. Finally, the compounds which obey our restrictive rules but which were not obtained directly from carbon disulfide were also omitted from our considerations on the reactivity.

2 STRUCTURE OF CARBON DISULFIDE COMPLEXES

2.1 GENERAL CLASSIFICATION OF THE STRUCTURES

In this section we shall describe the results of about thirty structural analyses of transition metal complexes with CS_2, which are summarized in **Tables 1** and **2.** Some detailed discussion of the geometrical features of the earlier complexes can be also found in previous review papers (ref. 2-5).

For practical purposes the structures were separated into two broad classes. Class 1 (**Table 1**) is related to the complexes containing an η^2-C,S bonded CS_2 group (hereafter η^2-CS_2 complexes). Class 2 (**Table 2**) includes the compounds with an η^1-C bonded CS_2 group (hereafter η^1-CS_2 complexes).

Class 1 is the broadest and may be divided into five different categories:

A) Classical mononuclear η^2-CS_2 complexes containing the triatomic in its intact form.

B) Binuclear complexes where the exocyclic sulfur atom of one η^2-CS_2 unit acts as a donor ligand toward a second metal fragment.

C) A series of dimers obtained by combining two M-η^2-CS_2 units, whose exocyclic sulfur atoms reciprocally donate a σ lone pair to the metal center of the other unit.

D) An unique example of M-η^2-CS_2 moiety, which chelates a second metal through both sulfur atoms.

E) η^2-Dithioalkyl ester, CS_2R, compounds. Their inclusion as carbon disulfide complexes ensues from the similarity with the compounds of type 1B, by taking for granted the isolobal analogy between alkyl groups and certain ML_n fragments (ref. 9).

Complexes of class **2**, containing the triatomic in its intact form, have not been so far isolated in the solid state. Thus, all the compounds summarized in **Table 2** are characterized by one or two linkages ensuing from one or both sulfur atoms of the η^1-CS_2 group. Class 2 has been subdivided into five categories:

A) Trinuclear compounds where the two sulfur atoms of the M-η^1-CS_2 unit act as σ ligands toward two distinct metal centers.

B) Binuclear compounds where the M-η^1-CS_2 unit acts as a bidentate 1,1-dithio ligand toward another metal fragment.

C) Binuclear compounds where the CS_2 grouping bridges two similar metal centers by using the carbon atom on one side, and one sulfur atom on the other side.

Table 1. Structural parameters of η^2-CS_2 complexes

	M–C(Å)	M–S_1(Å)	C–S_1(Å)	C–S_2(Å)	S_1–C–S_2(°)	Fig.	ref.
A. mononuclear (M–C(=S_2)–S_1 ring)							
$Pt(CS_2)(PPh_3)_2$	2.06(5)	2.33(2)	1.72(5)	1.54(5)	136(4)	1	12
$Pd(CS_2)(PPh_3)_2$	2.00(3)	2.31(1)	1.65(3)	1.63(3)	140(2)		13
$Co(CS_2)(Cp)(PMe_3)$	1.89(1)	2.24(0)	1.68(1)	1.60(1)	141.2(7)	2	15
$Fe(CS_2)(CO)_2(PPh_3)(PMe_3)$	1.983(8)	2.334(2)	1.676(7)	1.615(8)	138.9(1)	3	16
$[Rh(CS_2)(np_3)]BPh_4$	2.09(1)	2.387(5)	1.65(2)	1.54(1)	143(1)	4	17
$Co(CS_2)$(triphos)	1.88(1)	2.206(4)	1.68(1)	1.62(1)	133.8(8)	5	18
$Ni(CS_2)$(triphos)	1.86(1)	2.197(3)	1.63(1)	1.61(1)	136.1(7)		8
$V(CS_2)(Cp)_2$	2.075(4)	2.432(2)	1.667(4)	1.618(4)	137.5(3)	6	19
$Nb(CS_2)(Cp)_2(Me)$	2.206(8)	2.503(4)	1.68(1)	1.61(1)	137.7(5)	7	20
$Nb(CS_2)(Cp)_2(\eta^1$-$C_3H_5)$	2.24(2)	2.518(9) 2.514(9)[a]	1.72(2)	1.57(2)	137(1)		21

Table 1. continued

B. binuclear M(–C(–S_2–M′)–S_1)							
$(PMe_2Ph)_2(CO)_2Fe(CS_2)Mn(Cp)(CO)_2$	1.939(6)	2.325(2)	1.658(6)	1.642(2)	139.7(3)	8	22
$(triphos)Co(CS_2)Cr(CO)_5$	1.87(2)[b]	2.180(8)	1.65(3)	1.63(2)	137(1)	9	18
	1.86(2)	2.220(7)	1.67(3)	1.62(3)	136(1)		
$(Cp)_2(Bu)Nb(CS_2)W(CO)_5$	2.11(8)	2.52(3)	1.52(8)	1.67(8)	135(5)	10	23
C. binuclear M(S–C(S)M′)(C–S_2)S_1							
$[Ni(CS_2)(PPh_3)]_2$	1.81(1)	2.157(4)	1.63(1)	1.68(1)	137.3(9)	13	26
$[Pd(CS_2)(P(t-Bu)_2Ph)]_2$	1.980(5)	2.316(1)	1.643(5)	1.650(5)	135.9(3)		27
$[Pt(CS_2)(P(t-Bu)_2Ph)]_2$	1.970(6)	2.337(2)	1.674(6)	1.651(6)	133.3(4)		28

Table 1. continued

D. binuclear M(C S1 S2)M							
$[\{Co(triphos)\}_2(CS_2)](BPh_4)_2$	1.72(5)	2.27(1)	1.75(7)	1.72(7)	112(4)	14	18
E. dithioalkyl ester M(C S1)–S2–R							
$[Fe(CS_2CH_2Ph)(CO)_2(PMe_3)_2]PF_6$	1.890(3)	2.322(1)	1.634(3)	1.669(3)	136.2(0)	11	24
$[V(CS_2Me)(Cp)_2]I_3$	2.08(1)	2.452(4)	1.63(1)	1.66(1)	134.8(9)	12	19
$[Ru(CS_2Me)(CO)_2(PPh_3)_2]ClO_4$	2.03[c]	2.467	1.66	1.65	135		25

[a]The unit cell consists of two independent molecules of the complex. Values of the bond lenghts and angles within the CS_2 molecule were constrained in the refinement to be equal in both complex molecules.

[b]Values for two independent molecules of the complex.

[c]No estimated standard deviations reported.

Table 2. Structural parameters of η^1-CS_2 complexes

	M-C(Å)	C-S_1(Å)	C-S_2(Å)	S_1-C-S_2(°)	Fig.	ref.
A. trinuclear M-C(S_1—M')(S_2—M'')						
$(Cp)(CO)_2FeC(S)SFe(Cp)(CO)_2W(CO)_5$	1.975(4)	1.715(4)	1.670(5)	119.2(2)	15	31
B. binuclear M-C(S)(S)M'						
$[(PPh_3)_2ClPt(CS_2)Pt(PPh_3)_2]BF_4$	1.95(2)	1.71(2)	1.69(2)	109.9(9)	16	32
C. binuclear M-C(=S_1)—S_2-M						
$[PtCl(dppm)]_2(CS_2)$	2.18[a]	1.63	1.55	c	17	33

Table 2. continued

Compound						
D. dithioalkyl ester $M-C(=S_2)-S_1-R$ 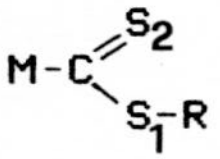						
$OsH(CS_2Me)(CO)_2(PPh_3)_2$	2.137(1)	1.724(5)	1.648(4)	121.4(3)	18	37
E. miscellaneous						
$Mn[CS_2C_2(CO_2Me)_2](CO)[P(OMe)_3](Cp)$	1.876(2)	1.732(2)	1.753(2)	108.7(2)	19	39
$Fe_3(CS_2C_2H_2)(CO)_8S_2$	1.948(3)	1.675(3)	1.683(3)	c		40
$Rh(C_2S_4)(PMe_3)(Cp)$	2.04(1)	1.68(2)	1.65(2)	115.2(9)	20	43
$Rh_2Cl_2(C_2S_4)(dppm)_2(CO)$	1.90(2)[b]	1.75(2)	1.67(2)	116(1)	21	44
	1.95(2)	1.75(2)	1.62(2)	119(1)		
$[(Cp)(CO)_2\overline{FeSC(Fp)S}CS(Fp)]SO_3CF_3$	1.67(3)	1.73(4)	1.88(4)	102(2)	22	45

Fp = $Fe(CO)_2(Cp)$

[a]Neither realistic e.s.d.'s nor bond angles within the bridging CS_2 fragment were reported. [b]Values for two independent molecules. [c]Not reported.

D) η^1-Dithioalkyl ester complexes. The latter can be considered the η^1 analogs of compounds 1E.

E) A number of complexes classified as miscellaneous in which CS_2 is a part of some cyclic unit, structured, case by case, in different manners.

2.2 STRUCTURES OF M-η^2-CS_2 COMPLEXES

In the survey of η^2-structures special attention will be devoted to the nature of the different ML_n fragments which support this type of coordination. A distinction in these terms will be extremely useful at a later stage, while discussing the electronic underpinnings of this coordination mode (see Section 4.2).

Although in many cases rather high standard deviations obscure a detailed comparison of the structures, it can be safely said that most of the 10 structures 1A exhibit quite similar geometries of the coordinated CS_2 molecule.

The mean value of the endocyclic C-S bond length where both the carbon and sulfur atoms are metal-bonded is 1.68(3) Å (range 1.63-1.72 Å), while the analogous value of the uncoordinated (exocyclic) C-S bond length is 1.60(3) Å (range 1.54-1.63 Å). The mean value of the S-C-S angle is 138(3)° (range 134-143°). Interestingly, the length of both C-S linkages is intermediate between that of the double bond in the linear, free CS_2 molecule (1.554 Å) (ref. 10) and the single bond (1.81 Å) (ref. 11) of the Me_2S molecule.

The ML_2 fragment appears in the isostructural complexes $M(\eta^2\text{-}CS_2)(PPh_3)_2$ (M = Pt (ref. 12), Pd (ref. 13)) (**Fig. 1**).

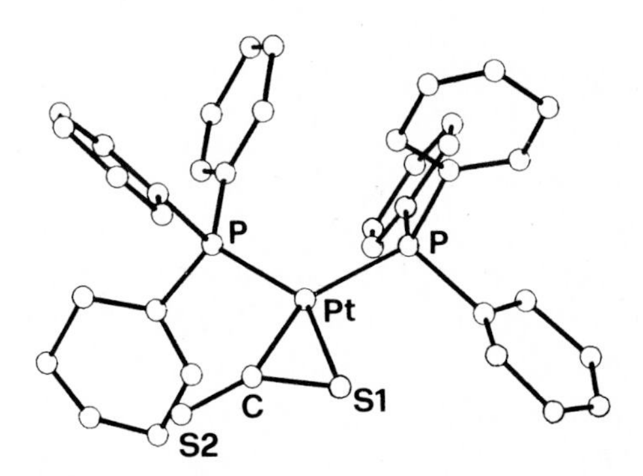

Fig. 1. $Pt(\eta^2\text{-}CS_2)(PPh_3)_2$ (CDFC)

The unusual asymmetry of the metal fragment in these complexes is worthy of some attention. In fact, the M-P bond <u>trans</u> to the carbon atom is about 0.1 Å longer than the other M-P bond. The same type of asymmetry is found also in the analogous carbon dioxide complex $Ni(\eta^2-CO_2)(PCy_3)_2$ (ref. 14). The origin of the effect is as yet unclear and the suggestion that the Pt-P bond <u>trans</u> to the carbon atom has a lower π-type character (ref. 12), is not verified. Notice that both structures mentioned above are affected by relatively high standard deviations so that any fine comparison with respect to the geometry of the $M-CS_2$ groupings is not appropriate.

Dihapto coordination of CS_2 is supported also by ML_4 fragments with pseudo C_{2v} symmetry. These include also M(Cp)L fragments if the well known equivalence of a cyclopentadienyl ring with three separated σ ligands is taken into account. The complex $Co(\eta^2-CS_2)-(Cp)(PMe_3)$ (ref. 15) is an example. The structure of this compound is presented in **Fig. 2.** The geometry of the $M-\eta^2-CS_2$ grouping compares well with that found in the complex $Fe(\eta^2-CS_2)(CO)_2(PPh_3)-$

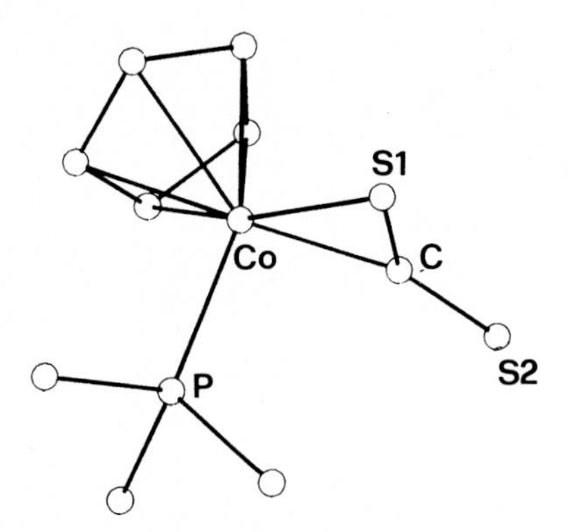

Fig. 2. $Co(\eta^2-CS_2)(Cp)(PMe_3)$ (CDFC)

(PMe_3) (ref. 16), shown in **Fig. 3.** The supporting metal fragment of the latter compound is of the canonical type ML_4. Both the coordinated and the uncoordinated C-S bond lengths as well as the S-C-S angles are almost identical in the two structures, in nice agreement with the assumption of equivalent electronic distributions at the two d^8 metal centers. It is worthwhile to notice that in the iron complex as well as in other 1B and 1E compounds formally derived from it (vide infra), the two carbonyl ligands lie in the same plane as CS_2. Finally, in the iron fragment of **Fig. 3** there is no evidence of the particular asymmetry found in the complexes $ML_2(\eta^2\text{-}CS_2)$. The two Fe-C bonds are almost equal and, ultimately, the slightly longer Fe-C bond (by 0.04 Å) is _trans_ to

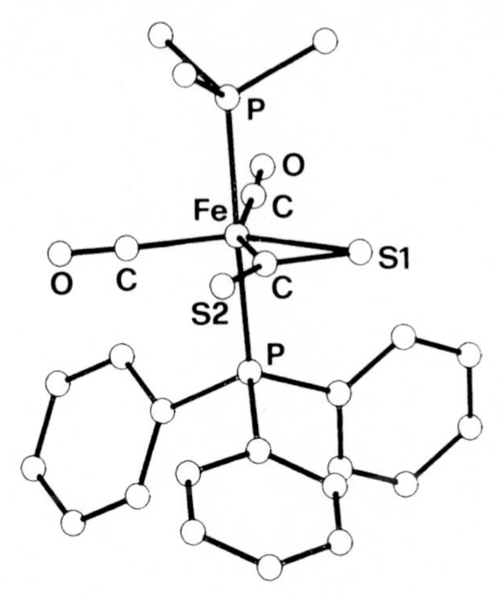

Fig. 3. $Fe(\eta^2\text{-}CS_2)(CO)_2(PMe_3)(PPh_3)$ (CDFC)

the sulfur rather than to the carbon atom.

A third example of $d^8\text{-}ML_4$ supporting fragment is found in the Rh(I) compound $[(np_3)Rh(\eta^2\text{-}CS_2)]BPh_4$ (ref. 17), shown in **Fig. 4.** The $M(np_3)$ fragment which usually has a C_{3v} symmetry, adapts here to an arrangement of the type ML_4 with pseudo C_{2v} symmetry. The

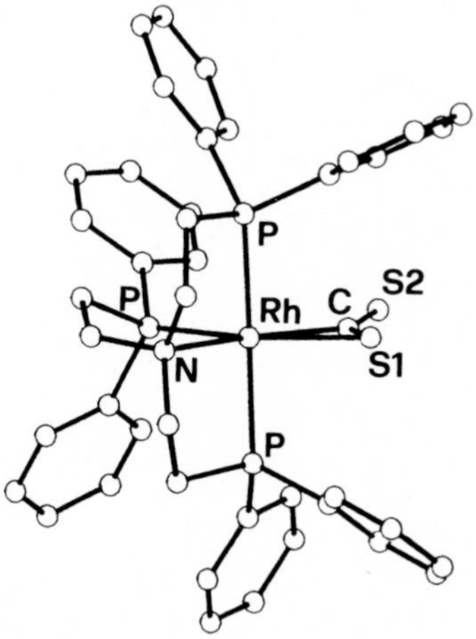

Fig. 4. $[(np_3)Rh(\eta^2\text{-}CS_2)]^+$

P(1) and P(3) atoms of **Fig. 4** are trans to each other as the P(1)-Rh-P(3) angle is 161.9(1)°, while the N and P(2) atoms are in a cis arrangement. The CS_2 molecule lies in the plane defined by the latter two atoms and by the metal. The geometrical parameters of the Rh-η^2-CS_2 moiety are slightly off the average values calculated for the complexes of class 1A. In particular, the structure exhibits the largest S-C-S angle of 143(1)° and the shortest C-S exocyclic bond (1.54(1) Å) amongst the monomeric η^2-CS_2 complexes. It will be shown in Section 4.4 that such a combination of values is indicative of a relatively poor metal-CS_2 π-backdonation.

A pyramidal ML_3 fragment appears in the two isostructural complexes (triphos)Co(η^2-CS_2) (**Fig. 5**) (ref. 18) and (triphos)Ni-(η^2-CS_2) (ref. 8). By assuming C_{3v} symmetry for the (triphos)M skeleton, the orientation of the CS_2 molecule is such that the MCS_2 grouping lies coplanar with the plane defined by the metal, one phosphorus atom and the threefold axis. Chemically, the two Co and Ni complexes differ for the total electron count. However

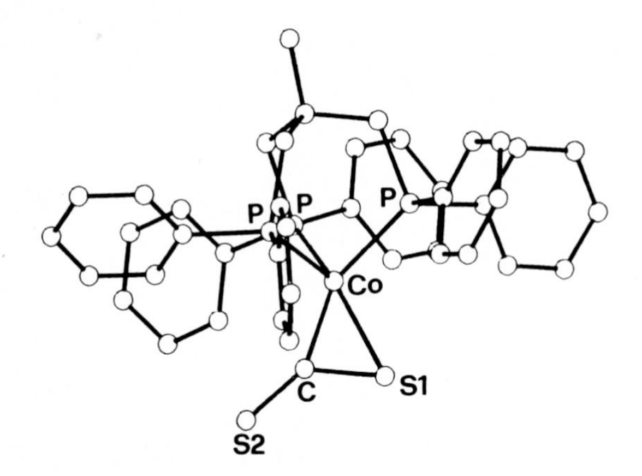

Fig. 5. (triphos)Co(η^2-CS_2)

the primary geometry barely changes in going from the paramagnetic cobalt to the diamagnetic nickel complex (the two structures are isomorphous) and only minor differences are observed for the two C-S linkages and for the S-C-S angle. The bonds lengths are both larger in the Co complex, in particular the coordinated C-S bond (1.68(1) vs. 1.63(1) Å) and also the CS_2 molecule is here more bent (133.8(8)° vs. 136.1(7)°). The latter differences are indicative of a larger backdonation from cobalt in spite of its d^{17} total electron count (see Section 4.4).

The series of structures of type 1A is completed by complexes of vanadium and niobium which contain $M(Cp)_2$ and $M(Cp)_2(R)$ fragments (R = methyl, allyl), respectively. The vanadium compound $V(\eta^2-CS_2)-(Cp)_2$ (ref. 19), which has an unpaired electron as the complex (triphos)$Co(\eta^2-CS_2)$, is shown in **Fig. 6**.

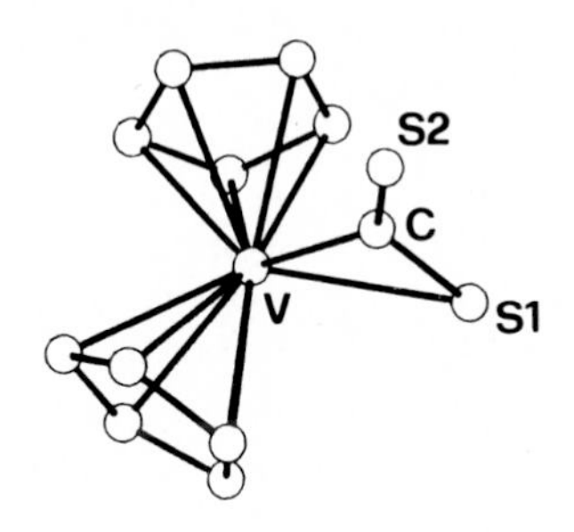

Fig. 6. $V(\eta^2-CS_2)(Cp)_2$ (CDFC)

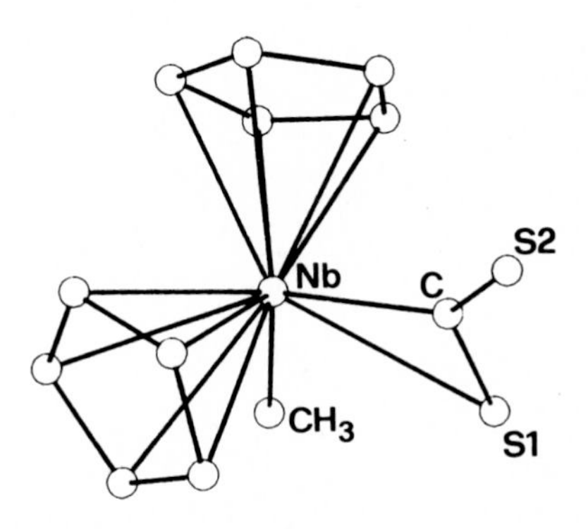

Fig. 7. $Nb(\eta^2-CS_2)(Cp)_2(Me)$ (CDFC)

Of the two niobium derivatives, $Nb(\eta^2-CS_2)(Cp)_2(Me)$ (ref. 20) and $Nb(\eta^2-CS_2)(Cp)_2(\eta^1-C_3H_5)$ (ref. 21) only the structure of the former is shown in **Fig. 7**. In all of these complexes the plane of CS_2 almost bisects the Cp-M-Cp angle.

There is enough evidence that, as a direct consequence of the dihapto coordination of CS_2 to a metal center, the exocyclic sulfur atom magnifies its σ donor capabilities (see Sections 3 and 4).

Thus, electron deficient transition metal fragments such as $M'(CO)_5$ (M' = Cr, W) or $Mn(Cp)(CO)_2$ which have an empty σ hybrid (ref. 9), react easily with M-η^2-CS_2 complexes to give adducts of class 1B. Three structures are known for this type of adducts, namely $(PMe_2Ph)_2Fe(CO)_2(\mu\text{-}CS_2)Mn(Cp)(CO)_2$ (ref. 22) (**Fig. 8**), (triphos)-Co(μ-CS_2)$Cr(CO)_5$ (ref. 18) (**Fig. 9**) and $(Cp)_2(Bu)Nb(\mu\text{-}CS_2)W(CO)_5$ (ref. 23) (**Fig. 10**). Remarkably, the primary geometry of the

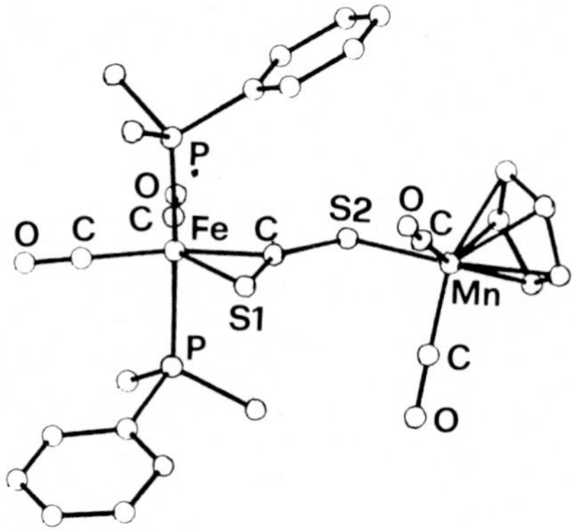

Fig. 8. $(PMe_2Ph)_2Fe(CO)_2(\mu\text{-}CS_2)Mn(Cp)(CO)_2$ (CDFC)

parent $ML_n(\eta^2\text{-}CS_2)$ fragments is barely changed in the adducts. However, the interesting geometrical aspects of the latter compounds need to be discussed together with those of compounds 1E. Actually, the binding of an alkyl group to the exocyclic sulfur atom of a M-η^2-CS_2 moiety reproduces the trends observable when the

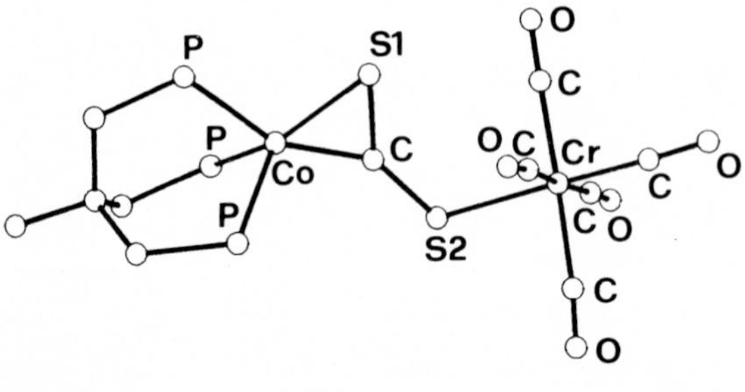

Fig. 9. (triphos)Co(μ-CS_2)$Cr(CO)_5$

latter moiety is linked to an electrophilic metal center (type 1B). Significant in this respect is the structure of the cation $[Fe(\eta^2-CS_2CH_2Ph)(CO)_2(PMe_3)_2]^+$ (ref. 24), shown in **Fig. 11**, which together with the structures in **Fig. 3** and **8**, completes a series whose basic framework is the unit $FeL_4(\eta^2-CS_2)$.

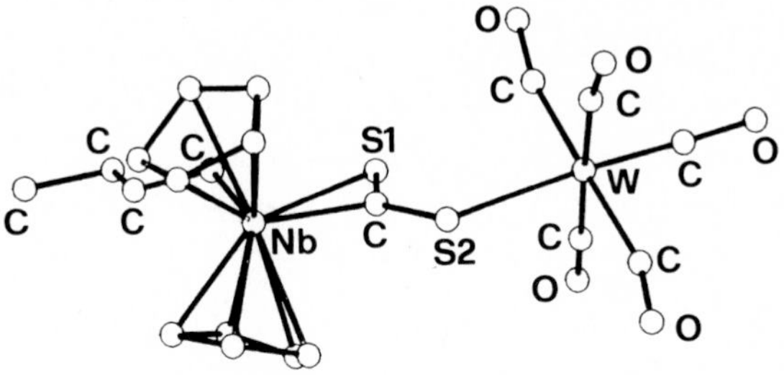

Fig. 10. $(Cp)(Bu)Nb(\mu-CS_2)W(CO)_5$ (CDFC)

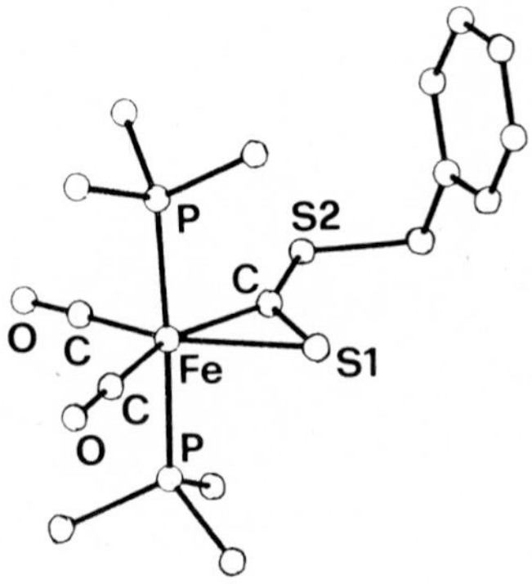

Fig. 11. $[Fe(\eta^2-CS_2CH_2Ph)(CO)_2(PMe_3)_2]^+$ (CDFC)

The primary geometry of the parent $FeL_4(\eta^2-CS_2)$ complex is only slightly changed in its manganese and alkyl adducts. Practically, the electrophilic center (E) of the two latter groups lies coplanar with the $M-\eta^2-CS_2$ unit and the C-S-E angles differ ± 7° from 112°. All this implies that no major rehybridization of the atoms of CS_2 has occurred. Also, the variations of the bond distances and angles of the $M-\eta^2-CS_2$ unit are not large. In general, there is only a combined shortening and lengthening of the coordinated and uncoordinated C-S linkages, respectively. Incidentally, the effect seems particularly large in the niobium complex shown in **Fig. 10** but is likely unrealistic on account of the large standard deviations reported. A rationale for these trends will be presented in Section 4.4. Concerning the geometrical variations within the M-C-S triangle, the M-S bond seems constant, whereas there is a progressive shortening of the M-C bond in compounds of 1B and 1E with respect to the monomers 1A. Compare in particular the M-C distance of 1.899(3) Å in the alkylated iron complex vs. 1.983(8) Å in the precursor $Fe(\eta^2-CS_2)(PMe_3)(PPh_3)(CO)_2$. The latter shortening has been attributed (ref. 24) to an increase of the Fe-C double bond character. The argument is critically discussed in Section 4.2 and 4.4. Also, notice that, in $V(\eta^2-CS_2)(Cp)_2$ (ref. 19) (**Fig. 6**) and in its methylated derivative of type 1E (ref. 19) (**Fig. 12**), there is no shortening of the V-C bond (2.07(4) vs. 2.08(1) Å). Significantly, however, the shortening of the C-S(endo) and the lengthening of the C-S(exo) bonds are observed also in this pair of vanadium structures. Notice that these two vanadium compounds represent one of the examples in which the engagement of

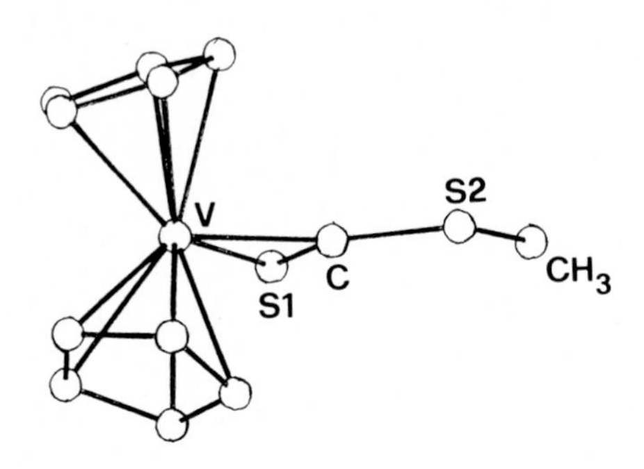

Fig. 12. $[V(\eta^2-CS_2Me)(Cp)_2]^+$ (CDFC)

the sulfur(exo) in bonding to electrophiles makes the C-S(endo) bond even shorter than its C-S(exo) analog (1.63(1) vs. 1.66(1) Å).

Another example of an alkylated $M-\eta^2-CS_2$ moiety is found in the complex cation $[Ru(\eta^2-CS_2Me)(CO)_2(PPh_3)_2]^+$ (ref. 25). The supporting metal fragment is of type ML_4, in complete analogy with that already found in the previous series of $Fe-\eta^2-CS_2$ complexes. Apparently, there are no unusual features in this structure although it must be properly considered that this represents the only known example of a $Ru-\eta^2-CS_2$ species.

The series of dimeric compounds $[M(PR_3)(CS_2)]_2$ (M = Ni (ref. 26), Pd (ref. 27), Pt (ref. 28); PR_3 = PPh_3, $PPh(t\text{-}Bu)_2$), are described as belonging to class 1C. Here, two $M(PR_3)$ fragments, related by a crystallographic center of inversion, are held together by two CS_2 groups in such a way as to form a six-membered M-S-C-M-S-C ring, see **Fig. 13**. The arrangement is such that each CS_2 molecule is η^2-bonded to one metal and σ-bonded to the other.

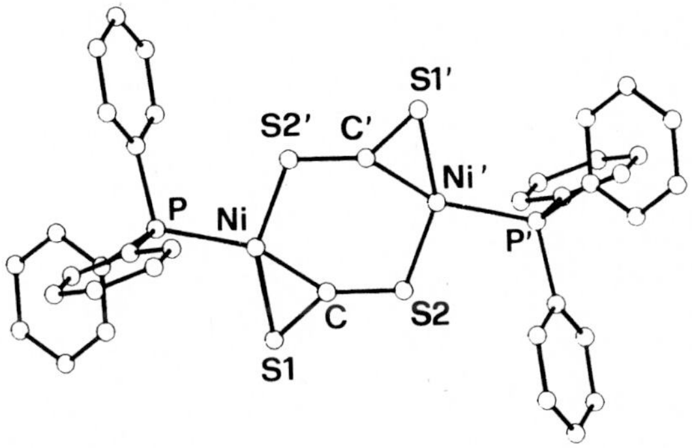

Fig. 13. $[Ni(PPh_3)(CS_2)]_2$

Substantially, these 1C compounds are not very different from those of Class 1B in the sense that the exocyclic sulfur atom of a $M\text{-}\eta^2\text{-}CS_2$ moiety is used as a σ ligand toward another metal, which, in turn, is able to stabilize a dihapto-coordination of CS_2. In going from nickel to platinum, there is both an elongation of the exocyclic C-S vector (range 1.63(1)-1.67(6) Å) and a shrinkage of the η^2-bonded analog (range 1.68(1)-1.65(6)Å). Also, the S-C-S angle bends progressively (range 137.3-133.4°). A rationale for the observed trends, based on the different basicity of the metal centers, is attempted in Section 4.4.

Finally, the compounds of class 1D are limited to an unique example, namely the complex dication $[(triphos)Co(\mu\text{-}CS_2)Co\text{-}$

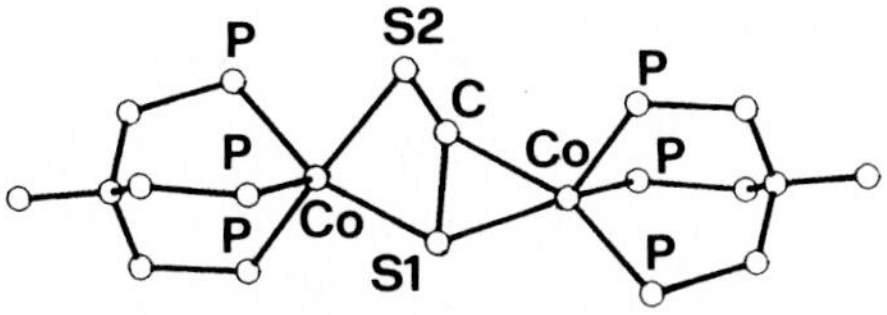

Fig. 14. $[(triphos)Co(\mu\text{-}CS_2)Co(triphos)]^{2+}$

(triphos)]$^{2+}$ (ref. 18), shown in **Fig. 14.** Here carbon disulfide, dihapto-bonded to one Co(triphos) fragment, chelates an identical fragment with its two sulfur atoms. The solution of the structure presented some problems due to the presence of a crystallographic center of inversion, midway between the two metals. Accordingly, a model, in which CS_2 is dihapto-bonded 50% to one metal and 50% to its centrosymmetric image, was used and successfully refined. This type of disorder has been recently reported also for an SCO grouping which bridges two centrosymmetrically related $Co_3(CO)_9C$ units (ref. 29).

The two S-Co bonds, which the sulfur atoms of the Co-η^2-CS_2 unit form with the other cobalt atom, are similar (2.27(1) vs. 2.31(1) Å). If the latter difference is reliable (recall the disorder problem), the shorter bond unexpectedly involves the sulfur atom which bridges directly the two metals. In conclusion, the residual sulfur bonding capabilities of M-η^2-CS_2 moieties appear to be almost equivalent. The planarity of the Co(μ-CS_2)Co bridged framework and the diamagnetism of the dimer, which results from the combination of the two paramagnetic centers (triphos)Co(η^2-CS_2) and (triphos)Co^{2+}, are both indicative of $\pi_\perp$ electron coupling effects through the CS_2 bridge. Qualitative MO arguments which support this viewpoint are presented in Section 4.2. Remarkable geometrical trends observed in the compound are the long C-S and short Co-S distances as well as the small S-C-S angle. The latter magnitude is typical of M-η^1-CS_2 compounds (see next Section), where the sulfur atoms are engaged in other bonds besides that with the carbon atom.

2.3 STRUCTURES OF M-η^1-CS_2 COMPLEXES

Although mononuclear η^1-CS_2 complexes are known in solution (see Section 3.5), no example of solid state structures containing a η^1-C-bonded carbon disulfide molecule in the intact form has been reported as yet.

Conversely, stable crystalline samples are obtainable when one or both sulfur atoms of the triatomic utilize the localized lone pair to make a new bond to another grouping. Simple VSEPR (ref. 30) arguments predict reduced repulsions between the two sulfur atoms when their lone pairs are not free but engaged in bonding interactions. Also, the qualitative MO theory (see Section 4.3) easily accounts for the fact. Practically, all of the complexes 2A–2E (**Table 2**) can be seen in this light, although only the species 2A–2C can be considered real coordination compounds of CS_2.

The origin of the compound $(Cp)(CO)_2FeC(\overline{S)SFe(Cp)(CO)_2W}(CO)_5$ (ref. 31) supports the latter observation. In fact, this trinuclear species, shown in **Fig. 15**, is obtained in a stepwise reaction starting from the mononuclear anion $(Cp)(CO)_2Fe(\eta^1\text{-}CS_2)^-$, which, in this form, was never crystallized (see Section 3.5).

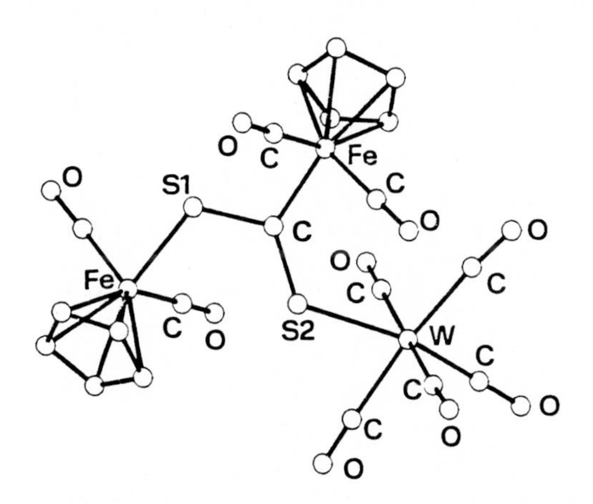

Fig. 15. $(Cp)(CO)_2FeC(\overline{S)SFe(Cp)(CO)_2W}(CO)_5$ (CDFC)

The most striking structural difference between $M\text{-}(\eta^1\text{-}CS_2)$ and $M\text{-}(\eta^2\text{-}CS_2)$ moieties is the pronounced bending of the S–C–S angle in the former species. In the trinuclear compound of **Fig. 15**, the angle measures 119.2(2)°, about 15–20° less than the corresponding angles in compounds 1A, 1B, 1C, 1E (only the dimer of class 1D represents a noticeable exception). In most of the other η^1 complexes, presented below, the difference is even more pronounced. On the other hand, the C–S bond lengths are only slightly elongated with respect to $M\text{-}\eta^2\text{-}CS_2$ compounds, so that they are still intermediate between double and single bond values. It is noteworthy that, in the compound of **Fig. 15**, the C–S bond, which leads to the $W(CO)_5$ fragment, is slightly shorter than the analogous bond connecting the second $Fe(Cp)(CO)_2$ fragment of the structure (1.715(4) vs. 1.670(5) Å). The authors suggest that this is a consequence of the better σ acceptor capabilities of $W(CO)_5$ with respect to those of the fragment $Fe(Cp)(CO)_2$.

Beside the fragment of type ML_5 which, in the previous compound, was found capable of supporting η^1-coordination of CS_2, planar ML_3 fragments have also the same capability. This is demonstrated by the structure of the binuclear cation $[(PPh_3)_2ClPt(\mu-CS_2)-Pt(PPh_3)_2]^+$ (ref. 32), shown in **Fig. 16**. Again, the sulfur atoms are both engaged in external bonds, since CS_2 chelates the metal fragment $Pt(PPh_3)_2$. At variance with what was found in the previous compound of type 2A, the two C-S linkages (1.71(2) vs. 1.69(2) Å) are practically equal on account of the equivalent donor functions exerted by the two sulfur atoms.

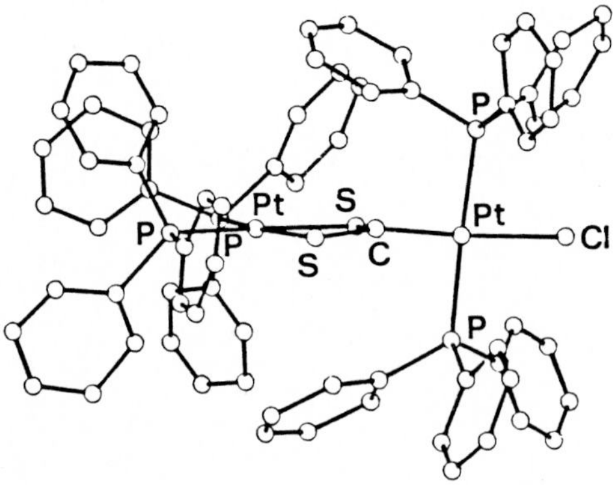

Fig. 16. $[(PPh_3)_2ClPt(\mu-CS_2)Pt(PPh_3)_2]^+$ (CDFC)

The binuclear compound $[PtCl(dppm)]_2(CS_2)$ (ref. 33), classified as 2C, can be considered an A-frame complex (ref. 34-36), whose apex is unusually occupied by a CS_2 molecule. As shown in **Fig. 17**, CS_2 uses the carbon and one sulfur atom to bridge the two metal centers, which ultimately are both square-planarly coordinated. Notice that the metal fragment, which supports in this case the η^1-mode, is again of the type ML_3 (planar), where the ligands L are represented by two P and one Cl atoms. The structure shows that the η^1-coordination mode can be stabilized also if one sulfur atom of CS_2 remains free from linkages with external groups. No fine detail of the geometry of the $Pt(\mu-CS_2)Pt$ framework can be critically discussed on account of the very bad quality of the reported crystallographic data. In addition, significant data, such as the S-C-S angle, are missing.

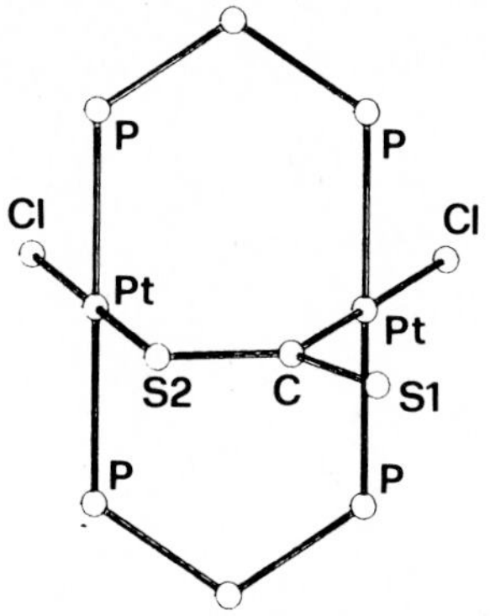

Fig. 17. $[PtCl(dppm)]_2(CS_2)$ (Approximate model reconstructed from available geometrical data)

The compound $OsH(\eta^1-CS_2Me)(CO)_2(PPh_3)_2$ (ref. 36), shown in **Fig. 18**, is included in the series of $M-\eta^1-CS_2$ complexes in spite of its metal-dithiomethyl ester nature. Again, we take into account the isolobal analogy between certain transition metal fragments and alkyl groups (ref. 9). As in the previous structure (**Fig. 17**), one sulfur atoms is in a terminal position. The bond formed by this

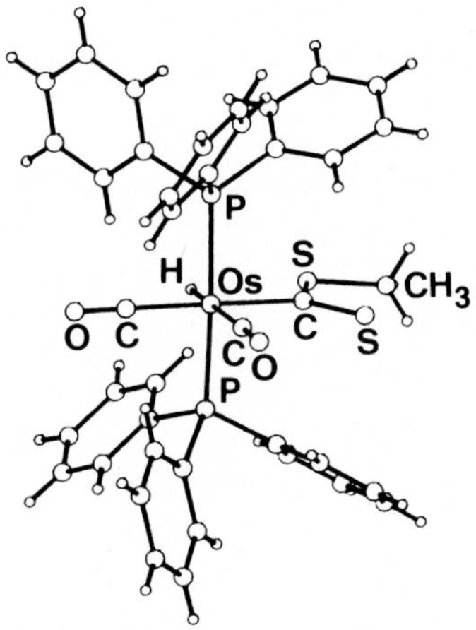

Fig. 18. $OsH(\eta^1-CS_2Me)(CO)_2(PPh_3)_2$ (CDFC)

free sulfur atom with carbon is substantially shorter than the C-S bond leading to the alkyl group (1.648(4) vs. 1.724(5) Å). Accordingly, a larger percentage of double bond character can be assigned to the former C-S linkage. The M-C distance of 2.137(5) Å is, in any case, longer than the corresponding distances observed in some Os-carbene structures (ref. 38-39). The latter may be taken as an indication that the M-C double bond character is less important in these M-CS_2 carbenoid species.

Finally, we report the compounds 2E, which only indirectly can be referred to as CS_2 complexes. In fact, the triatomic here is part of heterocycles which may or may not contain also a metal atom. In a sense, the addition of an activated alkyne to an M-CS_2 moiety, to give a η^1-$\overline{\text{C-S-C(R)-C(R)-S}}$ cycle, can be envisaged in the same light as the addition of alkyl groups to give the compounds 1E or 2D. By using the same standards, other types of cycles, which contain the M-η^1-CS_2 moiety, have been also introduced in our list.

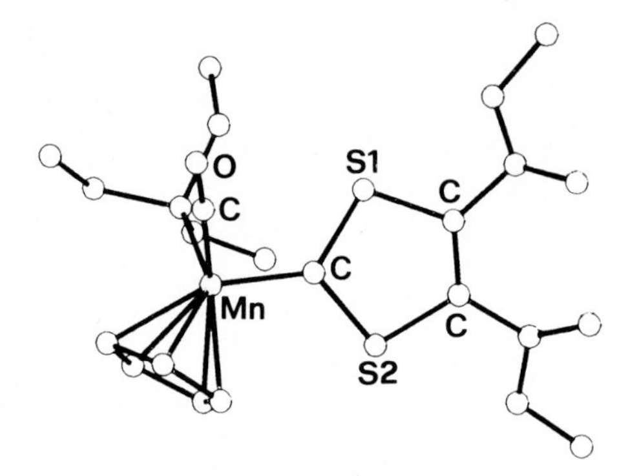

Fig.19. Mn[$CS_2C_2(CO_2Me)_2$](CO)[P(OMe)$_3$](Cp) (CDFC)

Let's describe these miscellaneous CS_2 containing heterocycles in some detail.

The structure of Mn[$CS_2C_2(CO_2Me)_2$](CO)[P(OMe)$_3$](Cp) (ref. 39), shown in **Fig. 19**, contains a 1,3-dithiol-2-ylidene ligand. Incidentally, the same coordination mode of an equivalent cycle is found in the structure of the cluster $Fe_3(CS_2C_2H_2)(CO)_8S_2$ (ref. 40). However, the latter compound will not be discussed here, since it was not obtained directly from a metal-CS_2 precursor. Conversely, the compound of **Fig. 19** is obtainable from the reaction of the complex Mn(η^2-CS_2)(CO)[P(OMe)$_3$](Cp) with dimethylacetylene dicarboxylate. The Mn-η^2-CS_2 precursor (ref. 41) has not been structurally characterized, but there is little doubt that the η^2-coordination mode enhances the 1,3-dipole character of CS_2 so that the reaction displays the typical character of 1,3-dipolar cycloadditions. This point will be discussed in Sections 3.2 and 4.5. Here, we like to point out some geometrical features of the $M\text{-}\eta^1\text{-}CS_2$ grouping and of the whole C-S-C-C-S cycle as well. Both the C-S bonds are significantly elongated (1.732(2) and 1.753(2) Å) with respect to any other structurally characterized $M\text{-}CS_2$ grouping in any coordination mode (at least relative to crystallographically reliable results). The latter values, almost indicative of C-S single bonds, together with a C-C bond length of 1.343(3) Å, typical of C-C double bond, seem to exclude any aromaticity of the five-membered ring. This would allow some Mn-C double bond character in the complex. The length of the Mn-C bond (1.876(2) Å) matches perfectly the values reported for several Mn-carbene complexes of the fragment $Mn(CO)_2(Cp)$ (ref. 42). Partial Mn-C double bond character has also been predicted for the latter complexes because the metal fragment in question is known for its strong π donor capabilities when compared with those of other $d^6\text{-}ML_5$ fragments (ref. 42). Notice that, on account of its favorable π-donor capabilities, $Mn(CO)_2(Cp)$ is also able to support η^2 coordination of CS_2. This is an important point, since 1,3-dithiol-2-ylidene complexes appear to be supported only by metal fragments which typically form $\eta^2\text{-}CS_2$ complexes (see Section 4.5).

The two rhodium compounds $Rh(C_2S_4)(PMe_3)(Cp)$ (ref. 43), **Fig. 20**, and $Rh_2Cl_2(CO)(C_2S_4)(dppm)_2$ (ref. 44), **Fig. 21**, share the common

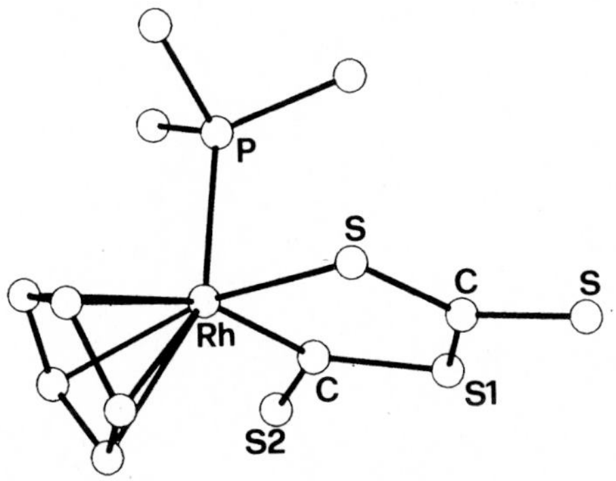

Fig. 20. $Rh(C_2S_4)(PMe_3)(Cp)$ (CDFC)

feature of containing the five-membered $\overline{Rh-C-S-C-S}$ metallocycle, which results from a head-to-tail dimerization of CS_2. A Rh-η^1-CS_2 moiety can be envisaged within this cycle. One sulfur atom of the latter moiety is linked to the carbon of a second CS_2 molecule. Finally, the ring is closed by a Rh-S linkage. In $Rh(C_2S_4)(PMe_3)$-(Cp), two exocyclic sulfur atoms of the $Rh(C_2S_4)$ grouping are not involved in other bonds. Conversely, one exocyclic sulfur atom acts as a ligand toward a second metal center in $Rh_2Cl_2(CO)(C_2S_4)$-$(dppm)_2$. In this case, the η^1-coordinated CS_2 molecule is both, part of the metallo-ring and a bridge between two Rh metals. The bridging mode is quite similar to that observed in the compound $[PtCl(dppm)]_2(CS_2)$ (ref. 33) (category 2C). Notice that the metal fragment, supporting η^1-coordination of CS_2, attains an ML_5 character through the formation of a Rh-Rh bond and through the action of one sulfur atom of the C_2S_4 unit, which acts as a ligand.

An ML_5 fragment is also recognizable in the other rhodium species of **Fig. 20**. In order to check if some π electron delocalization occurs within the $\overline{Rh-C-S-C-S}$ ring, the M-η^1-CS_2 moiety in these rhodium compounds can conveniently be compared with that of other non-cyclic systems. For instance, in $Rh(C_2S_4)$-$(PMe_3)(Cp)$, the exocyclic C-S(2) bond (1.65(2) Å) is practically equal to that (1.648(4) Å) found in the complex OsH(η^1-CS_2Me)-

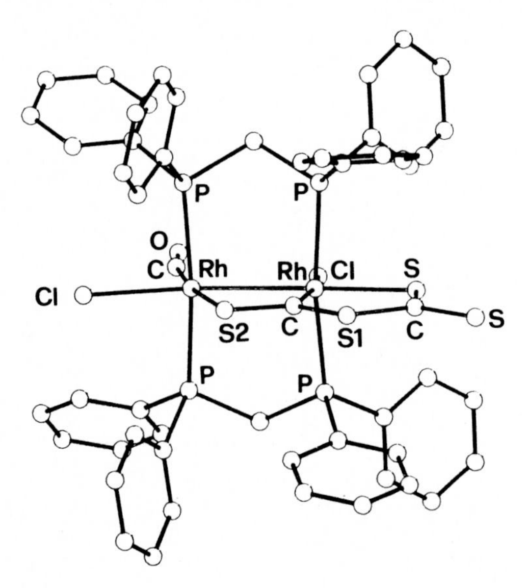

Fig. 21. $Rh_2Cl_2(C_2S_4)(dppm)_2(CO)$ (CDFC)

$(CO)_2(PPh_3)_2$ (ref. 36) (**Fig. 18**), whereas the adjacent C-S(1) bond is shorter when part of the metallocycle (1.68(2) vs. 1.724(5) Å). By contrast, in $Rh_2Cl_2(CO)(C_2S_4)(dppm)_2$, the C-S(1) bond in the metallocycle is the longest found (1.75(2) Å). However, the latter rhodium structure contains two independent complex cations in the asymmetric unit, whose structural features are somewhat different. Note, for example, that the equivalent Rh-C and the C-S(2) bonds of the two molecules (**Table 2**) differ by more than the associated standard deviations. Under these circumstances, any conclusion can be hardly drawn about the amount of π delocalization in $\overline{\text{Rh-C-S-C-S}}$ cycles.

The last structure of our survey (**Fig. 22**) represents an unique example of a trithioanhydride complex of transition metals. The synthetic strategy to obtain the cation $[(Cp)(CO)_2\overline{\text{FeSC(Fp)SCS}}\text{-}(Fp)]^+$, Fp = $Fe(CO)_2(Cp)$ (ref. 45), is reported in Section 3.6. Here, we like to mention the nature of the important precursor, namely the species $Fe(CO)_2(Cp)CS_2ML_n$, which, not known structurally, is very likely comparable to the complexes of type 2D having an alkyl group in place of the electrophilic ML_n fragment. The cation of **Fig. 22** is characterized by a C_2S_3 skeleton acting as tetradentate ligand toward three different iron centers. Presently, we are mainly interested in the η^1-coordination mode of CS_2 supported by the fragment $Fe(CO)_2(Cp)$, a situation already found in

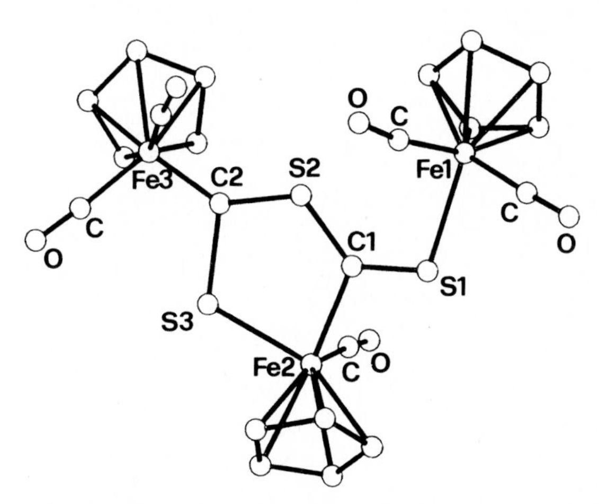

Fig. 22. $[(Cp)(CO)_2\overline{\text{FeSC(Fp)SCS}}(Fp)]^+$ **(Coordinates kindly supplied by the authors (ref. 45))**

the prototype compound 2A. Interestingly, the inner five-membered cycle C-S-C-Fe-S resembles the structure of 1,3-dithiol-2-ylidene rings, if the alkyne is thought of as being substituted for a more complex binuclear organometallic fragment. Unfortunately, a direct comparison based on geometrical parameters is not possible due to the very poor quality of the structure of $[(Cp)(CO)_2FeSC(Fp)SCS(Fp)]^+$.

3 REACTIVITY OF CARBON DISULFIDE METAL COMPLEXES

3.1 η^2-C,S BONDED CS_2 COMPLEXES

With very few exceptions, dihapto-CS_2 complexes are synthesized by reaction of CS_2 with metal-ligand fragments containing a nucleophilic metal center. Generally, low valent transition metal complexes with phosphines, phosphites or cyclopentadienyl as additional ligands, are suitable starting compounds to form η^2-CS_2 derivatives. An alternative synthetic route to η^2-CS_2 compounds is the reduction with Na or $NaBH_4$ of high valent metal complexes containing bidentate 1,1-dithio ligands such as trithiocarbonates, CS_3^{2-} and CS_2R^-, or phosphoniodithiocarboxylates, R_3PCS_2. This method, however, has so far been used to prepare only two η^2-CS_2 complexes, namely (triphos)Co(η^2-CS_2) (ref. 46,47), and (triphos)Ni(η^2-CS_2) (ref. 48,49).

As a consequence of the dihapto-coordination, every atom and bond of the metal-η^2-CS_2 moiety becomes a potentially reactive site. Indeed, the reaction possibilities of η^2-CS_2 metal complexes are so numerous and various that it is difficult to classify them. This is particularly true when considering the great variety of products obtained from the reactions of η^2-CS_2 complexes. For this reason, we prefer to classify the reactions of these compounds according to the type of reagent employed. As a warning to the reader, in most of the formulas used throughout this section, the bonds are simply meant to represent the connectivity between the atoms. In general, the high electron delocalization in these molecules requires specific considerations, case by case. The argument is developed in some detail in Section 4.2.

3.2 REACTIONS WITH ELECTROPHILES

3.2.1 Reactions with alkylating agents

The uncoordinated sulfur atom of the η^2-CS_2 group is a nucleophilic center susceptible to attack by electrophiles. Alkylation of the exocyclic sulfur atom may be accomplished by using any sort of alkylating agents and, generally, results in the formation of cationic η^2-dithioalkyl ester ligands (ref. 19,23,24, 50-61) (1). η^2-CS_2R complexes are characterized by a strong or medium IR absorption between 1100 and 1150 cm^{-1}, assigned to ν(C=S). A small decrease of the C=S stretching frequency is usually observed on going from η^2-CS_2 complexes to the corresponding η^2-CS_2R derivatives, which, usually, do not show the low

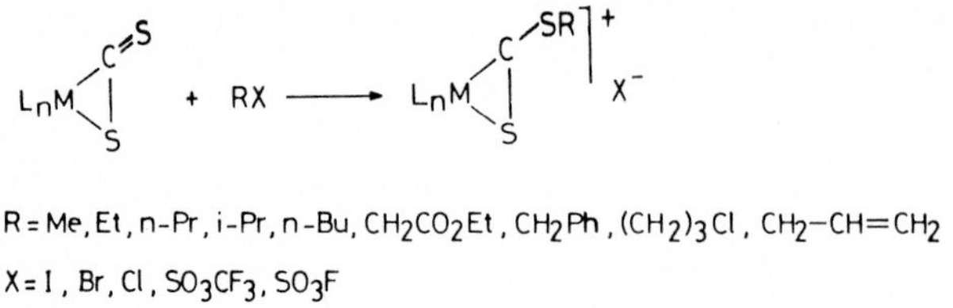

R = Me, Et, n-Pr, i-Pr, n-Bu, CH_2CO_2Et, CH_2Ph, $(CH_2)_3Cl$, CH_2–CH=CH_2
X = I, Br, Cl, SO_3CF_3, SO_3F

(1)

ν(C–S) stretching vibration in the range 650–620 cm^{-1}, always present in the IR spectra of η^2-CS_2 complexes. X-ray crystal structures have been reported for the complexes [Ru(η^2-CS_2Me)-$(CO)_2(PPh_3)_2$]ClO_4 (ref. 25), [V$(Cp)_2$(η^2-CS_2Me)]I_3 (ref. 19) and [Fe(η^2-CS_2CH_2Ph)$(CO)_2(PMe_3)_2$]PF_6 (ref. 24). A particular case is the methylation by $MeOSO_2F$ of the two complexes (np_3)Co(η^2-CS_2) (ref. 62) and (np_3)Ni(η^2-CS_2) (ref. 63). The formation of the η^2-dithiomethyl ester ligand is accompanied by the rearrangement of the np_3 coligand. As a result, the previously uncoordinated nitrogen atom approaches the metal center, while a phosphine group of np_3 attacks the carbon atom of the η^2-CS_2Me ligand to give an η^2-phosphoniodithiomethyl ester ligand (2). The crystal structure of the nickel derivative has been established by X-ray methods (ref. 63). The geometric requirements of the np_3 ligand should play

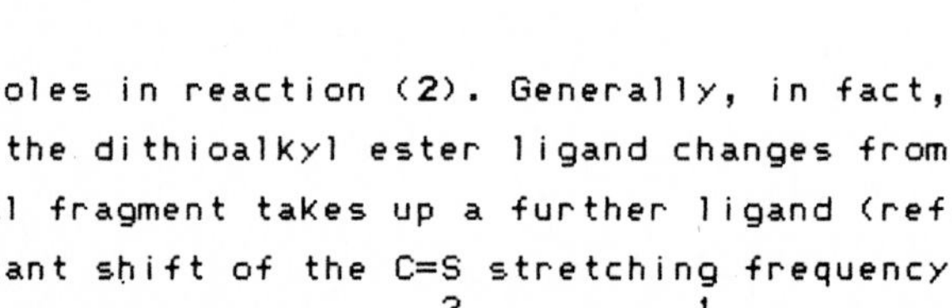

(2)

unique roles in reaction (2). Generally, in fact, the coordination mode of the dithioalkyl ester ligand changes from η^2 to η^1 while the metal fragment takes up a further ligand (ref. 25,52,55). A significant shift of the C=S stretching frequency to ca. 1000 cm^{-1} is observed on going from η^2-CS_2R to η^1-CS_2R. A few examples of η^1-CS_2R complexes obtained through this route have been reported by Roper (ref. 25,52) (3), and Werner (ref. 4) (4).

Both η^2-CS_2R and η^1-CS_2R compounds can function as precursors to thiocarbonyl complexes. The conversion to thiocarbonyls is achieved via alkylthiol elimination by reaction of dithioalkyl

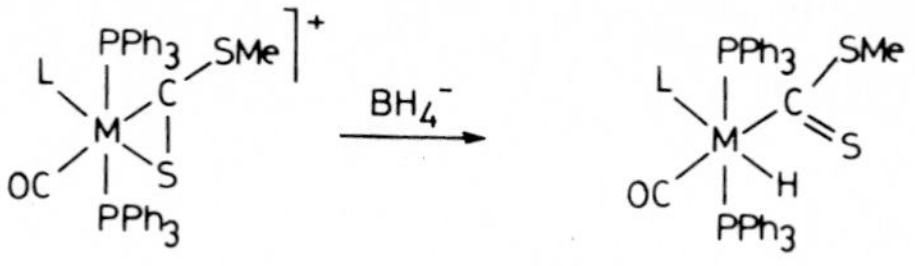

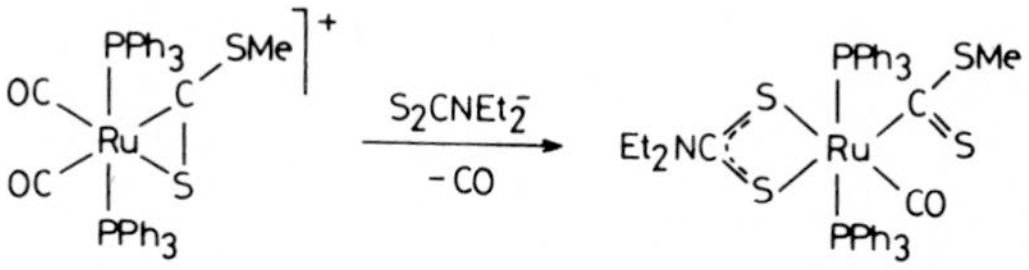

M = Os, L = CO; M = Ir, L = Cl

(3)

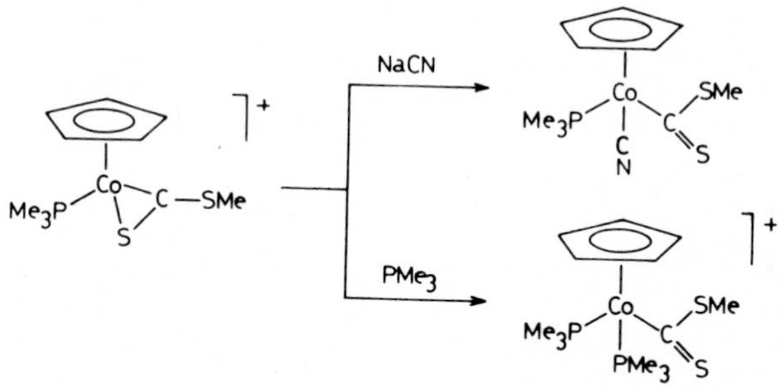

(4)

ester complexes with an acid (ref. 25,51) (5) or hydride ion (ref. 55,52) (6).

$$[Ru(\eta^2\text{-}CS_2Me)(CO)_2(PPh_3)_2]^+ \xrightarrow{HCl} RuCl_2(CS)(CO)(PPh_3)_2 + MeSH$$

$$Ru(\eta^1\text{-}CS_2Me)(S_2CNEt_2)(CO)(PPh_3)_2 \xrightarrow{HCl} Ru(CS)(S_2CNEt_2)(CO)(PPh_3)_2 + MeSH$$

(5)

$$Pd(\eta^2\text{-}CS_2Me)(PMe_3)_2 \xrightarrow{NaH} Pd(CS)(PMe_3)_2 + MeSH$$

$$OsH(\eta^1\text{-}CS_2Me)(CO)_2(PPh_3)_2 \xrightarrow{\Delta} Os(CS)(CO)_2(PPh_3)_2 + MeSH$$

(6)

Interestingly, the thiocarbonyl complex $[(np_3)Co(CS)]BPh_4$ is obtained by treatment of the η^2-phosphoniodithiomethyl ester complex $[(np_3)Co(CS_2Me)]BPh_4$ with $NaBH_4$ (ref. 64) (see Section 3.3.2). Alkylation of the exocyclic sulfur atom is, therefore, an important, although not essential, step for the conversion of η^2-CS_2 to CS. Other routes to thiocarbonyls from η^2-CS_2 complexes, which do not require the preliminary alkylation of the exo-sulfur atom, are described in Section 3.3

The intermediacy of a dithiomethyl ester complex has been found to occur in the metal-promoted conversion of η^2-CS_2 to η^2-H_2CS reported below (ref. 65) (7).

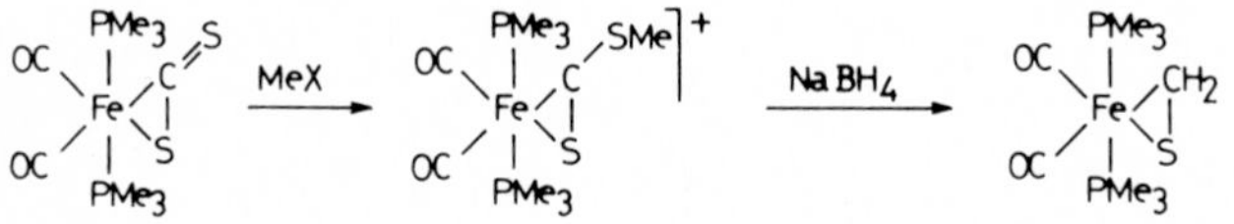

(7)

Dithioalkyl esters are not the only compounds obtainable from the reaction of η^2-CS_2 complexes with alkylating agents. When an excess of the latter reagent is used, the double alkylation of the metal-η^2-CS_2 moiety may occur (ref. 25,51,66). As a result, dithiocarbene complexes are obtained (8). The IR spectra exhibit absorptions in the range 990-940 cm^{-1}. Examples are given in (9)

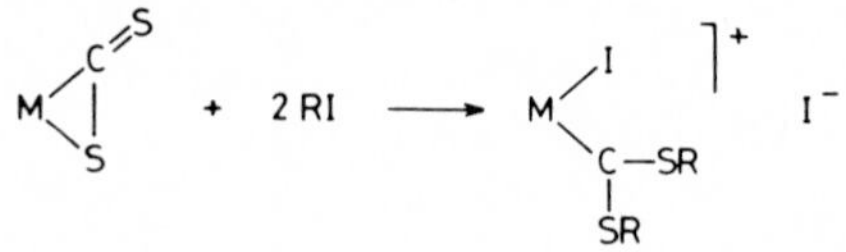

(8)

(ref. 51) and in (**10**) (ref. 66) (R = Me, Et). This route to

$$Os(\eta^2\text{-}CS_2)(CO)_2(PPh_3)_2 + 2\ RI \longrightarrow [OsI\{C(SR)_2\}(CO)_2(PPh_3)_2]I$$

(9)

$$Pt(\eta^2\text{-}CS_2)(PPh_3)_2 + 2\ RI \longrightarrow [PtI\{C(SR)_2\}(PPh_3)_2]I$$

(10)

dithiocarbene ligands has been considered to proceed through electrophilic attack by alkyl halides at both sulfur atoms of the metal-η^2-CS_2 moiety (ref. 66). Such an interpretation could implicitly suggest that not only the uncoordinated sulfur atom, but also the one involved in dihapto bonding to the metal must be considered as nucleophilic centers. Although no alkylation product of type **A** or **B** has been isolated, the binuclear complex cation (**11**) (ref. 18), where both the sulfur atoms of CS_2 coordinate the rightmost cobalt atom, could indirectly support this hypothesis. However, the electronic structure of (**11**) is exceptional (see Section 4.4), and no definite conclusions about the nucleophilic character of the sulfur atom involved in dihapto bonding should be drawn. By contrast, it has been clearly demonstrated that the

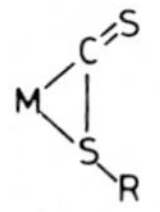

A 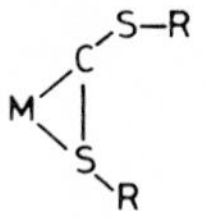 B 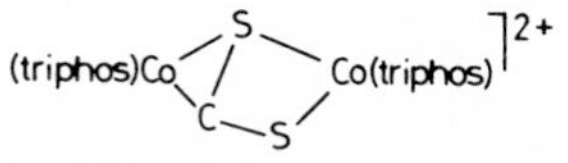 (11)

alkylation of η^1-CS_2R ligands to give dithiocarbene $C(SR)_2$ ligands is an easy process (ref. 25). Since the products of reactions (**9**) and (**10**) contain the iodide ligand, a reasonable reaction mechanism for their formation could involve the alkylation of the exocyclic sulfur, followed by coordination of the iodide ion to the metal to give an η^1-CS_2R intermediate. Further alkylation of the latter group should produce the dithiocarbene ligand. A stepwise reaction of this type has been exploited to synthesize the dithiocarbene complex $[Ru\{C(SMe)_2\}(CO)(PPh_3)_2(S_2CNEt_2)]^+$ from $Ru(CO)_2(PPh_3)_2$-(η^2-CS_2) (ref. 25) (**12**).

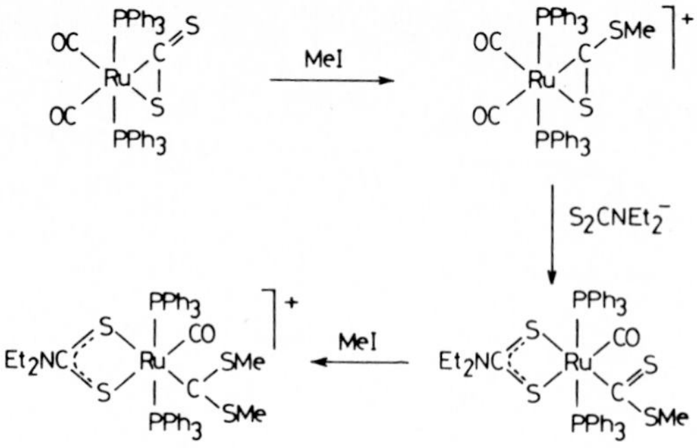

(12)

Dithiocarbene derivatives are obtained by reaction of $Os(\eta^2-CS_2)(CO)_2(PPh_3)_2$, $Os(\eta^2-CS_2)(CO)(CNR)(PPh_3)_2$, or $Ru(\eta^2-CS_2)(CO)(CNR)(PPh_3)_2$ (R = p-tolyl) with 1,2-dibromoethane (ref. 67). Also, it has been suggested that these reactions proceed via η^1-dithioalkyl ester intermediates. The corresponding (1,3-dithiol-2-ylidene)derivatives have been isolated as ClO_4^- salts. The stepwise reaction sequence shown in (13) has been proposed.

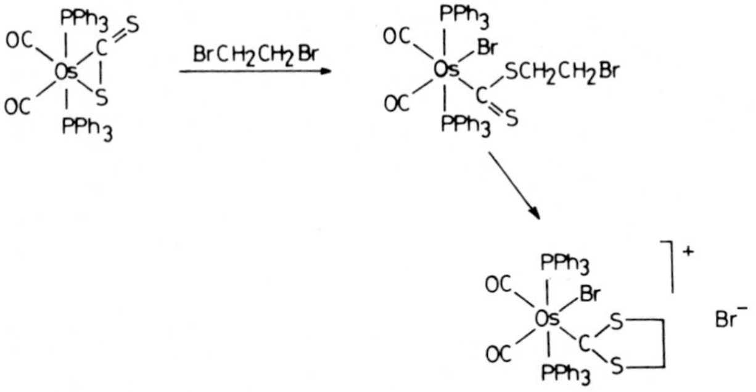

(13)

The formation of dithioheterocyclic ligands is not the only possible reaction for dihaloalkanes with η^2-CS_2 complexes. Depending on the length of the alkane chain and on the reaction conditions, different products are obtained. For example, $Ru(\eta^2-CS_2)(CO)_2(PPh_3)_2$ reacts with 1,2-dibromoethane to give the neutral dithiocarbene complex $RuBr_2(CO)(PPh_3)_2(CSCH_2CH_2S)$, whereas with methylene bromide, 1,3-dibromopropane or 1,4-dibromobutane, the monoalkylated neutral complexes $RuBr(CO)(PPh_3)_2(CS_2(CH_2)_nBr)$ (n = 1, 3 or 4), are formed (ref. 3).

Bridging bis(dithioalkyl ester)complexes can be obtained by reacting bifunctional alkyl halides with η^2-CS_2 compounds, as recently reported by Amaudrut et al. (ref. 61). The dimer $(Cp)_2(Bu)Nb(C(S)SCH_2S(S)C)Nb(Bu)(Cp)_2$ (**14**) is formed by reacting $Nb(Cp)_2(Bu)(\eta^2$-$CS_2)$ with diiodomethane.

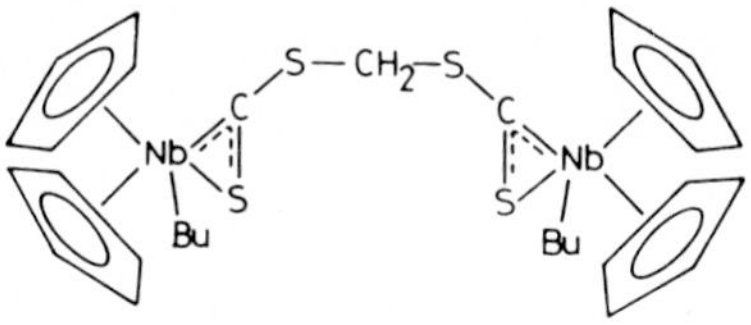

(**14**)

Finally, it is interesting to report the multifarious behaviour of $Pt(\eta^2$-$CS_2)(PPh_3)_2$ with alkylating agents. As already mentioned, this complex is doubly alkylated after treatment with an excess of MeI (ref. 66). When a stoichiometric amount of the latter is used, the formation of a bridging-CS_2 dimer occurs (ref. 68) (**15**).

$$\mathbf{Pt(\eta^2\text{-}CS_2)(PPh_3)_2 + MeI \longrightarrow [(PPh_3)_2IPt\text{-}C(S)_2Pt(PPh_3)_2]I}$$

(**15**)

Analogously, the reaction of a stoichiometric amount of 1,2-diiodoethane also gives a μ-CS_2 complex (ref. 68) (**16**).

$$\mathbf{Pt(\eta^2\text{-}CS_2)(PPh_3)_2 + C_2H_4I_2 \longrightarrow (PPh_3)_2IPt\text{-}C(S)_2PtI(PPh_3)}$$

(**16**)

The great stability of platiniodithiocarboxylato-(S,S′)platinum type complexes has been suggested to be the driving force for the existence of these compounds. Since their generation requires halide transfer to the CS_2-carrying platinum atom, any halogen source (MeI or EtI_2) promotes the dimerization. Monoalkylation of $Pt(\eta^2$-$CS_2)(PPh_3)_2$ to the η^2-CS_2R derivative is, conversely, achieved by action of alkylating agents with non-coordinating anions ($[OEt_3]BF_4$, $MeOSO_2CF_3$) (ref. 68).

3.2.2 Reactions with activated alkynes and olefins

Depending on the nature of the metal fragment and of the coligands, acetylenes with electron-withdrawing substituents (electrophilic acetylenes) can add to the metal-η^2-CS_2 moiety in two different ways (17).

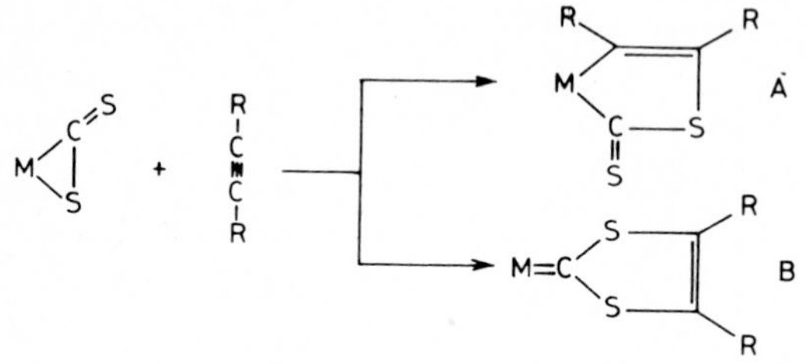

(17)

Only two examples of heterocyclic five-membered metallo-ring complexes of type **A** are known, $\overline{RhC(CO_2Me)C(CO_2Me)SC}(S)(Cp)(PPh_3)$ (ref. 69), and $\overline{FeC(CO_2Me)C(CO_2Me)SC}(S)L_2(CO)_2$ (L = PMe_2Ph, PMe_3) (ref. 70), obtained by reacting the corresponding η^2-CS_2 compounds with dimethyl acetylenedicarboxylate. Many more are the (1,3-dithiol-2-ylidene)complexes of type **B**. At least five different η^2-CS_2 complexes have been found to react with activated alkynes yielding dithiocarbenes: $Fe(\eta^2\text{-}CS_2)(CO)_2L_2$ (L = $P(OMe)_3$, PPh_3, PMe_2Ph, PMe_3) (ref. 70-72), $Mn(\eta^2\text{-}CS_2)(CO)L(C_5H_4R)$ (L = $P(OMe)_3$, PMe_2Ph, PMe_3; R = H, Me) (ref. 39), (triphos)$Ni(\eta^2\text{-}CS_2)$

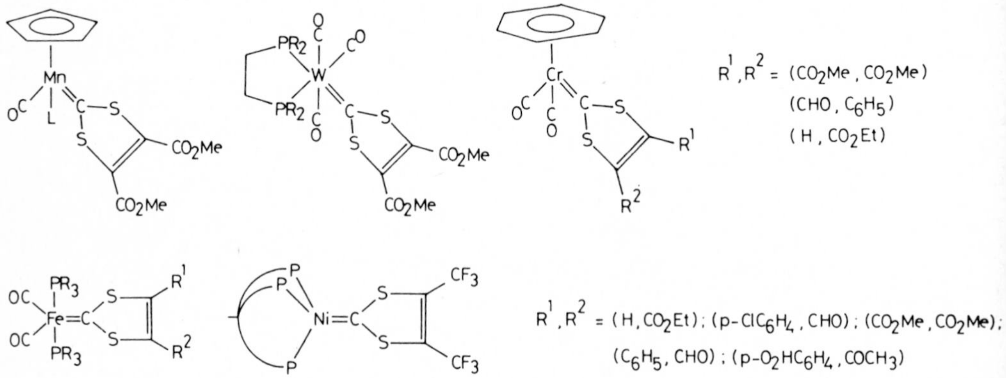

R^1,R^2 = (CO_2Me, CO_2Me)
(CHO, C_6H_5)
(H, CO_2Et)

R^1,R^2 = (H,CO_2Et); (p-ClC_6H_4, CHO); (CO_2Me, CO_2Me); (C_6H_5, CHO); (p-$O_2HC_6H_4$, $COCH_3$)

(18)

(ref. 73,74), $Cr(\eta^2\text{-}CS_2)(\eta^6\text{-arene})(CO)_2$ (arene = mesytilene, anisole, o-xylene, m-xylene) (ref. 75), and $W(CO)_3(R_2PC_2H_4PR_2)$-$(\eta^2\text{-}CS_2)$ (R = Me, Ph) (ref. 54) (**18**). Of the (1,3-dithiol-2--ylidene)compounds shown below, only the structure of the Mn derivative with L = $P(OMe)_3$ has been elucidated by means of an X-ray analysis (ref. 39).

It is likely that the energies of the two types of rings (**A** and **B**) are comparable. Clear-cut support for this hypothesis comes from the reaction of dimethyl acetylenedicarboxylate with $Fe(\eta^2\text{-}CS_2)(CO)_2L_2$ reported by Dixneuf et al. (ref. 70,76) (**19**).

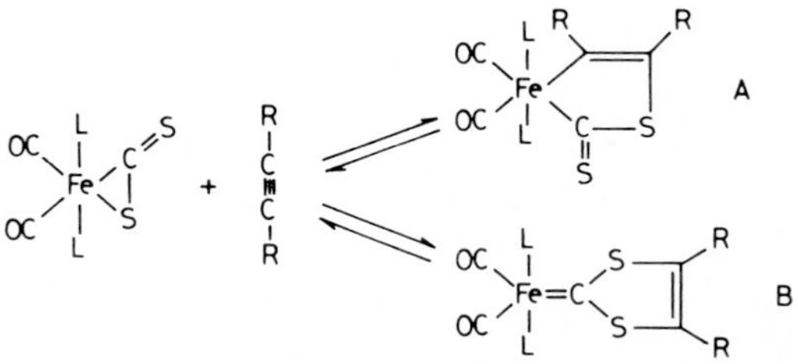

(**19**)

The carbene complex **B** is stable when L is $P(OMe)_3$ or PPh_3 but isomerizes into the heterometallocycle **A** when L is a stronger donor group such as PMe_3. An equilibrium between forms **A** and **B** is observed when L is PMe_2Ph. The equilibrium constant and the rate of the isomerization of **B** increases with the electron-donating capability of the ligand L. The authors, by assuming a 1,3-cyclo-addition-type reactivity, propose that both isomers **A** and **B** can be formed on account of a double 1,3-dipole character of the $M\text{-}\eta^2\text{-}CS_2$ moiety. Accordingly, the A⇀B isomerization would proceed by initial retrocycloaddition from **B** to the $\eta^2\text{-}CS_2$ iron complex, followed by 1,3-cycloaddition of the alkyne to the $\eta^2\text{-}CS_2$ moiety (**20**).

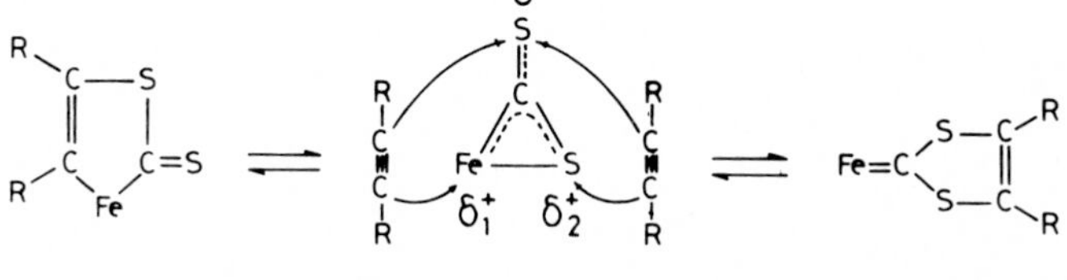

(**20**)

Actually, the CS_2 molecule can itself behave as a 1,3-dipole and it has been shown that, under particular conditions, it can add a dipolarophile, such as a substituted acetylene to give 1,3-dithiocarbene complexes (ref. 77). In light of the facile formation of type **B** rings, it ensues that the 1,3-dipole character of CS_2 is enhanced by η^2-coordination. As an extention, 1,3-dipole character has also been assigned to the M-C-S(exocyclic) grouping to explain the formation of type **A** metallo-rings (ref. 21) (**21**). In Section

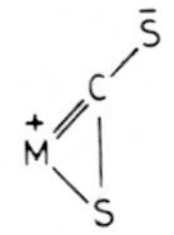

(**21**)

4.2, the validity of the latter Lewis structure is questioned. Moreover, 1,3-dipolar cycloadditions, well understood in organic chemistry (ref. 78), are "frontier-orbital" controlled, and the π-type electron distribution of the reactants must conform to precise requirements. In conclusion, we believe that it is somewhat hazardous to classify the reactions yielding metallo-rings of type **A** as classical 1,3-dipolar cycloadditions without any detailed mechanistic study.

The addition of activated acetylenes to η^2-CS_2 metal complexes to give dithiocarbene derivatives has found interesting potential applications in the synthesis of sulfur-containing products as shown in (**22**).

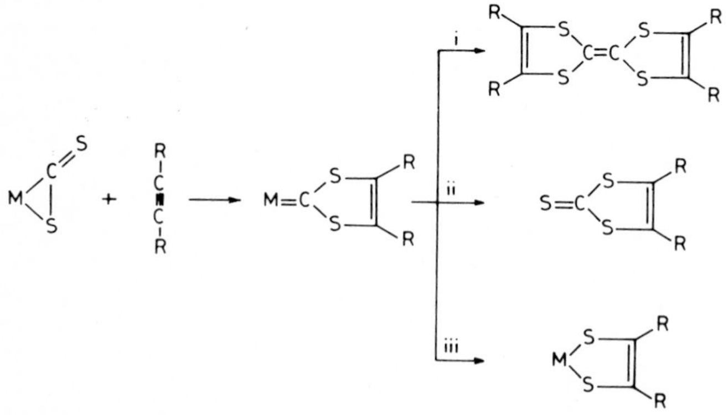

(**22**)

i. Tetrakis(trifluoromethyl)tetrathiafulvalene has been obtained in 50-60% yields through the reaction of (triphos)Ni(η^2-CS_2) and $F_3CC{\equiv}CCF_3$, followed by CO addition, or through direct reaction by using an excess of hexafluorobut-2-yne (ref. 73,74) (**23**).

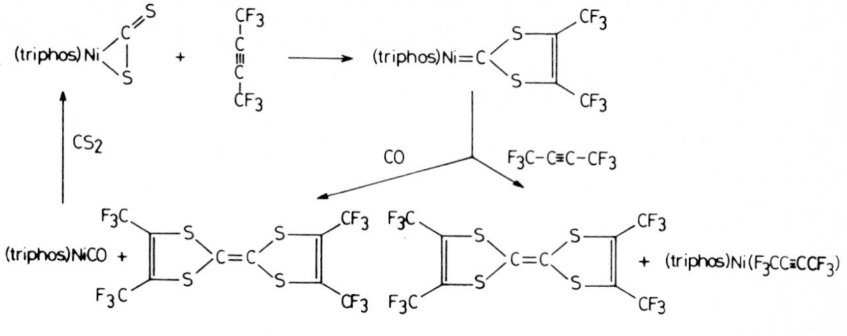

(**23**)

The possibility of regenerating the starting η^2-CS_2 complex by treatment of the carbonyl (triphos)Ni(CO) with CS_2 is an interesting feature of this route to tetrathiafulvalene derivatives.

Addition of iodine or thermolysis has also been used to synthesize in 55-85% yields tetrathiafulvalene derivatives from (1,3-dithiol-2-ylidene)iron complexes (ref. 79) (**24**).

Excellent yields of tetrakis(trifluoromethyl)tetrathiafulvalene (150-200%) have been obtained by reacting CS_2 and $F_3CC{\equiv}CCF_3$ in the

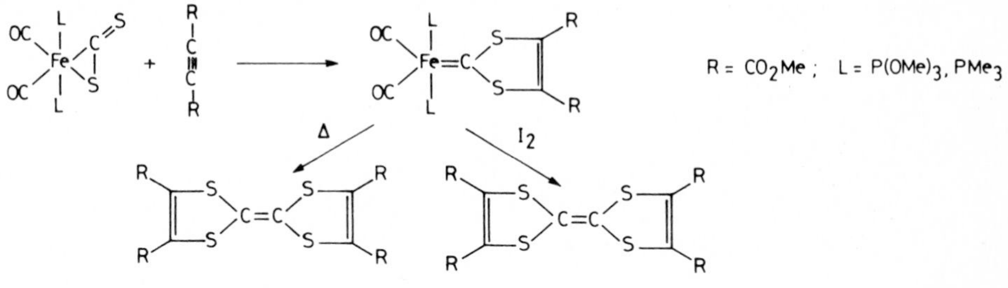

R = CO_2Me ; L = $P(OMe)_3$, PMe_3

(**24**)

presence of the nickel(0) complex Ni(np_3), which is known to form (np_3)Ni(η^2-CS_2) with CS_2 (ref. 74). A 'catalytic' mechanism depicted in (**25**) has been suggested to explain the observed yields. The low turn-over was attributed to the poisoning of the catalyst Ni(np_3).

ii. 1,3-dithiole-2-thiones containing functional groups can be

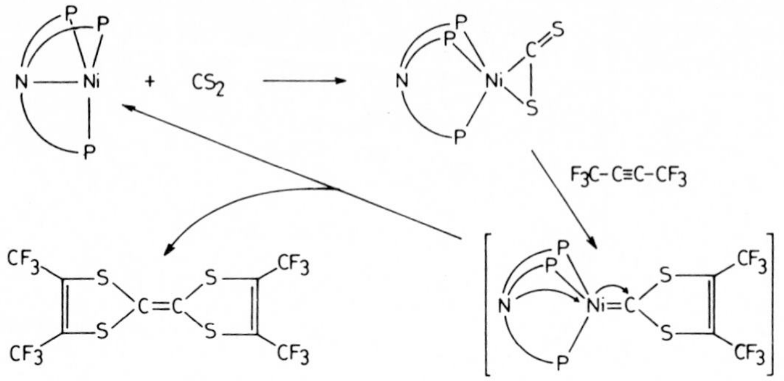

(**25**)

obtained in a one-step reaction by refluxing, in an excess of sulfur, mixtures of $Fe(\eta^2-CS_2)(CO)_2L_2$ (L = PPh_3, $P(OMe)_3$) and $R^1C{\equiv}CR^2$ (R^1, R^2 = CO_2Me, CO_2Et, Ph, CHO, COMe) (ref. 80) (**26**).

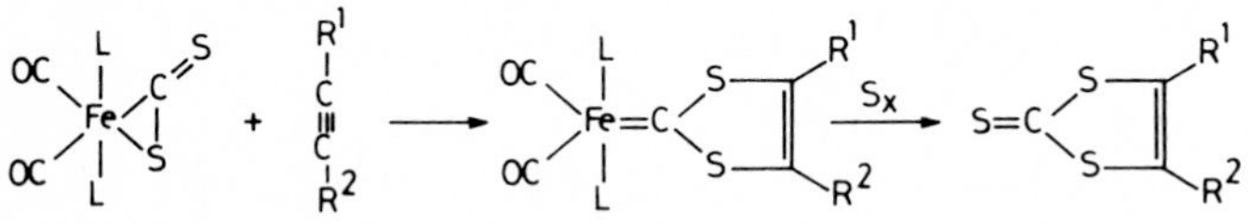

(**26**)

iii. Decomposition in air of solutions of some (1,3-dithiol-2-ylidene)complexes, besides yielding tetrathiafulvalene derivatives produces dithiolene complexes. Examples are given in (**27**) (ref. 81) and (**28**) (ref. 74). These reactions represent a

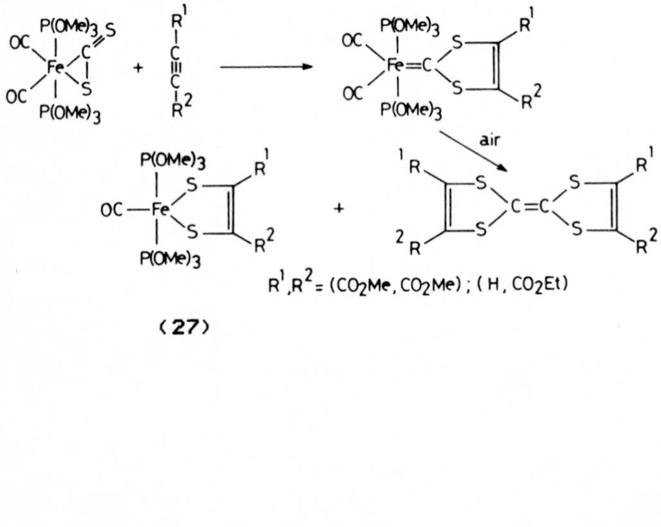

R^1,R^2 = (CO_2Me, CO_2Me); (H, CO_2Et)

(**27**)

novel and potentially useful route to unsymmetrical dithiolene complexes since the usual methods of synthesis lead only to symmetrical dithiolates.

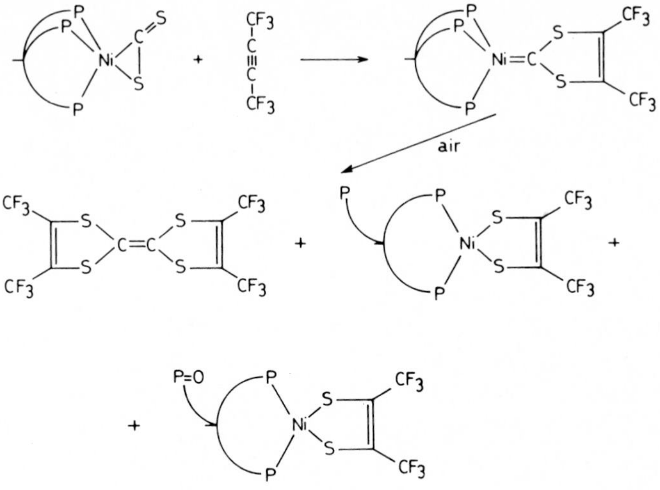

(28)

Addition of two acetylene molecules to a metal-η^2-CS_2 moiety has been recently observed in the reaction of (triphos)RhCl(η^2-CS_2) with $F_3CC{\equiv}CCF_3$ yielding the complex (triphos)ClRhC(CF_3)=C(CF_3)CSC-(CF_3)=C(CF_3)S (ref. 82). The X-ray structure of the latter compound is shown in (**29**). The intermediacy of the (1,3-dithiol-2-ylidene) complex, (triphos)RhCl[$CS_2C_2(CF_3)_2$], has been proposed to explain the formation of the rhodacyclobutene complex (**30**).

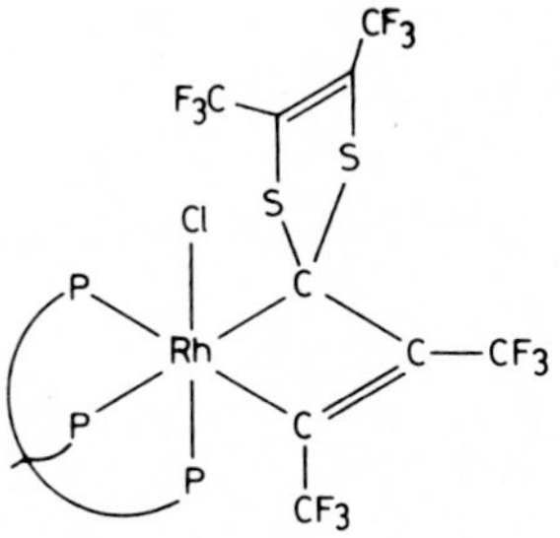

(29)

A very unusual reaction between η^2-CS_2 complexes and alkynes has been described by Dixneuf et al.. In alcohols Fe(η^2-CS_2)$(CO)_2L_2$

(L = $P(OMe)_3$, PPh_3, $PMePh_2$, $Ph_2PC{\equiv}CPh$, $Ph_2PC{\equiv}C$-t-Bu) react with phosphinoalkynes of formula $Ph_2PC{\equiv}CR$ (R = Ph, t-Bu) to give

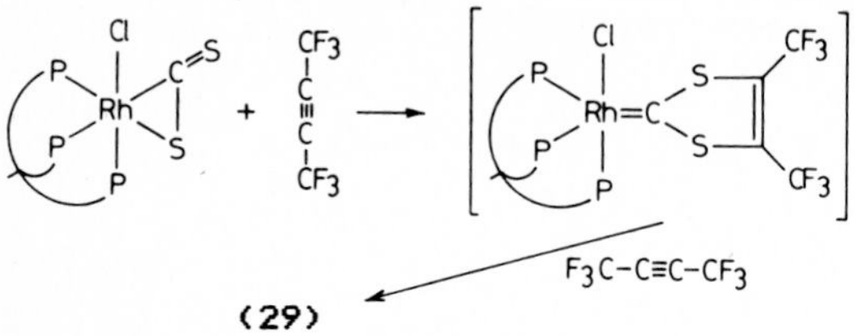

(29)

(30)

$Fe(CO)L[Ph_2PCH{=}C(R)S][CS(OR')]$ (R'= Me, Et, t-Bu) via intramolecular coupling of η^2-CS_2 and $Ph_2PC{\equiv}CR$ groups, followed by CS_2 fragmentation (ref. 83,84) (**31**). The expected alkyne addition to the η^2-CS_2 moiety to give the (1,3-dithiol-2-ylidene)ligand does not take place probably due to the low electrophilicity of phosphinoalkynes. The following stepwise mechanism for the formation

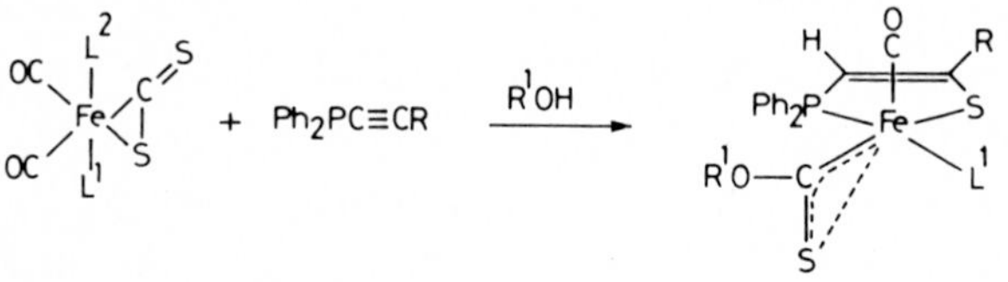

(**31**)

of the final η^2-alkoxythiocarbonyl compounds has been suggested on

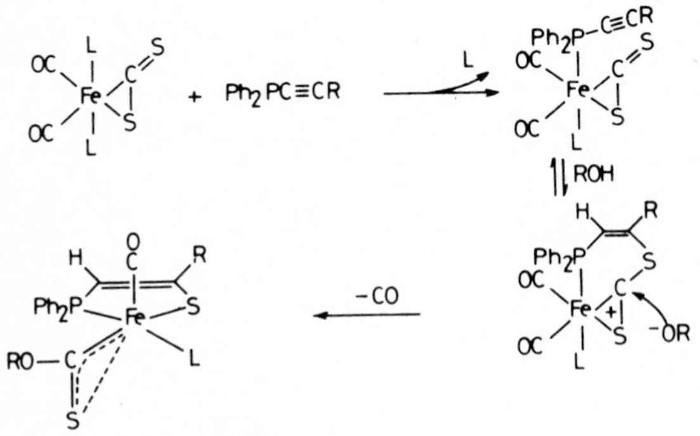

(**32**)

the basis of chemical and spectroscopic evidence (**32**). These (η^2-alkoxythiolato)iron complexes can add a basic phosphine ligand, such as PMe_3 or PMe_2Ph, to give the corresponding (η^1-alkoxythiolato)derivatives (ref. 85) (**33**). Both η^2- and η^1-alkoxythiocarbonyl complexes of iron may find useful applications in organometallic chemistry. In particular, they can function as starting materials to synthesize (2-phosphonioalkenethiolato) iron derivatives (ref. 85).

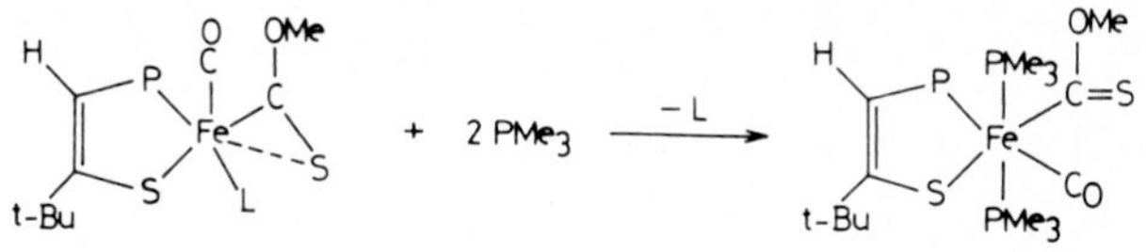

(**33**)

Notwithstanding the ease by which electrophilic alkynes can add to η^2-CS_2 complexes, cycloaddition with activated olefins has not been so far reported. Also, in organic chemistry, cycloadditions of acetylenes to electron rich 1,3-dipoles are known to be more efficient than those involving the corresponding alkenes. The rationale for this difference in behavior is implicit in the frontier MO theory for these reactions (ref. 86).

The reactions of η^2-CS_2 compounds with olefins are very rare, being limited to the regio- and stereoselective addition of $Fe(\eta^2\text{-}CS_2)(CO)_2L_2$ complexes (L = PPh_3, PMe_2Ph) to α,β-unsaturated carbonyls (ref. 87). In the presence of strong acids, the η^2-CS_2 iron complexes can be incorporated into α,β-unsaturated aldehydes, ketones, and esters such as acrolein, methyl vinyl ketone, chalcone, and methyl acrylate to give cationic complexes having a bulky sulfur-containing organometallic group linked to C3 of the carbonyl substrate. An example of these reactions is given in (**34**). The final cationic products result from the addition of the

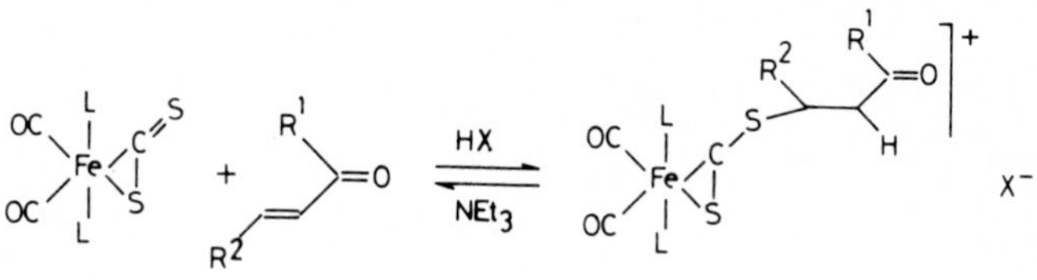

(**34**)

nucleophilic uncoordinated sulfur atom of the η^2-CS_2 complex to the protonated, activated olefin. They are stable in acidic or neutral media. By contrast, in the presence of a base, abstraction of the proton at the α-position of the carbonyl occurs and retroaddition takes place.

3.2.3 Reactions with electrophilic metal fragments and complexes

The nucleophilic behavior of the η^2-coordinated CS_2 ligand has been widely exploited to synthesize homo- and heterometal CS_2-bridged complexes. To accomplish this goal, η^2-CS_2 complexes are reacted either with electrophilic metal fragments (route I), or with metal complexes containing weakly bound ligands (route II). In this way, dimers with CS_2 bonding modes **A**, **B**, and **C** have been obtained (**35**). In particular, compounds of types **A** and **C** can be

M—C(S)(S)M   M(C,S)—S—M   M(C,S)S—M

A   B   C

(**35**)

synthesized through route I, whereas those of type **B** may be obtained through both routes. To classes **C** and **A** belong the complexes $[(triphos)Co(\mu\text{-}CS_2)Co(triphos)](BPh_4)_2$ (ref. 18) (**36**), and $[(PPh_3)_2ClPt(\mu\text{-}CS_2)M(PPh_3)_2]BF_4$ (M = Pt, Pd) (ref. 88,89) (**37**).

$$\mathbf{(triphos)Co(\eta^2\text{-}CS_2) + [(triphos)Co]^{2+} \longrightarrow [(triphos)Co(\mu\text{-}CS_2)Co(triphos)]^{2+}}$$

(**36**)

$$\mathbf{Pt(\eta^2\text{-}CS_2)(PPh_3)_2 + 1/2\ [PtCl(PPh_3)_2](BF_4)_2 \longrightarrow [(PPh_3)_2ClPt\text{-}C(S)(S)Pt(PPh_3)_2]BF_4}$$

(**37**)

The cobalt complex is fluxional. In solution a single ^{31}P NMR resonance for the two triphos ligands is observed even at low temperatures (-80°C) (ref. 90). A plausible step in the formation of the platinum complex has been suggested to be electrophilic attack at the exocyclic sulfur atom by the chloroplatinum species to give an $\eta^2(C,S):\eta^1(S)$ intermediate (**38**), essentially a dimer of type B. Support for this mechanism is provided by the formation of $[(CO)_2L_2Fe(\mu\text{-}CS_2)PtCl(PPh_3)_2]BF_4$ (L = PPh_3, PMe_2Ph) (**39**) from

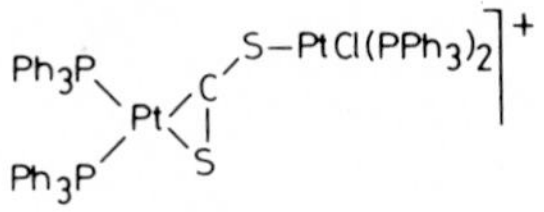

(**38**)

the reaction of $Fe(\eta^2\text{-}CS_2)(CO)_2L_2$ with 1/2 $[PtCl(PPh_3)_2]_2(BF_4)_2$ to give the following compound (ref. 68).

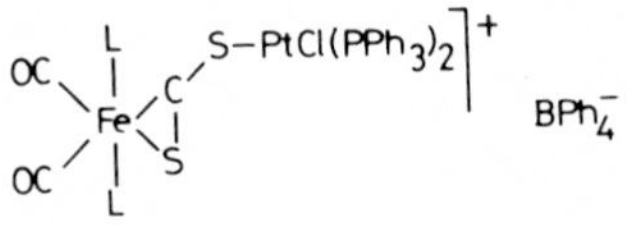

(**39**)

A particular example of $\eta^2(C,S):\eta^1(S)$ complex is the double CS_2-bridged complex $(PPh_3)Ni(\mu\text{-}CS_2)_2Ni(PPh_3)$ (**40**). This is obtained by reacting CS_2 with a variety of triphenylphosphine-nickel(0) complexes like $(PPh_3)_2Ni(CO)_2$ (ref. 91), $(PPh_3)_2Ni(C_2H_4)$ (ref. 92), and $[(PPh_3)_2Ni(C_3Ph_3)]BPh_4$ (ref. 26). The crystal structure determinations of (**40**), and those of very similar Pd and Pt derivatives have been recently reported (see Section 2.2).

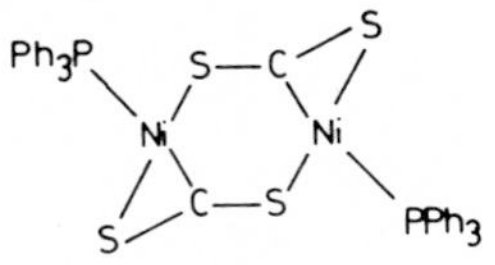

(**40**)

The formation of the $(\mu\text{-}CS_2)_2$ dimer (**40**) is likely to involve the intermediate mononuclear complex $(PPh_3)_2Ni(\eta^2\text{-}CS_2)$. The conversion of the latter complex to the final dimeric product could proceed through the mutual displacement of one phosphine ligand by the uncoordinated sulfur atom of two η^2-CS_2 complex molecules. This hypothesis is supported by the fact that, just after the addition of CS_2, exposure to air of the reaction mixture yields the dithiocarbonate complex $(PPh_3)_2Ni(S_2CO)$ (ref. 90) (**41**). Oxidation by atmospheric oxygen to dithiocarbonate complexes is, in fact, a well known chemical property of η^2-CS_2 compounds (see Section 3.4).

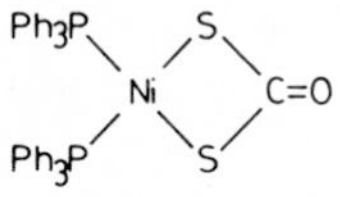

(**41**)

The CS_2-bridged complexes of type B are relatively numerous. They are generally prepared by reacting η^2-CS_2 compounds with tetrahydrofuran adducts of metal carbonyls. In a few cases however, the formation of the bridging CS_2 complex is achieved by spontaneous degradation in solution of the starting η^2-CS_2 complex (ref. 41). The following combinations have been reported: $Mn(CS_2)Mn$ (ref. 41), $Cr(CS_2)Cr$ (ref. 93), $W(CS_2)W$ (ref. 54), $Co(CS_2)M$ (M = Cr, Mn) (ref. 15,18), $Pt(CS_2)M$ (M = Cr, W) (ref. 89), $Nb(CS_2)M$ (M = Nb, Cr, Mo, W, Fe) (ref. 23), $Rh(CS_2)M$ (M = Cr,Mn) (ref. 43) , $Fe(CS_2)M$ (Mn, Mo) (ref. 22), $V(CS_2)M$ (M = Cr,Mo, Mn, Fe) (ref. 94). Crystal structures are available for (triphos)$Co(\mu$-$CS_2)Cr(CO)_5$ (ref. 18), $(PMe_2Ph)_2(CO)_2Fe(\mu$-$CS_2)Mn(CO)_2(Cp)$ (ref. 22), and $(Bu)(Cp)_2Nb(\mu$-$CS_2)W(CO)_5$ (ref. 23) (**42**).

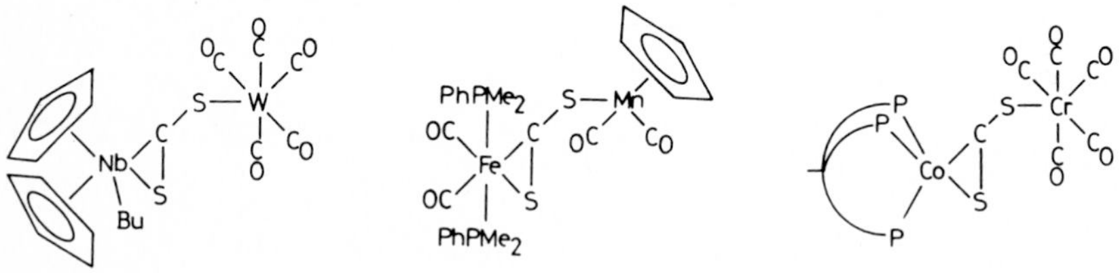

(**42**)

Polynuclear species of formula $PdLCS_2$ which on the basis of spectroscopic data are assigned a polymeric structure of type (**A**) or (**B**), have been recently reported (ref. 95) (**43**). One phosphine ligand is abstracted from $Pd(\eta^2$-$CS_2)L_2$ (L = PPh_3, $P(p\text{-}tol)_3$, PCy_3) complexes by reaction with $[Pd(OCMe_2)(bipy)(C_6F_5)]ClO_4$ (**44**).

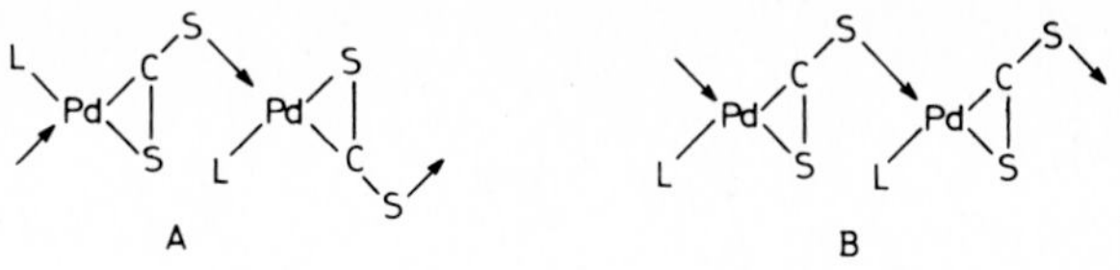

A B

(43)

$$Pd(\eta^2\text{-}CS_2)L_2 + [Pd(OCMe_2)(bipy)(C_6F_5)]ClO_4 \xrightarrow{Me_2CO/CS_2} \text{'}PdLCS_2\text{'} + [PdL(bipy)(C_6F_5)]ClO_4$$

(44)

Alternatively, 'Pd(PCy_3)CS_2' can be obtained by reacting an excess of CS_2 with the corresponding η^2-CS_2 complex (45).

$$Pd(\eta^2\text{-}CS_2)(PCy_3)_2 + CS_2 \longrightarrow \text{'}Pd(PCy_3)CS_2\text{'} + Cy_3PCS_2$$

(45)

3.2.4 Reactions with protic acids

In most instances the reaction of protic acids on η^2-CS_2 complexes gives sulfur-free products (ref. 68,96). For example, when Pt(η^2-CS_2)(PPh_3)$_2$ in tetrahydrofuran is treated with HCl, PtH(Cl)(PPh_3)$_2$ is formed (ref. 68). However, in some cases, the reactions occur specifically at the metal-η^2-CS_2 moiety giving a variety of sulfur-containing ligands (46).

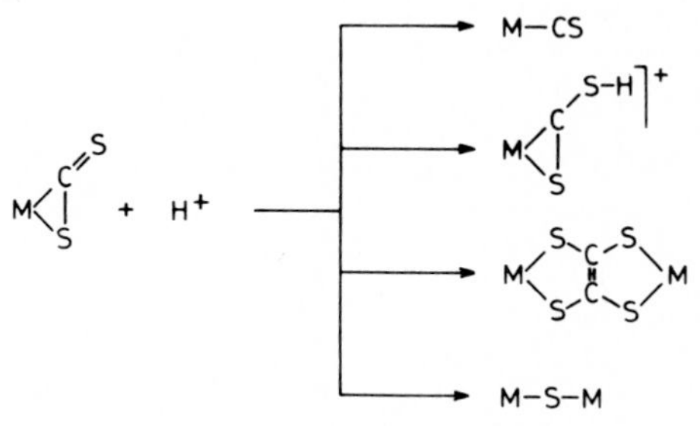

M—CS

M—S—M

(46)

The first reaction of an η^2-CS_2 complex with a protic acid was reported by Roper et al. in 1975. By reacting HCl with Os(η^2-CS_2)-

$(CO)_2(PPh_3)_2$, the corresponding thiocarbonyl is obtained and H_2S evolves (ref. 51) (**47**).

$$Os(\eta^2\text{-}CS_2)(CO)_2(PPh_3)_2 + 2\ HCl \longrightarrow OsCl_2(CO)(CS)(PPh_3)_2 + H_2S$$

(**47**)

Recently, Fehlhammer et al. have described the reaction of HBF_4 with $Fe(\eta^2\text{-}CS_2)(CO)_2L_2$ (L = PMe_2Ph, PPh_3). The protonation of the exocyclic sulfur atom occurs with subsequent formation of the $\eta^2\text{-}CS_2H$ derivative (ref. 68) (**48**). Although not very stable,

$$(OC)_2L_2Fe(\eta^2\text{-}CS_2) + HBF_4 \longrightarrow [(OC)_2L_2Fe(\eta^2\text{-}CS\text{-}SH)]^+ BF_4^-$$

(**48**)

the $\eta^2\text{-}CS_2H$ complex has been characterized by IR spectroscopy (νSH: 2490 cm^{-1}, L = PMe_2Ph; 2460 cm^{-1}, L = PPh_3). In light of Fehlhammer's results, a reasonable mechanism for reaction (**47**) is one which implies the protonation of the exocyclic sulfur to give an $\eta^2\text{-}CS_2H$ ligand. Subsequent action of H^+ on the $Os\text{-}\eta^2\text{-}CS_2H$ moiety should trigger the extrusion of H_2S and the formation of the thiocarbonyl complex. A similar reactivity pattern has previously been observed for the acid-promoted conversion of metal-$\eta^2\text{-}CS_2R$ to metal-CS and RSH (see Section 3.2.1).

The action of H^+ and other Lewis-acids on $\eta^2\text{-}CS_2$ complexes has been found to promote the head-to-head dimerization of CS_2 to C_2S_4. By treatment of the complexes (triphos)RhCl($\eta^2\text{-}CS_2$) (ref. 97) and (triphos)IrCl($\eta^2\text{-}CS_2$) (ref. 90) in tetrahydrofuran with H^+, $HgCl_2$, BF_3, or alkali cations, the corresponding ethenetetrathiolate binuclear derivatives are formed (**49**). These are isolated as BPh_4^-, Cl^-, or BF_4^- salts. The structure of the Rh/BPh_4^- derivative has been established by X-ray methods (ref. 97).

An important step for the formation of the $\mu\text{-}C_2S_4$ dimers is believed to be the opening of the MCS ring, which is favored by the action of the Lewis-acid at the uncoordinated sulfur atom.

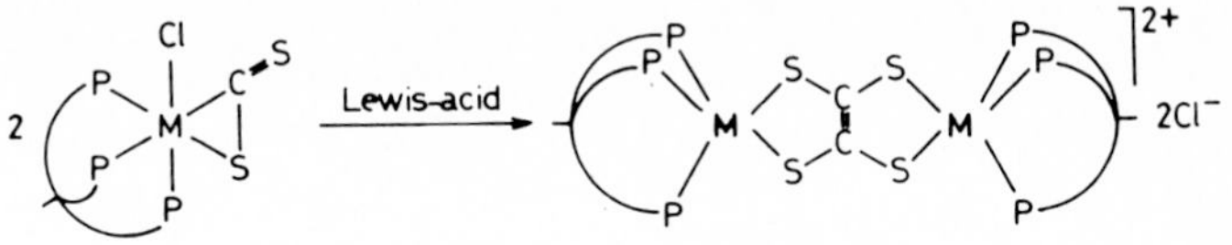

(49)

Fragmentation of the η^2-CS_2 group to give a bridging sulfido ligand upon reaction with HBF_4 has been also observed. The complex (triphos)Ni(η^2-CS_2) reacts with HBF_4 giving a μ-S dimer (ref. 98) (**50**). Interestingly, the same reaction on the isomorphous cobalt

$$\text{(triphos)Ni}(\eta^2\text{-}CS_2) \xrightarrow{HBF_4} [\text{(triphos)Ni-S-Ni(triphos)}](BF_4)_2$$

(50)

derivative (triphos)Co(η^2-CS_2) yields the μ-CS_2 complex [(triphos)Co(μ-CS_2)Co(triphos)]$(BF_4)_2$ (ref. 18).

3.2.5 Reactions with CS_2

Some metal-η^2-CS_2 complexes may add a second CS_2 molecule to give metallo-rings of the type $\overline{M\text{-}C(S)SC(S)S}$. This has been observed by Werner et al. who obtained the complex Rh(C_2S_4)(Cp)(PMe_3) by reacting Rh(η^2-CS_2)(Cp)(PMe_3) with CS_2 (ref. 43). According to the results of an X-ray analysis, the metallocycle is planar and possesses nearly identical C-S bond lengths. The chelating C_2S_4

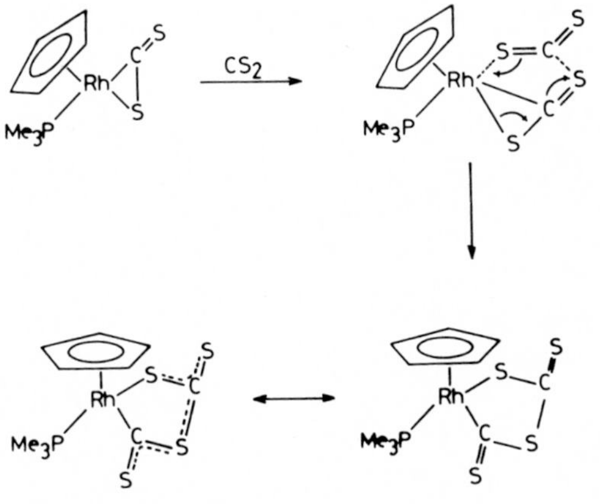

(51)

ligand forms a highly delocalized π-electron system. The authors suggest a mechanism, which involves nucleophilic attack by the exocyclic sulfur atom of the η^2-coordinated CS_2 on the electrophilic carbon atom of an incoming CS_2 molecule, followed by ring closure through the formation of a Rh-S bond (ref. 4) (**51**).

A C_2S_4 metalloheterocycle similar to that found by Werner, has been suggested by Thewissen to be an intermediate in the transformation of some rhodium η^2-CS_2 complexes into thiocarbonyls. The rhodium(I) complexes $Rh[X-C(Z)-Y](PPh_3)_2$, in which the X-C(Z)-Y uninegative unsaturated heteroallylic ligands (X, Y, Z = P ,S, or N) are coordinated to the metal by two of the three heteroatoms, react at elevated temperature with an excess of CS_2 to give the thiocarbonyl complexes $Rh[X-C(Z)-Y](CS)(PPh_3)$ (ref. 99). The stepwise mechanism shown in (**52**) has been proposed. The

$[Rh(CS_3)]$-complex

(**52**)

sequence $Rh(CS_2) \longrightarrow Rh(C_2S_4) \longrightarrow Rh(CS)(CS_3) \longrightarrow Rh(CS_3)$ is remarkable. Overall, this amounts to a metal-promoted disproportionation of CS_2 to CS and CS_3^{2-}.

The interpretations given by Werner and Thewissen for the formation of the $\overline{MSCSC}$ metallo-ring are somewhat in contrast with each other. The first author suggests that the electrophilic carbon atom of an incoming CS_2 molecule interacts with the exocyclic sulfur atom of the metal-η^2-CS_2 moiety, whereas Thewissen suggests the endocyclic sulfur atom as a candidate for nucleophilic attack. In other words, this would correspond to a different opening mode of the $\overline{MCS}$ cycle. It is also worth mentioning here a third possibility, proposed on the basis of a theoretical analysis (ref. 7), and schematically shown in (**53**). According to this work, the sulfur atom of a free CS_2 molecule attacks the carbon atom of coordinated CS_2 while the Rh-C linkage is unfastening.

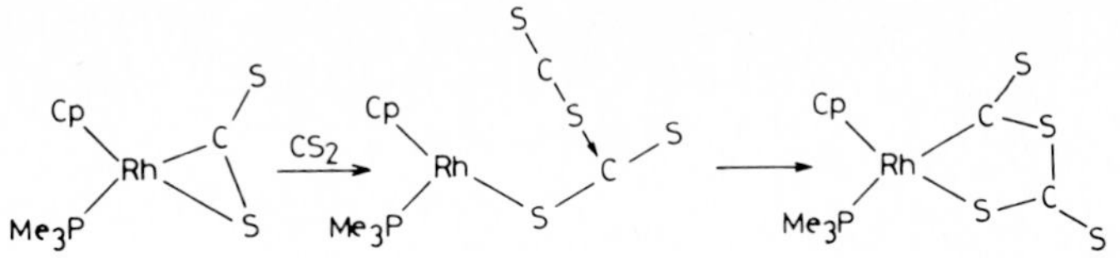

(53)

A C_2S_4 intermediate has been also invoked by Fortune and Manning to explain the thermal decomposition of $Co(Cp)(PPh_3)(\eta^2-CS_2)$ in methylene chloride at 40°C, which gives approximately equal amounts of $Co(Cp)(PPh_3)(CS)$ and $Co(Cp)(PPh_3)(CS_3)$ (ref. 100).

3.3 REACTIONS WITH NUCLEOPHILES

3.3.1 Reactions with tertiary phosphines

Desulfurization of the CS_2 ligand by alkyl- and arylphosphines to give thiocarbonyls and phosphine sulfides was first described by Wilkinson et al. in 1966 (ref. 101). The formation of trans-$RhX(CS)(PPh_3)_2$ (X = Cl, Br) complexes by the direct reaction of $RhX(PPh_3)_3$ with CS_2 was suggested to proceed via η^2-CS_2 intermediates (ref. 102). Since then, a number of thiocarbonyls have been synthesized either by treatment of η^2-CS_2 complexes with PR_3 (ref. 54,93,103) or through the reaction of metal complex/PR_3 mixtures with CS_2 (ref. 102,104-107). In the latter case, however, the presence of η^2-CS_2 ligands at a certain stage in the reaction has been often inferred by means of spectroscopic methods. Significant examples are reported below (ref. 54,93,106) (**54**).

$$Cr(arene)(CO)_2(\eta^2-CS_2) \xrightarrow{PMe_3} Cr(arene)(CO)_2(CS) + SPMe_3$$

$$Mn(Cp)(CO)_2THF \xrightarrow{CS_2} Mn(\eta^2-CS_2)(Cp)(CO)_2$$

$$Mn(Cp)(CO)_2THF \xrightarrow{PPh_3/CS_2} Mn(Cp)(CO)_2(CS) + SPPh_3$$

$$Mn(\eta^2-CS_2)(Cp)(CO)_2 \xrightarrow{PPh_3} Mn(Cp)(CO)_2(CS) + SPPh_3$$

$$W(CO)_2(\eta^2\text{-}CS_2)(diphos) \xrightarrow{PBu_3} W(CO)_2(CS)(diphos) + SPBu_3$$

(54)

The mechanism for these reactions has been suggested to involve nucleophilic attack by PR_3 at the endocyclic sulfur atom of the metal-η^2-CS_2 moiety (ref. 102,104) (**55**). Accordingly, the coordinated sulfur atom is considered an electrophilic site.

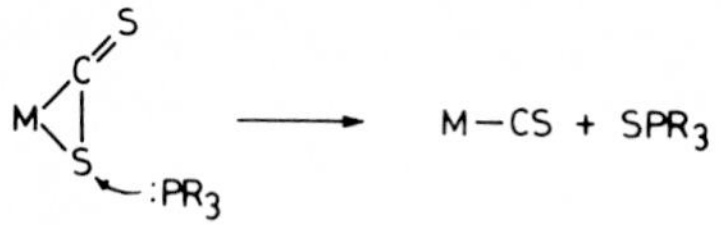

$$M\text{-}CS + SPR_3$$

(55)

The electrophilic character of the sulfur in $\overline{MCS}$ has been also invoked to explain the 1,3-cycloaddition reactions of alkynes to η^2-CS_2 complexes (ref. 70). In particular, the canonical form (**56**) has been considered important in describing the electronic nature of the metal-η^2-CS_2 moiety.

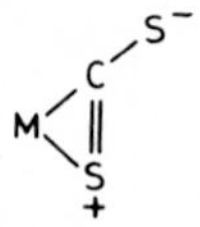

(56)

Recent results suggest a stepwise mechanism involving the intermediacy of phosphoniodithiocarboxylate complexes to explain the desulfurization reaction of η^2-CS_2 compounds to thiocarbonyls by tertiary phosphines. The addition of tributylphosphine to a methylene chloride solution of $W(CO)_2(\eta^2\text{-}CS_2)$(diphos) results, in fact, in the formation of a thiocarbonyl, but, under appropriate conditions, a phosponiodithiocarboxylate complex has been isolated by chromatographic methods (ref. 54) (**57**). Analogously, the reaction of triethylphosphine with (triphos)$RhCl(\eta^2\text{-}CS_2)$ gives the corresponding S_2CPEt_3 complex as a crystalline product (ref. 90) (**58**). Furthermore, $PdL_2(\eta^2\text{-}CS_2)$ (L = $PMePh_2$, PMe_3) react with trimethylphosphine to give binuclear complexes containing bridging S_2CPMe_3 zwitterions (ref. 55) (**59**). The bonding mode of the bridging zwitterions is still unknown. The presence of S_2CPR_3

(57)

(58)

$$2\ Pd(\eta^2\text{-}CS_2)(PR_3)_2 + 4\ PMe_3 \longrightarrow [(PMe_3)Pd(S_2CPMe_3)]_2 + 4\ PR_3$$

(59)

ligands in the products of reactions (57), (58), and (59) is confirmed by the presence of the same final products through the addition of the preformed zwitterions to the corresponding metal fragments. Within this context, it is noteworthy to report that the addition of S_2CPEt_3 to $Cr(C_6H_6)(CO)_2THF$ quantitatively gives the known thiocarbonyl $Cr(C_6H_6)(CO)_2(CS)$ and $SPEt_3$ through a stepwise mechanism which probably involves the intermediacy of a phosphoniodithiocarboxylate complex (ref. 90) (60). On the basis of these results, a reasonable mechanism for the formation of thiocarbonyls from $\eta^2\text{-}CS_2$ complexes is shown in (61), which makes use of nucleophilic attack by tertiary phosphines at the electrophilic $\overline{MCS}$ carbon atom. At the present time, the factors which govern the eventual evolution of the phosphoniodithioformate intermediates

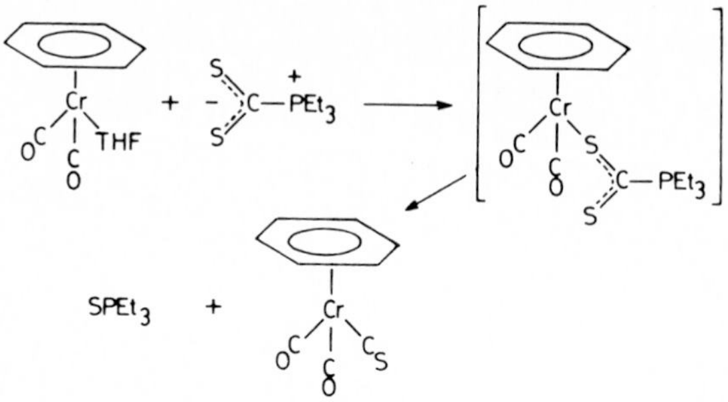

(60)

to thiocarbonyls are not completely understood.

It has been reported that the conversion of metal-η^2-CS_2 to metal-CS via reaction with PR_3, is affected by the nature of the coligands (ref. 93,103). In particular, the less basic these are,

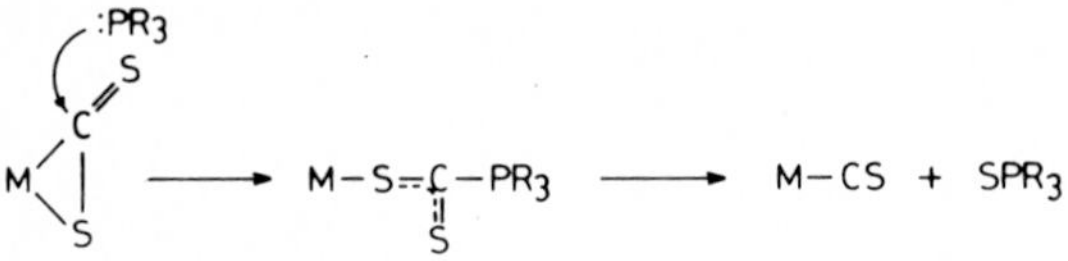

(61)

the more readily the desulfurization reaction occurs. In other words, the accumulation of charge at the metal obstructs the reaction. For example, $Fe(\eta^2$-$CS_2)(CO)_2(P(OPh)_3)_2$ is desulfurized by PBu_3; whereas, under the same conditions, $Fe(\eta^2$-$CS_2)(CO)_2(PR_3)_2$ (R = alkyl, phenyl) are not (ref. 103). Similarly, the conversion of $Cr(arene)(CO)_2(\eta^2$-$CS_2)$ to $Cr(arene)(CO)_2(CS)$ depends on the basicity of the arene ligands (ref. 93). Notice that the reaction pathway shown in (**61**) is in good agreement with the latter experimental results. In fact, according to mechanism (**61**), the desulfurization process would be facilitated when the charge density at the metal decreases, that is to say when the CS_2 carbon atom is more electrophilic. Within this context, it is reasonable that alkylphosphines are, generally, more efficient than arylphosphines to desulfurize η^2-CS_2 complexes. Only the former, in fact, are known to react directly with CS_2 giving zwitterions of the type S_2CPR_3 (ref. 108).

3.3.2 Reactions with H^- and N_3^-

Nucleophiles other than tertiary phosphines can react with η^2-CS_2 metal complexes. One of these is the hydride ion, which has been found to attack the metal-η^2-CS_2 moiety at the carbon atom. Reduction of <u>mer</u>-$W(CO)_3$(diphos)(η^2-CS_2) with $LiHBEt_3$ gives the dithioformate complex <u>fac</u>-[$W(CO)_3$(diphos)(η^1-SCHS)]$^-$ (ref. 109) (62).

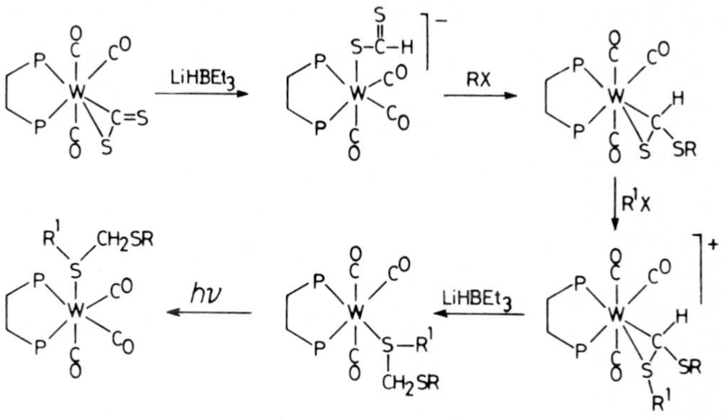

(62)

Interestingly, the dithioformate complex reacts with alkyl halides, RBr (R = Me, Et, CH_2Ph), in two steps via dithio ester complexes, <u>mer</u>-$W(CO)_3$(diphos)(η^2-SCHSR), to dithiocarbenium ion derivatives, <u>mer</u>-[$W(CO)_3$(diphos)(η^2-R′SCHSR)]X (R′= Me, CH_2Ph). These, in turn, react with $LiHBEt_3$ to give dithioacetal complexes, <u>mer</u>-$W(CO)_3$(diphos)(η^1-R′SCH_2SR), which isomerize photochemically to the corresponding <u>fac</u> compounds. The sequence to note is
η^2-CS_2 $\longrightarrow$ η^1-SCHS $\longrightarrow$ η^2-SCHSR $\longrightarrow$ η^2-R′SCHSR $\longrightarrow$ η^1-R′SCH_2SR.

Metal complexes containing SCHSR ligands have been also obtained by reacting dithioalkyl ester compounds with $NaBH_4$. [$Nb(Cp)_2$(Bu)-(η^2-CS_2R)]$^+$ (R = Me, Et, i-Pr, CH_2CO_2Et, CH_2Ph, $CH_2CH_2CH_2Cl$,

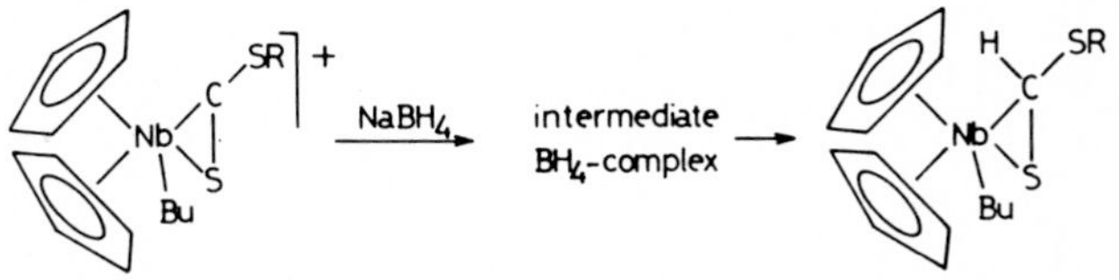

(63)

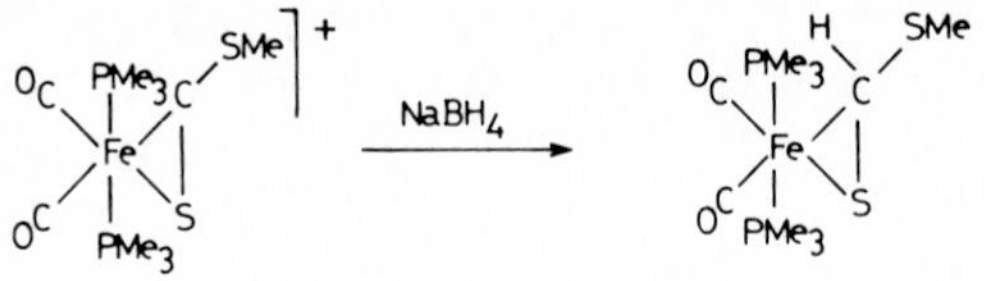

(**64**)

$CH_2CH{=}CH_2$) react with $NaBH_4$ in tetrahydrofuran to give $Nb(Cp)_2(Bu)(\eta^2\text{-}SCHSR)$ via tetraidroborato-metal intermediates (ref. 61) (**63**). Similarly, $[Fe(\eta^2\text{-}CS_2Me)(CO)_2(PMe_3)_2]^+$ reacts with $NaBH_4$ yielding the η^2-SCHSMe complex $Fe(\eta^2\text{-}SCHSMe)(CO)_2(PMe_3)_2$ (ref. 65) (**64**). Similarly, an intermediate containing the SCHSR ligand has been suggested to play an important role in the conversion of metal-η^2-CS_2 to metal-CS as shown below (**65**) (ref. 64).

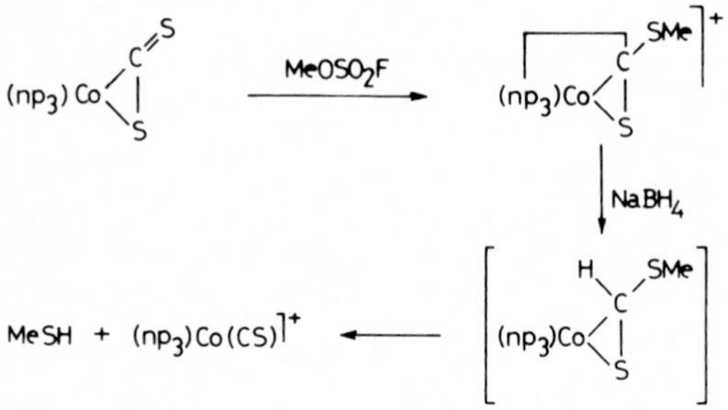

(**65**)

Another interesting reaction of η^2-CS_2 complexes is the cycloaddition of the azide ion, recently reported by Schenk et al. for the complexes $W(CO)_3(R_2PC_2H_4PR_2)(\eta^2\text{-}CS_2)$ (R = Ph, Me). These react with $(NEt_4)N_3$ in methylene chloride to give the 1,2,3,4--thiatriazole-5-thiolate complexes $W(CO)_3(R_2PC_2H_4PR_2)(N_3CS_2)$,

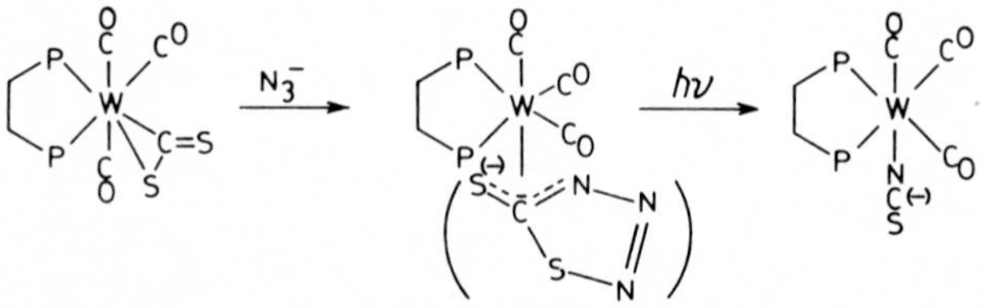

(**66**)

which decompose photolytically to the corresponding isothiocyanate derivatives (ref. 54) (**66**). Unfortunately, the coordination mode of the 1,2,3,4-thiatriazole-5-thiolate group in this compound remains rather obscure.

Interestingly, 1,2,3,4-thiatriazole-5-thiolate complexes can be synthesized also by reaction of CS_2 with azido complexes (ref. 110).

3.3.3 Reactions with electron-rich metal fragments

As previously described, the reaction of η^2-CS_2 compounds with electrophilic metal fragments can give CS_2-bridged complexes of types **A**, **B**, and **C**. A fourth type of CS_2-bridged compounds, **D** (**67**), is observed in some μ-CS_2 compounds obtained by reacting η^2-CS_2 complexes with electron-rich metal fragments.

M—C(—S—M)(=S)

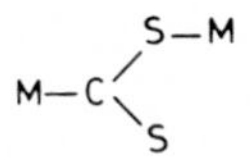

D

(**67**)

$Pt(\eta^2$-$CS_2)(PPh_3)_2$ reacts with $Pt(C_2H_4)(PPh_3)_2$ to give the μ-CS_2 dimer $(Ph_3P)_2Pt(\mu$-$CS_2)Pt(PPh_3)_2$ (ref. 111) (**68**). Regardless of the Pt-Pt linkage, the mode of coordination of CS_2 with respect to the two metals is just that depicted in **D**. Although Walker et al.

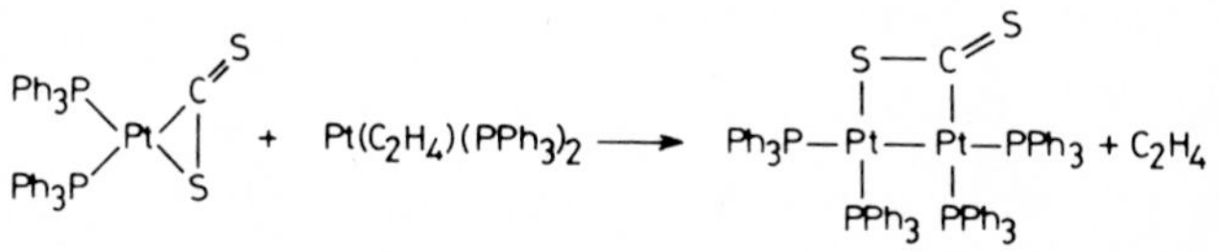

$$(Ph_3P)_2Pt(\eta^2\text{-}CS_2) + Pt(C_2H_4)(PPh_3)_2 \longrightarrow Ph_3P(PPh_3)Pt(\mu\text{-}S\text{—}C(=S))Pt(PPh_3)PPh_3 + C_2H_4$$

(**68**)

suggest that the basic $Pt(PPh_3)_2$ fragment attacks the electrophilic carbon atom of the η^2-CS_2 ligand, it is more likely that $Pt(PPh_3)_2$ adds oxidatively to the Pt-C linkage. The capability of cleaving chemical bonds through oxidative addition is a well known property of the metal fragment in question (ref. 112-114).

An intermediate of type **D** has also been hypothesized in the fragmentation of η^2-CS_2 to CS and S reported below (ref. 111) (**69**).

Interestingly, the fragmentation of the coordinated CS_2 group seems to be reversible; that is, addition of diphos to a suspension of the μ-S thiocarbonyl dimer yields a complex analogous to the μ-CS_2 intermediate, but containing two diphos molecules.

In light of Walker's results, a similar fragmentation of the η^2-CS_2 group to CS and S observed in the reaction of

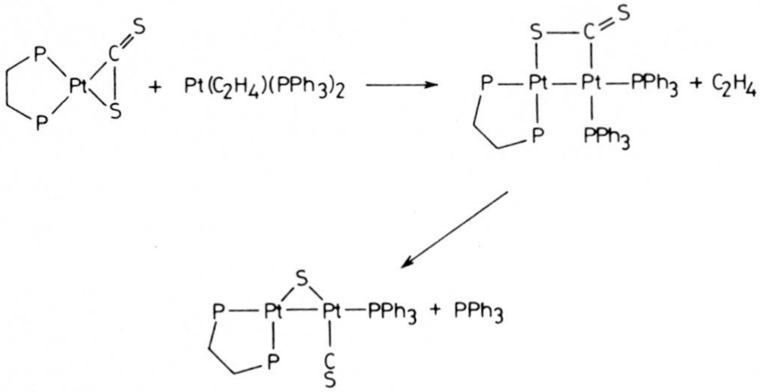

(69)

$Co(\eta^2\text{-}CS_2)(Cp)(PMe_3)$ with $(Cp)(PMe_3)Co(\mu\text{-}CO)_2Mn(CO)(C_5H_4Me)$ could proceed via a μ-CS_2 intermediate of type **D** (ref. 115,116) (**70**).

$Co(Cp)(PMe_3)(\eta^2\text{-}CS_2)$ + $(Cp)(PMe_3)Co(\mu\text{-}CO)(C_5H_4Me)$ ⟶

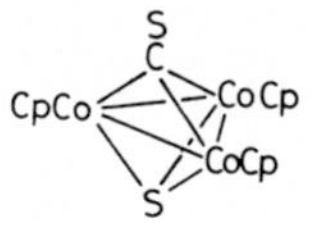

+ 3 PMe_3 + 2 $Mn(C_5H_4Me)(CO)_3$

(**70**)

Recently, Fortune and Manning have reported some alternative routes to the synthesis of Werner's trinuclear complex $Co_3(Cp)_3(\mu_3\text{-}S)$-($\mu_3$-CS) shown in (**70**). These include photolysis at room temperature, or thermolysis at temperatures >80°C of the CS_2 complexes $Co(\eta^2\text{-}CS_2)(Cp)L$ (L = tertiary phosphine or organoisocyanide), and reaction of $Co(\eta^2\text{-}CS_2)(Cp)(PMePh_2)$ with $Co(Cp)(CO)_2$ (mole ratio 1:2 in refluxing benzene) (ref. 107).

3.4 MISCELLANEOUS REACTIONS

3.4.1 Reactions with Group 6B elements

The paramagnetic complex (triphos)Co(η^2-CS_2) reacts with molecular oxygen at room temperature, cyclo-octasulfur or amorphous selenium at reflux temperatures, to form the corresponding dithiocarbonate, trithiocarbonate, and moselenodithiocarbonate derivatives, respectively (ref. 57) (71). Under the same conditions, oxygen reacts with the diamagnetic isomorphous complex

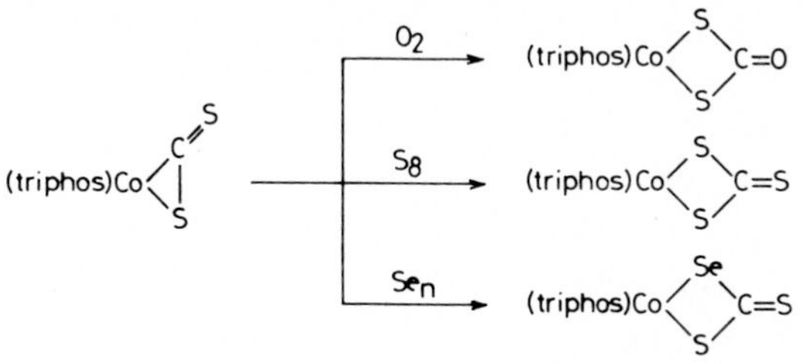

(71)

(triphos)Ni(η^2-CS_2) to give the dithiocarbonate derivative (triphos=O)Ni(S_2CO), whereas the action of sulfur or selenium

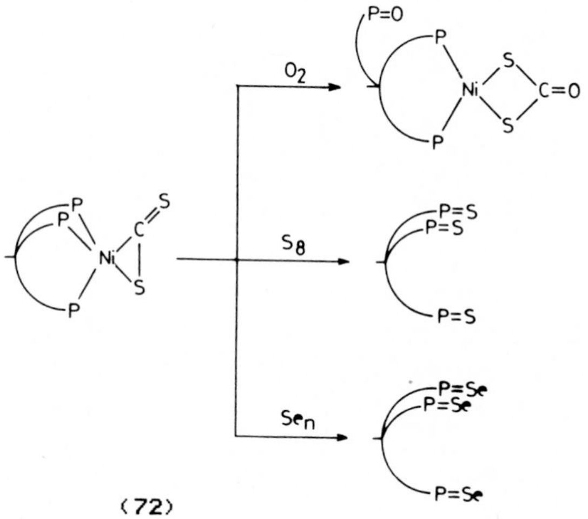

(72)

results in the destruction of the complex, after which the sulfurated, or seleniated triphos ligand is obtained (ref. 98)(**72**).

In at least two other cases dithiocarbonate ligands have been obtained from the reaction of η^2-CS_2 with molecular oxygen, namely the complex $Pt(\eta^2-CS_2)(PPh_3)_2$ (ref. 117) (73), and the intermediate $Ni(\eta^2-CS_2)(PPh_3)_2$ (see Section 3.2.3) (74)

$$Pt(\eta^2-CS_2)(PPh_3)_2 + O_2 \longrightarrow Pt(PPh_3)_2(S_2CO)$$

(73)

$$Ni(PPh_3)_2(C_2H_4) \xrightarrow{CS_2} [Ni(PPh_3)_2(\eta^2-CS_2)] \xrightarrow{O_2} Ni(PPh_3)_2(S_2CO)$$

(74)

Cyclo-octasulfur and amorphous selenium react also with the dithiomethyl ester complex $[(triphos)Co(\eta^2-CS_2Me)]BPh_4$ to give trithiocarbonate or monoselenodithiocarbonate derivatives of formulas $[(triphos)Co)(S_2CSMe)]BPh_4$ and $[(triphos)Co(SeSCSMe)]BPh_4$, respectively (ref. 57) (75).

$$(triphos)Co(\eta^2-CS_2Me)]^+ \xrightarrow{S_8} (triphos)Co(S_2C-SMe)]^+$$

$$(triphos)Co(\eta^2-CS_2Me)]^+ \xrightarrow{Se_n} (triphos)Co(SeSC{=}SMe)]^+$$

(75)

The reactions reported in (75) confirm that the metal-η^2-CS_2 moiety is electronically very similar to the η^2-CS_2 parent (see Section 4.2). Various hypotheses about the mechanism of these reactions have been made. None of them, however, has so far received experimental support. Also, Fischer-type carbenes are known to undergo similar reactions with Group VIB elements (ref. 118). The carbenoid character of the η^2-CS_2 and η^2-CS_2R moieties (see Section 4.5) is, thus, indirectly confirmed.

Interestingly, $Pt(PPh_3)_2(\eta^2-CS_2)$ can also be converted to the dithiocarbonate derivative by oxidation with SO_2 (ref. 117) (76). The coupling of two SO_2 molecules at the metal, followed by transfer of one oxygen atom to the η^2-CS_2 group, has been proposed to explain the formation of the dithiocarbonate ligand. The formation of S and SO_3 as products suggests the intermediacy of

$$Pt(\eta^2\text{-}CS_2)(PPh_3)_2 + SO_2 \longrightarrow Pt(PPh_3)_2(S_2CO) + S + SO_3$$

(76)

free or bonded S_2O_3 species in reaction (76). S_2O_3 is known, in fact, to disproportionate at room temperature to give S and SO_3.

3.4.2 Displacement reactions

Neutral molecules such as CO, SO_2, SCNPh, and CSe_2, or cations such as NO^+ and $(N_2Ph)^+$ can selectively displace the η^2-CS_2 ligand from its complexes. In most instances, the replacement occurs as a thermal reaction at ambient temperature. In other cases, UV irradiation of the solutions is necessary to have a complete conversion. A 'modus operandi' seems to be the use of a large excess of the replacing ligand.

The displacement of η^2-CS_2 by CO in $Fe(\eta^2\text{-}CS_2)(CO)_2(PR_3)_2$ complexes has been particularly investigated (ref. 53,119) (77). For R = Ph, the η^2-CS_2 group is slowly displaced by CO in toluene

$$Fe(\eta^2\text{-}CS_2)(CO)_2(PR_3)_2 + CO \longrightarrow Fe(CO)_3(PR_3)_2 + CS_2$$

(77)

solution at room temperature (ref. 53). Conversely, for R = Et or OMe, the reactions occur easily by UV irradiation of the complexes in toluene solution (ref. 119). Photolysis of $Fe(\eta^2\text{-}CS_2)(CO)_2(PR_3)_2$ (R = Et, OMe) in the presence of an excess of PEt_3 or $P(OMe)_3$ gives the corresponding $Fe(CO)_2(PR_3)_3$ derivatives (ref. 119). SO_2 and $(N_2Ph)BF_4$ react at room temperature with $Fe(CO)_2(PR_3)_2(\eta^2\text{-}CS_2)$ (R = Et, OMe, OPh) yielding $Fe(CO)_2(PR_3)_2(SO_2)$, and $[Fe(CO)_2(PR_3)_2\text{-}(N_2Ph)]BF_4$, respectively (ref. 53).

The displacement of the η^2-CS_2 ligand by NO^+ in $Fe(\eta^2\text{-}CS_2)(CO)_2\text{-}(PR_3)_2$ (R = Ph, Me) has been suggested to take place by initial interaction of the nitrosyl cation with the nucleophilic uncoordinated sulfur atom, rather than by substitution at the metal. Support for this reaction pathway comes from the selectivity of the substitution as well as from the inertness of the dithiomethyl ester derivatives $Fe(\eta^2\text{-}CS_2Me)I(CO)(PR_3)_2$ toward the addition of $NOBF_4$ (ref. 120) (78).

$$(OC)_2L_2Fe(\eta^2\text{-}CS_2) \xrightarrow{NOPF_6} [(OC)_2L_2Fe\text{-}NO]^+ \; PF_6^-$$

(78)

Carbon monoxide displaces carbon disulfide from (triphos)Ni-(η^2-CS_2). This reaction is reversible, and depends upon which ligand is in excess (ref. 121) (79).

$$\text{(triphos)Ni}(\eta^2\text{-}CS_2) \underset{CS_2}{\overset{CO}{\rightleftharpoons}} \text{(triphos)Ni(CO)}$$

(79)

Another molecule which is particularly suitable for the displacement of the η^2-CS_2 ligand is SO_2. In addition to the iron complexes reported above, (triphos)Ni(η^2-CS_2) (ref. 90) and $Pt(PCy_3)_2(\eta^2\text{-}CS_2)$ (ref. 122) easily undergo replacement of CS_2 by SO_2 to give (triphos)Ni(SO_2) and $Pt(PCy_3)_2(SO_2)$. The addition of CS_2 to the SO_2 complexes regenerates the η^2-CS_2 starting material (**80**). Unexpectedely, the reaction of (triphos)Co(η^2-CS_2) in

$$\text{(triphos)Ni}(\eta^2\text{-}CS_2) \underset{CS_2}{\overset{SO_2}{\rightleftharpoons}} \text{(triphos)Ni}(SO_2)$$

$$Pt(\eta^2\text{-}CS_2)(PCy_3)_2 \underset{CS_2}{\overset{SO_2}{\rightleftharpoons}} Pt(PCy_3)_2(SO_2)$$

(**80**)

tetrahydrofuran with SO_2 gives the sulfito complex (triphos)Co(SO_3) the structure of which has been established by X-ray methods (ref. 123). The SO_3^{2-} ligand is bonded to cobalt through two oxygen atoms. At present the mechanism for this reaction is not understood.

The reactions of the heteroallene molecules SCNPh (ref. 124), COS (ref. 90), CSe_2 (ref. 90), and CO_2 (ref. 121) with (triphos)M(η^2-CS_2) (M = Co, Ni) are particularly interesting. Although the two η^2-CS_2 compounds are isostructural, the one electron difference existing between the two is of critical importance in determining the reactivity toward the above heterocumulenes. The reactions take place much more readily for nickel than for cobalt, and, in general, different products are obtained (**81**). More specifically, phenylisothiocyanate reacts

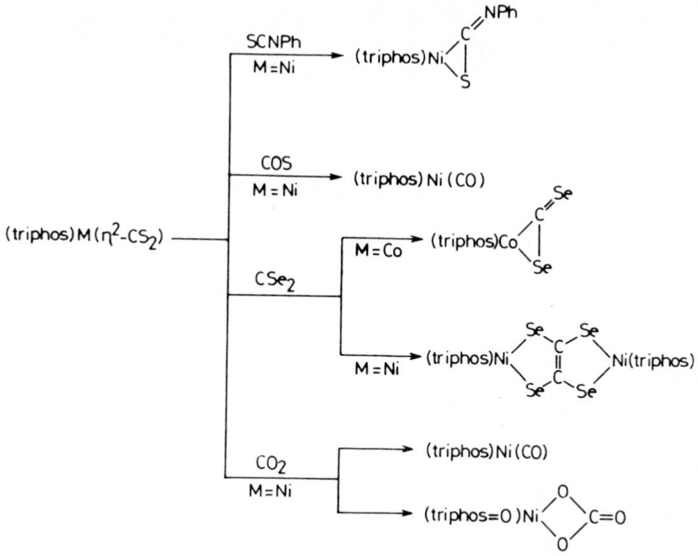

(**81**)

at room temperature with (triphos)Ni(η^2-CS_2) to give the η^2-S,C bonded SCNPh complex (triphos)Ni(η^2-SCNPh) (ref. 124), but does not react with (triphos)Co(η^2-CS_2) even at reflux temperatures. Similarly, COS does not react with the cobalt complex but reacts with the nickel complex to give the carbonyl (triphos)Ni(CO), probably through a η^2-COS intermediate (ref. 90). Carbon diselenide displaces the CS_2 ligand from both nickel and cobalt complexes but the reaction products vary with the nature of the metal: a simple replacement of CS_2 with CSe_2 to give (triphos)Co(η^2-CSe_2) is observed with cobalt, whereas a head-to-head dimerization of two CSe_2 molecules to produce C_2Se_4 is observed with nickel (ref. 90). Finally, CO_2 reacts with (triphos)Ni(η^2-CS_2), but not with the cobalt analog. Depending on the reaction conditions, the carbonyl (triphos)Ni(CO) or the carbonate complex, (triphos=O)Ni(CO_3), is formed (ref. 121).

Displacement of CS_2 by SCNPh to give the complex Co(Cp)(PPh_3)-(η^2-SCNPh) has also been reported (ref. 107).

3.4.3 Catalytic reactions

Reduction of $Co(acac)_3$ with triethylaluminium in the presence of buta-1,3-diene, followed by addition of CS_2 gives the η^2-CS_2 complex shown in (82) (ref. 125).

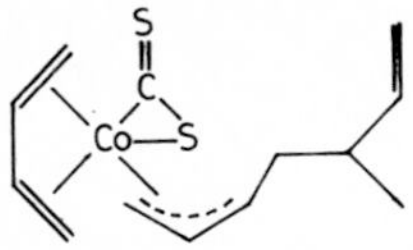

(82)

The latter compound catalyzes the formation of syndiotactic poly-1,2-butadiene. Syndiotactic 1,2 polymerization has been suggested to proceed under the influence of the η^2-CS_2 group, which enhances the reactivity between the terminal carbon atoms of butadiene and the C3 carbon of the π-allyl end. A key role of the catalytic process has been assigned to the nucleophilic character of the uncoordinated sulfur atom of η^2-CS_2, which would interact with organoaluminium.

3.5 η^1-C-BONDED CS_2 COMPLEXES

η^1-C-Bonded CS_2 complexes are obtained by reacting CS_2 with carbonyl monoanions (**83**). Although none of the above anionic complexes could be isolated even in the presence of bulky cations

$$L_nM^- + CS_2 \longrightarrow L_nM\text{-}C\begin{smallmatrix}S\\S\end{smallmatrix}\ -$$

ML_n = $Fe(CO)_2(\eta\text{-dienyl})$ (dienyl = C_5H_5, C_5H_4Me, C_5Me_5),
$Ru(CO)_2(Cp)$, $Re(CO)_5$, $Mn(CO)_4(PCy_3)$

(83)

such as NEt_4^+, $AsPh_4^+$ or $N(PPh_3)_2^+$, enough spectroscopic and chemical evidence for the existence of these dithiocarboxylate anions in solution has been provided (ref. 126-129).

The chemistry of C-bonded CS_2 metal anions is much less extensive than that of η^2-CS_2 complexes. It is limited to electrophilic attack at one or both sulfur atoms. In particular, η^1-C-bonded CS_2 complexes can displace halide ion from alkyl halides (ref. 126,128-130), $SnPh_3Cl$ (ref. 126), $SnMe_3Cl$ (ref. 131), and $SiMe_3Cl$ (ref. 131) to give η^1-dithio ester complexes (**84**).

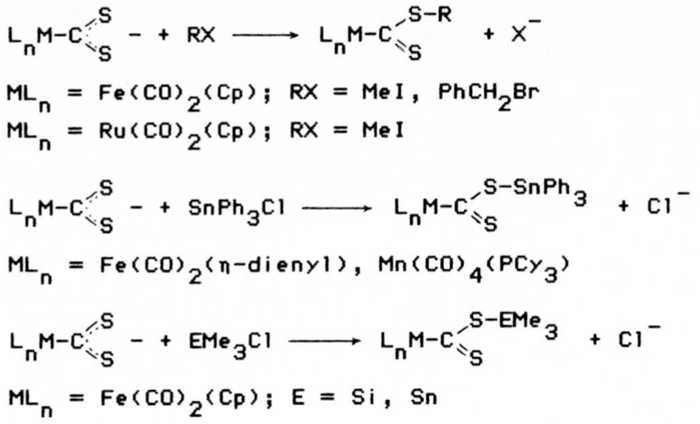

$$L_nM-C(=S)S^- + RX \longrightarrow L_nM-C(=S)S-R + X^-$$

ML_n = $Fe(CO)_2(Cp)$; RX = MeI, $PhCH_2Br$
ML_n = $Ru(CO)_2(Cp)$; RX = MeI

$$L_nM-C(=S)S^- + SnPh_3Cl \longrightarrow L_nM-C(=S)S-SnPh_3 + Cl^-$$

ML_n = $Fe(CO)_2$(η-dienyl), $Mn(CO)_4(PCy_3)$

$$L_nM-C(=S)S^- + EMe_3Cl \longrightarrow L_nM-C(=S)S-EMe_3 + Cl^-$$

ML_n = $Fe(CO)_2(Cp)$; E = Si, Sn

(84)

The uncoordinated sulfur atom of the dithioester ligand in the complex $(Cp)(CO)_2FeC(=S)SR$ can react either with alkyl fluorosulfonates to give the corresponding cationic dithiocarbene derivatives (ref. 130) or with 16-electron metal fragments to form heterodinuclear methyldithiocarboxylato-bridged complexes (ref. 131) (**85**). Interestingly the complex $(Cp)(CO)_2FeC(=S)SMe$

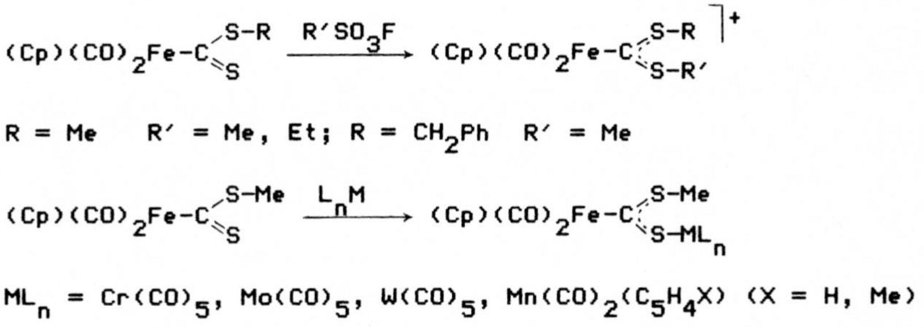

$$(Cp)(CO)_2Fe-C(=S)S-R \xrightarrow{R'SO_3F} [(Cp)(CO)_2Fe-C(S-R)(S-R')]^+$$

R = Me R′ = Me, Et; R = CH_2Ph R′ = Me

$$(Cp)(CO)_2Fe-C(=S)S-Me \xrightarrow{L_nM} (Cp)(CO)_2Fe-C(S-Me)(S-ML_n)$$

ML_n = $Cr(CO)_5$, $Mo(CO)_5$, $W(CO)_5$, $Mn(CO)_2(C_5H_4X)$ (X = H, Me)

(**85**)

reacts with dicobaltoctacarbonyl with formation of a carbido cluster containing the MeSC fragment located on top of the cobalt triangle (ref. 131) (**86**).

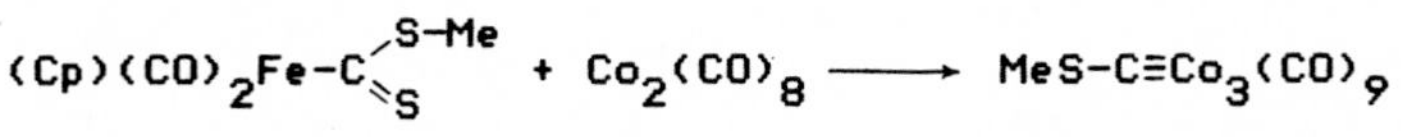

$$(Cp)(CO)_2Fe-C(=S)S-Me + Co_2(CO)_8 \longrightarrow MeS-C{\equiv}Co_3(CO)_9$$

(**86**)

The $(Cp)(CO)_2FeCS_2^-$ anion reacts also with organometallic derivatives producing complexes with structure I or II (**87**), in which the metallodithiocarboxylato fragment behaves as a uni- or bidentate ligand by means of one or both sulfur atoms (ref. 31,89, 126,131,132) (88).

$(Cp)(CO)_2Fe-C(S)S-ML_n$ (I) $(Cp)(CO)_2Fe-CS_2ML_n$ (II)

(87)

A perusal of IR spectra in the $\nu(CS_2)$ region indicates a substantial difference in the absorption pattern between complexes

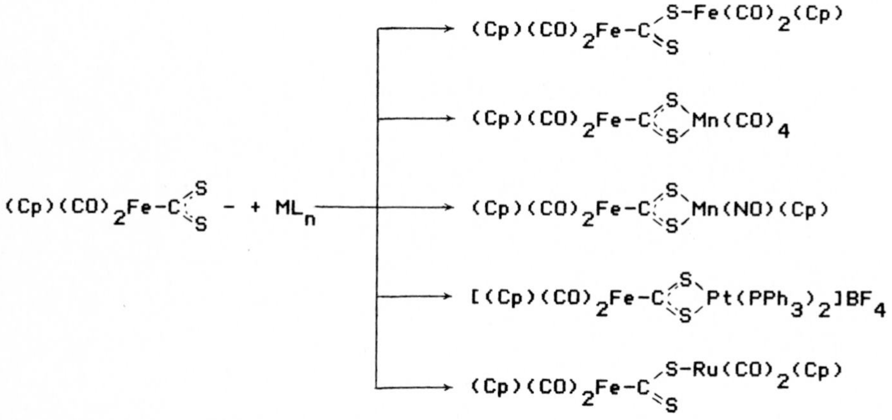

$(Cp)(CO)_2Fe-CS_2^- + ML_n \rightarrow$
$(Cp)(CO)_2Fe-C(S)S-Fe(CO)_2(Cp)$
$(Cp)(CO)_2Fe-CS_2Mn(CO)_4$
$(Cp)(CO)_2Fe-CS_2Mn(NO)(Cp)$
$[(Cp)(CO)_2Fe-CS_2Pt(PPh_3)_2]BF_4$
$(Cp)(CO)_2Fe-C(S)S-Ru(CO)_2(Cp)$

(88)

of structure I and II. While complexes of the former type exhibit only one main band at ca. 1000 cm^{-1} (ref. 126,131,132), those of structure II show two bands at ca. 870 and 920 cm^{-1} attributable to ν_{sym} and $\nu_{asym}(CS_2)$ stretching vibration modes (ref. 31,89,132).

Depending on the reaction conditions, both the chelato-complex

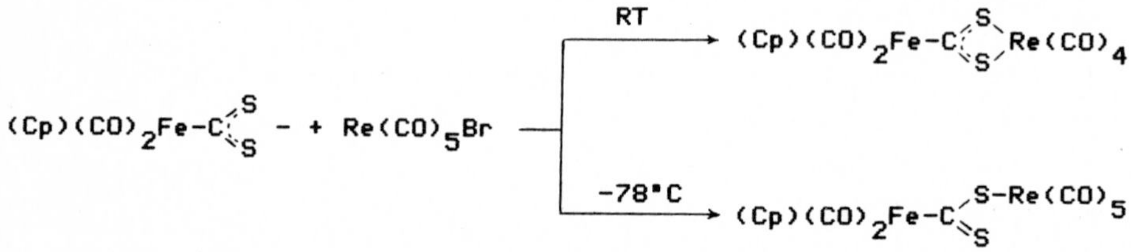

$(Cp)(CO)_2Fe-CS_2^- + Re(CO)_5Br$
RT $\rightarrow (Cp)(CO)_2Fe-CS_2Re(CO)_4$
$-78°C \rightarrow (Cp)(CO)_2Fe-C(S)S-Re(CO)_5$

(89)

$(Cp)(CO)_2FeC(=S)SRe(CO)_4$ and the unidentate $(Cp)(CO)_2FeC(=S)S$-$Re(CO)_5$ derivative have been obtained by Busetto et al. from the reaction of $(Cp)(CO)_2FeCS_2^-$ with $Re(CO)_5Br$ (ref. 132) (**89**).

A trinuclear compound containing two ferriodithiocarboxylato-chelate ligands has been synthesized by Fehlhammer by reacting $(Cp)(CO)_2FeCS_2^-$ with $PtCl_2$ in the molar ratio 2:1 (ref. 89) (**90**).

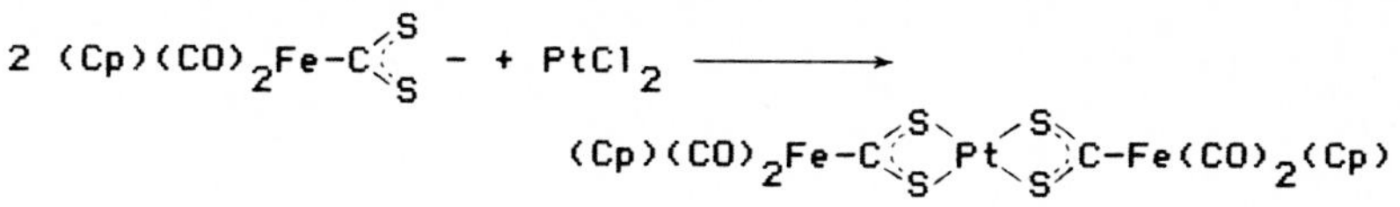

$$2\ (Cp)(CO)_2Fe{-}C{\overset{S}{\underset{S}{\lessdot}}}^{-} + PtCl_2 \longrightarrow (Cp)(CO)_2Fe{-}C\langle S_2\rangle Pt\langle S_2\rangle C{-}Fe(CO)_2(Cp)$$

(**90**)

Protonation of the thione sulfur atom of the dithiocarboxylate complex anion $(Cp)(CO)_2FeCS_2^-$ by protic acids has been recently reported by Fehlhammer (ref. 68,133) (**91**). The neutral complex

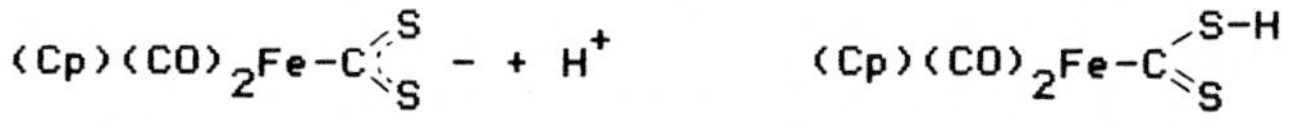

$$(Cp)(CO)_2Fe{-}CS_2^- + H^+ \qquad (Cp)(CO)_2Fe{-}C(=S){-}S{-}H$$

(**91**)

$$Fe{-}C\langle S{-}Os,\ S{-}Os\rangle\ (Os{-}Os)$$

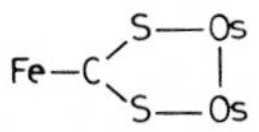

(**92**)

$(Cp)(CO)_2FeCS_2H$ represents the first example of a complex containing a C-bonded CS_2H unit. Reaction of $(Cp)(CO)_2FeCS_2H$ with $Os_3(NCMe)_2(CO)_{10}$ gives a μ_3-CS_2 trinuclear compound which contains an unusual dimetalloheterocycle (ref. 133) (**92**).

3.6 CS_2-BRIDGED COMPLEXES

In this section we describe the reactivity of those complexes where a CS_2 ligand bridges two metal centers. Beginning with the synthesis of $K_6[(CN)_5Co(CS_2)Co(CN)_5]$ (ref. 134), some fifty complexes of this type have been reported in the literature. The synthetic work almost exclusively exploits the reactivity of η^2-CS_2 metal complexes (see Section 3.2.3) or C-bonded CS_2 complex anions towards electrophilic metal fragments (ref. 31,89,126,131,132).

Only a few examples of synthesis by direct reaction of CS_2 with metal species have been reported (ref. 18,26,27,33,111,122, 134,135). In most cases, the structures of CS_2 bridged complexes have been assigned on the basis of spectroscopic data. Complete X-ray structure determinations have been reported only for nine complexes (see Section 2, Table 1 and 2); the following bridging modes **A**, **B**, **C**, and **D** have been identified:

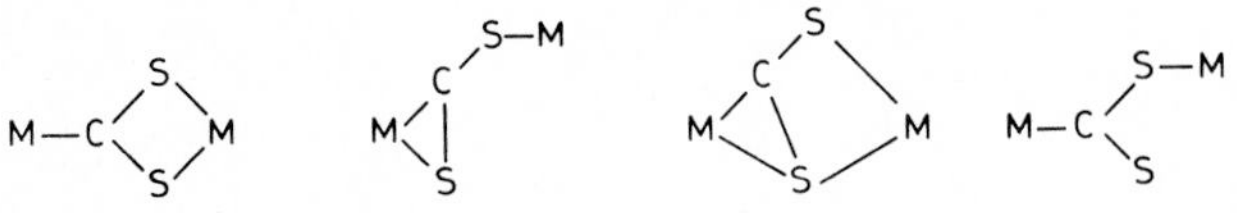

A B C D

IR spectroscopy is a very useful diagnostic tool for discriminating among these four different bonding modes of CS_2. In particular, in the $\nu(CS_2)$ region, complexes of type **B** and **D** show only one band at 1110-1170 and 930-1020 cm^{-1}, respectively, while complexes of type **A** show in general two bands at ca. 870 and 920 cm^{-1}. It is noteworthy that no ν(C=S) bands were observed for the type **C** complex [(triphos)Co(μ-CS_2)Co(triphos)]$(BPh_4)_2$ (ref. 18).

Generally, in bonding modes **A**, **B** and **C** the CS_2 group is inert towards both electrophilic and nucleophilic reagents. This is likely due to the engagement in the bonding framework of all three atoms of the CS_2 molecule. By contrast, the thione sulfur atom of μ-CS_2 complexes of type **D** has a strong nucleophilic character. Accordingly, a general chemical property of metallodithioesters is the ease by which the uncoordinated sulfur is capable of reacting with various electrophiles.

The metallodithioester-metal complexes (Cp)$(CO)_2$FeC(=S)SRe$(CO)_5$, (Cp)$(CO)_2$FeC(=S)Fe$(CO)_2$(Cp) and (Cp)$(CO)_2$FeC(=S)Ru$(CO)_2$(Cp) undergo alkylation at the thione sulfur atom by alkyl-trifluoromethane sulfonate producing stable cationic dithiocarbene complexes (ref. 131,132,136) (**93**).

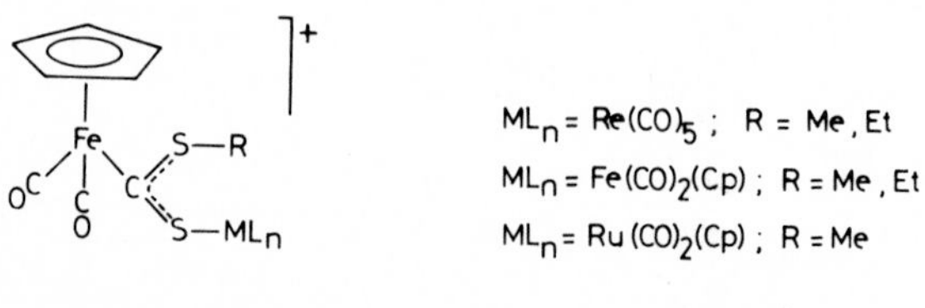

ML_n = $Re(CO)_5$; R = Me, Et
ML_n = $Fe(CO)_2(Cp)$; R = Me, Et
ML_n = $Ru(CO)_2(Cp)$; R = Me

(**93**)

Depending on the reaction conditions, the dinuclear complex $Pt_2Cl_2(\mu\text{-}CS_2)(\mu\text{-}dppm)_2$ undergoes methylation of one or both sulfur atoms. The cationic dithiomethyl ester-bridged complex $[Pt_2Cl_2(\mu\text{-}CS_2Me)(\mu\text{-}dppm)_2]^+$ is obtained by reacting the $\mu\text{-}CS_2$ dimer with $MeOSO_2CF_3$ at room temperature, whereas the cationic dithiocarbene $[Pt_2Cl_2I\{C(SMe)_2\}(\mu\text{-}dppm)_2]^+$ is formed when the reaction is carried out in MeI at reflux temperature (ref. 33) (**94**). The formation of the dithiocarbene complex likely occurs through a stepwise mechanism. In the first step, the uncoordinated sulfur atom of $\mu\text{-}CS_2$ is methylated by MeI to give the cationic $\mu\text{-}CS_2Me$ complex. Second, the presence of iodine ions in the reaction mixture could induce the formation of a $Pt\text{-}\eta^1\text{-}CS_2Me$ moiety by

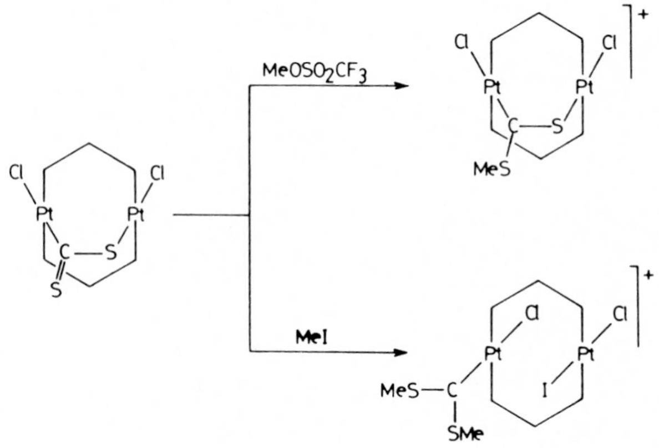

(**94**)

displacing the sulfur ligand from the other platinum atom. Subsequent methylation of the free sulfur atom of the $\eta^1\text{-}CS_2Me$ group should give the final product. An analogous reaction pathway has been proposed to explain the formation of dithiocarbenes from the reaction of some $\eta^2\text{-}CS_2R$ complexes with alkyl halides (see Section 3.2.1).

The thione sulfur atom in metallodithioester complexes $(Cp)(CO)_2FeC(=S)SFe(CO)_2(Cp)$ and $(Cp)(CO)_2FeC(=S)SRe(CO)_5$ reacts rapidly at room temperature with $M(CO)_5THF$ (M = Cr, Mo, W), $Mn(CO)_2(Cp)THF$ or $[Fe(CO)_2(Cp)THF]BF_4$ giving hetero- (ref. 31,133) (**95**) or homo-trinuclear complexes (ref. 133) (**96**) in which a CS_2 molecule acts as a triply bridging ligand. In the $\nu(CS)$ region the $\mu_3\text{-}CS_2$ bridged derivatives reveal a remarkably constant two band pattern at ca. 770 and 950 cm^{-1} attributable to the $\nu_{sym}(CS_2)$ and $\nu_{asym}(CS_2)$ vibrations, respectively (ref. 31).

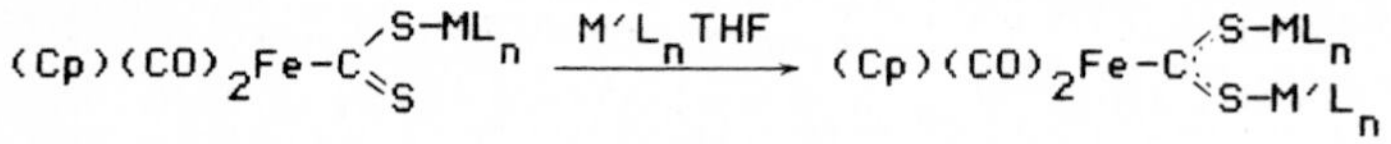

$$(Cp)(CO)_2Fe-C\begin{smallmatrix}S-ML_n \\ S\end{smallmatrix} \xrightarrow{M'L_nTHF} (Cp)(CO)_2Fe-C\begin{smallmatrix}S-ML_n \\ S-M'L_n\end{smallmatrix}$$

ML_n = $Fe(CO)_2(Cp)$, $Re(CO)_5$

$M'L_n$ = $Cr(CO)_5$, $Mo(CO)_5$, $W(CO)_5$, $Mn(CO)_2(Cp)$

(95)

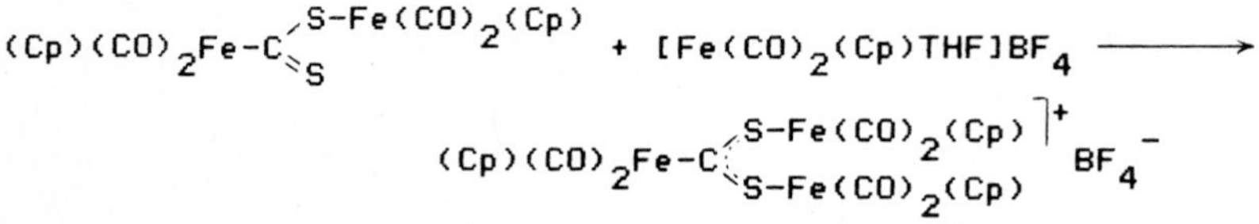

$$(Cp)(CO)_2Fe-C\begin{smallmatrix}S-Fe(CO)_2(Cp) \\ S\end{smallmatrix} + [Fe(CO)_2(Cp)THF]BF_4 \longrightarrow (Cp)(CO)_2Fe-C\begin{smallmatrix}S-Fe(CO)_2(Cp) \\ S-Fe(CO)_2(Cp)\end{smallmatrix}^{+} BF_4^{-}$$

(96)

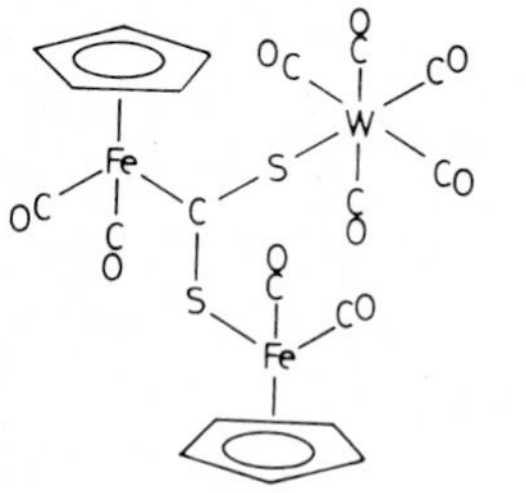

(97)

The structure of the complex $(Cp)(CO)_2FeC(=S)SFe(CO)_2(Cp)W(CO)_5$ (97) has been determined by X-ray methods. This is the first structure of a triply bridged-CS_2 complex (ref. 31).

Generally, μ_3-CS_2 derivatives are air stable in the solid state; however, they decompose in chlorinated solvents, even under

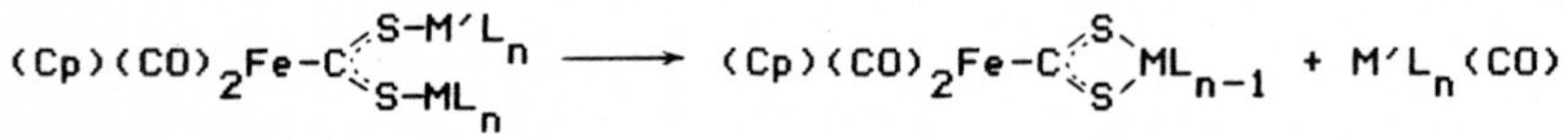

$$(Cp)(CO)_2Fe-C\begin{smallmatrix}S-M'L_n \\ S-ML_n\end{smallmatrix} \longrightarrow (Cp)(CO)_2Fe-C\begin{smallmatrix}S \\ S\end{smallmatrix}ML_{n-1} + M'L_n(CO)$$

ML_n = $Fe(CO)_2(Cp)$, $Re(CO)_5$

$M'L_n$ = $Cr(CO)_5$, $Mo(CO)_5$, $W(CO)_5$, $Mn(CO)_2(Cp)$

(98)

nitrogen, to give dithiocarboxylato-metal complexes (98). Although in low yields and under drastic reaction conditions, the same dithiocarboxylate derivatives can be obtained directly from the dinuclear μ-CS_2 complexes $(Cp)(CO)_2Fe(=S)SFe(CO)_2(Cp)$ (99), and $(Cp)(CO)_2FeC(=S)SRe(CO)_5$ (100) (ref. 31). The ease with which reaction (98) takes place compared to reactions (99) and (100) has been tentatively explained by Busetto (ref. 31). In particular,

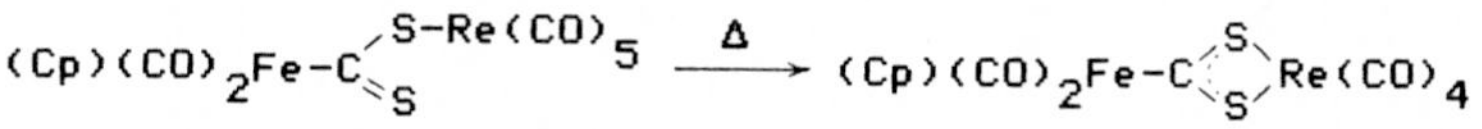

$$(Cp)(CO)_2Fe{-}C(=S){-}S{-}Re(CO)_5 \xrightarrow{\Delta} (Cp)(CO)_2Fe{-}C(S)_2Re(CO)_4$$

(99)

$$(Cp)(CO)_2Fe{-}C(=S){-}S{-}Fe(CO)_2(Cp) \xrightarrow{h\nu \text{ or } \Delta} (Cp)(CO)_2Fe{-}C(S)_2Fe(CO)(Cp)$$

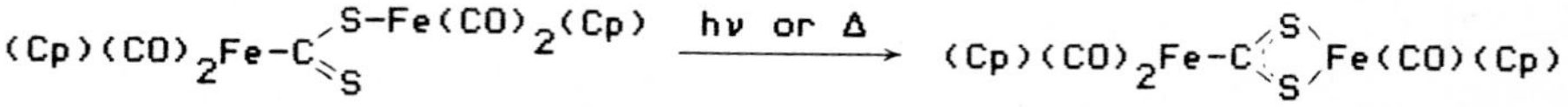

(100)

the further attachment of molecular Lewis-acids at the uncoordinated sulfur atom in the unit Fe(=S)SM seems to have the effect of increasing the rate of the kinetically controlled metallodithiocarboxylate formation rather than the labilizing of the carbon-sulfur bonds of the CS_2 molecule.

The complex $(Cp)(CO)_2FeC(=S)SFe(CO)_2(Cp)$ is easily protonated by CF_3SO_2OH to give the dithiocarbene triflate salt $[(Cp)(CO)_2FeC(SH)$-

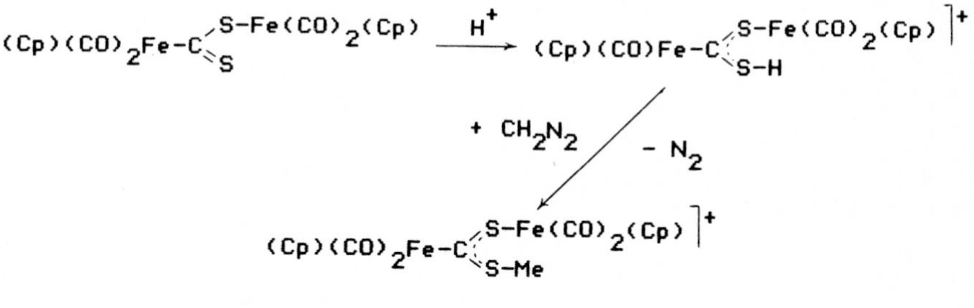

$$(Cp)(CO)_2Fe{-}C(=S){-}S{-}Fe(CO)_2(Cp) \xrightarrow{H^+} [(Cp)(CO)Fe{-}C(S{-}H){-}S{-}Fe(CO)_2(Cp)]^+$$

$$\xrightarrow{+\ CH_2N_2,\ -\ N_2} [(Cp)(CO)_2Fe{-}C(S{-}Me){-}S{-}Fe(CO)_2(Cp)]^+$$

(101)

$SFe(CO)_2(Cp)]SO_3CF_3$ (ref. 131). Interestingly, the latter complex reacts with diazomethane to yield, quantitatively, the homodinuclear methyldithiocarboxylato-bridged complex $[(Cp)(CO)_2FeC(SMe)SFe(CO)_2(Cp)]SO_3CF_3$ (ref. 131) (101).

Reactions of the most stable metallodithioester-metal complex $(Cp)(CO)_2FeC(=S)SFe(CO)_2(Cp)$ with other Lewis acids such as BF_3,

$AlCl_3$ and HgX_2 (X = Cl, Br, I) also proceed with formation of S-adducts as shown by their separation from the reaction mixture or by the IR spectra of the solutions; no stable (BF_3, $AlCl_3$) or pure (HgX_2) products can be isolated, however (ref. 131).

A further interesting point in evidence for the nucleophilic character of the thione sulfur atom in complexes of the type $(Cp)(CO)_2FeC(=S)SML_n$ is provided by their reactions with the cationic complex $[(Cp)(CO)_2Fe(CS)]SO_3CF_3$ (ref. 45). Five-membered metallacycles, $[(Cp)(CO)\overline{FeSC\{Fe(CO)_2(Cp)\}SC}SML_n]SO_3CF_3$, are obtained via nucleophilic addition of the basic thione sulfur atom at the thiocarbonyl carbon atom (**102**). An X-ray structure

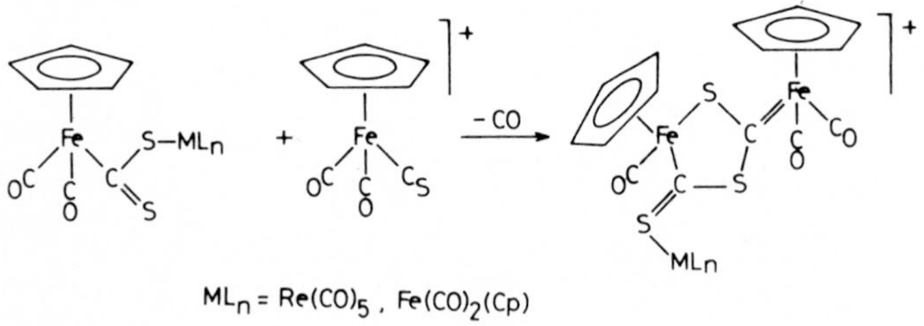

$ML_n = Re(CO)_5,\ Fe(CO)_2(Cp)$

(**102**)

determination, undertaken for the complex with ML_n = $Fe(CO)_2(Cp)$, has shown the presence of a novel trithioanhydride grouping acting as a tetradentate ligand (ref. 45).

Bridged CS_2 complexes of type **A**, **B**, and **C** do not show particular reactivity in the CS_2 group with the exception of the dimeric cobalt(I) complex $[(triphos)Co(\mu\text{-}CS_2)Co(triphos)](BPh_4)_2$ (**103**).

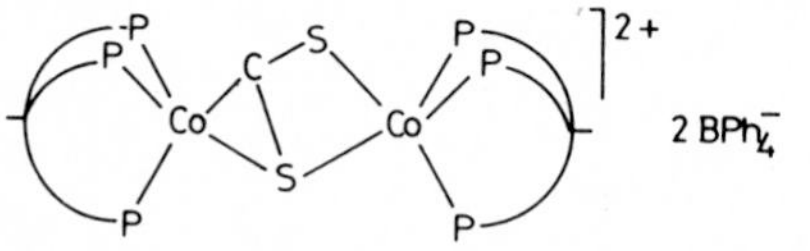

(**103**)

Depending on the reducing agent employed, $NaC_{10}H_8$, or $NaBH_4$, the cobalt(0) complex $(triphos)Co(\eta^2\text{-}CS_2)$ or the 1:1 solid solution formed by $(triphos)Co(\eta^2\text{-}CS_2)$ and $(triphos)Co(\eta^1\text{-}BH_4)$ can be isolated (ref. 18) (**104**). The formation of the $\eta^2\text{-}CS_2$ complex is uniquely due to electron transfer, which induces the rupture of the two cobalt-sulfur bonds. The formation of the solid solution can be attributed both to the reducing and coordinating capabilities of the anion BH_4^-, which can trap the outgoing (triphos)Co moiety to give the η^1-BH_4 complex.

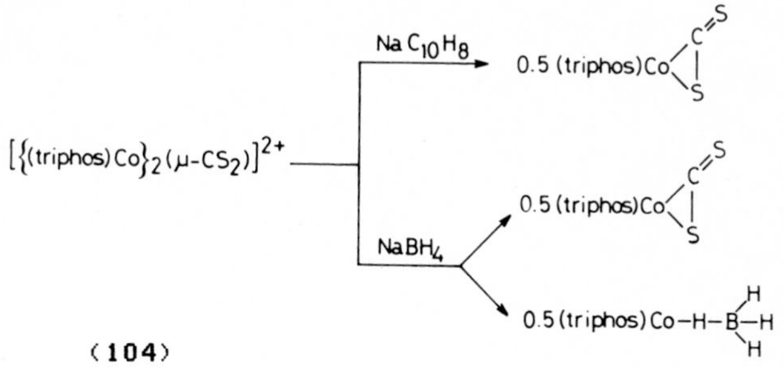

(104)

Finally, it is interesting to describe the synthesis of the μ-C_2S_4 complex $Rh_2Cl_2(CO)(C_2S_4)(dppm)_2$ reported by Cowie and Dwight (ref. 44). This complex can be prepared by reacting CS_2 either with trans-$[RhCl(CO)(dppm)]_2$, or with $Rh_2Cl_2(\mu$-$CO)(dppm)_2$. The structure

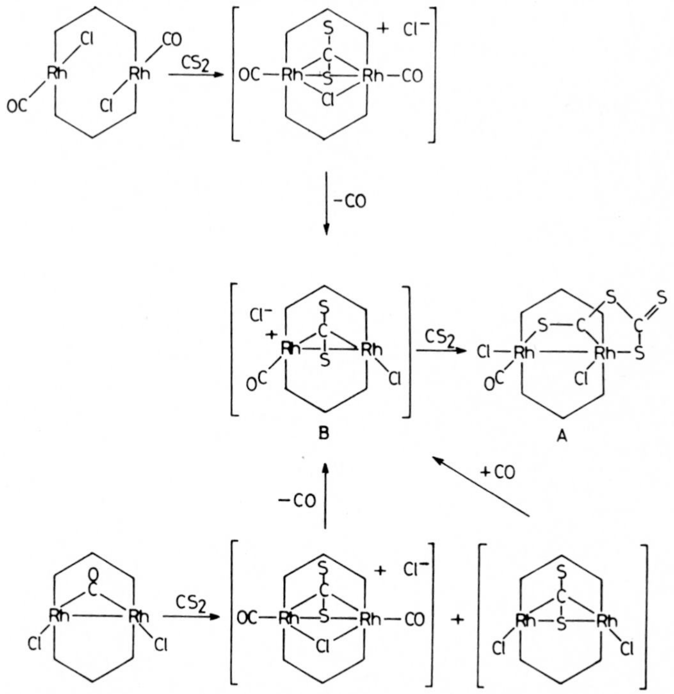

(105)

has been established by X-ray methods. By monitoring the stepwise addition of CS_2 using $^{31}P\{^1H\}$ and $^{13}C\{^{31}P\{^1H\}\}$ NMR and IR spectroscopy, the authors suggest a reaction sequence which involves intermediate dimeric species containing a μ-CS_2 molecule C-bonded to both rhodium centers (**105**). Conversion of intermediary **B** to the final product **A** is believed to proceed by electrophilic attack at one of the sulfur atoms of the μ-CS_2 grouping by another CS_2 molecule. Carbon-sulfur and rhodium-sulfur bond formation, and rhodium-carbon bond rupture are necessary steps to produce the bridging C_2S_4 ligand. The bonding mode of the μ-CS_2 group in intermediates of pathway (**105**) has not been yet clarified. Tentatively, we suggest that the CS_2 bridge in these rhodium dimers is similar to that found in $Pt_2Cl_2(\mu$-$CS_2)(\mu$-dppm$)_2$ prepared and X-ray characterized by Cameron et al. (ref. 33) (**106**). Nucleophilic attack by the thione sulfur atom of the μ-CS_2 group at the carbon atom of an incoming CS_2 molecule would explain the formation of the C_2S_4 ligand.

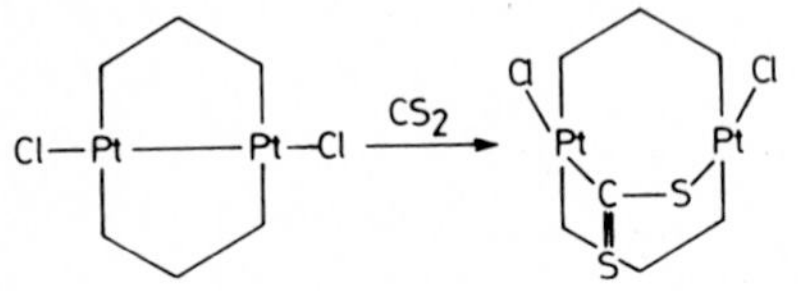

(**106**)

4 ELECTRONIC STRUCTURE OF CARBON DISULFIDE COMPLEXES

4.1 POSSIBLE COORDINATION MODES

The X-ray structures of CS_2 complexes were divided in two main classes (Section 2) depending on the coordination mode of the M-CS_2 unit. The most common mode is η^2 (side-on), schematically shown in (**107**), whereas the η^1-mode, shown in (**108**), is definitely stabilized only when one or both sulfur atoms are engaged in bonding to other groups. Incidentally, η^1-carbon dioxide complexes, although rare, have been isolated in the solid state (ref. 137-139).

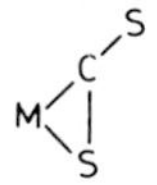

M—S—C—S

(**107**) 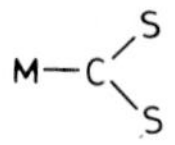 (**108**) 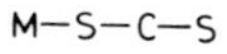 (**109**)

Before discussing the electronic factors which stabilize either one of the two main coordination modes of CS_2, a brief mention must be made relative to a third possibility, namely the end-on coordination mode shown in (**109**). This has been proposed only for a few compounds, such as $Rh(PPh_3)_2(\eta^2\text{-}CS_2)(CS_2)Cl$ (ref. 91) and $[Pd(PMe_3)_2(CS_2)(C_6H_5)(S_2CPMe_3)]BPh_4$ (ref. 140), on the basis of some tenuous spectroscopic arguments. In the end-on mode the linear CS_2 molecule would donate a sulfur σ-lone pair to the metal. In turn, this electron flow away from CS_2 would make the central carbon atom untolerably electrophilic. EHMO calculations have indicated, under certain conditions, a small total energy difference between side-on and end-on coordination of CS_2 as well as a low energy barrier for the interconversion of the two forms (ref. 7). This result is in part justified because the EHMO method both overestimates the repulsion between the sulfur lone pairs on bending CS_2 and underestimates the instability of an excessively positive central carbon atom. On the other hand, for the isoelectronic carbon dioxide molecule, even sophisticated ab-initio methods (ref. 141), confirm that the energetics of side-on and end-on forms are comparable in d^{10}-$L_2M(CO_2)$ complexes. The latter calculations agree well with the experimental η^2-coordination of CO_2 found in $(PCy_3)_2Ni(CO_2)$ (ref. 14), but predict an end-on mode for the unknown isoelectronic cation $[(PH_3)_2Cu(CO_2)]^+$. However, given the lack of significant experimental evidence, the analysis of mode (**109**) will be not developed further in the present context.

The general principles which govern stereochemical control of the bonding modes need to be discussed now.

4.2 M-η^2-CS_2 COORDINATION MODE

Two reasonable models have been proposed (ref. 16,53) to describe the η^2-coordination of CS_2:

i) The Dewar-Chatt-Duncanson model for metal-olefin bonding (ref. 142).

ii) The metallocyclopropane bonding pattern.

In the first case, one π C-S linkage participates with its bonding and antibonding combinations in donor-acceptor interactions with the metal. Conversely, the second point of view considers three independent M-C, M-S and C-S bonds which form a triangle. The coordinated carbon and sulfur atoms act as independent ligands toward the metal, which can be taken as formally oxidized by two electrons. Qualitative MO analysis shows (vide infra) that the actual electron distribution in the M-C-S cycle ranges between the limits i and ii.

VB theory uses the structures (**110**) and (**111**) to describe the M-η^2-CS_2 linkage in terms of the C-D-D model.

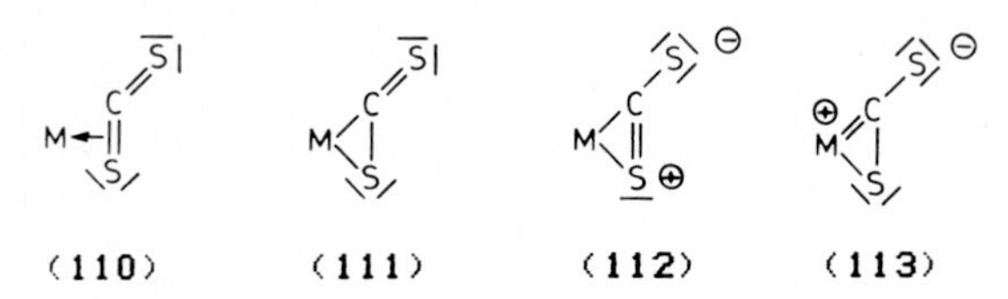

Model (**112**), given its 1,3-dipole character, would explain the facile cycloaddition of acetylenes to these complexes to yield five-membered metallocycles (see Section 4.5). Finally, model (**113**) has been proposed (ref. 16) on the basis of the carbene-type reactivity of η^2-CS_2 complexes, and on account of the short M-C distances associated with these compounds. We believe that the validity of model (**113**) is questionable since any metal d_π orbital, suited to form a $\pi_\perp$ bond with CS_2, projects into a node of the proper CS_2 $\pi_\perp^*$ orbital, halfway between the coordinated C and S atoms. This point is strongly supported by EHMO calculations, which dismiss the possibility of double bond between the metal and the carbon atom (ref. 7). On the other hand, the carbenoid reactivity exhibited by some η^2-CS_2 complexes can be rationalized, in most instances, without invoking any M-C double bond character (see Section 4.5).

Given the delocalization of both π and $\pi_\perp$ interactions over the three atoms of CS_2, the important point to be made is that both the donation and the backdonation to and from the metal cannot be restricted to one C=S linkage. For this reason, VB theory proposes a few resonance structures, but models such as (**110-112**) do not fully illustrate the nature of the electrons involved (either stick bonds or lone pairs). In fact, it is well known that single Lewis formulas are not well suited to represent systems such as CO_2, CS_2, SO_2, etc. and, ultimately, the difficulties extend to the metal complexes of these 16 and 18 electron triatomics. Apparent "octet rule" violations in the latter molecules were accounted for with the introduction of the concept of "fractionality" of certain LMOs (ref. 143) connected to a single center. However, no method other than MO theory describes better the highly delocalized nature of these systems. Applied to M-CS_2 complexes, the MO analysis offers a deep insight of the bonding features, with the advantage that it also permits a better evaluation of the VB structures, most familiar to the chemist.

A successful way of performing a general MO analysis involves the partitioning of the complexes into metal-ligand (L_nM) and CS_2 fragments (ref. 144-146). Let us consider the FMOs of CS_2 first.

The qualitative MO structure of the triatomics AB_2 are known in detail. The pioneering ideas of Mulliken and Walsh (ref. 147,148), developed in fine detail by Buenker and Peyerimhoff (ref. 149), are now beautifully summarized in the textbook of B.M. Gimarc (ref. 150).

For a qualitative understanding of the bonding with the metals, the CS_2 MOs, which need to be focused on, have p_π character. The contribution of the AOs of type s, which is of some importance (ref. 6), can be neglected in first approximation. In the linear geometry ($D_{\infty h}$), the p_π orbitals of each atom combine to give three doubly degenerate sets. These, depicted at left in **Fig. 23**, are classified in ascending order of energy as $1\pi_u$, π_g and $2\pi_u$. The first set is S-C-S overall bonding with π and $\pi_\perp$ components, the set $2\pi_u$ being its antibonding counterpart. Finally π_g is considered to be the equivalent of two lone pairs on the sulfur atoms having π and $\pi_\perp$ character, respectively.

For a 16 electron molecule, such as CS_2, the antibonding level $2\pi_u$ is vacant so that an overall double bond character can be correctly assigned to the three centers S=C=S linkage.

The qualitative MO theory explains the origin of each stick in the VB structure (**114**). Two C-S bonds and two lone pairs on sulfur atoms have a σ character which is implicit in the nature of four low MOs not reported in **Fig. 23.** The low energy of the two lone

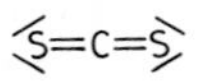

(**114**) 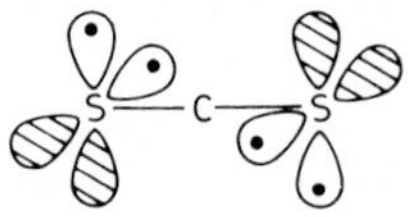 (**115**)

pairs, and some residual C-S σ bonding character in them, are indicative of their poor σ-donor capabilities. The second stick bond at each C-S linkage has actually 50% π and 50% $\pi_{\perp}$ character. The same holds for the second set of lone pairs in (**114**). Practically, the latter are only formal lone pairs since their electrons lie on two different p_{π} orbitals of each sulfur atom, as schematically shown in (**115**). The lack of well developed lone pairs on terminal atoms agrees well with the lack of any definite end-on complex of linear carbon disulfide.

The evolution of the above six π orbitals on bending the triatomic is also shown in **Fig. 23.** The overlap rule easily accounts for the trends of the energy variations. Since both the components of the π_g level (HOMO for 16 electron molecules) are destabilized on bending, the linear ground state is rationalized. The other feature of particular interest in the Walsh-type diagram is the fast energy decrease of LUMO, $2a_1$. This occurs because adjacent lobes on parallel in-plane p_{π} AOs, which are C-S antibonding, become less so and ultimately even C-S bonding. Were $2a_1$ populated by electrons, a bent geometry would be stabilized. Naturally, this is the case of 18 electron molecules such as SO_2, but even CS_2 bends when receiving electron density from a metal upon coordination.

In order to achieve M-η^2-CS_2 complexation the metal fragment must carry one empty σ and one filled π hybrid, both projecting toward CS_2. These hybrids are schematically shown in (**116**) and (**117**) respectively.

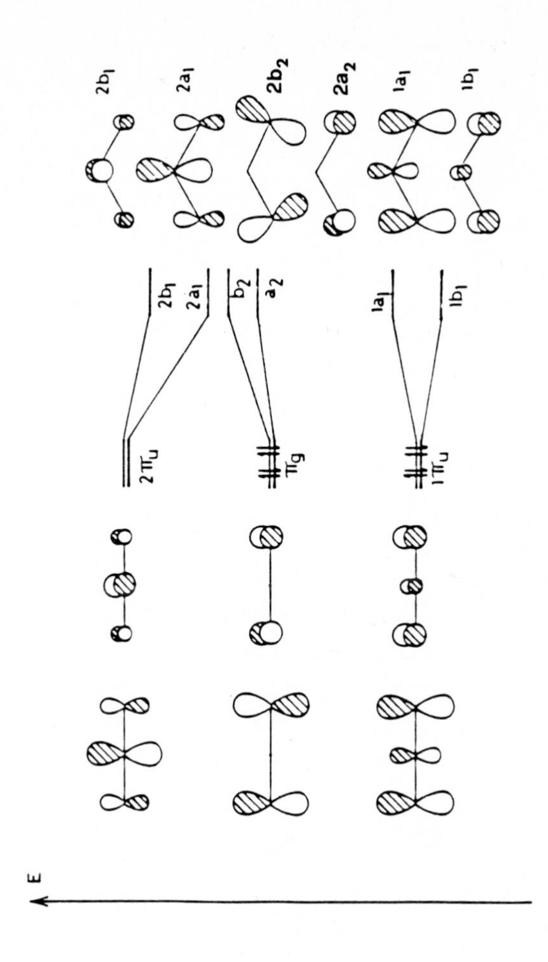

Fig. 23. Qualitative MO correlations for linear and bent CS_2

The frontier orbitals of the most typical transition metal fragments are now known in detail mainly thanks to the cataloging

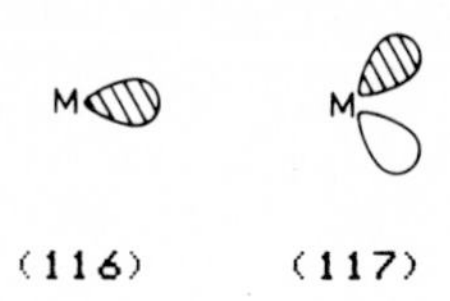

(116) (117)

work done by Roald Hoffmann and his school (ref. 9,144,145).

Fig. 24 presents the orbital templates for the metal fragments which are found in η^2-CS_2 complexes, namely d^{10}-L_2M, d^8-L_4M, d^{10}-L_3M or d^9-L_3M, d^3-$(Cp)_2M$ or d^2-$(Cp)_2(R)M$.

The fragments d^{10}-L_2M and d^8-L_4M, both having C_{2v} symmetry, are essentially similar, as their HOMO and the LUMO identify with the levels b_2, π, and a_1, σ, respectively. The L_3M fragment is different only because the π orbital used for π-backdonation, e_s, is degenerate with a perpendicular $\pi_\perp$ orbital, e_a. In general, the latter orbital has only a minor perturbative effect on the bonding to CS_2. In practice, e_a remains a metal centered, non-bonding, level. The structures (triphos)Ni(η^2-CS_2) (ref. 8) and (triphos)Co-(η^2-CS_2) (ref. 18), which, in first approximation, are superimposable, indirectly confirm that e_a can be indifferently populated by two or one electron. Also, EPR experiments confirm the identification of the HOMO with the π orbital e_a (ref. 8).

No MO calculations have been attempted for models of either $V(\eta^2$-$CS_2)(Cp)_2$ (ref. 19) or $Nb(\eta^2$-$CS_2)(Cp)_2(R)$ (ref. 20,21) complexes. However the frontier orbitals of $(Cp)_2M$ and $(Cp)_2(R)M$ fragments, presented at the right of **Fig. 24**, suggest bonding capabilities similar to those of the previous cases.

A d^3-Cp_2M fragment has a σ-acceptor orbital, $2a_1$, which lies higher in energy than the π orbital, b_2, suited for backdonation. Close to the latter, there is another orbital of a_1 symmetry. It may be reasonably assumed, also on the basis of a MO study performed for the olefin complex $Mo(C_2H_4)(Cp)_2$ (ref. 151), that this $1a_1$ level remains substantially non-bonding toward CS_2 and singly occupied (SOMO). EPR studies (ref. 94), performed on a series of derivatives of $V(\eta^2$-$CS_2)(Cp)_2$, where the exocyclic sulfur atom is linked to different electrophiles, show constant g and A_{iso} values. Indirectly, the assignement of the unpaired electron to a metal-centered orbital, non-bonding toward CS_2 is confirmed.

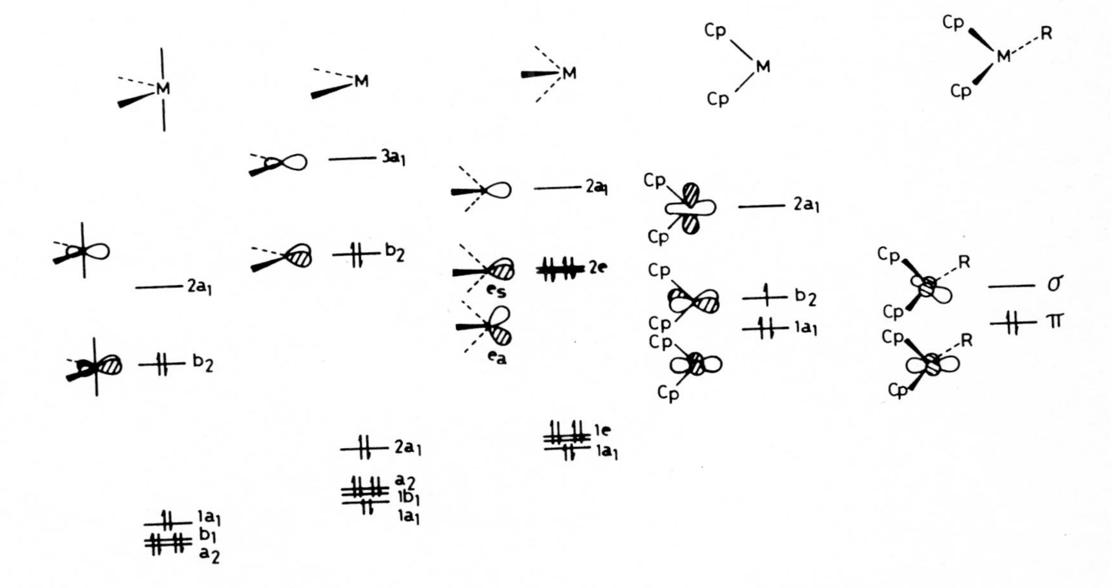

Fig. 24. Valence Molecular Orbitals of some metal fragments which support η^2-CS_2 coordination. The energy scale is arbitrary and meaningful to describe the order of the levels.

Ultimately, there is a strict analogy with the electronic character of the d^9 complex (triphos)Co(η^2-CS_2) (ref. 18).

Also, the d^2-$(Cp)_2(R)M$ fragment seems suited for dihapto-bonding to CS_2 on account of the σ acceptor and π donor character of its LUMO and HOMO levels. The interested reader may find the derivation of the important valence orbitals from those of the $(Cp)_2M$ fragment elsewhere (ref. 151,152). The olefin complex of formula $Nb(C_2H_4)(Cp)_2(R)$, analogous to the two Nb-CS_2 complexes in question, has been theoretically investigated by Eisenstein and Hoffmann (ref. 152).

Finally, it is worth mentioning that fragments, reducible to the type d^6-ML_5, also support η^2-CS_2 coordination in the complexes Mn(η^2-CS_2)(Cp)$(CO)_2$ (ref. 41) and Cr(η^2-CS_2)(η-arene)$(CO)_2$ (ref. 75), for which no X-ray structure is available. The orbital template of the ML_5 fragment is presented in **Fig. 26** of Section 4.3 together with those of other fragments typical of the η^1-CS_2 coordination. The amphotheric behavior of ML_5 depends, case by case, on the total number of electrons. Thus, in the anion $Fe(Cp)(CO)_2(CS_2)^-$ (ref. 126), which carries two electrons more than its Mn analog, the CS_2 molecule is η^1-coordinated. The point will receive further attention in the next section. Concerning the support for η^2-coordination, π-backdonation from the metal is provided by either one of the two members of a low lying e set, pure d orbitals. Significantly, the only ascertained examples of η^2-CS_2 coordination of ML_5 fragments involve Cr and Mn metals whose d orbitals are sufficiently expanded. On the other hand, the isolobal analogy between d^8-ML_4 and d^6-ML_5 fragments is well recognized (ref. 9). Before leaving this argument, it is worth mentioning that the fluxional behavior observed for Mn(η^2-CS_2)(Cp)$(CO)_2$ (ref. 41) is due to the presence of two perpendicular d_π orbitals, members of the e set, and suited for π-backdonation.

Fig. 25 summarizes the main interactions between the two selected metal orbitals **116** and **117** and the CS_2 π orbitals of **Fig. 23**.

On the right side, the CS_2 $\pi_\perp$ levels $1b_1$, a_2 and $2b_1$ remain basically unaffected. Only a_2 is slightly pushed up in energy by some lower filled metal $\pi_\perp$ level. Accordingly, the $\pi_\perp$ electron distribution within CS_2 (four electrons) is quasi unchanged after the interaction with the metal. As in free carbon disulfide, an

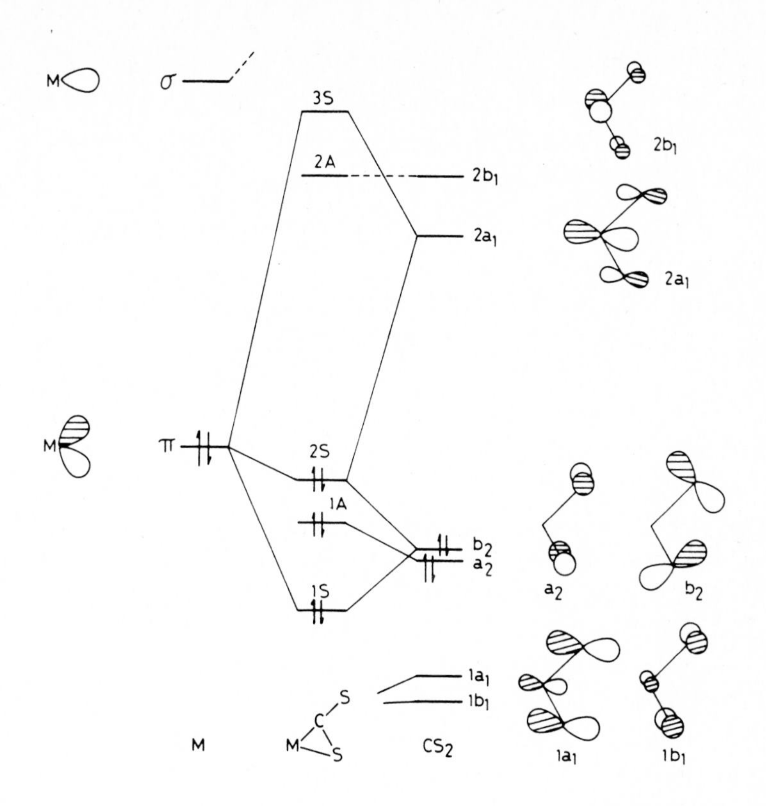

Fig. 25. Diagram of the important interactions between the valence FMOs of a typical metal fragment and of those of CS_2.

overall $\pi_{\perp}$ bond is distributed over the three centers and a $\pi_{\perp}$ lone pair is shared by the two sulfur atoms (see the VB structures (**111**) and (**112**)).

The important interactions between fragments occur in the plane of M-CS_2. Briefly, the CS_2 $1a_1$ level, more developed on sulfur than on carbon on account of electronegativity arguments (ref.150), donates electron density into the empty, high lying, σ metal hybrid, (**116**). Backdonation occurs from the metal π hybrid, (**117**), into the CS_2 in-plane π^* level, $2a_1$. On the basis of overlap and energy arguments which govern the orbital interactions at the second order (ref. 153), the π-backdonation is stronger, in nice agreement with the Lewis-acid nature commonly assigned to CS_2.

The diagram of **Fig. 25** unveils a critical point: the MO 2S, which, in first approximation, is the M-CS_2 bonding level corresponding to backdonation, is somewhat destabilized by a substantial mixing of the lower level b_2. Recall that, in free CS_2, b_2 is a non-bonding combination on the two sulfur atoms. The more b_2 is involved in bonding interactions to the metal, the less appropriate is the C-D-D model, which basically utilizes only the donor, $1a_1$, and the acceptor, $2a_1$, levels of CS_2.

In turn, the metallocyclopropane description ii) becomes more satisfying. The latter may be visualized as follows: the electron transfer from the metal to the CS_2 π^* level, $2a_1$, greatly reduces the in-plane overall π bonding of the triatomic. In the hypothetical case where $2a_1$ becomes fully populated (this corresponds to an oxidation of the metal) the overall S-C-S, in-plane π bonding, vanishes. After a redistribution of the in-plane carbon and sulfur p_π orbitals, a σ lone pair can be assigned to each atom of the triatomic. The lone pairs of carbon and coordinated sulfur atoms can be thought of as being either donated to, or shared with, the L_nM^{2+} fragment, whereas the third lone pair remains located on the exocyclic sulfur atom. In conclusion, the remarkable effect of metal coordination is the activation of a dormant nucleophilicity of the exocyclic sulfur atom. Notice, in fact, that free CS_2 can even be used as solvent in alkylation reactions (ref. 63). Eventually, the lone pair, on the exocyclic sulfur atom, is used to form a bond with an electrophilic reactant, without any further rehybridization. For example, **Table 1** shows quite similar geometrical features in $V(\eta^2\text{-}CS_2)(Cp)_2$ (**Fig. 6**) and its methylated derivative, $[V(\eta^2-CS_2CH_3)(Cp)_2]^+$ (**Fig. 12**). A similar relationship occurs in (triphos)Co(η^2-CS_2) (**Fig. 5**) and (triphos)Co(η^2-CS_2)-$Cr(CO)_5$ (**Fig. 9**).

The capability of the exocyclic sulfur atom to act as a strong ligand toward a second metal atom, which is apparent in the latter compound, is confirmed by the formation of certain binuclear (**118**) (ref. 26-28) and polynuclear (**119**) (ref. 95) derivatives of the unit $M(\eta^2-CS_2)(PR_3)$ (M = Ni, Pd, Pt). Formally, the systems derive from classical $M(\eta^2-CS_2)(PR_3)_2$ complexes by replacing one phosphine ligand with the exocyclic sulfur atom of another M-η^2-CS_2 unit.

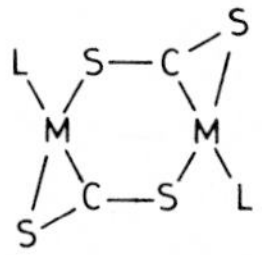

(**118**) 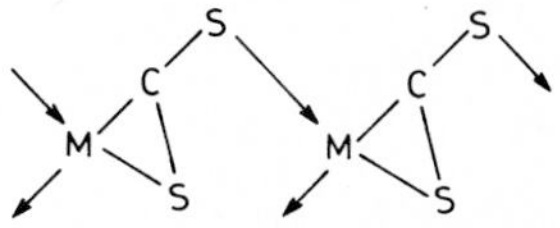 (**119**)

The diamagnetic dimer $[(triphos)Co(\mu-CS_2)Co(triphos)]^{2+}$ (ref.18) whose skeleton is shown in (**120**), has some unique features.

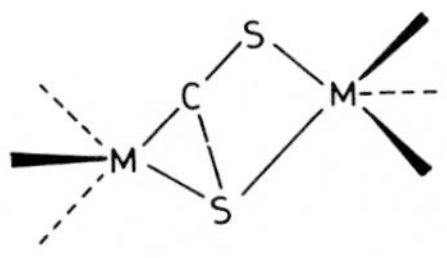

(**120**)

Surprisingly, the two sulfur atoms of a CS_2 molecule, dihapto bonded to a CoL_3 fragment, appear to have residual, almost equal, bonding capabilities toward a second metal fragment, as shown by the close similarity of the newly formed Co-S bonds. The bridging CS_2 molecule forms with the two metals two coplanar, condensed, three-membered and four-membered rings. The whole system is highly conjugated as shown by the quenching of the paramagnetism on the two metal centers. Consider, in fact, that a way of obtaining this μ-CS_2 dimer is to react two paramagnetic centers, namely the compound (triphos)Co(η^2-CS_2) and a d^7-$(triphos)Co^{2+}$ fragment.

Initially, the intriguing point is that CS_2 has a total of three in-plane p_π atomic orbitals available, so that it is not so clear how they comply with the four sticks, drawn to describe the connectivity between the metals and the atoms of CS_2.

EHMO calculations (ref. 154) shed some light on the real nature of the bonding. A qualitative analysis can be performed on the basis of a fragmentation of the dimer between M_2L_6 (terminal) and CS_2 (bridging) groups. Indeed, the bent CS_2 molecule uses the three in-plane π FMOs ($1a_1$, b_2, $2a_1$ in **Fig. 23**) to make positive overlaps with three opportune combinations of the metal orbitals (**116**) and (**117**). Given the very low symmetry of the system (a pseudo-mirror plane, at best), it is not possible to correlate any of the three overall bonding interactions with three specific, localized M-CS_2 stick bonds in (**120**). As implied by the cyclopropane model ii), the $ML_3(\eta^2\text{-}CS_2)$ part of the dimer can be ideally subdivided into bipositive ML_3^{2+} and binegative CS_2^{2-} fragments. In this case the CS_2 bridge is capable of donating a total of six electrons to two identical d^7-ML_3 fragments. At this stage, each metal attains a formal d^{16} electron count.

Eventually, the fourth bond between CS_2 and the metals, as well as the formal 18 electron count of the latter, are provided by $\pi_\perp$ interactions. The character of the LUMO and HOMO levels reveals where the extra stability of the dimer comes from (**121**).

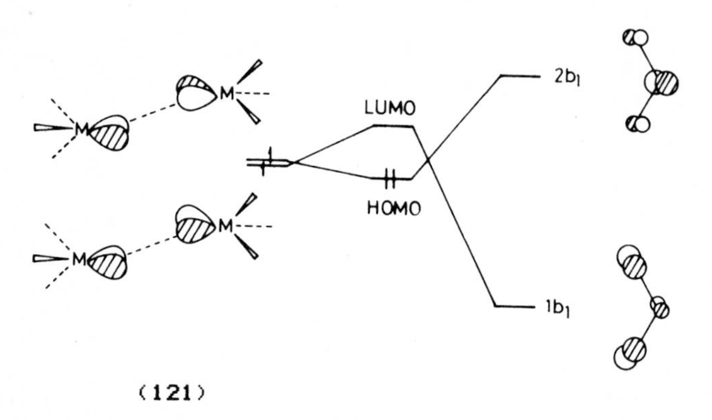

(**121**)

Thus, an in-phase combination of two ML_3 $\pi_\perp$ hybrids interacts, in antibonding fashion, with the filled, S-C-S overall bonding level $1b_1$. In this manner, a fourth CS_2 lone pair, $\pi_\perp$, is formally donated to the two metals. The corresponding out-of-phase metal $\pi_\perp$ level is stabilized by $2b_1$. In spite of the small overlap between the $\pi_\perp$ fragment orbitals, the latter interactions result in a HOMO-LUMO gap of ca. 0.8 eV, sufficiently large to account for the stabilization of the dimer. Moreover, the unpaired electrons of the metals couple together in the HOMO to complete the formal 18 electron count.

The described $\pi_{\perp}$ interactions also help to interpret some structural details of the dimer. Although the structure of $[(triphos)Co(\mu-CS_2)Co(triphos)]^{2+}$ is affected by high standard deviations, the observed trends for long C-S and short Co-C bonds agree with a sort of synergistic $\pi_{\perp}$ donation-backdonation effect. This reduces the overall S-C-S $\pi_{\perp}$ bonding and reenforces the Co-C linkage, by allowing some double bond character for it.

In summary, the bonding pattern depicted for the binuclear compound is due to unique electronic features. Notice that the monomer $(triphos)Ni(\eta^2-CS_2)$ (ref. 8), geometrically equivalent to $(triphos)Co(\eta^2-CS_2)$ (ref. 18), does not form analogous binuclear compounds. The obvious reason seems to be the abscence of favorable $\pi_{\perp}$ interactions, especially due to the impossibility of coupling unpaired spins.

4.3 η^1-C COORDINATION MODE

All of the known structural examples of η^1-coordination of carbon disulfide were enumerated in Section 2.3. Again, it is remarkable that no X-ray structure, where an intact CS_2 molecule is η^1 bonded to a metal has been reported as yet.

The electronic underpinnings of η^1-coordination of CS_2, can be related to those of carbene transition metal complexes. Many theoretical studies, even highly sophisticated ones, have been carried out for the latter (ref. 155-160). It goes beyond the purposes of the present article to review in detail the findings of these calculations, but a few concepts, which adapt well to η^1-CS_2 carbenoid compounds, are conveniently summarized. Two main bonding interactions occur between the CX_2 unit and the metal fragment. These, schematically shown in (**122**) and (**123**), are common to all carbenes but an important distinction is to be made between

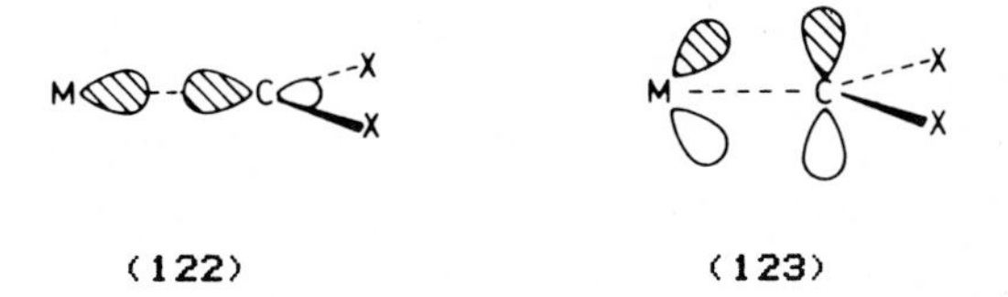

(**122**) (**123**)

Fisher-type and Schrock-type carbenes (ref. 160). The former may be viewed as singlet state CX_2 units donating a lone pair to the metal from the hybrid sp^2 carbon orbital, with a corresponding amount of backdonation from metal to an empty π orbital. Conversely, Schrock-type compounds are viewed as triplet-state carbenes spin-coupled to two electrons on an appropriate metal center (usually early trensition metals (ref. 160)). Obviously, the triplet state requires the CX_2 σ and π orbitals to be very close in energy.

The CS_2 grouping is peculiar in that the σ and π FMOs of interest, namely the levels $2a_1$ and $2b_1$ of **Fig. 23**, are quite high in energy on account of their antibonding character between carbon and sulfur atoms. An EHMO calculation (ref. 154) places both levels at an energy which is as much as 2 eV higher than that of the corresponding CH_2 levels. Thus, for $M-CS_2$, the metal hybrid of the σ interaction (**122**), lies lower in energy than the FMO $2a_1$. In accordance with the known perturbation theory rules which govern second-order interactions between orbitals (ref. 153), the M-C σ bond can be taken as dative toward carbon. Even in carbene complexes, the carbon atom is commonly believed as positively charged, although recent ab-initio calculations (ref. 159) indicate a shifting of the M-C linkage electron density toward the carbon atom.

In agreement with the above viewpoint, which assigns two electrons to the metal σ hybrid, the typical metal fragments of η^1-CS_2 coordination are the square pyramidal d^8-ML_5 (or $MCpL_2$, alternatively) and the planar $d^{10}-ML_3$ fragments. The corresponding orbital templates are shown at the left of **Fig. 26**.

It was emphasized in Section 4.2 that, whereas the fragment $Fe(Cp)(CO)_2$ stabilizes an anionic η^1 complex of CS_2 (ref. 126), the isostructural fragment $Mn(Cp)(CO)_2$ forms an uncharged η^2 species (ref. 41). Clearly, the different behavior is a consequence of the different total electron count of the CS_2 adducts. The a_1 level of the ML_5 fragment can be taken as populated in the iron complex, so that it acts as a σ donor toward CS_2, properly oriented for η^1-C coordination. By contrast, the electron deficiency at the Mn metal activates the dihapto bonding mode by which CS_2 can exert donor, beside acceptor, capabilities. The number of electrons, rather than the metal nature, activates the η^1-C coordination of CS_2. An explicit example is provided by the existence of L_5M dithiocarbene complexes of the chromium triad such as $W(CO)_5[CS(SR)]^-$ and $W(CO)_5[C(SR)_2]$. Notice that the latter are not directly obtainable

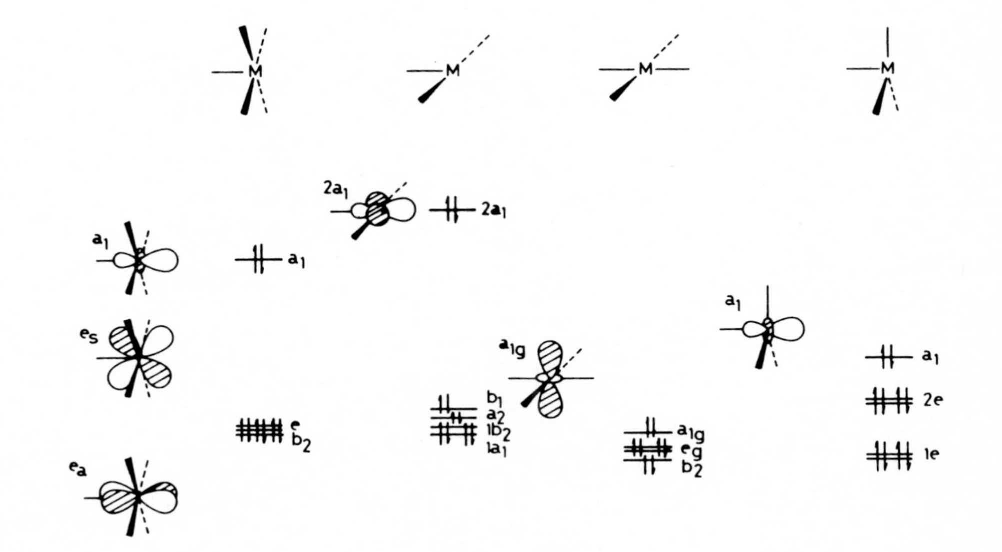

Fig. 26. Valence Molecular Orbitals of some metal fragments suitable to support η^1-CS_2 coordination. The energy scale is arbitrary and meaningful only to describe the order of the levels.

from carbon disulfide derivatives (ref. 161). In particular, $W(CO)_5[CS(SR)]^-$ has the same formal electron count as $Fe(Cp)(CO)_2$-[CS(CMe)] (ref. 162), which, in turn, can be taken as the formal methylated derivative of the anion $Fe(Cp)(CO)_2(\eta^1-CS_2)^-$ (ref. 126).

Concerning the π-backdonation in carbene complexes, Peter Hofmann has pointed out (ref. 155) that there is a competition, for $\pi_\perp$ bonding with carbon, between the metal and the other adjacent groupings. The competition, in η^1 dithiocarbenoid complexes, is well represented by the three VB resonance structures (**124-126**).

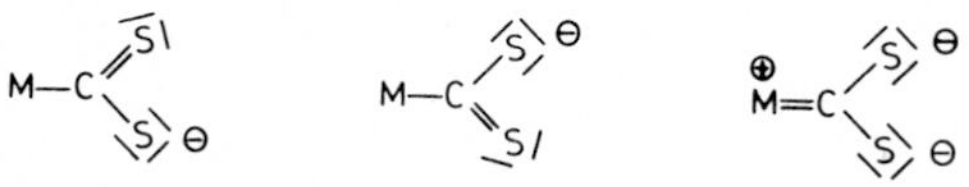

(**124**) (**125**) (**126**)

The models (**124**) and (**125**) are the η^1 equivalents of the η^2 structures (**111**) and (**112**), with the difference that an in-plane lone pair of one sulfur atom is not used for σ bonding to the metal. The structure (**126**) corresponds to a prevalent $\pi_\perp$ back-donation from the metal. However, given the strong donor character of the S or SR substituents, the weight of structure (**126**) cannot be large. In addition, recall that the metal is already donating σ electron density to CS_2. Experimentally, the elongation of the Os-C bond in the complex $OsH(\eta^1-CS_2Me)(PPh_3)_2(CO)_2$ (ref. 36) with respect to other osmium carbenes (ref. 42) (see Section 2.3) confirms the minor M-C double bond character of dithiocarbenes.

Somewhat surprisingly, the two C-S linkages of M-η^1-CS_2 derivatives are, on the average, only slightly elongated with respect to M-η^2-CS_2 complexes. In fact, given the C-S antibonding character of the $2a_1$ and $2b_1$ CS_2 levels, σ backdonation from the metal, as well as π (if any), should both stretch the C-S linkages. In addition, an increasing electron density in $2a_1$ favours the bending of CS_2. Since the latter effect is clear-cut in η^1-CS_2 complexes with S-C-S angles 15-30° smaller than in η^2 complexes, we conclude that the σ interaction is predominant.

Other candidates for η^1-type coordination are the square planar d^8-ML_4 complexes and the trigonal pyramidal d^{10}-L_4M fragments (at the right of **Fig. 26**). In these electronic configurations, the HOMOs of the fragments are σ hybrids which offer the electrons for the M-C linkage. Although the existence of η^1-coordinated complexes of CS_2 or CS_2R has not been demonstrated for these fragments, there are known examples of the corresponding CO_2 (ref. 137) and CO_2R (ref. 163) derivatives. In addition, in the known trigonal bipyramidal L_4M(carbene) complexes, the carbene occupies the apical position (ref. 42) and the preference is supported by MO calculations (ref. 159).

Whereas in the M-η^2-CS_2 complexes the metal is formally taken as accepting a lone pair from one sulfur atom, none of the fragments of **Fig. 26** has such an acceptor capability. Consequently, the transformation of η^2 into η^1 complexes is possible, only provided that the supporting metal fragment undergoes also an appropriate geometrical and electronic adjustment. Usually, as shown for some of the reactions reported in Section 3.1.1, a new ligand enters the coordination sphere at the metal fragment, while the CS_2 or CS_2R grouping adapts to the new coordination mode.

The compound $[PtCl(dppm)]_2(CS_2)$ shown in **Fig. 17** of Section 2.3 (ref. 33), can be classified among the η^1-CS_2 species. Actually the dimer in question belongs to the class of molecular "A-frame" complexes, which have small molecules as the apex bridge (ref. 34, 35). Usually this role is played by CO or SO_2 molecules which use two localized hybrids (formed by a σ-donor and a π-acceptor orbital) to connect the two metal atoms. In the present case, the CS_2 molecule, in the particular bridging disposition, projects two distinct orbitals (in principle, one acceptor on carbon and one donor on sulfur) toward the two different metal atoms. Ultimately, the similarity of the square planar environments at the two metal centers indicate that the bonding capabilities of the carbon and sulfur atoms are almost even. Unfortunately, the poor quality of the crystal structure in question does not allow any subtle consideration in this sense.

The class of 1,3-dithiol-2-ylidene complexes, of which Mn(Cp)-(CO)$[P(OMe)_3](CS_2C_2R_2)$ (ref. 39) has been structurally characterized (**Fig. 19**), is assigned to the category of ML_5-η^1-CS_2 systems. These compounds are obtainable from the reaction of η^2-CS_2 complexes with alkynes (see Section 3.1.1). The mechanism is likely that of 1,3-cycloadditions, although it has not been studied in detail (a brief discussion will be given in section 4.5). Preliminary EHMO calculations (ref. 154) show that the HOMO and LUMO levels of the uncharged five-membered cycle are quite similar to the $2a_1$ and $2b_1$ empty MOs of CS_2 (**Fig. 23**). The presence of two electrons in $2a_1$ suggests that the viewpoint, which initially assigns to the metal the σ electron pair used for M-C bonding, is not realistic in this case. Moreover, the σ and π levels in question are close in energy, so that the free cycle can exist in a triplet ground state. Significant, in this respect, is the unique possibility of forming tetrathiafulvalene derivatives from these 1,3-dithiol-2-ylidene complexes, likely through a diradical association mechanism (ref. 73,74,79) (see Section 3.2.2). Moreover, all of the complexes containing the 1,3-dithiol-2-ylidene cycle, η^1- coordinated, are supported by metal fragments typical of η^2-CS_2 coordination, i.e. unsaturated 16 electron species. On the basis of all of these observations, an analogy between 1,3-dithiol-2-ylidene complexes and Schrock-type carbenes can be proposed.

4.4 GEOMETRICAL FEATURES OF M-η^2-CS_2 AND M-η^1-CS_2 MOIETIES

In M-η^2-CS_2 complexes the coordinated C-S bond is generally longer than the corresponding uncoordinated linkage, whereas in M-η^2-CS_2E complexes (E = inorganic or organic electrophile) the trend is sometimes reversed (see **Table 1**). In general, however, the differences are not large and, in any case, the average C-S values (1.68 and 1.60 Å for endo and exo C-S bonds) are intermediate between double and single bonds. Since the metal little affects the $\pi_\perp$ system of CS_2, the $\pi_\perp$ bonding interactions should be equally distributed over the two C-S linkages. At the same time, the overall in-plane S-C-S π bonding is vanishing on account of the electron donation from the metal. Indeed, EHMO calculations (ref. 7) for models using equal C-S lengths yield practically equal C-S overlap populations. Accordingly, the same weight could be assigned to the VB structures (**111**) and (**112**). In conclusion, the semiempirical method does not explain whether certain experimental C-S asymmetries originate from π or $\pi_\perp$ interactions.

Charge distribution arguments (not properly evaluated by the EHMO method) suggest that the localization of one in-plane lone pair on the uncoordinated sulfur atom may be counterbalanced by a heavier presence, on the coordinated sulfur atom, of the $\pi_\perp$ lone pair. Recall the presence of four electrons in the $\pi_\perp$ system. In this case a greater weight can be assigned to the VB structure (**111**). Conversely, the lengthening of the exocyclic C-S linkage in

$M-\eta^2-CS_2R$ complexes could indicate that the electron flow toward the electrophile favors the VB structure (**112**). In this case, the $\pi_{\perp}$ lone pair is shifted more toward the exocyclic sulfur atom and more double bond character is assigned to the coordinated C-S linkage. The important point is that there is no major rehybridization of the system in passing from $M-\eta^2-CS_2$ to $M-\eta^2-CS_2R$ compounds so that the charge compensation argument should be a good one.

In η^2-CS_2 compounds, the bending of the angle S-C-S can be taken as an indirect parameter of the basicity of the metal fragment. By referring to **Fig. 23**, the bending of CS_2 is governed by the energy and population of the level $2a_1$. A pertinent comparison can be made between the isostructural complexes (triphos)Co(η^2-CS_2) (ref. 18) and (triphos)Ni(η^2-CS_2) (ref. 8). The S-C-S angular value of 133.8(8)°, in the former, vs. 136.1(7)°, in the latter, indicates a larger donation from cobalt. In addition, the angle in the cobalt complex is the smallest within the whole series of $M-\eta^2-CS_2$ complexes (see **Table 1**). This is indicative, for (triphos)Co-(η^2-CS_2), of the strongest $M-CS_2$ interaction. Such a feature, together with the presence of the unpaired electron, which confers a radical character to the complex, accounts for certain unique reactivity patterns (see Sections 3.1.1 and 3.1.2).

The S-C-S angular parameter can be monitored also within the series of complexes with formula $[M(\eta^2-CS_2)(PR_3)]_2$ (M = Ni, Pd, Pt; PR_3 = PPh_3, $PPh(t-Bu)_2$) (ref. 26-28), and interesting conclusions can be drawn. First of all, it can be reasonably assumed that the basicity of the metal increases from Ni to Pt, not only on account of their different nature, but also on account of the increased donor capabilities of the terminal phosphine molecule ($PPh(t-Bu)_2$ > PPh_3). The latter observation agrees well with a progressively larger bending of CS_2 in the order Ni < Pd < Pt (the S-C-S angle varies between the values 137.3(9)° and 133.4(4)°). Remarkable are also the trends for the C-S bond lengths (see **Table 1**). On account of the nature of the orbital $2a_1$, antibonding for both the C-S linkages, a strong π-backdonation (as in platinum) should elongate the two C-S bonds by almost an equal amount. In actuality, the observed trend is an elongation of one C-S bond (the dihapto-bonded one) and a shrinking of the other. Notice that the two vectors are almost equal in the Pd derivative. Again, the qualitative argument which works best is that of charge compensation. The electron pair being donated from the σ-donor exocyclic sulfur to the second metal center, are compensated by an increased localization of the $\pi_{\perp}$ lone pair on the same sulfur atom. As a consequence, the η^2 coordinated C-S linkage acquires a larger double bond character. Also, the proposed (ref. 26) partial aromaticity of the six-membered cycle $\overline{Ni-C-S-Ni-C-S}$ becomes questionable in the light of the recently reported structures of Pd and Pt derivatives. In fact, the geometry of the $M-\eta^2-CS_2$ fragment in the dimers is substantially similar to that of the precursor mononuclear compounds $M(\eta^2-CS_2)(PPh_3)_2$ (M = Pd, Pt (ref. 12,13).

The relatively short M-C bond, generally observed in $M-\eta^2-CS_2$ complexes, (**Table 1**) has been attributed to a partial double bond by taking for granted the validity of the VB structure (**113**) (ref. 16). Actually, the argument is based on the close similarity of the Fe-C bond length (1.983(8) Å) in $Fe(\eta^2-CS_2)(PMe_3)(PPh_3)(CO)_2$ (ref. 16) with those reported for iron carbene complexes (see ref. 42 for an exhaustive list). In addition, a very similar value of 1.975(4) Å is also found in the $Fe-\eta^1-CS_2$ compound $(Cp)(CO)_2Fe-C\overline{(S)SFe(Cp)(CO)_2W}(CO)_5$ (ref. 31). The diagram of **Fig. 25** excludes any major $\pi_{\perp}$ bonding interaction between the metal and the carbon atom in $M-\eta^2-CS_2$ complexes. The short M-C bond could be favored by particularly good overlap and energy match of the in-plane interacting orbitals (ref. 7), but certainly the problem is not clearly understood. Even more intriguing, at the moment, is the progressive shortening of the Fe-C distance (range 1.983(8)-1.899(3) Å) observed in $Fe(\eta^2-CS_2)(PR_3)_2(CO)_2$ (ref. 16) and its $Mn(Cp)(CO)_2$ (ref. 22) and CH_2Ph (ref. 24) derivatives (see **Fig. 3,22,24** respectively).

The C-S distances in η^1-CS_2 moieties, although longer (on the average) than in η^2-CS_2 analogs, are still significantly shorter than the typical C-S single bond. This fact suggests that the weight of structure (**126**) cannot be large. In the complex $OsH(\eta^1-CS_2Me)(PPh_3)_2(CO)_2$ (ref. 36), shown in **Fig. 18**, the C-S linkage involving the free sulfur atom is significantly shorter than the other one (1.648(4) Å vs. 1.724(5) Å). The X-ray structure does not suggest any major rehybridization of the in-plane orbitals. Once again, the argument which works best is that the electron flow toward the alkyl group is compensated by a greater localization of the $\pi_{\perp}$ lone pair on the sulfur atom linked to it. In turn, a larger double bond character is assigned to the opposite

C-S linkage. Obviously, in the η^1 complexes with two substituents at the sulfur atoms (see **Table 2**), the asymmetry of the two C-S bonds is significantly reduced.

Finally, a remarkable feature of η^1 complexes is the large bending of the CS_2 moiety. This may be caused by two main factors: a) the η^1-coordination, rather than η^2, favors the accumulation of electron density in the CS_2 $2a_1$ level which is stabilized on bending (see **Fig. 23**), b) the engagement of one or both sulfur atoms in bonding with external groups (unfortunately no plain M-η^1-CS_2 structure is known) reduces the repulsion between the lone pairs (VSEPR argument (ref. 30)). The latter argument becomes more tenuous after a comparison between η^2- and η^1-CS_2R complexes. Although in principle the repulsions between the two sulfur atoms are comparable in the two cases, the compounds of Class 1E are significantly less bent than the corresponding compounds of Class 2D (see Section 2). Also, direct experimental evidence shows that the level $2a_1$ is more populated in η^1-CS_2R complexes. It is in fact well known that the C=S stretching frequency is higher by ca. 100-150 cm^{-1} in η^2- than in η^1-dithioalkyl esters (ref. 36).

4.5 THEORETICAL IMPLICATIONS FOR THE REACTIVITY OF CARBON DISULFIDE COMPLEXES

Coordination to a metal enhances the reactivity of carbon disulfide toward both electrophilic and nucleophilic reagents. The main trends are schematically summarized in (**128**). The exocyclic sulfur atom is highly activated toward electrophiles (E) and the carbon atom toward nucleophiles (Nu). Electrophilic character is also attributed to the endocyclic sulfur atom. Finally, the triatomic shows the character of 1,3-dipole and may react with dipolarophiles (D). In turn, η^1 complexes are nucleophilic at both sulfur atoms (**129**), whereas the carbon atom has a remarkable electrophilicity when both sulfur atoms are engaged in bonding with other groups (ref. 161) (**130**).

(**128**) (**129**) (**130**)

Unfortunately, very little is known about the various reactivity patterns because of the lack of information on many important intermediates. For this reason, the theoretical investigation in the field is scarce. By exploiting information on the electronic distributions in the basic η^1- and η^2-coordination modes, useful hints for typical reactions may be provided.

The reactions which are better understood, are those of M-η^2-CS_2 complexes with electrophiles (Section 3.2). The backdonation from the metal, which fills with electrons the CS_2 overall antibonding $2a_1$ level, ultimately favors the localization of an in-plane lone pair at the exocyclic sulfur atom. If the latter lone pair is used for binding an electrophile, there is an overall decrease of electron density at CS_2. Part of the donor capabilities of the triatomic toward the metal is lost and ultimately the η^1 mode is preferred. However, the η^2-coordination of the CS_2E grouping can still be stable, especially if the supporting metal fragment has no possibility of rearrangement. This usually happens when no extra ligands are present in solution or when the metal fragment, highly constrained, has little capability of making extra room for an incoming ligand, e.g. (triphos)M. With a noticeable exception discussed below, $\eta^2 \rightarrow \eta^1$ interconversion is paralleled by $ML_n \rightarrow ML_{n+1}$ interconversion. The reactions **(3)**,**(4)** and **(15)**,**(16)** are significant in this respect. Notice, for example, that in **(15)** the $(PPh_3)_2Pt$ fragment adds an iodide ligand and transforms into a new fragment of type L_3M (planar with C_{2v} symmetry). In some cases, when an excess of electrophilic reagent is used, the attack proceeds also at the second sulfur atom of CS_2, but after $\eta^2 \rightarrow \eta^1$ interconversion (see reactions **(8)**-**(10)** and **(12)**-**(13)**). This is also confirmed by the fact that η^1 complexes, which contain a free terminal sulfur atom, may undergo an electrophilic attack (see Sections 3.5 and 3.6).

Conversion from η^2- to η^1-coordination is also observed during the reaction of M-η^2-CS_2 complexes with activated alkynes to give five-membered cycles, either of type **A** or **B**, see (**131**). During the formation of metallo-rings of type **A**, the η^1-coordination of CS_2 is attained, while the ML_n fragment transforms into ML_{n+1}. In this case, one carbon atom of the alkyne acts as a new ligand toward the metal. Conversely, in 1,3-dithiol-2-ylidene derivatives of type **B**, the metal fragment preserves its nature, typically supportive of the η^2 mode.

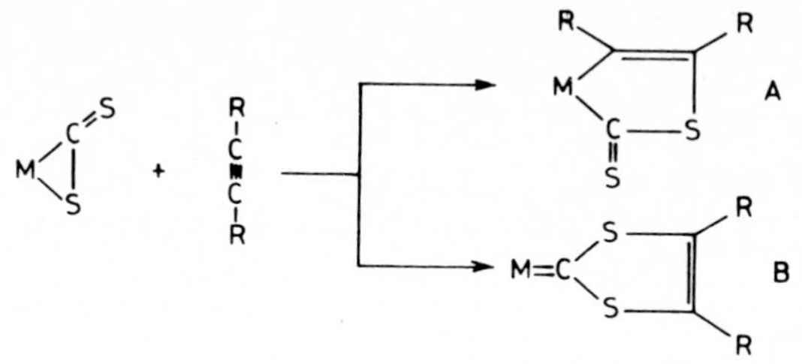

(131)

In Section 4.3 a certain analogy between the $\overline{C-S-C(R)-C(R)-S}$ cycle and the CH_2 group was already pointed out. Notice that the CH_2 ligand characterizes Schrock-type carbenes. Some authors have suggested (ref. 70) that the formation of compounds of type **B** corresponds to a typical 1,3-dipolar cycloaddition (ref. 78,86, 164,165). A 1,3-dipole is characterized by four π electrons, distributed over three centers (ref. 164). In the η^2-coordination mode, the four electron $\pi_\perp$ system of CS_2 is not perturbed by major interactions with the metal, so that the possibility of a concerted cycloaddition which involves the $\pi_\perp$ electrons cannot be dismissed. Ultimately, there are several conceivable ways through which the alkyne can approach an η^2-bound CS_2 complex. These extreme geometries are depicted in (**132**)-(**133**). In the former the alkyne

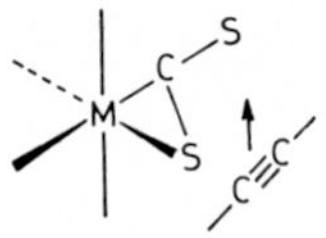

(132) 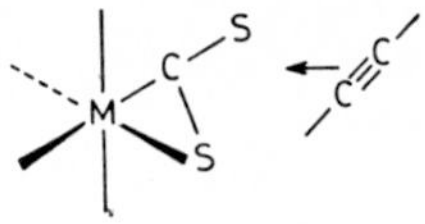 (133)

follows a route reminiscent of a suprafacial addition, a mechanism proposed long ago for typical 1,3-cycloadditions (ref. 165); in the latter, the addition and the eventual cyclization occur in the metallocycle plane. Also, the equilibrium between structures **A** and **B** of (**131**) was demonstrated by Dixneuf et al. in the case of the complex $Fe(CO)_2(PMe_2Ph)_2[CS_2C_2(CO_2Me)_2]$ (ref. 70). The authors propose that the dipolarophilic alkyne molecule may attack the η^2-CS_2 complex in two different ways in as much as they attribute 1,3-dipole character to the sequence Fe-C-S(exo) (VB structure (**113**)) as well as to the sequence S-C-S (VB structure (**112**)).

Obviously, the formation of 1,3-dithiol-2-ylidene complexes and of the equilibrium between the two possible five-membered cycles need specific theoretical investigation. A computational work is in progress to define the energetically favorable pathway, (**132**) or (**133**), to the cycle of type **B** and the mechanism of its interconversion into the cycle of type **A** (ref. 166).

In Section 3.2.4 the reactions of η^2-CS_2 complexes with protic acids were presented and summarized in (**46**). In a variety of cases, products which are very different from the starting materials were obtained. Even fragmentation of CS_2 to CS is observed. Indirectly, the variety of products obtainable points out to a crucial problem which must be faced whenever dealing with the general reactivity of η^2-CS_2 compounds; that is, the opening modes of the $\overline{M-C-S}$ triangle. Lewis acids, but Lewis bases as well, seem able to promote this opening. Unfortunately, any attempt of theoretical rationalization of this process is greatly complicated by the lack of symmetry in the triangle. In any event, it is predictable that the nature of the metal fragment, and the presence of electrophilic substituents at the exocyclic sulfur atom, variously affect the overall electronic distribution over the triangle. Accordingly, the site of attack of the acid (or the base) may change, case by case, and ultimately the same reagent can behave differently toward $\overline{M-C-S}$ cycles.

It is noteworthy that a modest activation energy (less than 10 kcal/mol) is required to unfasten the M-C linkage in η^2 complexes and to convert CS_2 back to linearity (ref. 7). This process is shematically shown in (**134**).

(**134**)

A clear-cut understanding of the reactivity of η^2-CS_2 complexes requires not only the identification of the side which is cleaved, but also the mode of rupture. If, for example, the bond in question is M-C, an electron pair may remain localized on the metal. As a consequence, the carbon atom becomes extremely electrophilic, with two MOs, $2a_1$ and $2b_1$, susceptible to attack by nucleophiles. Another possibility is the homolytic M-C cleveage in analogy with the thermolysis of the cyclopropane ring (ref. 167). In this case, the reaction may proceed through a diradical mechanism. In this light, the Lewis-acid-promoted head-to-head dimerization of η^2-CS_2 was interpreted (ref. 168) to give the ethenetetrathiolate bridging ligand in the complex cation $[(triphos)Rh(\mu$-$C_2S_4)Rh(triphos)]^{2+}$ (see Section 3.2.4, (**49**)). A qualitative MO analysis points out the existence of an overall $\pi_\perp$ delocalization through the $Rh(\mu$-$C_2S_4)Rh$ framework. Implicitly, the arrangement offers to the metals the possibility of coupling their unpaired spins (**135**).

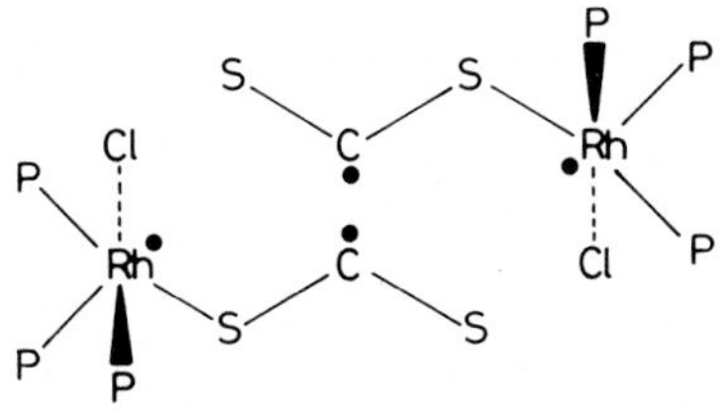

(**135**)

Very recently, the homolytic cleveage of a Fe-C bond has been proposed to explain the formation of a tetrathiooxalate ligand in the complex $Fe(CO)_2(PPh_3)(C_2S_4Me_2)$ obtainable from $[Fe(\eta^2$-$CS_2Me)$-$(CO)_2(PPh_3)_2]^+$ (ref. 169).

η^2-CS_2 complexes are susceptible to nucleophilic attack. The carbon atom is commonly accepted as the most probable site of the attack by nucleophiles such as H^-, PR_3, and N_3^- (see Section 3.3). On the other hand, the endocyclic sulfur atom has also been suggested as an alternative electrophilic center. The latter hypothesis is based on the formation of thiocarbonyls and phosphine sulfides upon attack by tertiary phosphines on η^2-CS_2 complexes (see Section 3.3.1). In this context, significant is the reaction of PPh_3 with the η^2-C,Se bonded CSSe complex $Co(\eta^2$-$CSSe)(Cp)(PMe_3)$, which yields the thiocarbonyl species $Co(CS)(Cp)(PMe_3)$ and $SePPh_3$ (ref. 4). Qualitatively, the nucleophilic attack is supported by a relevant weight to be assigned to the VB structure (**112**), where the sulfur atom, engaged in the dihapto linkage, is positively charged. Although useful in many respects, an interpretation of the reactivity of the compounds in terms of "charge control" is questionable. Notice that for carbene complexes the importance of the latter mechanism has been minimized on the basis of MO arguments (ref. 158,159). Recently very reliable ab-initio calculations (ref. 159) have shown that the M-C bond is indeed polarized, but, in contrast to a common belief, the negative charge is on the carbon atom. In conclusion, nucleophilic attacks on carbene complexes are "frontier controlled" and this is likely to hold, even to a larger extent, for carbon disulfide complexes. According to the frontier orbital theory of Fukui (ref. 170), the primary features of nucleophilic reactions are governed by the LUMO of the reactant. The site of the reaction is the AO whose coefficient is the largest in the LUMO. With reference to the MO diagram of **Fig. 125**, the situation is particularly clear for η^2-CS_2 complexes. In analogy with carbene complexes, the LUMO (2A) is strongly centered on the carbon atom because it coincides practically with the CS_2 MO $2b_1$, which has C-S $\pi_{\perp}^*$ character. On the other hand, a residual possibility of nucleophilic attack on one of the two sulfur atoms cannot be disregarded, although it seems less probable.

There are no examples in which, upon attack of a nucleophile on η^2-CS_2 complexes, the C-S linkage remains dihapto coordinated to the metal, but this is possible for η^2-CS_2R complexes. This has been observed for at least three different reactions (**2**), (**63**), and (**64**). For instance, the structure of $[(np_3)Ni(CS_2Me)]^+$ (see reaction (**2**)), indicates that the CS_2 plane has rotated by ca. 90° with respect to its η^2-CS_2R precursor. Notice that the carbon atom is only slightly pyramidalized and ultimately the interaction of the new fragment with the metal can be envisaged as that of a planar thioketone unit through its C=S linkage. In this case, the metal can still provide some π-backdonation. On the other hand, the close similarity of the LUMOs in η^2-CS_2 and η^2-CS_2R complexes suggests that the former complexes may also undergo an initial attack at the carbon atom by nucleophiles. However, the final reaction product cannot be a species of the type shown in (**136**), which should be unstable on account of a highly negative terminal sulfur atom. One possibility which is offered is the definitive cleveage of the M-C bond. See for instance the reactions of H^- and PR_3 with some η^2-CS_2 compounds, which yield complexes containing unidentate S-bonded dithioformate or phosphoniodithiocarboxylate ligands (see (**57**), (**58**) and (**62**) in Section 3.3). Remarkably, the metal loses the possibility of backdonation with the consequence that, in some cases, a too electron-rich metal is another source of instability. Other pathways may then be sought that ultimately may lead to the cleveage of one C-S bond.

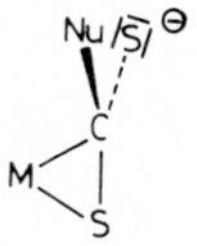

(**136**)

With reference to **Fig. 125**, the MO 3S of η^2-CS_2 complexes is largely centered on the carbon atom, since it is the antibonding level which corresponds to the metal-CS_2 π-backdonation. When the scheme (**134**) for the unfastening of CS_2 from the metal is monitored by MO calculations (ref. 7), the level 3S drops in energy and acquires more CS_2 character. Ultimately, the level 3S becomes almost equivalent to 2A, a pure CS_2 $\pi_\perp^*$ level. Eventually, the unfastened CS_2 molecule offers two sites for nucleophilic attack on the carbon atom. In conclusion, the formation of phosphine sulfides as products of the desulfurization reaction of η^2-CS_2 complexes (see Section 3.3.1) does not necessarily imply a direct attack of the phosphine on the coordinated sulfur atom. Reaction (**60**) may be interpreted along these lines.

A final comment is devoted to the very unique reactivity of the paramagnetic compound (triphos)Co(η^2-CS_2), reported in (**71**). The latter is able to form dithiocarbonate, trithiocarbonate and monoselenodithiocarbonate complexes from the reactions with elemental oxygen, sulfur and selenium, respectively. Such a capability might derive from the presence of one unpaired electron in a metal π hybrid of the starting complex (ref. 8). The predictable pathway goes through an excited state, where the unpaired spin occupies the MO 3A (see **Fig. 125**), which largely derives from the CS_2 $2b_1$ level. If so, the triatomic assumes a radical character which in turn can promote the homolytic cleveage of O-O, S-S or Se-Se linkages. Also the process, which eventually leads the metal to the oxydation state +2, is initiated. Remarkably, the analogous (triphos)Ni(η^2-CS_2) complex has no unpaired electrons but transforms also to a dithiocarbonate compound with oxygen, whereas totally different compounds are obtainable from sulfur and selenium (**72**). Our suggestion, in this case, is that the preliminary action of oxygen yields a cationic η^2

complex equivalent to the cobalt species, thereby allowing the reaction to proceed along the same pathway.

The reader may have found some of the proposals of the last Section somewhat speculative. On the other hand, most of the ideas are based on the knowledge of the basic electron distributions derived from the correlations between chemistry and structure of the complexes. Our suggestions, therefore, should be taken as interpretational hints, challenging both for the chemist and the theoretician.

Acknowledgements

We are indebted to Mr. F. Cecconi for his skillful work in making the drawings of this paper. We also thank Professor Pierre Dixneuf and Doctor Jerome Silvestre for a critical reading of the manuscript.

REFERENCES

1 M. C. Baird and G. Wilkinson, J. Chem. Soc., Chem. Commun., (1966) 514.
2 I. S. Butler and A. E. Fenster, J. Organomet. Chem., 66 (1974) 161.
3 P. V. Yaneff, Coord. Chem. Rev., 23 (1977) 183.
4 H. Werner, Coord. Chem. Rev., 43 (1982) 165.
5 J. A. Ibers, Chem. Soc. Rev., 11 (1982) 57.
6 S. Sasaki, H. Kudow and A. Ohyoshi, Inorg. Chem., 16 (1977) 202.
7 C. Mealli, R. Hoffmann and A. Stockis, Inorg. Chem., 23 (1984) 56.
8 C. Bianchini, D. Masi, C. Mealli and A. Meli, Inorg. Chem., 23 (1984) 2838.
9 R. Hoffmann, Angew. Chem., Int. Ed. Engl., 21 (1982) 711.
10 A. H. Guenther, J. Chem. Phys., 31 (1959) 1093.
11 W. Maier, Angew. Chem., 73 (1961) 120.
12 R. Mason, A. I. M. Rae, J. Chem. Soc. (A), (1970) 1767.
13 T. Kashiwagi, N. Yasuoka, T. Ueki, N. Kasai, M. Kakudo, S. Takahashi and N. Hagihara, Bull. Chem. Soc. Japan, 41 (1968), 296.
14 M. Aresta, C. F. Nobile, V. Albano, E. Forni and M. Manassero, J. Chem. Soc., Chem. Commun., (1975) 636.
15 H. Werner, K. Leonhard and C. Burschka, J. Organomet. Chem., 160 (1978) 291.
16 H. Le Bozec, P. H. Dixneuf, A. J. Carty and N. J. Taylor, Inorg. Chem., 17 (1978) 2568.
17 C. Bianchini, D. Masi, C. Mealli, A. Meli and M. Sabat, Organometallics, (1985) in press.
18 C. Bianchini, C. Mealli, A. Meli, A. Orlandini and L. Sacconi, Inorg. Chem., 19 (1980) 2968.
19 G. Fachinetti, C. Floriani, A. Chiesi-Villa and C. Guastini, J .Chem. Soc., Dalton Trans., (1979) 1612.
20 P. R. Mercier, J. Douglade, J. Amaudrut, J. Sala-Pala and J. Guerchais, Acta Cryst., B36 (1980) 2986.
21 M. G. B. Drew and L. Sun Pu, Acta Cryst., B33 (1977) 1207.
22 T. G. Southern, U. Oehmichen, J. Y. Le Marouille, H. Le Bozec, D. Grandjean and P. H. Dixneuf, Inorg, Chem., 19 (1980) 2976.

23 J. Amaudrut, J. Sala-Pala, J. E. Guerchais, R. Mercier and J. Douglade, J. Organomet. Chem., 235 (1982) 301.

24 D. Touchard, H. Le Bozec, P. H. Dixneuf, A. J. Carty and N. J. Taylor, Inorg. Chem., 20 (1981) 1811.

25 G.R. Clark, T. J. Collins, S. M. Jones and W. R. Roper, J. Organomet. Chem., 125 (1977) C23.

26 C. Bianchini, C. A. Ghilardi, A. Meli, S. Midollini and A. Orlandini, J. Chem. Soc., Chem. Commun., (1983) 753.

27 D. H. Farrar, R. R. Gukathasan and K. Won, J. Organomet. Chem., 275 (1984) 263.

28 D. H. Farrar, R. R. Gukathasan and S. A. Morris, Inorg. Chem., 23 (1984) 3258.

29 G. Gervasio, R. Rossetti and P. L. Stanghellini, J. Chem. Soc. Dalton Trans., (1983) 1613.

30 R. J. Gillespie, "Molecular Geometry", Van Nostrand-Rheinhold, London (1972).

31 L. Busetto, M. Monari, A. Palazzi, V. Albano, F. Demartin, J. Chem. Soc., Dalton Trans., (1983) 1849.

32 J. M. Lisy, E. D. Dobrzynski, R. J. Angelici and J. Clardy, J. Am. Chem. Soc., 97 (1975) 656.

33 T. S. Cameron, P. A. Gardner and K. R. Grundy, J. Organomet. Chem., 212 (1981) C19.

34 a) C. P. Kubiak, R. Eisenberg, J. Am. Chem. Soc., 99 (1977) 6129. b) Inorg. Chem., 19 (1980) 2726.

35 D. M. Hoffman and R. Hoffmann, Inorg. Chem., 20 (1981) 3543.

36 J. M. Waters and J. A. Ibers, Inorg. Chem., 16 (1977) 3273.

37 S. M. Boniface, G. R. Clark, J. Organomet. Chem., 208 (1981) 253.

38 C. M. Jensen, T. J. Lynch, C. B. Knobler, H. D. Kaesz, J. Am. Chem. Soc., 104 (1982), 4679.

39 J. Y. Le Marouille, C. Lelay, A. Benoit, D. Grandjean, D. Touchard, H. Le Bozec and P. Dixneuf, J. Organomet. Chem., 191 (1980) 133.

40 A. Benoit, J. Y. Le Marouille, C. Mahe and H. Patin, J. Organomet. Chem., 233 (1982) C51.

41 M. Herberhold, M. Süss-Fink and C. G. Kreiter, Angew. Chem., Int. Ed. Engl., 16 (1977) 193.

42 U. Shubert in: "Transition Metal Carbene Complexes", Verlag Chemie, Weinheim, 1983, p. 73 and references therein.

43 H. Werner, O. Kolb, R. Feser and U. Schubert, J. Organomet. Chem., 191 (1980) 283.
44 M. Cowie and S. K. Dwight, J. Organomet. Chem., 214 (1981) 233.
45 V. G. Albano, D. Braga, L. Busetto, M. Monari and V. Zanotti, J. Chem. Soc., Chem. Commun., (1984) 1257.
46 C. Bianchini, A. Meli and A. Orlandini, Inorg. Chem., 21 (1982) 4166.
47 C. Bianchini and A. Meli, J. Chem. Soc., Dalton Trans.,(1983) 2419.
48 C. Bianchini, P. Innocenti and A. Meli, J. Chem. Soc., Dalton Trans., (1983) 1777.
49 C. Bianchini, C. Mealli, A. Meli and G. Scapacci, Organometallics, 2 (1983) 141.
50 G. W. A. Fowles, L. S. Pu and D. A. Rice, J. Organomet. Chem., 54 (1973) C17.
51 K. R. Grundy, R. O. Harris and W. R. Roper, J. Organomet. Chem., 90 (1975) C34.
52 T. J. Collins, W. R. Roper and K. G. Town, J. Organomet. Chem., 121 (1976) C41.
53 P. Conway, S. M. Grant and A. R. Manning, J. Chem. Soc., Dalton Trans., (1979) 1920.
54 W, A. Schenk, T. Schwietzke, and H. M ller, J. Organomet. Chem., 232 (1982) C41.
55 W. Bertleff and H. Werner, Z. Naturforsh., 37b (1982) 1294.
56 O. Kolb, Diploma Thesis, University of Würzburg, 1979.
57 C. Bianchini and A. Meli, J. Chem. Soc., Chem. Commun., (1983) 156.
58 C. Bianchini, C. A. Ghilardi, A. Meli and A. Orlandini, J. Organomet. Chem., 246 (1983) C13.
59 J. Fortune and A. R. Manning, Organometallics, 2 (1983) 1719.
60 H. Stolzemberg and W. P. Fehlhammer, J. Organomet. Chem., 246 (1983) 105.
61 J, Amaudrut, A. Kadmiri, J. Pala-Sala and J. E.Guerchais, J. Organomet. Chem. 266 (1984) 53.
62 C. Bianchini, A. Meli and G. Scapacci, Organometallics, 2 (1983) 1834.
63 C. Bianchini, C. A. Ghilardi, A. Meli and A. Orlandini, J. Organomet. Chem., 270 (1984) 251.
64 C. Bianchini, D. Masi, C. Mealli, A. Meli, M. Sabat and G. Scapacci, J. Organomet. Chem. 273 (1984) 91.

65 D. Touchard, P. H. Dixneuf, R. D. Adams and B. E. Segmüller, Organometallics, 3 (1984) 640.
66 D. H. Farrar, R. O. Harris and A. Walker, J. Organomet. Chem., 124 (1977) 125.
67 T. J. Collins, K. R. Grundy, W. R. Roper and S. F. Wong, J. Organomet. Chem., 107 (1976) C37.
68 H. Stolzemberg, W. P. Fehlhammer and P. H. Dixneuf, J. Organomet. Chem., 246 (1983) 105.
69 Y. Wakatsuki, H. Yamazaki and H. Iwasaki, J. Am. Chem. Soc., 95 (1973) 5781.
70 H. Le Bozec, A. Gorgues and P. H. Dixneuf, Inorg. Chem., 20 (1981) 2486.
71 H. Le Bozec, A. Gorgues and P. H. Dixneuf J. Am. Chem. Soc., 100 (1978) 3946.
72 A. Gorgues and A. Le-Coq, Tetrahedron Lett., 50 (1979) 4829.
73 C. Bianchini and A. Meli, J. Chem. Soc., Chem. Commun., (1983) 1309.
74 C. Bianchini, A. Meli and G. Scapacci, Organometallics, 4 (1984) 264.
75 C.C. Frazier, N. D. Magnussen, L. N. Osuji and K. O. Parker, Organometallics, 1 (1982) 903.
76 H. Le Bozec, A, Gorgues and P. Dixneuf, J. Chem. Soc., Chem. Commun., (1978) 573.
77 H. D. Hartzler, J. Am. Chem. Soc., 92 (1970) 1412.
78 I. Fleming, Frontier Orbitals and Organic Chemical Reactions, John Wiley & Sons, New York, 1976, pp. 148-161.
79 H. Le Bozec and P. H. Dixneuf J. Chem. Soc., Chem. Commun., (1983) 1462.
80 M. Ngounda, H. Le Bozec and P. Dixneuf, J. Org. Chem., 47 (1982) 4000.
81 A. J. Carty, P. H. Dixneuf, A. Gorgues, F. Hartstock, H. Le Bozec and N. J. Taylor, Inorg. Chem., 20 (1981) 3929.
82 C. Bianchini, C. Mealli, A. Meli and M. Sabat, Organometallics, 4 (1985) 421.
83 A. J. Carty, F. Hartstock, N. J. Taylor, H. Le Bozec, P. Robert and P. H. Dixneuf, J. Chem. Soc., Chem. Commun., (1980) 361.
84 P. Robert, H, Le Bozec, P. H. Dixneuf, F. Hartstock, N. J. Taylor and A. J. Carty, Organometallics, 1 (1982) 1148.
85 P. Robert, B. Demerseman and P. H. Dixneuf, Organometallics, 3 (1984) 1771.

86 K. N. Houk, J. Sims, R. E. Duke, Jr., R. W. Strozier and J. K George, J. Am. Chem. Soc., 95 (1973) 7287.
87 D. Plusquellec and P. Dixneuf, Organometallics, 1 (1982) 1401.
88 W. P. Fehlhammer and H. Stolzemberg, Inorg. Chim. Acta, 44 (1980) L151.
89 W. P. Fehlhammer, A. Mayr and H. Stolzemberg, Angew. Chem., Int. Ed. Engl., 18 (1979) 626.
90 C. Bianchini and A. Meli unpublished results.
91 M. C. Baird and G. Wilkinson, J. Chem. Soc. A, (1967) 865.
92 E. Uhlig and W. Poppitz, Z. Chem., 19 (1979) 191.
93 M. Herberhold and M. Suss-Fink, Chem. Ber., 111 (1978) 2273.
94 C. Moise, J. Organomet. Chem., 247 (1983) 27.
95 R. Uson, J. Fornies and M. A. Uson, Inorg. Chim. Acta, 81 (1984) L27.
96 H. Le Bozec, A. Gorgues and P. Dixneuf, J. Organomet. Chem., 174 (1979) C24.
97 C. Bianchini, C. Mealli, A. Meli and M. Sabat, Inorg. Chem., 23 (1984) 4125.
98 C. Bianchini, C. A. Ghilardi, A. Meli and A. Orlandini, J. Organomet. Chem., (1985) in press.
99 D. H. M. Thewissen, J. Organomet. Chem., 188 (1980) 211.
100 J. Fortune and A. R. Manning, Organometallics, 2 (1983) 1719.
101 M. C. Baird and G. Wilkinson, J. Chem. Soc., Chem. Commun., (1966) 267.
102 M. C. Baird, G. Hartwell, Jr. and G. Wilkinson, Inorg. Phys. Theor., (1967) 2037.
103 P. Conway, A. R. Manning and F. S. Stephens, J. Organomet. Chem. 186 (1980) C64.
104 A. E. Fenster and I. S. Butler, Inorg. Chem. 13 (1974) 915.
105 G. Jaonen and R. Dabard, J. Organomet. Chem., 72 (1974) 377.
106 I. S. Butler, N. J. Coville and D. Cozak, J. Organomet. Chem., 133 (1977) 59.
107 J. Fortune and A. R. Manning, J. Organomet. Chem., 190 (1980) C95.
108 C. Bianchini, C. A. Ghilardi, A. Meli and A. Orlandini, Inorg. Chem., 22 (1983) 2188, and references therein.
109 W. A. Schenk and T. Schwietzke, Organometallics, 2 (1983) 1905.
110 G. La Monica, G. Ardizzone, S. Cenini and F. Porta, J. Organomet. Chem., 273 (1984) 263 and references therein.

111 W. M. Hawling, A. Walker and M. A. Woitzik, J. Chem. Soc., Chem. Commun., (1983) 11.

112 E. D. Jemmis and R. Hoffmann, J. Am. Chem. Soc., 102 (1980) 2570.

113 M. Di Vaira, S. Moneti, M. Peruzzini and P. Stoppioni, J. Organomet. Chem., 226 (1984) C8.

114 A. P. Ginsberg, W. E. Lindsell, C. R. Sprinkle. K. W. West and R. L. Cohen, Inorg. Chem., 21 (1982) 3666.

115 H. Werner and K. Leonhard, Angew. Chem., Int. Ed. Engl., 18 (1979) 627.

116 H. Werner, K. Leonhard, O. Kolb, E. Rottinger and H. Vahrenkamp, Chem. Ber., 113 (1980) 1654.

117 I. Ghatak, D. M. P. Mingos, M. B. Hursthouse and P. R. Raithby, Trans. Met. Chem., 1 (1976) 119.

118 E. O. Fischer and s. Riedmuller, Chem. Ber., 107 (1974) 915.

119 C. C. Frazier, R. F. Kline and D. D. Borck, Inorg. Chem., 20 (1980) 4009.

120 D. Touchard, H. Le Bozec and P. Dixneuf, Inorg. Chim. Acta, 33 (1979) L141.

121 C. Bianchini, C. Mealli, A. Meli and M. Sabat, Inorg. Chem., 23 (1984) 2731.

122 J. M. Ritchey, D. C. Moody and R. P. Ryan, Inorg. Chem., 22 (1983) 2276.

123 C. A. Ghilardi, A. Orlandini and S. Midollini, private communication.

124 C. Bianchini, D. Masi, C. Mealli and A. Meli, J. Organomet. Chem., 247 (1983) C29.

125 H. Ashitaka, K. Jinda and H. Ueno, J. Polym. Sci., Polym. Chem. Ed., 21 (1983) 1989.

126 J. E. Ellis, R. W. Fennel and E. A. Flom, Inorg. Chem., 15 (1976) 2031.

127 L. Busetto, U. Belluco and R. J. Angelici, J. Organomet. Chem., 18 (1969) 213.

128 D. B. Dombek and R. J. Angelici, Inorg. Synth. 17 (1977), 100.

129 T. A. Wnuk and R. J. Angelici, Inorg. Chem., 16 (1977) 1173.

130 F. B. Mc Cormick and R. J. Angelici, Inorg. Chem., 18 (1979) 1231.

131 H. Stolzenberg, W. P. Fehlhammer, M. Monari, V. Zanotti and L. Busetto, J. Organomet. Chem., 272 (1984) 73.

132 L. Busetto, A. Palazzi and M. Monari, J. Chem. Soc., Dalton Trans., (1982) 1631.
133 H. Stolzenberg and W. P. Fehlhammer, J. Organomet. Chem., 235 (1982) C7.
134 M. C. Baird, G. Hartwell and G. Wilkinson, J. Chem. Soc. A, (1967) 2037.
135 J. M. Ritchey and D. C. Moody, Inorg. Chim. Acta, 74 (1983) 271.
136 L. Busetto, A. Palazzi and M. Monari, J. Organomet. Chem., 228 (1982) C19.
137 S. Gambarotta, F. Arena, C. Floriani and P. F. Zanazzi, J. Am. Chem. Soc., 104 (1982) 5082.
138 J. C. Calabrese, T. Herskovitz and J. B. Kinney, J. Am. Chem. Soc., 105 (1983) 5914.
139 R. Alvarez, E. Carmona, E. Gutierrez-Puebla, J. M. Marìn, A. Monge and M. L. Poveda, J. Chem. Soc., Chem. Commun., (1984) 1326.
140 H. Werner and W. Bertleff, Chem. Ber., 113 (1980) 267.
141 S. Sakaki, K. Kitaura and K. Morokuma, Inorg. Chem., 21 (1982) 760.
142 a) H. J. Dewar, Bull. Soc. Chim. Fr., 18 (1951) C79. b) J. Chatt and L. A. Duncanson, J. Chem. Soc., (1953) 2939.
143 T. A. Halgren, L. D. Brown, D. A. Kleier and W. N. Lipscomb, J. Am. Chem. SOc., 99 (1977) 6793.
144 R. Hoffmann, Science, 211 (1981) 995.
145 R. Hoffmann, T. A. Albright and D. L. Thorn, Pure Appl. Chem., 50 (1978), 1.
146 T. A. Albright, Tetrahedron, 38 (1982) 1339.
147 R. S. Mulliken, Rev. Mod. Phys., 14 (1942) 204; Can. J. Chem., 36 (1958) 10.
148 A. D. Walsh, J. Chem. Soc. (1953) 2260.
149 R. J. Buenker and S. D. Peyerimhoff, Chem. Rev., 74 (1974) 127.
150 B. M. Gimarc, Molecular Structure and Bonding, Academic Press, New York, 1979, p. 153.
151 J. W. Lauher and R. Hoffmann, J. Am. Chem. Soc., 98 (1976) 1729.
152 O. Eisenstein and R. Hoffmann, J. Am. Chem. Soc., 103 (1981) 4308.
153 R. Hoffmann, Accts. Chem. Res., 4 (1971) 1.
154 C. Mealli, unpublished results.

155 P. Hofmann in: Transition Metal Carbene Complexes, Verlag Chemie, Weinheim (1983).

156 R. J. Goddard, R. Hoffmann and E. D. Jemmis, J. Am. Chem. Soc., 102 (1980) 7667.

157 D. Spangler, J. J. Wendeloski, M. Dupuis, M. M. L. Chen and H. F. Schaefer III, J. Am. Chem. Soc., 103 (1981) 3985.

158 N. M. Kostic and R. F. Fenske, Organometallics, 1 (1982) 974.

159 H. Nakatsuji, J. Ushio, S. Han and T. Yonezawa, J. Am. Chem. Soc., 105 (1983) 426.

160 T. E. Taylor and M. B. Hall, J. Am. Chem. Soc., 106 (1984) 1576.

161 R. A. Pickering and R. J. Angelici, Inorg. Chem., 20 (1981) 2977.

162 F. B. McCormick and R. J. Angelici, Inorg. Chem., 20 (1981) 1111.

163 C. Bianchini, C. Mealli, A. Meli and M. Sabat, manuscript in preparation.

164 R. G. Pearson, Symmetry Rules for Chemical Reactions, John Wiley and Sons, New York, 1976, p. 364.

165 R. Huisgen, Angew. Chem., Int. Ed. Engl., 2 (1963) 633.

166 C. Mealli, C. Bianchini, J. Silvestre and R. Hoffmann, work in progress.

167 R. G. Bergmann in: Free Radicals, John Wiley and Sons, New York, 1973, vol. I.

168 C. Bianchini, C. Mealli, A. Meli and M. Sabat, XVII Congresso Nazionale Chimica Inorganica, Cefalù, Italy, 1984, p. 187.

169 D. Touchard, J. L. Fillaut, P. Dixneuf, C. Mealli, M. Sabat and L. Toupet, submitted for publication.

170 K. Fukui, Theory of Orientation and Stereoselection, Splinger-Verlag, Berlin, 1975.

Chapter 4

STEREOCHEMISTRY OF BAILAR INVERSION AND RELATED METAL ION SUBSTITUTION REACTIONS

W.G. JACKSON

Department of Chemistry, Faculty of Military Studies, University of New South Wales, Duntroon, A.C.T. 2600, Australia

1 INTRODUCTION

Optical inversion at chiral carbon centres is a familiar process to organic chemists. The direct conversion of one enantiomer totally into the other is of course thermodynamically prohibited; only racemization is possible. However, by a series of substitution reactions, each involving net chemical change, the restriction is removed. Walden (ref. 1) first accomplished this feat using malic acid,

$$\underset{(-)\text{-malic acid}}{\overset{*}{C}H(OH)CO_2H - CH_2CO_2H} \xrightarrow{PCl_5} \underset{(+)\text{-chlorosuccinic acid}}{\overset{*}{C}H(Cl)CO_2H - CH_2CO_2H} \xrightarrow[OH^-]{Ag^+} \underset{(+)\text{-malic acid}}{\overset{*}{C}H(OH)CO_2H - CH_2CO_2H} \quad (1)$$

Nowadays the term Walden inversion is applied to any single reaction in which inversion of configuration occurs, and where reactant and product generally differ in chemical composition. Furthermore, the term is normally reserved for reactions where the degree of inversion exceeds 50%, i.e. the majority of the substitution product has an absolute configuration opposite to that of the reactant.

At the outset, it needs to be made clear that inversion refers to a substantial molecular rearrangement, and not merely to a reaction where the inversion is *apparent* because the priority rules (ref. 2) require a change in labelling. Conversely, a chiral reactant and product carrying the same designation for the absolute configuration do not necessarily imply optical retention. The labels need to be translated into standard stereochemical diagrams such as Fischer projections (ref. 3) in order to assess the true stereochemical outcome of a substitution reaction. To illustrate the point, the chiral tetratertiary arsine ligand (**R**,**R**)-tetars becomes the formally labelled (**S**,**S**)-tetars merely through coordination to a metal ion, yet there is no true inversion at either of the chiral As centres on complexation (ref. 4).

Optical activity, recognized originally as arising from compounds containing tetrahedral carbon, is of course not confined to this element. The Nobel laureate Alfred Werner first vividly demonstrated this by resolving the purely inorganic (non-carbon containing) species $Co\left((OH)_2Co(NH_3)_4\right)_3^{6+}$ into its enantiomorphs (ref. 5). Several analogous compounds such as *cis*-$Rh(NHSO_2NH)_2(OH_2)_2^-$ (ref. 6) and $Pt(S_5)_3^{2-}$ (ref. 7) have since emerged. Furthermore, shortly after the introduction of the concept of asymmetric carbon, tetrahedron-based nitrogen, sulfur, phosphorus, arsenic and other analogs were recognized, e.g. $(Me)(Et)(CH_2CO_2H)S^+$ (ref. 3). There are now many more non-metallic elements which feature as the nuclear atom in compounds of this kind, although Walden inversion reactions at all these centers are not yet well documented.

The direct inorganic analogs of chiral tetrahedral carbon, M(abcd), where M is a transition metal ion,* are even today a rare breed, and little is known of their substitution stereochemistry. However, chiral pseudo-tetrahedral organometallic complexes, e.g.

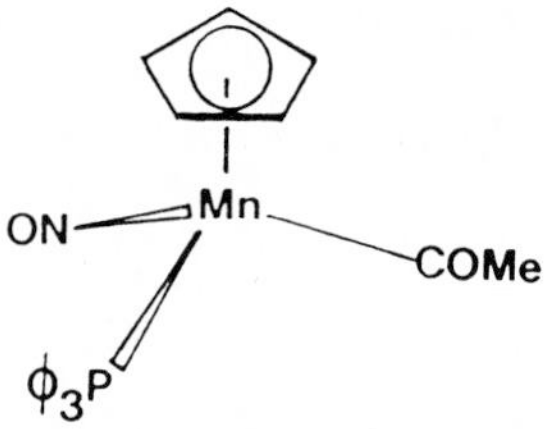

(2)

many of which have sufficient optical stability to be resolved and investigated, are a relatively recent development (ref. 9,10); their substitution stereochemistry is briefly dealt with later.

Inorganic complexes of the kind *cis*-$Co(en)_2AX^{n+}$ (3) were the first to be resolved (ref. 11). Although stereochemically these were not comparable to tetrahedral carbon, these classic cobalt (III) coordination complexes prepared by Werner were obvious candidates for the first attempts to observe optical inversion reactions at metal ion centers. They possessed moderate to high optical stability, facilitating the determination of relative configurations

* Chiral main group metal M(abcd) compounds [e.g. M = Sn(IV)] are known (ref. 8), and Walden inversions documented.

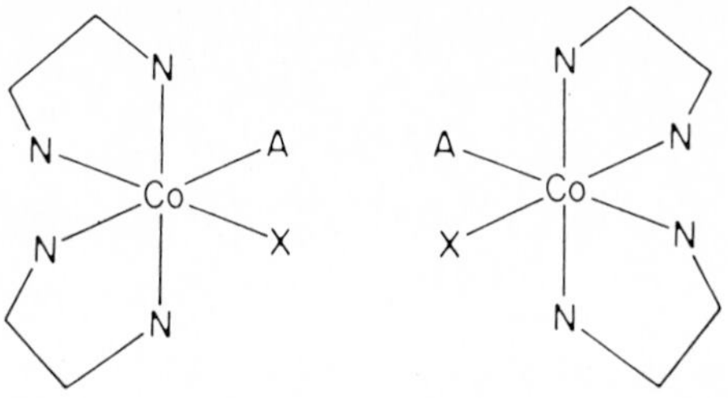

$$\Lambda\text{-}cis\text{-}Co(en)_2AX^{n+} \qquad \Delta\text{-}cis\text{-}Co(en)_2AX^{n+} \qquad (3)$$

of reactant and substitution product using conventional polarimetry. Thus, in 1934, Bailar and Auten reported the first Inorganic equivalent of Walden inversion (ref. 12). Their conclusion that this was a genuine inversion process has survived the passage of time, although few additional examples have since surfaced, despite effort. Apart from the *cis*-$Co(en)_2AX^{n+}$ examples (ref. 13–15), only the reactions between CN^- and $Fe(II)(diimine)_3^{2+}$ species [e.g. $Fe(phen)_3^{2+}$] appear to involve a path with net optical inversion (ref. 16–18):

$$\Lambda\text{-}Fe(diimine)_3^{2+} \xrightarrow[H_2O]{2CN^-} \Delta\text{-plus }\Lambda\text{-}Fe(diimine)_2(CN)_2 + diimine$$

$$\Delta > \Lambda \qquad (4)$$

We should mention that this and related processes are mechanistically controversial for quite different reasons, because of the prospect of CN^- (or OH^-) attack at the ligand imine function (ref. 19); the protracted debate over this issue is adequately covered elsewhere (ref. 20).

It remains true that, of the handful of substitution reactions of octahedral metal ion complexes which proceed with net inversion, none are mechanistically well understood, despite two reviews on the subject (ref. 21,22). The problem is two-fold, one a lack of data for a range of metal ion centers, and two, a lack of detailed and mechanistically discriminatory experiments for a *particular* Bailar inversion reaction, as such a process has properly become to be called. One of the purposes of this article therefore will be an attempt to expose areas of special attention, rather than go over well trampled ground.

In particular, the statistically unlikely single step "double substitution" reaction,

$$MX_2 \xrightarrow{2Y} MY_2 \qquad (5)$$

is purported to merit consideration, in a general context, as a viable alternative to stepwise substitution,

$$MX_2 \xrightarrow{Y} MXY \xrightarrow{Y} MY_2. \tag{6}$$

Indeed, evidence is reviewed which unambiguously establishes double substitution as a reality. It is now certain that this process is operative for the classic Bailar inversion reaction of Λ-$Co(en)_2Cl_2^+$ in H_2O containing OH^- (ref. 23),

$$\Lambda\text{-}Co(en)_2Cl_2^+ \xrightarrow[H_2O]{2OH^-} \Lambda\text{- and } \Delta\text{- and } \textit{trans}\text{-}Co(en)_2(OH)_2^+ + 2Cl^- \tag{7}$$

and the suggestion (ref. 23) that *trans*-$Co(en)_2Cl_2^+$ might also undergo double substitution has now been realized. The difficulty in the past has been devising not only the optimal experimental conditions to observe the simultaneous loss of two leaving groups, but also sufficiently sensitive detection techniques, since normal stepwise substitution (6) is inevitably competitive with double substitution.

So-called inversion reactions (ref. 24,25) of complexes such as *cis*-α-(**R**,**R**)-$Co(trien)Cl_2^+$, (trien = $H_2N(CH_2)_2NH(CH_2)_2NH(CH_2)_2NH_2$),

$$\Lambda\text{-}\textit{cis}\text{-}\alpha\text{-}(\mathbf{R},\mathbf{R})\text{-}Co(trien)Cl_2^+ \xrightarrow[H_2O]{OH^-} \Delta\text{-}\textit{cis}\text{-}\beta\text{-}(\mathbf{R},\mathbf{R})\text{- or } (\mathbf{S},\mathbf{R})\text{-}Co(trien)Cl(OH)^+ + Cl^- \tag{8}$$

will not be included in this article as such. They are not genuine optical inversions, in the sense that the possible substitution products are diastereomeric rather than enantiomeric, and hence such reactions are not fundamentally different from those giving geometrically isomeric products where questions of chirality are not involved, e.g.

$$\textit{cis}\text{-}Co(NH_3)_4Cl_2^+ \xrightarrow{H_2O} \textit{cis}\text{-}Co(NH_3)_4Cl(OH_2)^{2+} \text{ and } \textit{trans}\text{-}Co(NH_3)_4Cl(OH_2)^{2+} \tag{9}$$

Moreover, it need be borne in mind that topological rearrangements such as (8) can be coupled with the chirality at the nitrogen centers (ref. 26). These are fixed and optically stable in acid solution, but able to invert in the presence of OH^- provided the nitrogen arrangement is "planar" (meriodinal); inversion at "bent" (facial) nitrogen centres is stereochemically prohibitive.

Reactions such as

$$\Lambda\text{-}cis\text{-}\alpha\text{-}(\mathbf{S},\mathbf{S})\text{-Co}\Big((\mathbf{R})\text{picpn}\Big)\text{Cl}_2^+ \xrightarrow{C_2O_4^{2-}} \Delta\text{-}cis\text{-}\alpha\text{-}(\mathbf{R},\mathbf{R})\text{-Co}\Big((\mathbf{R})\text{picpn}\Big)\text{ox}^+ + 2\text{Cl}^- \qquad (10)$$

can also be viewed simply as isomerization.* Certainly there might be net inversion about the cobalt centre (ref. 27), but again this is coupled with inversion at the chiral nitrogen centres. Moreover, the reaction (10) must occur in several steps, and the overall inversion could arise merely because the (**R**)picpn ligand is chiral and optically stable, and imparts both kinetic and thermodynamic preferences for one (of several) isomers, in this case the isomer having the cobalt configuration inverted with respect to the reactant. Thus cobalt inversion can be realized because the basic conditions permit N-inversion but not racemization at the chiral carbon center of the (**R**)picpn ligand. We add that, while this Λ-α-(**S**,**S**) to Δ-α-(**R**,**R**) rearrangement has been termed a "new form of octahedral inversion" (ref. 27), in the context of this article it is better labelled a new form of geometric isomerization.

Finally, to put Bailar inversion reactions in proper perspective, it is appropriate that there is some coverage of the stereochemistry of octahedral metal ion substitution reactions in general. They are also of importance in their own right. The state of the art here appears to have been well explicated in recent reviews (ref. 28,29), one devoted exclusively to stereochemical aspects (ref. 30). However, there have been sufficient important developments in the interim to warrant inclusion of this material here. The review is intended to be critical rather than a complete survey of the literature, and hopefully it is a balanced selection to reveal as much of the stereochemical features as possible.

2 THE CLASSIC BAILAR INVERSION REACTION

The reaction depicted below is representative of all those investigated for $Co(en)_2AX^{n+}$ complexes and has received the most attention.

---

* (**R**)-picpn = (3**R**)-3-methyl-1,6-di(2-pyridyl)-2,5-diazahexane.

$$\Lambda\text{-}cis\text{-}Co(en)_2Cl_2^+ \xrightarrow[H_2O]{2OH^-} \Delta\text{-}cis\text{-} + \Lambda\text{-}cis\text{-} + trans\text{-}Co(en)_2(OH)_2^+ + 2Cl^-$$

$$\Delta > \Lambda$$

(11)

Originally, various reagents were used in attempts to observe inversion about cobalt, including Ag^+, CO_3^{2-}, and OH^-, separately or in combination (ref. 12–14,31,32). Also, most of these reactions appeared to be carried out heterogeneously. Indeed, the Bailar inversion was first formulated essentially* as follows:

$$d\text{-}cis\text{-}Co(en)_2Cl_2^+ \xrightarrow[H_2O\ paste]{Ag_2CO_3} \ell\text{-}Co(en)_2(CO_3)^+$$

$$HCl \uparrow\downarrow aq.K_2CO_3 \qquad\qquad HCl \downarrow\uparrow aq.K_2CO_3$$

$$d\text{-}Co(en)_2CO_3^+ \qquad\qquad \ell\text{-}cis\text{-}Co(en)_2Cl_2^+$$

(12)

Much of the speculation about the mechanism of the inversion, and the uncertainties in the conclusions, revolved around the identification of the effective species, e.g. "Ag_2CO_3" or "AgOH" (ref. 2,13,33,34). A recent paper has clarified many aspects of the reaction, and it is now known (ref. 23) that OH^- alone is sufficient to effect inversion, and this can occur in homogeneous solution. However, the ultimate development of *unequivocal* evidence for this and other conclusions regarding the processes involved is far from straightforward. We deal systematically with this in the next few sections.

2.1 Historical background—normal substitution?

It was first believed (ref. 13,31) that the optical inversion occurred in the first step of the expected sequence

$$\Lambda\text{-}cis\text{-}Co(en)_2Cl_2^+ \xrightarrow{OH^-} Co(en)_2Cl(OH)^+ \xrightarrow{OH^-} Co(en)_2(OH)_2^+$$

(13)

because, in separate experiments, it was shown that Λ-*cis*-$Co(en)_2(OH)Cl^+$ reacted with OH^- with predominant retention (ref. 13,31,35), whereas overall there was inversion. The absolute configuration of all the cis complexes involved in (13) are known (ref. 36), but even so, there can be no question

* There was some early confusion over the signs of optical rotation (d, ℓ) at the Na_D line.

that a genuine optical inversion arises, for the following reason. The activity of the final products, measured directly as $Co(en)_2(OH)_2^+$, or indirectly as $Co(en)_2(OH_2)_2^{3+}$ or $Co(en)_2OCO_2^+$, obtained through subsequent chemical interconversion, actually changes sign when experiments are carried out for a range of $[OH^-]$ (ref. 31). We return to this point later, but it suffices to note now that this result *requires* net inversion, at either low or high $[OH^-]$ (and thus net optical retention at the other extreme). This is so because the activity in the products arises from a *single* active cis species.* Furthermore, the known absolute configurations establish that it is the high $[OH^-]$ end of the scale where inversion predominates over retention.

The stereochemistry for the first step of (13) was determined first in 1962 by Tobe and Chan (ref. 35). They found predominant optical retention, under conditions of low $[OH^-]$. Thus Boucher, Kyuno and Bailar were led in 1964 (ref. 31) to conclude that the steric course for this first hydrolysis step must be $[OH^-]$ dependent, to accommodate their observations of overall inversion at high $[OH^-]$. Dwyer *et al.* had concluded a year earlier that inversion occurred in the first step for the Ag^+/OH^- reaction (ref. 13). In 1968 Dittmar and Archer set out to test the first proposal (ref. 33). Although they found a steric course which apparently depended upon $[OH^-]$, they reported a $\sim 97\%$ retained configuration at low $[OH^-]$, but essentially racemic *cis*-$Co(en)_2(OH)Cl^+$ at the highest $[OH^-]$ examined. Clearly the latter could not accommodate the observation of overall inversion under comparable conditions. Also, they confirmed the predominance of optical retention for the relevant component of the second step of (13), Λ-*cis*-$Co(en)_2(OH)Cl^+ + OH^-$, as had Tobe and Chan somewhat earlier.

It was on this basis that Dittmar and Archer first suggested a radical reaction pathway (ref. 33), namely a *single* step process in which two Cl^- ions were replaced by OH^-,

$$\Lambda\text{-}cis\text{-}Co(en)_2Cl_2^+ \underset{H_2O}{\overset{2OH^-}{\longrightarrow}} \Delta\text{-}cis\text{-}Co(en)_2(OH)_2^+ \qquad (14)$$

If their experimental data were correct, this conclusion seems enforced. As will be shown, this unique mechanism was substantiated by later experiments, although the facts upon which the proposal were first based have proved incorrect (ref. 23). Archer and Kwak (ref. 34) later presented additional results

* A change in the relative amounts of *two* or more active species can give rise to a change in sign.

supporting some of the original claims, although Tobe, Page and Farrago (ref. 37) have discredited important details of the original article.

We reproduce below the essence of a very recent publication (ref. 23) which resolves the major problems raised:

i) The apparent $[OH^-]$ dependence of the steric course of "normal" base hydrolysis of Λ-*cis*-$Co(en)_2Cl_2^+$; no other $Co(en)_2AX^{n+}$ complex seems to show such a dependence.

ii) The origin of the clear but anomalous $[OH^-]$ dependence of the overall steric course, i.e. the isomer distribution for $Co(en)_2(OH)_2^+$, whether arising in the normal stepwise fashion (13) or via the radical double-substitution (14) processes.

iii) The proposal of double-substitution. What are the general implications, given that double-substitution can be proved to be a viable substitution pathway?

2.2 Steric course of base hydrolysis of Λ-*cis*-$Co(en)_2(OH)Cl^+$

Clearly the second step of (13) need be understood before the first can be interpreted. This is straightforward, since the $Co(en)_2(OH)_2^+$ products are unreactive. Acid quenching generates $Co(en)_2(OH_2)_2^{3+}$ isomeric mixtures with retention, and the cis and trans components have vastly different electronic spectra (ϵ_{492}(cis) 79.2, ϵ_{492}(trans) 17.9). Also, HCO_3^- quenching generates the intensely optically rotating ion Λ-$Co(en)_2OCO_2^+$ (e.g. $[M]_{365}^{20} - 6271$ for Λ(+) enantiomer), and this reaction occurs with $100 \pm 1\%$ optical retention (ref. 13,23,36). Further, while Λ-*cis*-$Co(en)_2(OH_2)_2^{3+}$ generates Λ-$Co(en)_2OCO_2^+$, *trans*-$Co(en)_2(OH_2)_2^{3+}$ gives *trans*-$Co(en)_2(OCO_2)_2^-$ on HCO_3^- quenching, and these species also have greatly different visible spectra (ref. 36). Thus the agreement of the analysis in terms of both the aqua and carbonate ions is reassurance of its accuracy. The abbreviated results (Table 1) confirm the predominance of optical retention claimed by Archer *et al.* (ref. 33,34) in two separate works, but refute the data of Chan and Tobe (ref. 35) which suggested a significant loss of activity (75%, cf. our 7.5% at 0°C). Nonetheless, it remains clear that optical retention prevails.

It is significant that the stereochemical result (95% cis, 92.5% activity, 5% trans; 0°C) is independent of $[OH^-]$ over the range 0.1–1.0 M. We need add that the general rate law $k(obsd) = k_S + k_{OH}[OH^-]$ prevails for the substitution reactions of $Co(en)_2AX^{n+}$ complexes (ref. 28,30); the individual terms correspond to the normal aquation process ($[OH^-]$ independent) and the normal S_N1cb process ($[OH^-]$ dependent). The published rate data* (Table 2)

* Data for $\mu \sim 0$. For 0.1–1.0 M $[OH^-]$, k_{OH} values are probably a little lower with k_S values about the same.

TABLE 1

Stereochemistry of base hydrolysis of Λ-(+)-$Co(en)_2(OH)Cl^+$

$[OH^-]$,M	% cis[c]	% activity[c,d]	% Λ(+)	% Δ(−)	% trans
			25°C		
0.1	94.5	+ 90.6	92.5	2.0	5.5
0.1	94.0[a,b]				6.0
0.1	92.5				7.5
0.5	94.6	+ 90.7	92.7	2.0	5.4
			0°C		
0.1	94.8	+ 92.1	93.4	1.3	5.2
1.0	95.7	+ 92.8	94.3	1.5	4.3

[a] Racemic reactant.
[b] Acid quenching; product analysis for $Co(en)_2(OH_2)_2^{3+}$.
[c] HCO_3^- quenching; product analysis for carbonato species, except where noted.
[d] Plus sign means net optical retention (Λ > Δ) and does not relate to the signs of optical rotation in the $Co(en)_2OCO_2^+$ product.

show that the hydrolysis reaction occurs exclusively via the $[OH^-]$ dependent pathway above 0.1 M OH^- (ref. 35). The constant stereochemical result confirms this; the steric courses for the spontaneous (84.5% cis, 84.5% activity; 15.5% trans; 25°C) (ref. 36) and base catalyzed hydrolyses are different.

2.3 Steric course for the first step—base hydrolysis of Λ-*cis*-$Co(en)_2Cl_2^+$

The published rate data (Table 2) indicate that the loss of the first Cl^- from *cis*-$Co(en)_2Cl_2^+$ is much faster (~ 40-fold, 0°C) than Cl^- loss from *cis*-$Co(en)_2(OH)Cl^+$. Base hydrolysis of *trans*-$Co(en)_2(OH)Cl^+$ is slower again. The use of stoichiometric quantities of cobalt complex and OH^- suffer from the problem of rapid loss of Cl^- from $Co(en)_2(OH)Cl^+$ via the spontaneous pathway (25°C; cis, $t_{\frac{1}{2}}$ 0.96 min; trans, $t_{\frac{1}{2}}$ 7.2 min) (ref. 38), before *all* Λ-*cis*-$Co(en)_2Cl_2^+$ is consumed. High $[OH^-]$ lead to a significant contribution from a parallel reaction pathway, discussed ahead. Thus the products from the first step should be observed cleanly by using 2 to 3 equivalents of NaOH at low [Co].

This expectation was realized. Product cis/trans isomer distributions

TABLE 2

Aquation and base hydrolysis rate data[a] for $Co(en)_2ACl^+$

	T,°C	trans		cis	
		10^5 k_S	10^2 k_{OH}	10^5 k_S	10^2 k_{OH}
$Co(en)_2Cl_2^+$	25	4.2	310,000	24	70,000
	0	0.08	8,500	0.78	1,500
$Co(en)_2(OH)Cl^+$	25	160	56	1,200	1,100
	0	3.2	1.7	30	37

[a] From ref. 35 ($\mu \sim 0$ M).
[b] $k(obsd) = k_S + k_{OH}[OH^-]$; k_S, sec^{-1}; k_{OH}, M^{-1} sec^{-1}.

were analyzed spectrophotometrically in terms of both $Co(en)_2(OH_2)Cl^{2+}$ (H^+ quenching) and carbonato species (H^+/HCO_3^- quenching, *trans*-$Co(en)_2(OCO_2)Cl$ and $Co(en)_2OCO_2^+$). The activity was determined for (+)-$Co(en)_2OCO_2^+$ arising from the (+)-*cis*-$Co(en)_2(OH_2)Cl^{2+}$. Under the conditions, there is $\leq$ 10% secondary hydrolysis of *cis*-$Co(en)_2(OH)Cl^+$ prior to quenching, where there is negligible further reaction of *trans*-$Co(en)_2(OH)Cl^+$. Although there is good agreement, the results for the HCO_3^- quenching experiments are judged better (Table 3), because *cis*-$Co(en)_2(OH)Cl^+$ rearranges little on further reaction (95% cis product; 7.5% racemization—section 2.2, Table 1), even if it were completely consumed, and *cis*-$Co(en)_2(OH)Cl^+$ and *cis*-$Co(en)_2(OH)_2^+$ give the same species ($Co(en)_2OCO_2^+$) with H^+/HCO_3^-, with complete geometric and optical retention (ref. 36).

The activity arising in the first step was also checked by allowing time for *all* of the active *cis*-$Co(en)_2(OH)Cl^+$ to hydrolyze through to $Co(en)_2(OH)_2^+$, and then correcting the measured activity for the known loss (Table 1) in this second step. Subsequent racemization of Λ-*cis*-$Co(en)_2(OH)_2^+$ prior to quenching is negligible (ref. 39). The result obtained in this way is in very good agreement with that deduced above (Table 3).

It should be noted that because *cis*-$Co(en)_2Cl_2^+$ base hydrolyses with considerable rearrangement (77%, 25°C) to trans product, small absolute errors become much larger proportional errors for the % cis product. The accuracy of the Λ/Δ ratio thus suffers, despite the very well-defined % activity from which this is derived. It was for this reason that we devoted attention to measuring the cis/trans product distribution by alternate routes. The agree-

TABLE 3

Steric course for the first step of base hydrolysis of Λ-*cis*-$Co(en)_2Cl_2^+$
[Co] (1.5–2.5)10^{-3} M, $[OH^-]$ 5 $\times 10^{-3}$ M

T,°C	% cis	% Λ[a]	% Δ[b]	% trans	% activity[c]
25	23	15.8	7.5	77	8.5
	23.5[f]	–	–	76.5[f]	–
	19[d]	–	–	81[d]	–
	–	–	–	–	8.3[e]
0	18.5	13.6	4.9	81.5	8.7

[a] % Λ-*cis* = (% cis + % activity)/2.
[b] % Δ-*cis* = % cis – % Λ-*cis*.
[c] Measured as active $Co(en)_2OCO_2^+$; % activity = 10^2 [(% Λ-*cis* – %Δ-*cis*)/(% Λ-*cis* + % Δ-*cis* + % trans)].
[d] Analysis for H^+ quenched reaction; all other data for HCO_3^- quenching.
[e] For complete reaction through to $Co(en)_2(OH)_2^+$.
[f] Measured for racemate.

ment between the results so obtained (Table 3) suggests that the numbers are accurate to the relatively small error limits quoted.

Chan and Tobe (ref. 35) recorded a much higher % cis product for the base hydrolysis of *cis*-$Co(en)_2Cl_2^+$ (37%, 0°C), and lower activity (5%; cf our 8.7%). The literature data translate to a 21% Λ, 16% Δ, 63% trans product distribution. The contrast to our result 13.5% Λ, 5.0% Δ, 81.5% trans emphasizes the point made in the previous paragraph. The discrepancies are consistent with the use of a higher $[OH^-]$ by Chan and Tobe; some contribution from the special inversion parallel pathway discussed ahead accommodates both their lower activity and higher % cis product.

Archer and Dittmar (ref. 33) reported 23% cis product (0°C) for low $[OH^-]$, close to our result, but claimed the cis product was 94% active. Particularly alarming however are their data indicating that the Λ-/Δ-*cis*-$Co(en)_2(OH)Cl^+$ product ratio is dependent on $[OH^-]$: 97/3 at 0.03 M, 83/17 at 0.1 M and close to 50/50 (racemic) at 0.3 M$[OH^-]$. Also, the *cis-/trans*-$Co(en)_2(OH)Cl^+$ ratio was reported to be $[OH^-]$ dependent. This remarkable result is in conflict with the generally accepted S_N1cb mechanism of base hydrolysis, which predicts a steric course *independent* of $[OH^-]$.

Archer and Dittmar (ref. 33) employed a method of analysis which circumvented the problem of a parallel double-substitution pathway (14). They made no assumptions about the composition of the products. The now routine matrix methods were used to resolve multicomponent product spectra into their components, and the *cis-/trans*-$Co(en)_2(OH_2)Cl^{2+}$ ratios were determined for mixtures containing $Co(en)_2(OH_2)_2^{3+}$ species also, whether arising from double substitution and/or secondary base hydrolysis. However, there were fundamental deficiencies in such analyses, noted in detail elsewhere (ref. 36), and alluded to ahead.

We sought a new and precise measurement of the steric course of hydrolysis for the first step of the base hydrolysis of Λ-*cis*-$Co(en)_2Cl_2^+$ at higher $[OH^-]$, free from the difficulty of analyzing multicomponent mixtures. There were three phases in achieving this goal.

First, recognizing the possibility of *direct* $Co(en)_2(OH)_2^+$ formation (14), there was a need to measure the *cis-/trans*-$Co(en)_2(OH)Cl^+$ product distribution directly. This was achieved by acid-quenching, and ion-exchange chromatography at 0°C which cleanly separated any unreacted Λ-*cis*-$Co(en)_2Cl_2^+$ from $Co(en)_2(OH_2)Cl^{2+}$ and, in turn, from Λ-*cis*-$Co(en)_2(OH)_2^+$ (this approach was employed by Archer and Kwak (ref. 34), with partial success).

Second, there was a need for a device to correct for any loss of $Co(en)_2(OH)Cl^+$ through secondary hydrolysis (13); the cis isomer is much more reactive than the trans and thus any appreciable secondary reaction alters the observed *cis-/trans*-$Co(en)_2(OH_2)Cl^{2+}$ ratio. We took advantage of the fact that secondary hydrolysis depletes Λ- and Δ-$Co(en)_2(OH)Cl^+$ to identical extents, so that the % activity of *cis*-$Co(en)_2(OH)Cl^+$ remains unchanged. Thus, since first formed *trans*-$Co(en)_2(OH)Cl^+$ does not react significantly in the time scales involved while *cis*-$Co(en)_2(OH)Cl^+$ does, the loss in activity, per mole of cobalt found for $Co(en)_2(OH_2)Cl^{2+}$, is a direct measure of the consumed cis isomer. This deduction assumes that the % activity, if there were no secondary hydrolysis, is known. The accurately determined value for low $[OH^-]$ was used, and we are forced to assume it does not depend upon $[OH^-]$. However, there is no inbuilt assumption about the $[OH^-]$ dependence of the *cis-/trans*-$Co(en)_2(OH)Cl^+$ product ratio, which provides an acid test for the analysis. Moreover, reaction times were such as to minimize secondary loss of *cis*-$Co(en)_2(OH)Cl^+$, yet providing *variable and small* corrections to be made. The constancy of the corrected % activity for the immediate $Co(en)_2(OH)Cl^+$ product then provides a second test.

The third phase was the minimization or elimination of multicomponent mixtures, by utilizing the reactions (ref. 13,23,36) shown in (15). Hence cobalt concentrations for the three separate bands from the chromatography were determined for the common (single) species $Co(en)_2OCO_2^+$ (ϵ_{510} 133). Isomer ratios were then precisely and separately determined for $Co(en)_2(OH_2)Cl^{2+}$ and $Co(en)_2(OH_2)_2^{3+}$, each only two-component mixtures.

Finally, optical activities for each of the two bands were measured, again for the common (single) species, Λ-$Co(en)_2OCO_2^+$. This ion is inherently

$$\Lambda\text{-}cis\text{-}Co(en)_2(OH_2)Cl^{2+} \underset{\text{rapid}}{\overset{HCO_3^-}{\longrightarrow}}$$

$$\Lambda\text{-}cis\text{-}Co(en)_2(OCO_2)Cl \underset{\text{rapid}}{\longrightarrow} \Lambda\text{-}Co(en)_2OCO_2^+$$

$$trans\text{-}Co(en)_2(OH_2)Cl^{2+} \underset{\text{rapid}}{\overset{HCO_3^-}{\longrightarrow}} trans\text{-}Co(en)_2(OCO_2)Cl \overset{\Delta}{\longrightarrow} Co(en)_2OCO_2^+$$

$$\Lambda\text{-}cis\text{-}Co(en)_2(OH_2)_2^{3+} \underset{\text{rapid}}{\overset{HCO_3^-}{\longrightarrow}} \Lambda\text{-}Co(en)_2OCO_2^+$$

$$trans\text{-}Co(en)_2(OH_2)_2^{3+} \underset{\text{rapid}}{\overset{HCO_3^-}{\longrightarrow}} trans\text{-}Co(en)_2(OCO_2)_2^- \overset{\Delta}{\longrightarrow} Co(en)_2OCO_2^+$$

$$\Lambda\text{-}cis\text{-}Co(en)_2Cl_2^+ \underset{\Delta}{\overset{HCO_3^-}{\longrightarrow}} Co(en)_2OCO_2^+ \tag{15}$$

much more active (5–10-fold) than either Λ-*cis*-$Co(en)_2(OH_2)Cl^{2+}$ or Λ-*cis*-$Co(en)_2(OH_2)_2^{3+}$ (ref. 13,36). Previously, analyses were in terms of the latter two ions, moreover as an unseparated or partly separated mixture (ref. 33,34).

The steric course data for 1.0 M OH^- (0°C) are given in Table 4. The results corrected as above show 23% and 21% *cis*-$Co(en)_2(OH_2)Cl^{2+}$ product in repetitive experiments, compared to 18.5% cis observed for low $[OH^-]$ ($\sim 5 \times 10^{-4}$ M, 0°C; Table 3), and obtained by direct spectrophotometric measurements. The corresponding cis activities are 38% and 41%, respectively, compared to 47% at low $[OH^-]$. The agreement between the low and high $[OH^-]$ data is excellent, especially since we are dealing with quite low % cis products. Indeed, both the magnitude and sense of the discrepancies (~ +4% cis, −8% activity) are consistent with a trace of *trans*-$Co(en)_2(OH_2)Cl^{2+}$ isomerization during the chromatography (~ 1 h, ~ 5°C). This is clear from preliminary data obtained for (±)-*cis*-$Co(en)_2Cl_2^+$, where the chromatography took longer and was carried out at a slightly higher temperature. The isomerization merely increases the apparent % cis product, and hence decreases its apparent activity, since cis produced by this route must be racemic. This point may be clearer when it is recalled that the correction for loss of *cis*-$Co(en)_2(OH)Cl^+$ through secondary hydrolysis is based on the activity *expected* (8.7%) for the $Co(en)_2(OH_2)Cl^{2+}$ product. The shortfall (4.4% and 6.2%, Table 4) certainly reflects 30–50% secondary hydrolysis, but adjustment of each to 8.7% is based on *total* cobalt in the $Co(en)_2(OH_2)Cl^{2+}$ band from the column, independent of any subsequent trans to cis rearrangement.

TABLE 4

Stereochemistry of the $Co(en)_2(OH)Cl^+$ products for the base hydrolysis of Λ-*cis*-$Co(en)_2Cl_2^+$ at high $[OH^-]$, 0°C

	observed[c]			corrected[a]		
$[OH^-]$	% cis	%cis activ.[b,e]	% activ.[e]	% cis	% cis activ.[b,e]	% trans
1.0	23.7[d]	–	–	–	–	–
	26.0[d]	–	–	–	–	–
	14.0	+31.4	+4.4	22.7	+38.3	77.3
	16.0	+38.8	+6.2	21.1	+41.2	78.9
0.33	15.5	+34.8	+5.4	22.2	+39.2	77.8
	16.0	+38.1	+6.1	21.4	+40.7	78.6

[a] For secondary reaction of *cis*-$Co(en)_2(OH)Cl^+$, using the figure of +8.7% activity for first formed $Co(en)_2(OH)Cl^+$ (ref. 23); compare column 3.
[b] 10^2 (% activity/% cis).
[c] $Co(en)_2(OH_2)_2^{3+}$ (22–46% of total cobalt) and unreacted Λ-*cis*-$Co(en)_2Cl_2^+$ (31–3%) removed by ion-exchange chromatography at 0–5°C; analysis for $Co(en)_2(OH_2)Cl^{2+}$ (43–54%) after H^+/HCO_3^- quenching.
[d] Results for racemic reactant; chromatography at 5–10°C.
[e] % activity = 10^2 (Λ-*cis* – Δ-*cis*)/total Co; + sign denotes net retained Λ configuration.

Table 4 includes data for another $[OH^-]$, for reasons given in the next section (2.4). Since the 10^{-3} M and 1.0 M $[OH^-]$ data agree, it occasions no surprise that the steric course for 0.33 M OH^- is the same, but the result serves to reinforce the conclusion that the steric course of *stepwise* base hydrolysis of Λ-*cis*-$Co(en)_2Cl_2^+$ is independent of $[OH^-]$.

In summary, the data of Dittmar and Archer (ref. 33), suggesting a steric course where the cis/trans and Λ/Δ product ratios vary with $[OH^-]$, appear to be incorrect. The predictions of the accepted S_N1cb mechanism thus seem firm.

2.4 Steric course for the overall reaction

On the basis of the results described in sections 2.2 and 2.3, overall optical retention should predominate, since it does for each step of (13). Additionally, this should be independent of $[OH^-]$. For stepwise loss of Cl^- from Λ-*cis*-$Co(en)_2Cl_2^+$, the product distribution 95% *cis*- and 5% *trans*-$Co(en)_2(OH)_2^+$ is predicted, given time to consume all of the less reactive of the first formed $Co(en)_2(OH)Cl^+$, *trans*-$Co(en)_2(OH)Cl^+$. The final % optical activity predicted is $0.905 \times 8.5 = 7.7\%$ at 25°C ($0.925 \times 8.7 = 8.0$ at 0°C); here of course the complete consumption of the achiral *trans*- $Co(en)_2(OH)Cl^+$ ion is irrelevant.

An examination of the $[OH^-]$ dependence of the activity of the $Co(en)_2(OH)_2^+$, after complete hydrolysis, reveals agreement with prediction only at the low $[OH^-]$ end of the scale (Table 5).

TABLE 5

Dependence on $[OH^-]$ of activity of $Co(en)_2(OH)_2^+$ derived from complete base hydrolysis of Λ-*cis*-$Co(en)_2Cl_2^+$

	% activity[a,b]	
$[OH^-]$,M[e]	25°C	0°C
0.005	+8.3	+8.7
0.01	+7.4	
0.05	+4.7	
0.10	+0.6	
0.25	−1.5	
0.50	−8.6 (−6.7[d])	
0.75	−12.5	
1.0	−14.8	−20.3[c] (−17.2[d])

[a] 10^2 (Λ-*cis* - Δ-*cis*)/total Co; + denotes Λ > Δ, − denotes Δ > Λ.
[b] [Co], (1.6–62) 10^{-3} M; $[OH^-] \gg$ [Co] except for lowest $[OH^-]$.
[c] Average (±0.4%) of four values, [Co] = (1.6–14)10^{-3} M; all others are at least duplicate measurements.
[d] Slow addition of aqueous OH^-; all others, rapid.
[e] NaOH; μ not constant.

Clearly the overall result is strongly $[OH^-]$ dependent, but equally remarkable is the fact that above ~ 0.1 M $[OH^-]$ there is net inversion. The predominance of optical retention for each step of the normal stepwise hydrolysis, coupled with the $[OH^-]$ independence, leads to *the inescapable conclusion that* *there exists a path from Λ-cis-(+)-Co(en)$_2$Cl$_2^+$ to Δ-cis-(−)-Co(en)$_2$(OH)$_2^+$ which does NOT involve the intermediacy of Λ- or Δ-cis-Co(en)$_2$(OH)Cl$^+$.* Moreover, this unusual path is, in some unprecedented way, promoted by OH^-.

We sought the precise steric course for this double Cl^- loss path, corrected for the contribution from the normal route. Also an obvious question was the [OH^-] dependence of the rate of this double Cl^- loss reaction. Finally, if the contributions from the normal and double Cl^- substitutions could be separated at several [OH^-], the [OH^-] dependence of its stereochemistry (if any) could be determined. The following experiments were designed to answer all these questions.

Λ-Co(en)$_2$Cl$_2^+$ was base hydrolyzed under conditions that maximized the contribution to Co(en)$_2$(OH)$_2^+$ from the double Cl^- loss route, i.e. high [OH^-], but which minimized the contribution to Co(en)$_2$(OH)$_2^+$ from the second step of the normal hydrolysis route, i.e. very short reaction times. In principle it is possible to have, at high [OH^-], all the Λ-Co(en)Cl$_2^+$ but a negligible amount of the Co(en)$_2$(OH)Cl$^+$ reacted. Several experiments were conducted at 1 M [OH^-], 0°C with the desired result. The products were cleanly separated by ion-exchange chromatography at ~ 0°C, after H^+ quenching at 2–3 s reaction time, followed by HCO_3^- treatment (section 2.3). The data were corrected for any subsequent reaction of *cis*-Co(en)$_2$(OH)Cl$^+$ as described earlier. The results (Table 6) verify that appreciable (~ 40%) Co(en)$_2$(OH)$_2^+$ is derived directly from Λ-Co(en)$_2$Cl$_2^+$. This conclusion is not contingent upon the accuracy of corrections for secondary hydrolysis of *cis*-Co(en)$_2$(OH)Cl$^+$, since even *complete* secondary reaction of this ion contributes only ~ 12% (absolutely) to total cobalt in Co(en)$_2$(OH)$_2^+$. This arises because most (77%, 0°C) of first formed Co(en)$_2$(OH)Cl$^+$ comprises the relatively unreactive trans isomer.

The stereochemistry found for Co(en)$_2$(OH)$_2^+$ via the double Cl^- loss path is 82% cis (70% Δ, 12% Λ) and 18% trans at 0°C; the agreement between independent experiments, requiring different corrections for secondary hydrolysis, is good (Table 6). A further (and critical) test is the reconciliation of these data (Table 6), for separated components, with those of Table 5, for direct measurements on final mixtures. This is done as follows. From Table 6 it can be seen that there is a 41% contribution to the overall reaction (1 M [OH^-], 0°C) from double Cl^- loss. Using the values for the active product arising via both routes (+8.0%, stepwise; −60%, double Cl^- path), the figure of $0.59(8.0) + 0.41(-60) = -19.9(\pm 1)\%$ activity is calculated for the complete hydrolysis of Λ-Co(en)$_2$Cl$_2^+$ in 1.00 M [OH^-] at 0°C. This agrees remarkably well with that directly observed (−20.3 ± 0.4%, Table 5), and this agreement seems to secure the analyses.

TABLE 6

Product analysis for controlled base hydrolysis of Λ-*cis*-$Co(en)_2Cl_2^+$ at 0°C

		observed[a]			corrected[b]	
		recovd reactant	$(OH)Cl^+$	$(OH)_2^+$	$(OH)Cl^+$	$(OH)_2^+$
1.0 M OH^-						
% total Co	(i)[d]	11.1	43.4	45.6	–	–
	(ii)	8.1	50.5	41.4	–	–
	(iii)	3.4	50.4	46.2	56.0	40.6
	(iv)	4.8	53.5	41.7	57.0	38.2
corrd % total Co[c]	(i)	–	48.7	51.3	–	–
	(ii)	–	55.0	45.0	–	–
	(iii)	–	52.2	47.8	58.0	42.0
	(iv)	–	56.2	43.8	59.9	40.1
% cis[f]	(i)	100	23.7	80.0	–	–
	(ii)	100	26.0	86.1	–	–
	(iii)	100	14.0	82.5	22.7	80.8
	(iv)	100	16.0	84.5	21.1	83.5
% activ.[e]	(iii)	100	+4.4	−45.2	(+8.7)	−57.6
	(iv)	100	+6.2	−49.3	(+8.7)	−58.1
% Λ-*cis*[f]	(iii)	100	9.2	18.7	15.7	11.6
	(iv)	100	11.1	17.6	14.9	12.7
% Δ-*cis*[f]	(iii)	0	4.8	63.9	7.0	69.2
	(iv)	0	4.9	66.9	6.2	70.8
0.33 OH^-						
% total Co[d]	(v)	17.3	53.9	28.8	58.5	24.2
	(vi)	31.3	46.2	22.5	49.3	19.4
corrd % total Co[c]	(v)	–	65.2	34.8	70.7	29.3
	(vi)	–	67.2	32.8	71.8	28.2
% cis[f]	(v)	100	15.5	86.1	22.2	84.4
	(vi)	100	16.0	87.5	21.4	86.3
% activ.[e]	(v)	100	+5.4	−37.8	(+8.7)	−53.4
	(vi)	100	+6.1	−43.8	(+8.7)	−57.7
% Λ-*cis*[f]	(v)	100	10.5	24.1	15.5	15.5
	(vi)	100	11.1	21.9	15.1	14.3
% Δ-*cis*[f]	(v)	0	5.0	62.0	6.7	68.9
	(vi)	0	4.9	65.6	6.3	72.0

[a] Products and unreacted complex separated chromatographically at $\leq 5°C$.
[b] For secondary reaction of *cis*-$Co(en)_2(OH)Cl^+$, using the figure of +8.7% activity for first formed $Co(en)_2(OH)Cl^+$ (ref. 23).
[c] Results immediately above normalized to 100%, after allowing for unreacted dichloro complex.
[d] (i), (ii), ... are experiment numbers; preliminary experiments ((i) and (ii)) were performed on the racemate, with chromatography at 5–10°C.
[e] 10^2 (Λ-*cis* – Δ-*cis*)/(total Co in the relevant complex); + denotes net optical retention; –, inversion.
[f] % trans = 100 – % cis; % cis = % Λ-*cis* + % Δ-*cis*; % Λ-*cis* = (% cis + % activity)/2.

It remained to determine the $[OH^-]$ dependence of the steric course for the double Cl^- loss path. Two experiments, similar to those described for 1 M $[OH^-]$, were conducted using 1/3 M $[OH^-]$ at 0°C. The analysis of the products separated efficiently by ion-exchange chromatography are included in Table 6. As expected, the proportions of directly formed $Co(en)_2(OH)_2^+$ are lower, and consequently the corrections for secondary hydrolysis of *cis*-$Co(en)_2(OH)Cl^+$ are proportionately greater. Nonetheless, despite the necessarily lower accuracy of the steric course data, the $Co(en)_2(OH)_2^+$ stereochemistry is the same as at 1.0 M $[OH^-]$ (80.5% cis—70.5% Λ, 15% Δ—and 15.5% trans). This result not only further supports the analysis, but also demonstrates that the $[OH^-]$ dependence of the observed activity for the overall reaction (Table 5) results entirely from changes in the relative contributions of the two reaction pathways, stepwise and double Cl^- loss. Indeed, there is no intrinsic $[OH^-]$ dependence of the stereochemistry for any of the three basic reactions studied ((13), (14)).

The results (Table 5) are somewhat deceptive insofar as they indicate that net inversion is significant at quite low $[OH^-]$. However, the low percent activity (+8%) for the normal stepwise route is offset by the high percentage of opposite sign (–60%) for the double Cl^- loss path, and thus the overall result is especially sensitive to a contribution from the latter.

The temperature dependences (0, 25°C) for the stereochemistry of the various processes have been recorded also. A mild dependence is clear for the first base hydrolysis step of Λ-*cis*-$Co(en)_2Cl_2^+$, and also for Λ-*cis*-$Co(en)_2(OH)Cl^+$. More work is required to decide if the numerically larger activity at 0°C (–20.3%) for the overall process (cf –14.8%, 25°C) is a result of inherent temperature dependences for the contributing stepwise and double Cl^- loss pathways, or whether it arises from a higher activation enthalpy for the stepwise Cl^- loss contribution. Certainly the former can be temperature dependent, although established in only few instances, and we re-emphasize the need (ref. 40,41) for the precise determination of the temperature dependence of the steric course of ligand substitution.

2.5 Homogeneous or heterogeneous substitution?

The early Bailar inversion investigations (ref. 12–15) involved reagents in pastes or as an insoluble salt, e.g. Ag_2CO_3 in contact with an aqueous solution of the complex. Presumably reaction occurs at the surface of the solid phase, but a key question is: given the opportunity, could it also occur homogeneously? For the classic Bailar inversion reaction of Λ-*cis*-$Co(en)_2Cl_2^+$ in aqueous NaOH, this has been difficult to answer. The rate of base hydrolysis in 1 M OH^- is especially fast ($t_{1/2}$ $\simeq$ 1 ms, 25°C; $\simeq$ 50 ms, 0°C) (ref. 35), and mixing aqueous complex solutions and aqueous NaOH in a time short compared to the reaction time is obviously difficult. Archer *et al.* tried this (ref. 33,34), and it is uncertain whether mixing was sufficiently fast. We (and others (ref. 31)) have rapidly dissolved the solid complex directly in aqueous OH^-, where the rate of dissolution is the difficulty. Inversion is observed at high $[OH^-]$ under both sets of conditions, which removes the possibility that inversion *must* arise from a heterogeneous reaction between aqueous OH^- and the surface of the solid complex. Since *cis*-$Co(en)_2Cl_2^+$ very rapidly consumes OH^- and this can reduce the local $[OH^-]$ if mixing (liquid/liquid or solid/liquid) is inadequate, it needed to be established that the observed activities (Table 5) were independent of mixing time and [Co], in conditions where $[OH^-] \gg [Co]$. This was found to be so.

In summary, the reaction proceeds in *homogeneous* solution. This is supported by the clear fact that base hydrolysis can occur faster than the *complete* dissolution of the complex when larger samples are used, but provided adequate mixing is obtained, the results are the same.

2.6 Rate law

In a previous publication (ref. 23) the rate law for the reaction

$$cis\text{-}Co(en)_2Cl_2^+ \xrightarrow{2OH^-} Co(en)_2(OH)_2^+ \tag{16}$$

was deduced from the $[OH^-]$ dependence of the final optical rotations (Table 5). This was possible because the activities for each of the two (parallel) contributing reactions, namely stepwise and double Cl^- loss, were both very different and independent of $[OH^-]$, and also because the rate law for one of the paths, stepwise Cl^- loss, was known. Implicit in this analysis (although not stated) was the assumption that the stereochemical outcome of varied OH^- resided in the rate determining step for the double Cl^- loss path. An alternative prospect is considered ahead, but for the present we note that, on this basis, an approximate rate law for this reaction was deduced to be

$$-d/dt\ [CoCl_2^+] = -k_{xx}\ [OH^-]^2\ [CoCl_2^+] \tag{17}$$

Thus the complete rate law,

$$\begin{aligned} -d/dt\ [CoCl_2^+] &= -\{k_x\ [OH^-] + k_{xx}\ [OH^-]^2\}\ [CoCl_2^+] \\ &= k(obsd)\ [CoCl_2^+] \end{aligned} \tag{18}$$

accommodates both the exclusiveness of the normal base hydrolysis reaction at low $[OH^-]$ (first-order dependence), as well as the increased significance of the double Cl^- loss path as $[OH^-]$ is raised (second-order dependence).

A value for k_{xx} can be estimated from the data. At 1 M $[OH^-]$, 41% of the reaction occurs via double Cl^- loss, and thus 0.59 k(obsd) $= k_x\ [OH^-] = k_x$ and 0.41 k(obsd) $= k_{xx}\ [OH^-]^2 = k_{xx}$. Hence $0.41/0.59 = k_{xx}/k_x$. Given $k_x = 15\ (M^{-1}s^{-1})$ at 0°C, $k_{xx} = 10\ (M^{-2}s^{-1})$. Note first that this estimation of k_{xx} is only approximate, because the data (Table 5) do not refer to constant ionic strength, and the value of k_x used refers to $\mu \sim 0$), and second, that k_x is the second-order rate constant for normal base hydrolysis of *cis*-$Co(en)_2Cl_2^+$ (rather than of $Co(en)_2(OH)Cl^+$), albeit the formation of *cis*-$Co(en)_2(OH)_2^+$ via the stepwise Cl^- loss route is rate determined by the second step,

$$cis\text{-}Co(en)_2Cl_2^+ \xrightarrow{\text{fast}} Co(en)_2(OH)Cl^+ \xrightarrow{\text{slow}} Co(en)_2(OH)_2^+ \tag{19}$$

This arises because the division in reaction pathways occurs in the first step,

$$\Lambda\text{-}cis\text{-}Co(en)_2Cl_2^+ \xrightarrow{k_x\ [OH^-]} Co(en)_2(OH)Cl^+ \longrightarrow Co(en)_2(OH)_2^+ \quad \text{51.5\% }\Lambda\text{, 43.5\% }\Delta\text{, 5\% trans}$$

$$\Lambda\text{-}cis\text{-}Co(en)_2Cl_2^+ \xrightarrow{k_{xx}\ [OH^-]^2} Co(en)_2(OH)_2^+ \quad \text{70\% }\Delta\text{, 12\% }\Lambda\text{, 18\% trans} \tag{20}$$

Previous speculation on the mechanism of Bailar inversion has lacked information on the rate law. The $[OH^-]$ dependence of the rate derived from direct measurements has now become a crucial item, because it is conceivable that the role of OH^- does not appear in the rate determining step for the double Cl^- loss pathway. Indeed, the apparent simultaneous loss of two Cl^- could well be the more statistically reasonable sequential process (21), where the role of high $[OH^-]$ lies in re-directing the normal hydrolysis pathway in a faster non-rate determining process.

$$
Co(en)_2Cl_2^+ \xrightarrow[\text{slow}]{OH^-} I \begin{cases} \xrightarrow[OH^-]{\text{fast}} I^* \xrightarrow{\text{fast}} Co(en)_2(OH)_2^+ \\ \xrightarrow{\text{fast}} Co(en)_2(OH)Cl^+ \rightarrow Co(en)_2(OH)_2^+ \end{cases} \tag{21}
$$

Here I represents the normal intermediate in the base catalyzed process, in this case the $Co(en)(en\text{-}H)Cl_2$ 6-coordinate conjugate base, where en-H represents deprotonated ethylenediamine. This aspect is undisputed, although the sequence of the subsequent (and much faster) reactions of this conjugate base* remains contentious but need not concern us here (see section 8).

It is premature to discuss in detail the nature of the intermediate I* until the validity of (18) is checked by direct kinetic measurements. This is straightforward in principle, since for (21) the normal base hydrolysis rate law $k(obsd) = k[OH^-]$ will be strictly obeyed, for all $[OH^-]$, whereas for (18) a clearly discernible departure from linearity is predicted for a plot of k(obsd) *vs* $[OH^-]$. This important set of experiments is currently being undertaken (ref. 42).**

It is relevant to comment that the observed variation in the product distribution with varied $[OH^-]$, according to (14), is chemically reasonable if, e.g., OH^- effects a second (and irreversible) deprotonation of I ($pK_{a_1} \geq 16$, and precedent thus dictates $pK_{a2} \gg 16$) (ref. 43,44). Thus the role of OH^- could be focused on changing the spin state of I (from diamagnetic to paramagnetic) through either deprotonation by OH^-, and/or by ion-pairing of this *neutral* substrate with OH^- (and these processes, which are stoichiometrically equivalent, are very difficult to distinguish experimentally) (ref. 45). The focus

* I, on route to products by both paths shown in (21), could proceed with loss of Cl^- to pentacoordinate species. I* may be such an intermediate, derived from $Co(en)(en\text{-}H)Cl^+$ by *substitution* of the second Cl^- by OH^-.

** In NaOAc rather than the more usual $NaClO_4$ media, to maintain constant ionic strength. Acetate ion is a much poorer "nucleophile" than ClO_4^-, and ambiguities in interpretation of the data are much more readily resolved.

on spin state is to accommodate the enormous reactivity which I* (21) must possess, to offset its undoubted very low abundance, since high spin Co(III) is expected to be orders of magnitude more reactive than its normal low spin d^6 counterpart (ref. 46). It is also appropriate to comment that both normal and double Cl^- loss paths for the base catalyzed reactions of Λ-$Co(en)_2Cl_2^+$ do not seem to be accommodated by a *common* high spin Co(III) species; if one path goes via this species, the other would appear to find an alternate mechanism for acquiring the required reactivity.

2.7 Anion competition for the double-substitution pathway

The capture of anions during normal base hydrolysis

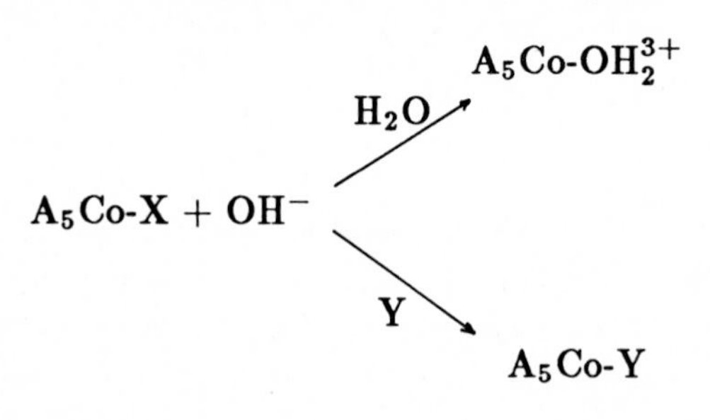

(22)

is a powerful mechanistic probe (ref. 28,30,47,48). Most experiments in this area have dealt with the extent rather than the stereochemistry of capture, although there are a few detailed studies of importance which are discussed in later sections.

The prospect of anion capture during the Bailar inversion reaction has been recognized (ref. 23), but no results have been forthcoming. However, some interesting preliminary data are now available for azide ion competition, and are presented ahead.

It has been suggested (ref. 23) that double-anion capture might be observable,

$$Co(en)_2Cl_2^+ \underset{OH^-}{\overset{2N_3^-}{\longrightarrow}} Co(en)_2(N_3)_2^+ \qquad (23)$$

Another possibility is capture of N_3^- and H_2O (or OH^-),

$$Co(en)_2Cl_2^+ \underset{OH^-}{\overset{H_2O,N_3^-}{\longrightarrow}} Co(en)_2(OH)N_3^+ \qquad (24)$$

Hydroxide ion, while certainly a much stronger base than N_3^-, is a decidely poorer competitor (nucleophile) in octahedral metal ion substitution reactions (ref. 49,50). For several OH^- catalyzed cobalt (III) substitution processes, it has been established that H_2O wins handsomely over OH^-, even at the 1 M level, in scavenging reduced coordination number intermediates. On a mole for mole basis, azide ion is a better nucleophile than H_2O, as can be gauged from the result (ref. 49),

$$(NH_3)_5CoCl^{2+} \xrightarrow[\text{pH } 10\text{–}14]{1.0\text{ M } N_3^-} (NH_3)_5CoN_3^{2+}\ (8.5\%) + (NH_3)_5CoOH^{2+}\ (91.5\%) \tag{25}$$

Clearly the observation of N_3^- capture ((23), (24)) in the context of Bailar inversion process would help clarify the special role that OH^- appears to play.

Apparent double-azide substitution (23) can arise sequentially,

$$\Lambda\text{-}cis\text{-}Co(en)_2Cl_2^+ \xrightarrow[OH^-]{N_3^-} Co(en)_2(N_3)Cl^+ \xrightarrow[OH^-]{N_3^-} Co(en)_2(N_3)_2^+ \tag{26}$$

Some estimate for the amount and stereochemistry of diazido product by this route can be arrived at from the results of N_3^- capture experiments for Λ-*cis*- and *trans*-$Co(en)_2(NH_3)X^{n+}$ ($n = 2,3$) in base (ref. 51,52). For the 2+ ions, there is ~ 25% azide capture in 1.0 M NaN_3, and ~ 30% for the 3+ substrates. The charge dependence is normal, and ~ 15% N_3^- capture could be expected for the 1+ ions *cis*-$Co(en)_2Cl_2^+$ and *cis*-$Co(en)_2(N_3)Cl^+$; the balance of product would be $Co(en)_2(OH)Cl^+$ and $Co(en)_2(OH)N_3^+$, respectively. Thus, overall, ~ 2–3% (.15 × 15) diazido product could be anticipated via the normal stepwise route (26). Moreover, this would comprise both cis and trans isomers, with the cis component largely racemized (~ 0.45 × 45 ≃20% activity for the cis diazido component), but with net optical retention (ref. 51,52). Trans isomer arising in the first step contributes to racemic cis formed in the second (26), and so the final activity of any *cis*-$Co(en)_2(N_3)_2^+$ would be minimal.

A similar analysis for azido-aqua product arising in a normal sequential fashion,

$$\Lambda\text{-}cis\text{-Co(en)}_2\text{Cl}_2^+ \xrightarrow{N_3^-} \text{Co(en)}_2(\text{N}_3)\text{Cl}^+ \xrightarrow[\text{OH}^-]{\text{H}_2\text{O}} \text{Co(en)}_2(\text{OH})\text{N}_3^+$$

$$\Lambda\text{-}cis\text{-Co(en)}_2\text{Cl}_2^+ \xrightarrow{\text{OH}^-} \text{Co(en)}_2(\text{OH})\text{Cl}^+ \xrightarrow{N_3^-} \text{Co(en)}_2(\text{OH})\text{N}_3^+ \qquad (27)$$

leads to an estimated 15% $Co(en)_2(OH)N_3^+$. Some 8% of this would be partly racemized cis, using the reported (ref. 53) steric course data for *cis*- and *trans*-$Co(en)_2(N_3)Cl^+$.

Duplicate experiments were carried out for Λ-*cis*-$[Co(en)_2Cl_2]Cl.H_2O$ reacted in 1.0 M NaN_3 containing 1.0 M NaOH, at 0°C. The reaction time of 2.0 min is sufficient to consume totally the intermediate cis species $Co(en)_2(OH)Cl^+$ and $Co(en)_2(N_3)Cl^+$, but little of the *cis*-$Co(en)_2(OH)N_3^+$ which is less reactive. Also, any *cis*- or *trans*-$Co(en)_2(N_3)_2^+$ product would be largely untouched (Table 7) (ref. 54). The acid (pH 3) quenched products were chromatographed on SP-Sephadex C-25 and the bands examined polarimetrically.

TABLE 7

Aquation and base hydrolysis rate data[a] for $Co(en)_2A(N_3)^+$

		trans		cis	
	T,°C	10^5 k_S	10^3 k_{OH}	10^5 k_S	10^3 k_{OH}
$Co(en)_2(N_3)_2^+$	25	[c]	11.3	[c]	6.6
$Co(en)_2(OH)(N_3)^+$	25	[c]	1.2	[c]	2.8
$Co(en)_2(N_3)Cl^+$	25	22	–	20	–
	0	0.62	410	0.70	170

[a] From ref. 54 ($\mu \sim 0$ M).
[b] $k(obsd) = k_S + k_{OH}[OH^-]$; k_S, sec^{-1}; k_{OH}, M^{-1} sec^{-1}.
[c] Negligibly small, pH $\geq$ 11.

No diazido complex was found. The limit of detection (spectrophotometric) was such that $\geq$ 1% would be observed. Second, the 2+ band off the column was appreciable and comprised largely $Co(en)_2(OH_2)N_3^{2+}$ (nitrosation tests); the rest was *trans*-$Co(en)_2(OH_2)Cl^{2+}$. The optical rotations were weak ($\sim$ 0.010°) but well above noise level ($\pm$ 0.001°). More significantly, the signs (589 nm, negative; 465 nm, positive) are relative magnitude ($\alpha_{589}/\alpha_{465}$) are diagnostic of Δ-$Co(en)_2(OH_2)(N_3)^{2+}$ (i.e. $\Delta > \Lambda$). The last band (3+)

off the column was $Co(en)_2(OH_2)_2^{3+}$, largely cis, and also having the net inverted configuration Δ.

At 1 M OH^-, the proportion of the reaction proceeding via the normal stepwise route is $\sim$ 60%, the balance of 40% occurring by double-chloride substitution. Thus the observation of no detectable diazido product ($\leq$ 1%) indicates an upper limit of about 2% by *either* pathway. Since this is a little less than that predicted (admittedly crudely) for the stepwise route (26) alone, it seems clear that two N_3^- ions are not captured in the double-substitution path to any significant extent (< 1%). This is an important result, for it suggests no special nucleophilic role for OH^-, since it is not mimicked by the better nucleophile N_3^-. An OH^- selective ion-pairing role also seems doubtful, unless the OH^-/N_3^- distinction vests in the ability of OH^- to abstract a proton from a coordinated amine, within the ion-pair. This amounts to a variation of the double deprotonation mechanism outlined earlier,

$$\begin{array}{ccccc}
(en)_2CoCl_2^+ & \underset{fast}{\overset{OH^-}{\rightleftharpoons}} & (en)(en\text{-}H)CoCl_2 & \underset{fast}{\overset{OH^-}{\rightleftharpoons}} & (en\text{-}H)(en\text{-}H)CoCl_2^- \\
 & & H_2O \big\downarrow k_1 & & 2H_2O \big\downarrow k_2 \\
 & & (en)_2Co(OH)Cl^+ & & (en)_2Co(OH)_2^+
\end{array} \tag{28}$$

This scheme is the simplest of several which accommodate the two term rate law (18). With OH^- functioning exclusively as a base, it also obviously accounts for the lack of *double* N_3^- ion capture. Indeed, if the path from the doubly deprotonated reactant through to dihydroxo product is actually sequential, involving intermediates of reduced coordination number which are likely unchanged or anionic, it is not conducive to N_3^- scavenging. Such a proposal implies attack of N_3^- one at a time, and 10% capture in each of two steps can yield only 1% diazido product.

The second result, the formation of significant amounts of *cis*-$Co(en)_2(OH_2)N_3^{2+}$ having the inverted Δ configuration, is important for two reasons. First, this is the first report of a Bailar inversion in cobalt (III) chemistry where the substitution involves an incoming anion. Second, it shows that *one* anion can be captured by this double-substitution route. Whether substitution by N_3^- and H_2O is synchronous or sequential (and in what order) remains undecided. Further experiments and detailed analysis are required to separate the two contributions ((24), (27)) to $Co(en)_2(OH)N_3^+$, in order to quantify the stereochemistry.

3 BASE CATALYZED HYDROLYSIS OF *trans*-$Co(en)_2Cl_2^+$: DOUBLE SUBSTITUTION, AND GENERAL IMPLICATIONS

The inversion reaction of the chiral *cis*-$Co(en)_2X_2^+$ complexes we have linked with the requirement for double substitution. However, the converse need not be true, i.e. achiral species such as *trans*-$Co(en)_2Cl_2^+$ can yield only racemic cis product, yet the prospect exists for direct $Co(en)_2(OH)_2^+$ formation. This suggestion (ref. 23) has now been tested.

Our new data (Table 8) (ref. 55) for the steric course of base hydrolysis of *trans*- $Co(en)_2(OH)Cl^+$, and of *trans*-$Co(en)_2Cl_2^+$ at low $[OH^-]$ (Table 9), are in reasonable agreement with those (ref. 35) of Tobe and Chan. The detailed analysis followed that described earlier for the corresponding cis isomers. The new feature is the study of the hydroxide ion dependence of the steric course. As expected, *trans*-$Co(en)_2(OH)Cl^+$ shows no dependence up to 1.0 M OH^-. This reaction has a high rate of background aquation, as does *cis*-$Co(en)_2(OH)Cl^+$ (although the cis rates are much larger, Table 2), and therefore a minimum $[OH^-]$ of 0.1 M is needed to drive the reaction essentially all via the base catalyzed pathway. The data of Table 9 are corrected for a little secondary reaction of *cis*- and *trans*-$Co(en)_2(OH)Cl^+$, in the time required for the hydrolysis experiments.

TABLE 8

Steric course of base hydrolysis of *trans*-$Co(en)_2(OH)Cl^+$

T,°C	$[OH^-]$,M	% cis	% trans
25	1.0	92.5[a]	7.5
		90.5[b]	9.5
	0.1	91[a]	9
0	1.0	93[a]	7
	(low)[d]	(94±2)[c]	(6±2)[c]

[a] H^+ quenched.
[b] H^+/HCO_3^- quenched.
[c] Literature values (ref. 35).
[d] Not specified in ref. 35.

Because *cis*- and *trans*-$Co(en)_2(OH)Cl^+$ hydrolyze in strong base (0.2–1.0 M) with almost an identical steric course (93% cis, 25°C), the final isomer distribution derived from *trans*-$Co(en)_2Cl_2^+$ by stepwise hydrolysis is independent of the steric course for the first step. Thus the first efforts to observe double-substitution for the *trans*-$Co(en)_2Cl_2^+ + OH^-$ reaction consisted simply in completely hydrolyzing *trans*-$[Co(en)_2Cl_2]Cl$ by direct dissolution in

TABLE 9

Stereochemistry of base hydrolysis of *trans*-$Co(en)_2Cl_2^+$

		$Co(en)_2(OH)Cl^+$ [b]		$Co(en)_2(OH)_2^+$ [a]	
T,°	$[OH^-]$,M	% cis	% trans	% cis	% trans
25	1.0	–	–	88	12
		–	–	89	11
	0.002	10.5	89.5[c]	–	–
		10.5	89.5[d]	–	–
0	0.002	8	92[c]	–	–
		9	91[d]	–	–
	(low)[e]	(5±2)	(95±2)[e]	–	–

[a] Observed product distribution for complete reaction through to $Co(en)_2(OH)_2^+$.
[b] Some correction for secondary hydrolysis.
[c] Analysis for aqua products; H^+ quenched.
[d] Analysis for carbonato products; HCO_3^- quenched.
[e] $[OH^-]$ not given; ref. 35.

1.0 M NaOH (25°C), and looking for a steric course other than 93% *cis*-$Co(en)_2(OH)_2^+$. We found 88.5% cis (Table 9), somewhat less, but unconvincing of a contribution from double-substitution. Thus the steric course for *trans*-$Co(en)_2Cl_2^+$ at high $[OH^-]$ (1.0 M) was examined more closely by H^+ quenching at very short reaction times (1–2 s, 0°C), and chromatography of the products, in a manner similar to that described for the cis isomer. In principle the steric course for the trans isomer is easier to determine; it reacts ~ 6-fold more rapidly than the cis form at 0°C (Table 2), and secondary reactions of $Co(en)_2(OH)Cl^+$ are thus less significant. Indeed, the primary hydrolysis of the trans-dichloro isomer is 230-fold faster than the hydrolysis of the faster reacting component *cis*-$Co(en)_2(OH)Cl^+$ of first formed $Co(en)_2(OH)Cl^+$, but nonetheless in the time necessary to get much of the *trans*-$[Co(en)_2Cl_2]Cl$ (2.0 g scale) dissolved in 1 M NaOH before acid quenching (1–2 sec), some secondary reaction is inevitable. Corrections for this secondary reaction were made on the basis of the cis/trans isomer ratio observed for the second (2+ ion) band after chromatography (0°C) of the products. The value of 92% trans/8% cis product, that determined for low $[OH^-]$ was used. We did not have the additional probe of activity provided by the chiral cis system, but nonetheless the corrected data (Table 10) are internally consistent.

TABLE 10

Product analysis for controlled base hydrolysis of *trans*-$Co(en)_2Cl_2^+$ in 1.0 M $[OH^-]$ at 0°C

		observed %			corrected %[a]	
		recovd reactant	$(OH)Cl^+$	$(OH)_2^+$	$(OH)Cl^+$	$(OH)_2^+$
% total Co	(i)[c]	44.4	38.7	16.9	41.9	13.7
	(ii)	56.7	31.8	11.5	33.3	10.0
corrd[b]	(i)	–	69.6	30.4	75.3	24.7
% total Co	(ii)	–	73.4	26.6	77.0	23.0
% trans	(i)	100	94	19.5[d],22.5[e]	(92)	24
	(ii)	100	91	21[e]	(92)	23

[a] Corrected for secondary hydrolysis of $Co(en)_2(OH)Cl^+$.
[b] Normalized to 100% after deducting % unreacted *trans*-$Co(en)_2Cl_2^+$.
[c] (i), (ii) represent separate experiments.
[d] H^+ quenched; isomeric analysis for aqua species.
[e] HCO_3^- quenched; isomeric analysis for carbonato species.

It would seem that the steric course of base hydrolysis of *trans*-$Co(en)_2Cl_2^+$ does not depend appreciably on $[OH^-]$, if at all (Table 9). This can be said because the corrections for secondary hydrolysis are small. Little *cis*-$Co(en)_2(OH)Cl^+$ (7%) is formed in the first step, although as much as half of this is lost in the 1–2 sec reaction time.

The more significant results are the amount and stereochemistry of the $Co(en)_2(OH)_2^+$ product. About 30% of the total cobalt is in this form, far too much to be accommodated by normal stepwise base hydrolysis. Indeed, allowing for the contribution from $Co(en)_2(OH)Cl^+$, the corrected value is 24 ± 2%, indicating clearly that double-substitution has occurred,

$$trans\text{-}Co(en)_2Cl_2^+ \xrightarrow[H_2O]{2OH^-} cis\text{- plus } trans\text{-}Co(en)_2(OH)_2^+ + 2Cl^- \quad (29)$$

The second point is the stereochemistry, 77 ± 3% cis. There is obviously substantial trans to cis rearrangement, contrasting with the high retention (93% trans) observed for the normal base hydrolysis process. An important observation is that, while the stereochemistries of substitution of *trans*- and

Λ-*cis*-$Co(en)_2Cl_2^+$ are quite different, and this is usual (see section 8.1), the stereochemistry for the double-substitution pathways are essentially the *same* (77 $\pm$ 3% and 82 $\pm$ 3% cis, respectively). This suggests that a common (chiral)* intermediate is involved. This point is taken further in section 4.

There seems little point in speculation upon the nature of any intermediate while the evidence for its existence is thin. However, it does now seem worthwhile to systematically investigate the stereochemistry of double-substitution for a range of Λ-*cis*- and *trans*-$Co(en)_2XY^{n+}$ complexes. The complexes need not have equivalent leaving groups, but two which are readily substituted (X,Y $= Cl^-$, Br^-, $(CH_3)_2SO$, $CF_3SO_3^-$, $CH_3SO_3^-$ and others). The synthetic chemistry continues to be developed (ref. 36,56,57), and provides many suitable complexes in this class, including the resolved cis ions. Interesting questions will be not only whether the product stereochemistry is generally independent of that of the reactant, but also the role of the *two* leaving groups. It is even conceivable that the stereochemical outcome is independent of the nature of these. The inference could be a common four coordinate intermediate, e.g.

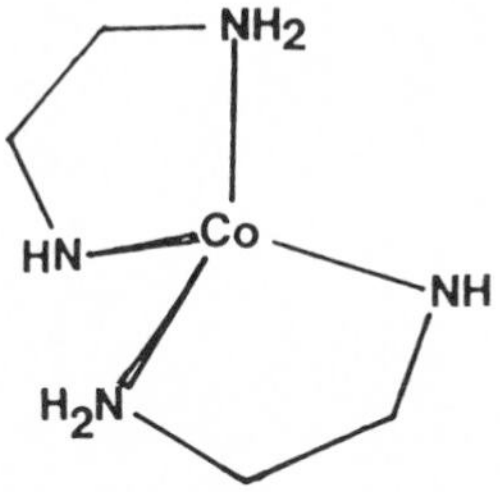

(30)

which is chiral by virtue of the two deprotonated amine centres. Highly unusual reactions can justify a radical proposal such as this one, especially if it can be tested. We have indicated one such test. Four coordinate cobalt(III) is not likely to be accepted readily, since even five coordinate intermediates in the normal octahedral metal ion substitution processes have resisted general recognition (ref. 58–60). It remains true, nonetheless, that a great deal of interesting inorganic chemistry has been generated from a contentious issue such as this one (ref. 28,30,47,61,62), and hopefully models for double-substitution will receive comparable attention with a similar result.

4 BAILAR INVERSION AND DOUBLE SUBSTITUTION IN OTHER SOLVENTS

It appears that the prospect of octahedral inversion in solvents other than water was recognized before other variables, such as a change in metal ion,

* The trans reactant must of course produce each hand of the chiral intermediate with equal probability.

were explored (ref. 15). Probably this reflects success with the early choice of liquid ammonia as solvent. In 1936, just two years following the discovery of Bailar inversion, it was reported that Λ-*cis*-$Co(en)_2Cl_2^+$ dissolved in liquid NH_3 leads to $Co(en)_2(NH_3)_2^+$ having predominantly the retained (Λ) or inverted (Δ) absolute configuration, depending upon the temperature (ref. 15). Certain aspects of this reaction were examined in more detail in a much later publication (Archer and Bailar, 1961, (ref. 63)).

In view of the double substitution pathway exposed for the hydrolysis of both Λ-*cis*-$Co(en)_2Cl_2^+$ and *trans*-$Co(en)_2Cl_2^+$ in H_2O/OH^-, and the fact that optical inversion occurs *only* by this route, it seems likely the "ammoniation" or ammoniolysis reaction follows suit. The facts recorded by Bailar *et al.* are not inconsistent with this (ref. 15,63). They reacted Λ-*cis*-$Co(en)_2Cl_2^+$ in $NH_3(\ell)$ at $-50°C$ and examined the $Co(en)_2(NH_3)_2^{3+}$ final products by optical rotatory dispersion and electronic spectroscopy. Also they reported qualitative observations on the *trans*-$Co(en)_2(NH_3)_2^{3+}$ content by selective precipitation with $S_2O_6^{2-}$. They found the *cis*-$Co(en)_2(NH_3)_2^{3+}$ to be predominantly Δ at $-50°$ but largely Λ at higher temperatures. They deduced that the inverted diammine product arose from Δ-*cis*-$Co(en)_2(NH_3)Cl^{2+}$ (with retention), formed by Bailar inversion in the first step, but they did not directly examine the activity of the first formed $Co(en)_2(NH_3)Cl^{2+}$ product,

$$\Lambda\text{-}cis\text{-}Co(en)_2Cl_2^+ \xrightarrow{NH_3} \underset{trans\,>\,\Delta\,>\,\Lambda}{Co(en)_2(NH_3)Cl^{2+}} \xrightarrow{NH_3} \underset{\Delta\,>\,\Lambda}{Co(en)_2(NH_3)_2^{3+}} \tag{31}$$

Separate experiments (ref. 63) established predominance of optical retention for the second phase of (31), supporting their conclusion, but clearly inversion could occur by direct double substitution,

$$\Lambda\text{-}cis\text{-}Co(en)_2Cl_2^+ \xrightarrow{2NH_3} \Delta\text{-}cis\text{-}Co(en)_2(NH_3)_2^{3+} + 2Cl^- \tag{32}$$

as in water.

Since the visible spectra of *cis*- and *trans*-$Co(en)_2(NH_3)_2^{3+}$ are similar, and the results obtained by selective trans precipitation are qualitative and variable, the precise steric course for each of the reactions in (31) and (32) must be considered uncertain, although overall inversion seems secure. Further, the ammoniolysis of *trans*-$Co(en)_2(NH_3)Cl^{2+}$ has been reported to be retentive (ref. 63), and this also could bear reinvestigation.

Some clear analogies between the $NH_3(\ell)$ and H_2O systems are apparent. The optical inversion appears to arise only by a base catalyzed pathway

(NH_3 or NH_2^-, and OH^-, respectively), and only if the parent complex has *two* good leaving groups which permit double substitution. This proposition seems worth testing further, not only for H_2O and $NH_3(\ell)$, but also for aprotic solvents such as dimethylsulfoxide (DMSO) where general rather than specific bases can be used. Base catalyzed substitution of $M(en)_2AX^{n+}$ type complexes in DMSO/OH^- has been demonstrated (ref. 64,65), but received little attention. Further, we have observed rapid substitution of species such as $(NH_3)_5Co(NC.CH_3)^{3+}$ in DMSO containing both nucleophilic and non-nucleophilic bases, e.g. 2,2',6,6'-tetramethylpiperidine (tmp) (ref. 66),

$$(NH_3)_5Co(NC.CH_3)^{3+} \xrightarrow[\text{rapid}]{\text{DMSO/tmp}} \begin{cases} (NH_3)_5CoOS(CH_3)_2^{3+} + CH_3CN \\ \xrightarrow{Y} (NH_3)_5CoY^{n+} + CH_3CN \end{cases} \quad (33)$$

Clearly such studies allow a handle on the relationship between the base term in the rate law and the stereochemistry of substitution, especially as the base is not the lyate ion of the solvent.

Although the kinetics of substitution reactions in liquid ammonia are reported (ref. 15,63) to be plagued by "catalysis problems" (i.e. irreproducible rates), Balt (ref. 67) has over recent years tidied up the general picture by using NH_4^+ (as NH_4ClO_4) to moderate the reactions. Unlike H_2O/OH^-, both the solvent lyate ion NH_2^- and the NH_3 solvent itself are capable of deprotonating aminecobalt(III) complexes to yield labile aminato species, the origin of the base catalysis. Both pathways are suppressed by NH_4^+ (the NH_2^- more so than the NH_3 deprotonation path), and it is possible to control by which predominant route the reaction proceeds. Base catalysis by solvent NH_3 is more usual. Also, effective base catalysis by the solvent tends to swamp the spontaneous ammoniolysis reaction, i.e. the $NH_3(\ell)$ equivalent of the two term rate law $k(obsd) = k_s + k_{OH}[OH^-]$ for H_2O/OH^- has a dominant second term. This is not as straightforward to demonstrate as it is in water, although it has been done (ref. 67), because it involves extrapolating rate data to $[NH_4^+]^{-1} = 0$.

Rate limiting deprotonation, akin to that for H_2O/OH^-, can be observed for certain aminecobalt(III) complexes in $NH_3(\ell)$ (ref. 67). Moreover, for a particular reaction, a change from rate limiting deprotonation to "normal" base catalyzed substitution (rapid pre-equilibrium conjugate base formation) can be effected by varying $[NH_4^+]$. The same prospect exists for solvent water but it is never realized because, except in one or two instances (ref. 68), the solvent is an ineffective base and at the $[OH^-]$ required to get significant reaction via the OH^- deprotonation route (over spontaneous hydrolysis), reprotonation of the reactive conjugate base is controlled by H_2O rather than H^+ addition. To some extent this is governed by the much lower acidities of aminecobalt(III) species in H_2O compared to $NH_3(\ell)$; conjugate base reprotonation by even H_2O is diffusion controlled.

The brief summary above of a number of papers dealing with the kinetics of substitution in $NH_3(\ell)$ seems necessary to appreciate the recent stereochemical data presented below. Balt *et al.* (ref. 69) have examined the substitution reactions of several *cis*- and *trans*-$Co(en)_2YX^{n+}$ complexes in $NH_3(\ell)$, at various $[NH_4^+]$ achieved by added NH_4ClO_4. Selected results appear in Table 11. The reaction of the *trans*-$Co(en)_2(DMF)Cl^{2+}$ ion (DMF = N,N'-dimethylformamide) illustrates the major point. In $NH_3(\ell)$ (−70°C) containing NH_4ClO_4(1 M) there is immediate reaction, 77% of the product comprising monosubstituted species (62% *trans*-$Co(en)_2(NH_3)Cl^{2+}$, loss of DMF; 15% *trans*-$Co(en)_2(NH_3)(DMF)^{3+}$, loss of Cl^-). The 23% balance of material is $Co(en)_2(NH_3)_2^{3+}$ (76% cis, 24% trans). Secondary reaction of the observed monosubstituted products is relatively slow, and thus most or all of the $Co(en)_2(NH_3)_2^{3+}$ must arise by direct double substitution,

$$trans\text{-}Co(en)_2(DMF)Cl^{2+} \xrightarrow{2NH_3} cis(76\%)\text{-} + trans(24\%)\text{-}Co(en)_2(NH_3)_2^{3+} \tag{34}$$

The data were obtained by *in situ* monitoring of the ^1H NMR spectrum at 250 MHz, using the technique of solvent peak saturation prior to data acquisition. The addition of a few mg of NaCl split the NH_2 resonances (attributable to en) sufficiently to make positive signal assignments and determine isomer ratios by conventional integration.

Other points to be called to attention are:

(i) The stereochemistry of the diammine product appears to be independent of the geometry of the starting material (e.g. *cis*- and *trans*-$Co(en)_2Cl_2^+$ each yield 28–30% *trans*-$Co(en)_2(NH_3)_2^{3+}$). This is a striking analogy to the aqueous chemistry (section 3). Experiments are underway on the resolved cis complex (ref. 70) to answer the question of the origin of the optical inversion raised earlier in this section, but with double substitution now established the result is predictable (optical retention for the *cis*-$Co(en)_2(NH_3)Cl^{2+}$ product, net inversion for *cis*-$Co(en)_2(NH_3)_2^{3+}$).

(ii) The 1+ ions *cis*- and *trans*-$Co(en)_2Br_2^+$ show the greatest amount of double substitution (65%). The other complexes show 25–35%. There

TABLE 11

Product analysis for controlled ammoniolysis[b] of $Co(en)_2YX^{n+}$

Reactant Geometry	Y	X	% $Co(NH_3)X$	% cis[f]	% $Co(NH_3)Y$	% cis[f]	% $Co(NH_3)_2^{3+}$	% cis[f]
cis	Cl	Cl[d]			≤ 71	0	≥ 29	72
trans	Cl	Cl[a]			71	0	29	70
cis	Br	Br[d]			≤ 35	0	≤ 65	78
trans	Br	Br			35	0	65	77
cis	Cl	Br	c	c	c	c	c	c
trans	Cl	Br	0		65	0	35	76
cis	Cl	DMSO[e]	c,g	c	68	0	32	79
trans	Cl	DMSO	10	0	63	0	27	80
cis	Cl	DMF[d]	c,g	c	72	0	28	76
trans	Cl	DMF	15	0	62	0	23	76
cis	DMSO	DMSO			65	34	35	81
trans	DMSO	DMSO			c	c	c	c

[a] Various T, -40 to $-70°C$; all others $-70°C$ except where noted otherwise.
[b] $[NH_4ClO_4]$, 10^{-3} to 1M; result independent of $[NH_4^+]$.
[c] Result not yet available.
[d] $-40°C$.
[e] $-50°C$.
[f] 10^2 % *cis*-$Co(NH_3)X$/total % $Co(NH_3)X$; 10^2 % *cis*-$Co(NH_3)Y$/total % $Co(NH_3)Y$; 10^2 % *cis*-$(NH_3)_2^{3+}$/total % $Co(NH_3)_2^{3+}$.
[g] Some possible contribution to diammine product from secondary reaction of *trans*-$(NH_3)X^{3+}$.

appears to be no systematic correlation with reactant geometry, charge on the complex, nor combination of leaving groups (although 2 Br^- appears to be the "best"). Results for the missing *cis*-$Co(en)_2ClBr^+$ and *trans*-$Co(en)_2(OS(CH_3)_2)_2^{3+}$ ions would be instructive.

(iii) The amount of double substitution, while leaving group dependent, seems to be independent of the starting geometry.

(iv) It is significant that a variation in the $[NH_4^+]$ does not affect the steric course of either of stepwise processes or the double substitution path, whereas the rates are suppressed by increased $[NH_4^+]$. Even more revealing is the $[NH_4^+]$ independence of the relative contributions to the overall reaction, from stepwise and double substitution. This strongly implies the same *form* for the rate laws for the two separate pathways, i.e. the typical $k(obsd) = k[NH_4^+]^{-1}$.

In water, the double substitution path appears to follow the rate law k(obsd) = $k_{xx}[OH^-]^2$ while the normal path is k(obsd) = $k_x[OH^-]$, and thus a variation in $[OH^-]$ affects the relative contributions. In liquid NH_3, the corresponding rate terms do not seem to be proportional to $[NH_2^-]^2$ and $[NH_2^-]$, respectively. However, more work is required to elucidate the detailed rate law for double substitution, as discussed in section 3.

Balt had earlier noted that some $Co(NH_3)_4(^{15}NH_3)_2^{3+}$ (10%) accompanied the ammoniolysis ($^{15}NH_3$) of *trans*-$Co(NH_3)_4Cl_2^+$ which yields primarily *trans*-$Co(NH_3)_4(^{15}NH_3)Cl^{2+}$ (ref. 71). Although not recognized at the time, it seems possible that the hexaammine arises by double substitution, although some of the pentaammine may have converted to hexaammine under work-up conditions. The stereochemistry of the minor $Co(NH_3)_4(^{15}NH_3)_2^{3+}$ product was not stated.

In summary, Bailar inversion and double substitution are not confined to water. Base catalysis remains a prerequisite, and there are both kinetic and stereochemical analogies between H_2O and NH_3, the only two solvents to receive close attention so far. Broadening of the solvent range as mentioned earlier seems a logical extension. In section 8.4 we briefly consider solvent effects in the context of the stereochemistry of *normal* base catalyzed substitution, with which this section is inevitably coupled.

5 BAILAR INVERSION FOR METAL IONS OTHER THAN COBALT(III)

It will be apparent that double-substitution, which has only been recognized recently (ref. 23), is not readily detected unless the correct conditions and methods of analysis are chosen. Moreover, careful and detailed experiments are necessary to separate the contribution from the normal stepwise hydrolysis pathway. Therefore, it seems fair to comment that the prospect has not been examined for any coordination complexes other than those of cobalt(III), and it represents an obvious area for fruitful research. The *trans*-$Rh(en)_2X_2^+/OH^-$ systems could represent a profitable start. On complete hydrolysis 100% *cis*-$Rh(en)_2(OH)_2^+$ is found (X = Cl^-, Br^-), and 50% cis/50% trans for X = I^-. The second step is faster than the first, but *trans*-$Rh(en)_2(OH)X^+$ (X = Br^-, Cl^-) have been shown to hydrolyze with retention (ref. 72,73). The steric change is highly unusual for Rh(III) substitution (ref. 28,30), and it could arise by double substitution.

The possibility of Bailar inversion, on the other hand, for *any* metal ion complex has been long recognized. Largely through the efforts of Bailar and his students (ref. 16–18,21,22,74), compounds of metal ions such as Cr(III), Fe(II), Ru(II), Rh(III), and Ir(III) have been examined, mostly for the conditions yielding inversion for Co(III) (H_2O and $NH_3(\ell)$* solvents). That no new examples have been found, Fe(II) chemistry aside (ref. 16–18), has been the disappointing result. The field is somewhat limited in scope by the resolvability (and optical stability) of reactant and product complexes, but it seems there is still potential, especially with the advances in rapid optical detection technology.

Given the apparent connection between optical inversion and double-substitution, we believe broader investigations in this area will help to understand if one is contingent upon the other. Even in normal octahedral metal ion substitution chemistry, stereochemical change appears to be confined to cobalt(III), and the reason is not well understood (ref. 30). Archer has focused on the factors which facilitate a transient spin change in diamagnetic d^6 systems, and has been successful in identifying a Bailar inversion for some Fe(II) systems which were chosen on this basis (ref. 16–18). While certainly non-singlet ground states, or accessible spin-multiplet excited states are well documented (ref. 75,76,77) for octahedral Co(III) and Fe(II) complexes (and for five coordinate Co(III) (ref. 78)), there remains no *direct* proof of the connection between this and stereochemical change in octahedral substitution.

It is becoming increasingly obvious that Bailar inversion and normal stepwise substitution are mechanistically distinct processes. A predominance of optical retention is *always* found for the latter (sections 7,8), and the examples are many. It is likely therefore that separate explanations pertain to the two processes. It has been seen that much still remains unknown about Bailar inversion to hold that optical inversion *always* requires double substitution, although this seems true for the few examples documented.

The additional requirement for base catalysis to observe Bailar inversion in Co(III) complexes affords a sharp contrast to Archer and Hardiman's Fe(II) inversion reaction (ref. 18), and this warrants special consideration. The resolved tris(diimine) Fe(II) species are optically stable and the absolute configuration of Λ-$Fe(phen)_3^{2+}$ (phen = 1,10-phenanthroline) is known from a single crystal X-ray study (ref. 79). Non-empirical exciton circular dichroism methods are especially useful for tris or cis-bis complexes of this kind which have exciton coupled chromophores, and for Λ-$Fe(phen)_3^{2+}$ the absolute configuration assigned on this basis is correct (ref. 80). This CD method was used for resolved $Fe(bipy)_3^{2+}$ (bipy = bipyridine), and also for the high field, low spin and optically stable $Fe(phen)_2(CN)_2$ and $Fe(bipy)(CN)_2$ derivatives for which X-ray structures have not been reported (ref. 18). All the tris

---

* Although the Λ-*cis*-$Rh(en)_2Cl_2^+$ and $Ir(en)_2Cl_2^+$ complexes react in $NH_3(\ell)/NH_2^-$ with optical retention (ref. 74), the question of double substitution still exists.

complexes react with aqueous CN^- in a single observable step,

$$Fe(diimine)_3^{2+} \underset{H_2O}{\overset{CN^-}{\longrightarrow}} Fe(diimine)_2(CN)_2 + diimine \tag{35}$$

The diimine ligands phen and bipy readily permit the cis configuration for the bis products. The achiral trans arrangement suffers severe steric interactions (ortho-H) and is not observed; if formed transiently, it must yield racemic cis product. Archer *et al.* found that the cis dicyano products were optically active (ref. 16–18), the sign and magnitude depending upon $[CN^-]$ as well as temperature. Only for the phen complex and its tris(3-methyl) analog were net inversion observed. However, for the bipy complex there were corresponding changes in the Λ/Δ $Fe(diimine)_2(CN)_2$ product ratio with $[CN^-]$ and temperature—more inversion at lower temperature and higher $[CN^-]$—only that here Δ never exceeded Λ product (when derived from Λ reactant). Finally, the introduction of a methyl group into each of the phen ligands of Λ-$Fe(phen)_3^{2+}$ creates three diastereoisomers. One isomer (the facial form) was obtained and resolved, and the Λ enantiomer was found to yield all three geometrically isomeric $Fe(mphen)_2(CN)_2$ products with CN^-, each having a net inverted (Δ) configuration, but only one strongly so (ref. 18).

An analysis of the results reveals competitive retention and inversion pathways to $Fe(diimine)_2(CN)_2$. The rate law coupled with $[CN^-]$ dependence of the Λ/Δ product distribution led to a method for decoupling the two paths. Previous work (ref. 81) has shown the presence of both $[CN^-]$ independent and dependent terms in the full rate law,

$$\begin{aligned} -d[\Lambda\text{-tris}]/dt &= \{k_0 + k_1[CN^-]\}[\Lambda\text{-tris}] \\ &= k(obsd)\ [\Lambda\text{-tris}] \end{aligned} \tag{36}$$

The k_0 term leads only to racemization (by two pathways), and k(obsd) *vs* $[CN^-]$ data treated in the usual way gives k_1. Curiously, $k_1[CN^-]$ is *not* linearly correlated with the extent of inversion, which *appears* to arise via this term. Archer *et al.* (ref. 18) explained this by assuming first that inversion arose in the rate determining step, (37). Next it was proposed that rapid racemization of first-formed $Fe(diimine)_2(OH_2)CN^+$ ($[CN^-]$ independent rate) was competitive with rapid CN^- anation, first-order in $[CN^-]$:

$$\Lambda\text{-Fe(diimine)}_3^{2+} \xrightarrow{H_2O,\ slow} \begin{cases} \xrightarrow{k_{ret}[CN^-]} \Lambda\text{-Fe(diimine)}_2(OH_2)CN^+ \\ \xrightarrow{k_{inv}[CN^-]} \Delta\text{-Fe(diimine)}_2(OH_2)CN^+ \end{cases} \quad (37)$$

$$\begin{array}{ccc} \Lambda\text{-Fe(diimine)}_2(OH_2)CN^+ & \xrightarrow{CN^-} & \Lambda\text{-Fe(diimine)}_2(CN)_2 \\ \downarrow\uparrow & & \\ \Delta\text{-Fe(diimine)}_2(OH_2)CN^+ & \xrightarrow{CN^-} & \Delta\text{-Fe(diimine)}_2(CN)_2 \end{array} \quad (38)$$

and thus k_{inv} and k_{ret} ($k_1 = k_{inv} + k_{ret}$ (36), (37)) could be deduced by an extrapolation procedure to high $[CN^-]$ ($1/[CN^-] \rightarrow 0$ in fact), where the competitive racemization was reduced to insignificance. Note that the $[CN^-]$ dependences of the rate (36) and product distribution are independent, since the CN^- substitution processes in (38) are faster than in (37), but the proposed competitive racemization (38) of the intermediate permits the connection to correlate the $[CN^-]$ dependences of both. However, it is by no means a unique account. The intermediacy of the unobserved* $Fe(diimine)_2(OH_2)CN^+$ is assumed, and obviously an independent investigation of this species would comment on the validity of the proposals.

An alternative mechanism is suggested by the Co(III) chemistry, double substitution by CN^- without the cyano-aqua intermediate.

$$\Lambda\text{-Fe(diimine)}_3^{2+} \xrightarrow[H_2O]{2CN^-} \Delta\text{- plus } \Lambda\text{-Fe(diimine)}_2(CN)_2 + \text{diimine}$$

$$\Lambda\text{-Co(en)}_2Cl_2^+ \xrightarrow[H_2O]{2OH^-} \Delta\text{- plus } \Lambda\text{- plus } \textit{trans}\ Co(en)_2(OH)_2^+ + 2Cl^- \quad (39)$$

However, even this process must be multistep since a term $k'[CN^-]^2$ does not appear in the rate law (36). This points to an important difference between the Fe(II) and Co(III) chemistry, namely the clear role of OH^- as a base for amine

* Also it does not seem to be a known species.

Co(III) species. The diimine ligands lack acidic protons.* Moreover, OH^- present in aqueous CN^- does not seem to bring about inversion in the Fe(II) systems, nor does it affect the CN^- results, proven in control experiments (ref. 16). It is possible that the apparently different rate laws for Co(III) and Fe(II) arise only because of an acid/base pre-equilibrium for Co(III), but to maintain the analogy would seem to demand a nucleophilic role for OH^- in the Co(III) inversions much like CN^- for Fe(II), but we have argued against this (section 2.7).

On the basis of the Fe(II) inversion mechanism embodied in (37) and (38), Archer and Hardiman deduced a dependence of k_{inv}/k_{ret} on temperature (ref. 18). They reported individual activation parameters for k_{inv} and k_{ret}, obtained in the usual way, but implicitly assuming separate reaction pathways. However, the observed temperature dependence of k_{inv}/k_{ret} could merely reflect the geometry or mode of CN^- attack for a single intermediate leading to $Fe(diimine)_2(OH_2)CN^+$, and which could arise in a single step (k_1, (37)); i.e. temperature dependence for the product distribution does not necessarily reside in the rate determining step leading to $Fe(diimine)_2(OH_2)CN^+$. There is an analogy here to the interpretation of the temperature dependence of the *cis-/trans-*$Co(en)_2(OH_2)Br^{2+}$ product distribution for the spontaneous aquation of *cis-*$Co(en)_2Br_2^+$ (ref. 40). If curvature in conventional Eyring plots of k(obsd) *vs* 1/T is not clear, there can be no strong case for assuming two separate rate-determining pathways. This is especially so if the temperature range examined is not large ($\Delta T = 15°C$ for the Fe(II) reactions, ref. 18). The significance of the values for $\Delta H^‡$, $\Delta S^‡$ derived separately for k_{ret} and k_{inv} is therefore questionable.

The quantitation for the analysis of the Fe(II) inversion reactions assumes the magnitude of the exciton CD for the optically pure $Fe(diimine)_2(CN)_2$ products.** Although not intrinsically invalidating any of the conclusions, a check of this assumption now seems possible. Yamagishi has described a method for raising (and hence testing) the optical purity of $Fe(phen)_2(CN)_2$, by selective removal of any racemic component using a colloidally dispersed clay (sodium montmorillonite), with and without doping by Λ-$Ni(phen)_3^{2+}$ (ref. 82). Optically pure Λ- and Δ-$Fe(phen)_2(CN)_2$ were apparently obtained, although the method was not successful for $Fe(bipy)_2(CN)_2$. Yamagishi also reported obtaining inverted Δ-$Fe(bipy)_2(CN)_2$ from the Λ-$Fe(bipy)_3^{2+}$ + CN^- reaction, and this is in conflict with the results of Archer *et al.* (ref. 18). To

* That the $Co(phen)_2Cl_2^+$ hydrolysis is not catalyzed by OH^- has been long known, a fact used to support the S_N1cb mechanism in its infancy (ref. 44) before more definitive proofs emerged.

** Theoretically 50% of the corresponding tris(chelate) values (ref. 18).

complete the record, the inversion for the Λ-$Fe(phen)_3^{2+}$ + CN^- reaction has been challenged by Nord (ref. 83), but it received later support from Gillard *et al.* (ref. 84). One can glean from the controversy that, at the very least, the reaction of either the phen *or* bipy complex is a Bailar inversion, and exciton CD assignments suggest that it is the $Fe(phen)_3^{2+}$ and $Fe(mphen)_3^{2+}$ + CN^- reactions where inversion predominates over retention.

6 INVERSION IN ORGANOMETALLIC COMPLEXES

There has been a flourishing activity in the chemistry of chiral organometallic species for some time, largely because of their potential in catalytic asymmetric synthesis. Recent reviews by Brunner (ref. 9,10) consider some reactions of pseudo-tetrahedral transition metal ions relevant to the present article. The essence of these are collected here; the reviews and later papers (ref. 85,86,87) should be consulted for details.

Complexes such as

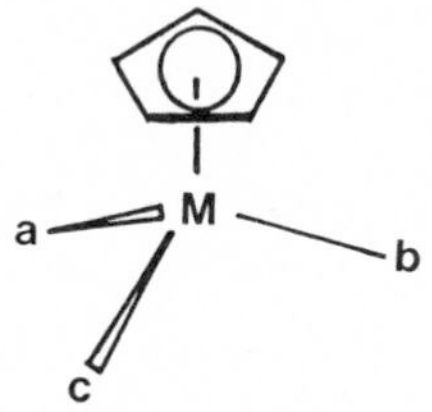

M a b c

(40)

could be regarded not as octahedral but rather as pseudo-tetrahedral species, having a formally chiral metal centre. Irrespective of the bonding of the polyhapto organic ligand, only two isomers (enantiomers) are possible. Because there are only four separate ligands arranged about M approximately tetrahedrally, similar considerations apply to the pseudo-five coordinate organometallics such as,

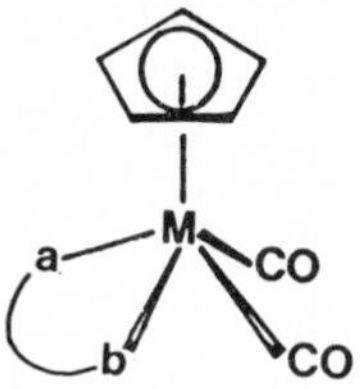

M a CO b CO

(41)

where the judicious choice of ligand types restricts the possible isomers and engenders a chiral metal atom; the only isomers are catoptromers.

A wide range of transition metal ion complexes exist for the kind (40), M = Mn, Fe, Ru, Cr, Mo, W, Co and (41), M = Mo, but a few have received considerably more attention than the others (ref. 9). Regardless of the semantics of the description of the stereochemistry of these complexes as tetrahedral or otherwise, substitution reactions phenomenologically akin to Walden inversion at tetrahedral C are known, and form the subject of the brief discussion below. Some mechanistic comparisons with substitution reactions leading to retained or racemic products are included, to emphasize important differences to traditional substitution stereochemistry at carbon and related non-transition metal tetrahedral centres.

6.1 Apparent inversion reactions

True chemical inversion involves substantial molecular rearrangement. For tetrahedral or pseudo-tetrahedral species, this corresponds to (i) the classical umbrella type inversion, and where, (ii) the handedness, defined by three of the four original groups attached to the central atom and which remain after reaction, is reversed.

First, the trivial cases of apparent inversion need be dismissed. The chirality labelling rules defining the metal centre as (**R**)- or (**S**)- of course take no account of mechanism (ref. 2). Thus a reaction entailing an (**R**)- to (**S**)- conversion or vice-versa is independent of both conditions (i) and (ii) above. In short, the nomenclature does not comment in any way on whether a reaction involving a chiral reactant and product is a true chemical inversion. More subtle is the relationship between chemical and optical inversion. True chemical inversion, as defined above, can only be inferred accurately from ORD and CD data when the relationship between the optical data (magnitude, and especially the sense) and the absolute configuration is well understood, and errors have been incurred because of deficiencies in this respect (ref. 85). Although the unambiguous interpretation of the ORD and CD spectra of chiral organometallics still remains difficult, the rapid accumulation of standards having absolute configurations known from single crystal X-ray studies is quickly diminishing the problem.

A less trivial example of apparent inversion is the reaction,

$(C_5H_5)Fe(CO)(P\phi_3)(C(O)OR) \xrightarrow{LiMe} (C_5H_5)Fe(CO)(P\phi_3)(C(O)Me)$ (42)

Clearly the handedness defined by the C_5H_5, CO and $P\phi_3$ groups has reversed, fulfilling condition (ii) for chemical inversion. This reaction also happens to

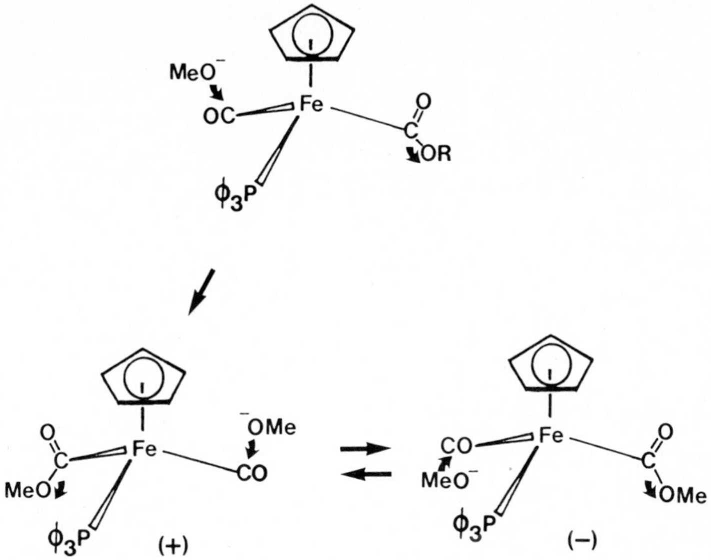

(43)

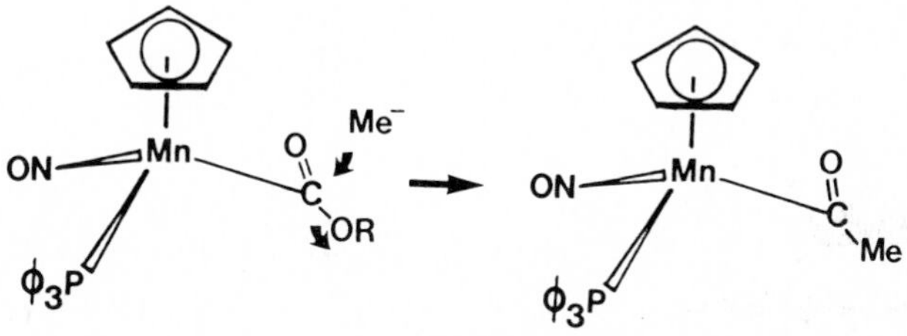

(44)

be an optical inversion (ref. 88). However, the mechanism does not involve substitution at the metal centre but rather attack at the ligand, leading to the interesting situation where the CO and RCO_2 groups have swapped roles.

An analogous mechanism (ref. 89) accommodates the formation of racemic product in the transesterification process, (43). The retention observed for the related reactions (44) of the isoelectronic Mn complex (ref. 90) sharply contrasts with the above. This arises merely because the CO.OR group rather than NO is attacked.

6.2 Genuine inversion reactions

Examples appear to be few, but the following, (45), (ref. 91) is a clear case. The rate is first order in the nucleophile PEt_3, and a transition state (46) similar to that for classical S_N2 substitution at tetrahedral C has been proposed.

The racemization process (47), more strictly epimerization (R = $\phi(CH_3)HC$), is mechanistically similar (ref. 92).

It requires the presence of free ligand which strongly catalyzes the reaction

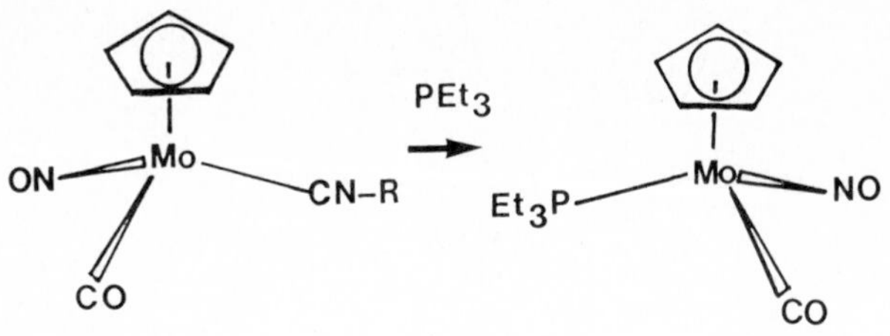

(45)

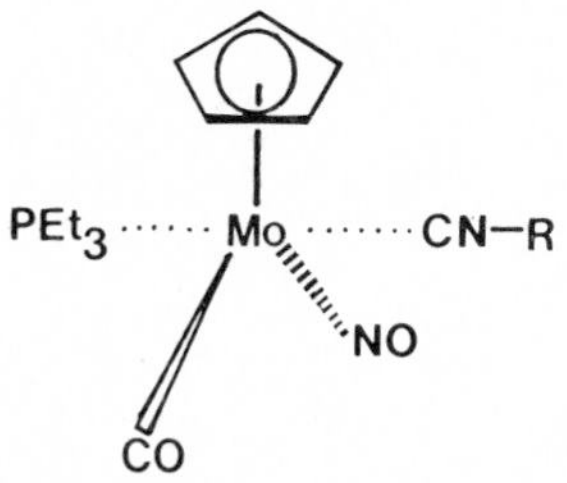

(46)

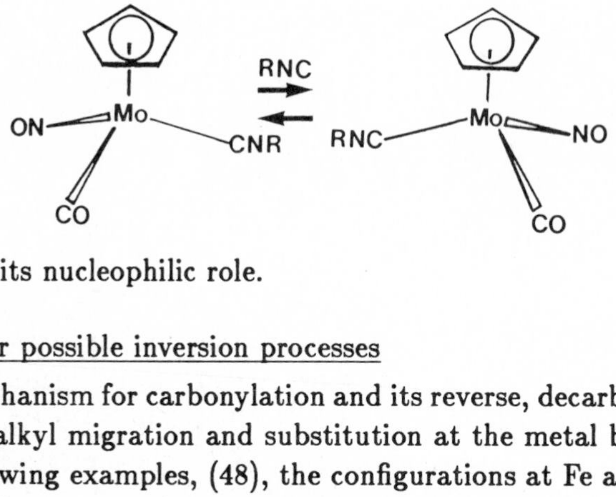

(47)

because of its nucleophilic role.

6.3 Other possible inversion processes

The mechanism for carbonylation and its reverse, decarbonylation, appears to involve alkyl migration and substitution at the metal by CO (ref. 93,94). In the following examples, (48), the configurations at Fe are inverted but the signposted

carbons show that this *seems* to be only apparent. In the carbonylation reaction, the mechanistic role of BF_3 is not fully understood (ref. 85), and the overall "retention" at Fe could arise by a succession of substitution processes involving either chemical retention or chemical inversion. A possible inversion sequence is shown below, (49).

The other interesting stereochemical feature of these reactions is that the chiral integrity of asymmetric alkyl groups is retained in the alkyl migration step to and from the coordinated carbonyl (ref. 96). However, for the formally analogous SO_2 insertion process, such a carbon centre is inverted while the metal centre retains its configuration (ref. 97,98). The mechanism of SO_2

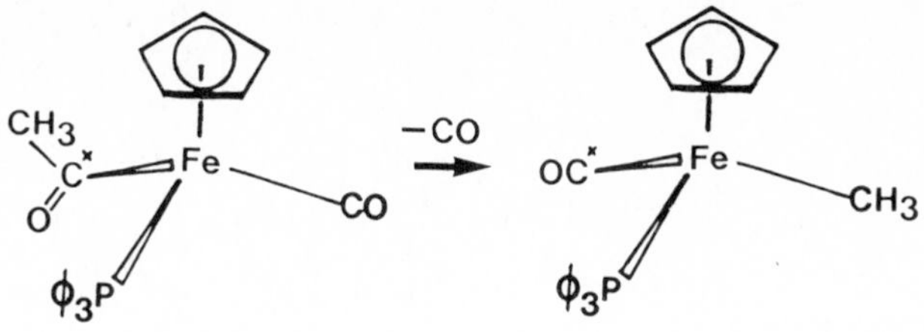

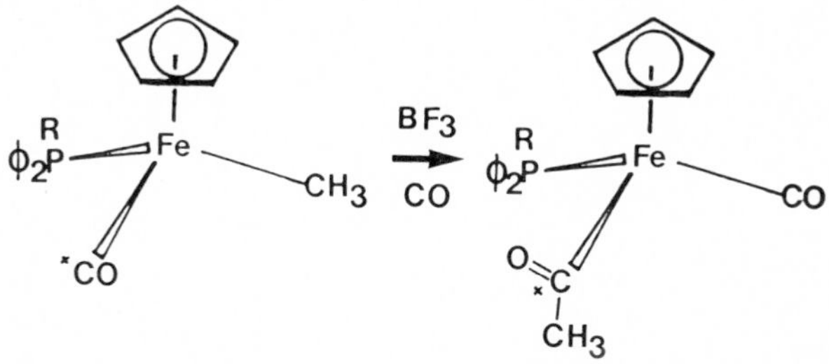

(48)

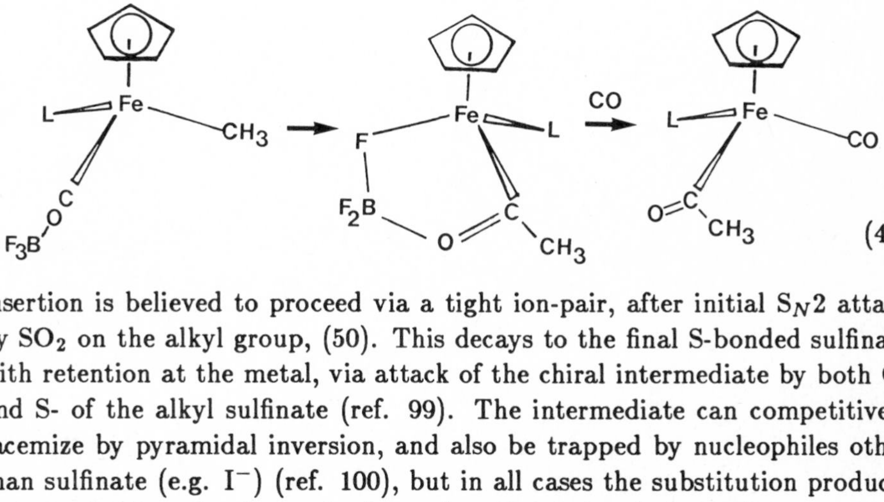

(49)

insertion is believed to proceed via a tight ion-pair, after initial S_N2 attack by SO_2 on the alkyl group, (50). This decays to the final S-bonded sulfinate with retention at the metal, via attack of the chiral intermediate by both O- and S- of the alkyl sulfinate (ref. 99). The intermediate can competitively racemize by pyramidal inversion, and also be trapped by nucleophiles other than sulfinate (e.g. I^-) (ref. 100), but in all cases the substitution products have predominantly the retained metal configuration (ref. 9,10).

6.4 Alternative modes of substitution

There appear to be more examples of metal substitution with retention than genuine Walden inversion reactions, for pseudo-tetrahedral organometallic complexes. This is opposite to the situation for carbon, but it is probably too early to generalize on this observation.

The dissociative reactions (ref. 101,102) of some Mn complexes, which show up the expected and fundamental mechanistic differences to carbon, merit consideration in the context of inversion.

The racemization process shown in (51) seems to be of the S_N1(lim) kind.

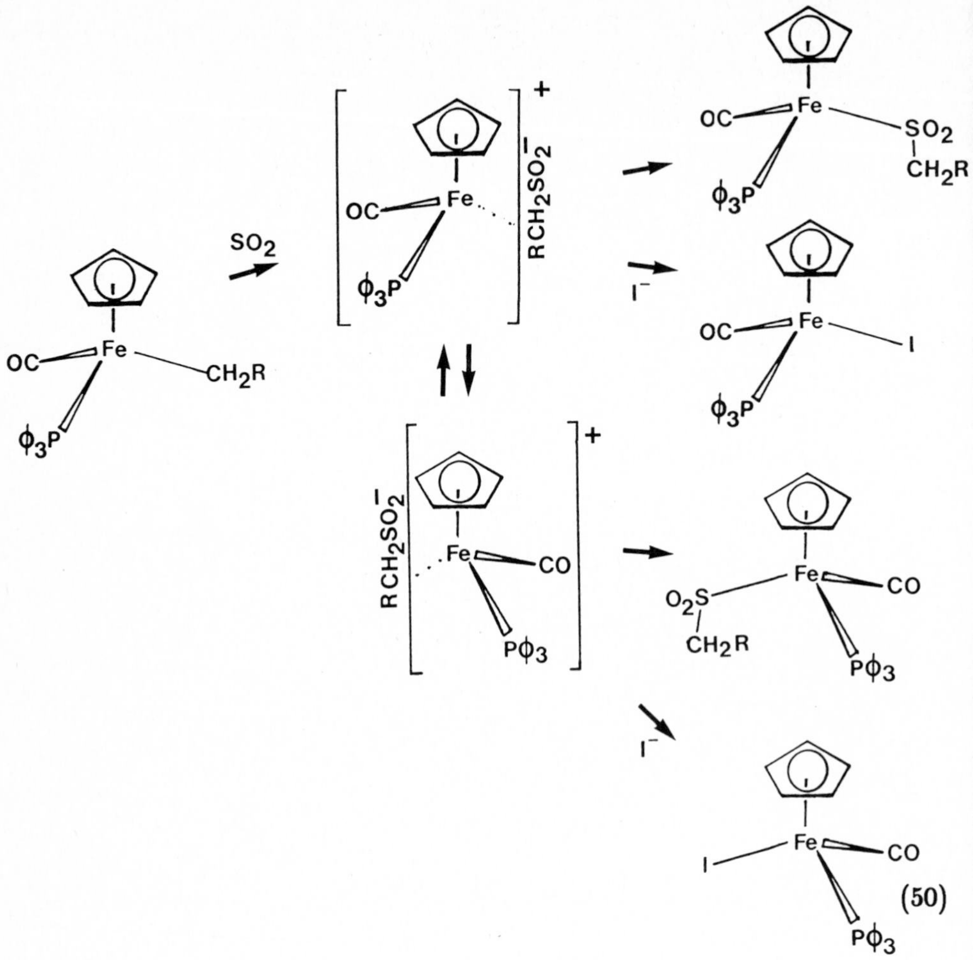

(50)

(51)

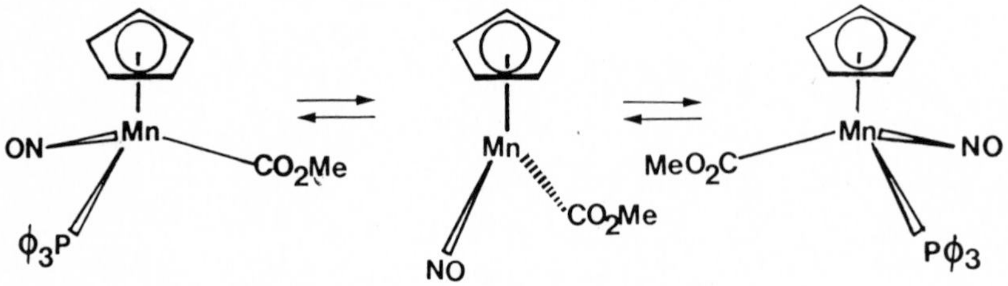

The rate is independent of $[P\phi_3]$, but the process is not intramolecular, but a substitution reaction since per-deuterated $P\phi_3$ is incorporated. Furthermore, this occurs for each act of substitution since the exchange and racemization rates are identical (ref. 101,102). A coordinately unsaturated intermediate which is stereochemically akin to the flat carbonium ions of carbon chemistry has been proposed. The general situation however is not quite this simple. For complexes having substituted acyl groups, mass-law retardation by dissociated $P\phi_3$ is observed (ref. 103,104). While clearly indicative of a limiting dissociative process, it appears re-entry of $P\phi_3$ does not give the racemate required by mechanism (51), but rather the regenerated reactant has largely the retained chirality. This is shown by the phosphine exchange rate *exceeding* the racemization rate, and also by trapping experiments (ref. 104) using the better ligand tris-*p*-anisylphosphine (52). Thus it was proposed that the intermediate was chiral, not flat as shown in (51). The observations were accommodated by competitive nucleophilic attack (with retention) and pyramidal inversion for the intermediate, leading to some racemic reactant and substitution product.

(52)

Calculations (ref. 105) support the proposal that the 16e$^-$ intermediate exists and is pyramidal. The flat "isomer" is believed not to exist, and in (52) is shown as the transition state for pyramidal inversion. Moreover, the proposal is not in conflict with mechanism (51) if here pyramidal inversion is faster than attack by the phosphine.

In summary, the work on chiral pseudo-tetrahedral organometallics has uncovered a lot of mechanistic detail of fundamental importance in its own right, as well as of relevance to classical Walden inversion. It is also germane to the mechanism of pyramidal inversion at non-metallic centres (ref. 8,30). Finally we comment that any attempted comparison with Bailar inversion at octahedral centres (sections 2–5) serves only to highlight the number of different issues involved.

7 STEREOCHEMISTRY OF OCTAHEDRAL AQUATION

The rates and stereochemistry of simple hydrolysis reactions such as

$$\textit{cis}\text{-Co(en)}_2\text{Cl}_2^+ \xrightarrow{H_2O} \textit{cis}\text{- plus } \textit{trans}\text{-Co(en)}_2(\text{OH}_2)\text{Cl}^{2+} + \text{Cl}^- \tag{53}$$

have been of interest since the days of Werner. The more recent developments are documented (ref. 30), and we need only consider here new facts which, hopefully, clarify the general picture or strengthen accepted thinking in areas where the data are limited.

It has become customary to talk of spontaneous and induced (or catalyzed) aquation (ref. 30,36,47). The former refers to normal hydrolysis, e.g. (53), the latter to that group of reactions where an especially good leaving group is generated *in situ* and consequently, where the hydrolysis is usually very rapid, e.g.

$$\begin{array}{l} (\text{NH}_3)_5\text{CoI}^{2+} + \text{Hg}^{2+} \rightleftharpoons (\text{NH}_3)_5\text{Co(IHg)}^{4+} \searrow \\ \qquad\qquad\qquad\qquad\qquad\qquad\qquad\qquad \text{H}_2\text{O} \rightarrow (\text{NH}_3)_5\text{CoOH}_2^{3+} \\ (\text{NH}_3)_5\text{CoN}_3^{2+} + \text{NO}^+ \rightleftharpoons (\text{NH}_3)_5\text{Co(N}_4\text{O)}^{3+} \nearrow \end{array} \tag{54}$$

This is a somewhat artificial distinction, emphasized by the recent synthesis of isolable reactive species such as *cis*-$\text{Co(en)}_2(\text{O}_3\text{SCF}_3)_2^+$ (ref. 56), *trans*-$\text{Co(en)}_2\text{Cl(O}_3\text{SCF}_3)^+$ (ref. 106) and $(\text{NH}_3)_5\text{CoOClO}_3^{2+}$ (ref. 107), which interface the once-clear two regions of reactivity—relatively slow or very fast. Acid-catalyzed aquation is in the same category. For example, H^+ induced loss of sulfamide from the $(\text{NH}_3)_5\text{CoNHSO}_2\text{NH}_2^{3+}$ ion is of course simple hydrolysis of the reactive protonated form (ref. 108),

$$(NH_3)_5CoNHSO_2NH_2^{2+} \overset{H^+}{\rightleftharpoons} (NH_3)_5CoNH_2SO_2NH_2^{3+} \xrightarrow{H_2O} (NH_3)_5CoOH_2^{3+} + NH_2SO_2NH_2 \qquad (55)$$

Clearly the only real difference between these various aquation processes resides in the nature of the group leaving from the reactive species. Also it should be noted that, for induced aquation, the precise nature of the leaving group is not always known.

7.1 Steric course of hydrolysis of Λ-*cis*- and *trans*-$Co(en)_2AX^{n+}$ ions

These classical reactions (ref. 44) continue to receive attention, largely because the accuracy of the early work has been called into question (ref. 36). Thus the main emphasis in recent work on the stereochemistry of hydrolysis of $Co(en)_2AX^{n+}$ ions has been, first, its *precise* definition, and second, efforts to vary X sufficiently widely so as to sensibly gauge this effect (ref. 109–113). Where once leaving groups were confined to Cl^- and Br^-, these have been extended to HgX^+, N_2 or N_4O, $CF_3SO_3^-$, $OP(OMe)_3$, DMSO, and many others. The spread in leaving groups now covers a substantial range in reactivity, size, charge, bonding properties, and geometry.

The basis for the effort to define the steric course of substitution more accurately, and for a very wide range of complexes, is simply this. If the reaction is limiting dissociative it involves a reduced coordination number intermediate, e.g.

$$cis\text{-}Co(en)_2AX \xrightarrow{slow} [Co(en)_2A] \rightarrow cis\text{-}Co(en)_2A(OH_2) \;/\; trans\text{-}Co(en)_2A(OH_2)$$

$$cis\text{-}Co(en)_2A(OH_2) \underset{H_2O,\ fast}{\rightleftharpoons} trans\text{-}Co(en)_2A(OH_2) \qquad (56)$$

and this requires the stereochemistry of the aqua products to be *strictly* independent of the nature of X. Obviously the strictness is contingent upon the error in the product distribution and what one considers to be an effective variation in X.

Early on it seemed that the steric course of aquation of $Co(en)_2AX^{n+}$ ions was dependent upon X. The very poor leaving groups of spontaneous aquation yielded a different stereochemical outcome to that resulting from the especially good ones of induced aquation (ref. 114). For the cis species at least, this difference has now largely dissolved (ref. 36,115). The steric course is, with one exception, independent of X at the $\pm$ 2% level of definition for each of the isomeric products. Moreover, the long-held belief that *cis*-$Co(en)_2AX^{n+}$ ions *always* spontaneously aquate with geometric and optical retention has gone.

Predictably, the steric course of aquation of the *trans*-$Co(en)_2AX^{n+}$ has come under extensive review, along the lines recorded for the cis substrates. Some of this new data have been published (ref. 109–111), but a more up-to-date record is given in Table 12 where we also include both recent (ref. 36,115) and new (ref. 112,113) cis results.

Before commenting further on these, some discussion on the new or revised data seems necessary. First, the closer examination (ref. 112) of the NO^+ induced aquation reactions of $Co(en)_2A(N_3)^{n+}$ complexes (ref.114,116) has uncovered the problem of surprisingly rapid (subsequent) nitrosation of the aqua products, which results in $Co(en)_2A(ONO)^{n+}$, and seriously interferes with product analysis. This has been overcome by N_3^- quenching which destroys excess HNO_2 and reverses nitrito complex formation. Second, *cis*-$Co(en)_2(OH_2)SO_3^+$ has been isolated and slow cis $\rightleftharpoons$ trans isomerization established (ref. 117). Thus for the first time the steric course of substitution of *trans*-$Co(en)_2(SO_3)X$, presumed retentive (ref. 118), has been directly measured and this aspect confirmed. Also, the first steric course data for a *cis*-$Co(en)_2(SO_3)X$ isomer have been obtained.

The entries for the aqua complexes $Co(en)_2(OH_2)_2^{3+}$ are derived from water exchange and cis/trans isomerization rate data (ref. 119), e.g.

$$\textit{trans-}Co(en)_2(^{18}OH_2)_2^{3+} \xrightarrow[H_2O]{} \begin{cases} \xrightarrow{k_{ex}} \textit{trans-}Co(en)_2(^{18}OH_2)(OH_2)^{3+} \\ \xrightarrow{k_{tc}} \textit{cis-}Co(en)_2(^{18}OH_2)(OH_2)^{3+} \end{cases} \quad (57)$$

Milking experiments (ref. 119) have established that isomerization is a substitution process rather than an intramolecular rearrangement, since each act of isomerization leads to incorporation of one labelled water molecule into the cis product. However, the previous figure of 63% trans product for reaction (57) appears to have involved the incorrect water exchange rate constant k_{ex} (ref. 116,123). The statistical factor of two is inbuilt into k_{tc} but not k_{ex}, since the usual measurement of k_{ex} gives the constant for the slowest step of the sequence,

$$M(sol^*)_n + sol \underset{sol}{\xrightarrow{nk_{ex}}} M(sol^*)_{n-1}(sol) + sol^* \underset{sol}{\xrightarrow{(n-1)k_{ex}}} \cdots (58)$$

TABLE 12

Stereochemistry of aquation of Λ-*cis*- and *trans*-$Co(en)_2AX^{n+}$

A	X	Inducing Agent	Λ-*cis* reactant % cis	Λ-*cis* reactant % Λ-*cis*	Λ-*cis* reactant Ref.	*trans* reactant % trans	*trans* reactant Ref.
Cl	Cl	spont.	75	75	36	74	110
	DMSO	spont.	(85)	–	120	70.5	109
	DMF	spont.	–	–		68	112
	DMA	spont.	(85)	–	120	–	
	Cl	Hg^{2+}	76	76.5	36	69	112
	Cl	$HgCl^+$	76	76.5	36	69	112
	N_3	NO^+	75	77.5	112	82	112
	N_3	Cl_2	76	77	113	–	
	N_3	HOCl	76	75	113	–	
	DMSO	Cl_2	75	75.5	36,115	69.5	109
	TMSO	Cl_2	75	–	115	–	
	O_3SCF_3	spont.	–	–	–	59.5	106
N_3	Cl	spont.	86	82	36	91	111
	Br	spont.	85	85.5	36	91	111
	DMSO	spont.	86	84	36	92	111
	DMF	spont.	85	–	121	92	111
	N_3	NO^+	84	82	114	87	112
	Cl	Hg^{2+}	83	83.5	36	86.5	112
	Br	Hg^{2+}	82	83.5	36	86	112
	N_3	Hg^{2+}	–	–		89	112
Br	Br	spont.,15°C	77	–	40	–	
		spont.,25°C	73.5	70	36,40	84.5	110
		spont.,40°C	65	–	40	–	
	Br	Hg^{2+}	61.5	60	36,40	78.5	112
	Br	$HgBr^+$	61.5	60	112	80	112
	N_3	NO^+	72.5	70.5	36	86	112
	DMSO	Cl_2	70.5	70.5	36	–	
OH	Cl	spont.	84	84.5	36	30	112
	Br	spont.	–	–		29	112
	$^{18}OH_2$	spont.	31[a]	(31[a])	119,122,123	~ 29	119,123
OH_2	$^{18}OH_2$	spont.	97.7	98[b]	119	77	119
	Cl	Hg^{2+}	95.5	93	36	65.5	112
	Br	Hg^{2+}	95	94	36	62	112
	N_3	NO^+	95	95	36	84.5	112
	DMSO	Cl_2	95.5	–	115	–	

Table 12 cont'd.

A	X	Inducing Agent	Λ-*cis* reactant % cis	% Λ-*cis*	Ref.	*trans* reactant % trans	Ref.
DMSO	Cl	Hg^{2+}	92	–	115	–	
	Br	Hg^{2+}	92	–	115	–	
	N_3	NO^+	92	–	115	–	
	DMSO	Cl_2	92	–	115	–	
NCS	Cl	spont.,25°C	100	–	112	58.5	112
		spont.,60°C	100	–	112	58	112
	Br	spont.,25°C	100	–	112	57	112
		spont.,60°C	100	–	112	56.5[57[c]]	112
	DMSO	spont.,25°C	100	–	112	55.5	112
		spont.,60°C	100	–	112	57	112
	N_3	NO^+,25°C	100	–	112	74.5	112
		NO^+,60°C	100	–	112	73	112
NO_2	Cl	spont.	100	100	112	100	112
	Br	spont.	100	100	112	100	112
	DMF	spont.	100	–	112	100	112
	DMSO	spont.	100	–	112	100	112
	CH_3CN	spont.	100	–	112	100	112
	Cl	Hg^{2+}	100	100	112	100	112
	Br	Hg^{2+}	100	100	112	100	112
	N_3	NO^+	100	–	112	100	112
	DMSO	Cl_2	100	–	112	100	112
	SO_4^{2-}	spont.	–	–		100	112
NH_3	Cl	Hg^{2+}	100	100	52	100	112
	Br	Hg^{2+}	100	100	52	100	112
	N_3	NO^+	100	100	52	100	112
	DMSO	Cl_2	100	100	51	–	
	$^{18}OH_2$	spont.	100	100	30,125	99	30,125
SO_3	N_3	H^+	some	–	117	100	117
	N_3	NO^+	some	–	117	100	117
	Cl	spont.	–	–		100	117

[a] k_{ex} used for this calculation (ref. 119) seems to be in error; however, % cis = % Λ-*cis* (see text).

[b] % cis = % Λ-*cis* because $k_r \simeq k_{ct}$ (see text).

[c] Value from ref. 124.

when $\geq$ 2 equivalent exchanging groups are involved (ref. 126,127). This has not always been recognized, although the point was made clear in the original publication on the diaqua complexes (ref. 119). The later error for $Co(en)_2(OH_2)_2^{3+}$ (ref. 116) was unfortunately perpetuated (ref. 123). The figure for the steric course for the cis isomer (Table 12) has been adjusted for the same reason.

Finally mention should be made of an effective way around the problem of rapid rearrangement following the primary substitution process, a situation which previously has made the steric course either very inaccurate or indeterminate. The hydrolysis of the very inert *trans*-$Co(en)_2(NCS)X^+$ ($X = Cl^-, Br^-$) illustrates the point. At 25°C, *trans*-$\rightleftharpoons$ *cis*-$Co(en)_2(OH_2)NCS^{2+}$ isomerization is 9-fold faster than hydrolysis of the chloro ion, and even faster (14-fold) at 60°C (ref. 112,124). Nonetheless, by allowing the complex to aquate for only ~ 5% reaction, the correction to the observed $Co(en)_2(OH_2)(NCS)^{2+}$ isomer distribution is actually quite small, and readily made, provided the swamping effect of unreacted *trans*-$Co(en)_2(NCS)Cl^+$ can be removed. This is easily achieved by ion-exchange chromatography, but it necessitates using gram quantities of reactant to provide sufficient aqua product for the usual spectrophotometric analysis. Also the success of the method is obviously demanding of the purity of the starting complexes, established here by blank experiments for zero aquation times. For the more reactive *trans*-$Co(en)_2(NCS)Br^+$ ion, where both the new and traditional analytical methods could be applied, the agreement is good (Table 12).

The steric course data for Λ-*cis*-$Co(en)_2(OH)X^+$ ($X = Cl^-, Br^-$) are supplemented by results for H_2O exchange in Λ-*cis*-$Co(en)_2(OH)OH_2^{2+}$. The exchange rate data are not sufficiently accurate to deduce the cis/trans product distribution to better than ±5–10%, but the optical purity (Λ/Δ ratio) of the cis product for this reaction can be assessed by comparing specific rates of racemization (k_r) and cis to trans isomerization (k_{ct}). If this ion follows the pattern for all other Λ-*cis*-$Co(en)_2AX^{n+}$ species, exchange of H_2O should give fully active Λ-*cis*-$Co(en)_2(OH)(OH_2)^{n+}$ product, irrespective of the proportion of co-formed trans. For the scheme below,

$$\Lambda\text{-}cis\text{-}Co(en)_2(OH)OH_2^{2+} + H_2^{18}O \begin{cases} \xrightarrow{k_{cc}} \Lambda\text{-}cis\text{-}Co(en)_2(OH)(^{18}OH_2)^{2+} \\ \xrightarrow{k_i} \Delta\text{-}cis\text{-}Co(en)_2(OH)(^{18}OH_2)^{2+} \\ \xrightarrow{k_{ct}} trans\text{-}Co(en)_2(OH)(^{18}OH_2)^{2+} \end{cases} \tag{59}$$

it is readily shown that $k_r = k_{ct} + 2k_i$. The data of Harrowfield and Sargeson (ref. 128) suggested $k_r > k_{ct}$ by a significant margin, implying the unique result that inversion competes with cis to trans rearrangement, i.e. $k_i \sim k_{ct}$. Note that $k_i \sim k_{ct}$ is a feature which nicely distinguishes spontaneous or induced substitution from the base catalyzed process, and since Λ-*cis*-$Co(en)_2(OH)OH_2^{2+}$ can conceivably react via an internal conjugate base pathway (see section 8), this result warranted closer scrutiny. The original article recognized the difficulty in extracting accurate rate data for specific ions (e.g. *cis*-$Co(en)_2(OH)OH_2^{2+}$) present in mixtures containing the two isomeric forms of the conjugate acids and bases, $Co(en)_2(OH_2)_2^{3+}$ and $Co(en)_2(OH)_2^+$, and where the distribution of the total of six species is governed by the pH, three cis/trans equilibrium and four acidity constants. We have since isolated optically pure Λ-*cis*-$[Co(en)_2(OH)(OH_2)]S_2O_6 \cdot H_2O$ and directly determined k_r and k_{ct} (25°C) in CO_2-free H_2O, in the usual ways, where these problems have been shown not to arise (ref. 122). Within experimental error (± 5%), $k_r = k_{ct}$, and hence this anomaly vanishes. Λ-*cis*-$Co(en)_2(OH)(OH_2)^{2+}$ racemizes exclusively by isomerization to the achiral trans form, in keeping with results (ref. 36,129) for other Λ-*cis*-$Co(en)_2A(OH_2)^{n+}$ ions (A = Cl^-, Br^-, N_3^- and probably H_2O). The strong implication is that if H_2O exchange gives appreciable cis product, this is fully active.

The now comprehensive data set (Table 12) for the steric course of hydrolysis of these classic Werner complexes provides further insight into the factors controlling the stereochemistry. The "rule" that cis complexes can never give less cis product than that derived from trans reactants, irrespective of A or X, remains valid (ref. 36,112), although this does not necessarily imply the validity of its basis (ref. 44). Also it remains true that A groups which cannot engage in L → M π bonding always lead to geometric and, where appropriate, optical retention. This list (NO_2^-, NH_3, CN^- (ref. 130,131) and others) includes simple σ-donors as well as e^- withdrawing and L ← M π bonding groups. Data for the temperature dependence of the steric course are still limited, although its independence for *trans*-$Co(en)_2(NCS)X^{n+}$ (25–60°C), slight dependence for Λ-*cis*-$Co(en)_2(OH)Cl^+$ (0–25°C), and appreciable dependence for *cis*-$Co(en)_2Br_2^+$ (25–40°C) suggest that a comprehensive study might prove fruitful. Vanquickenborne and Pierloot (ref. 41) have given a theoretical account for the portion of the results of Table 12 available to them, and this account included specific predictions of the sense of the temperature dependence of the cis/trans product distribution. So far the data are in accord with this (qualitative) prediction—more cis product at lower temperatures.

The steric change observed for $Co(en)_2AX^{n+}$ seems to depend very largely on the starting geometry and the nature of A. A recent re-examination (ref. 111) of the stereochemistry of aquation of *trans*-$Co(en)_2N_3X^{n+}$ (X = Cl^-,

Br^-, DMF, DMSO) has revealed that the claimed retention is incorrect; there is some rearrangement (~ 10%), comparable to that found (ref. 36) for the *cis*-$Co(en)_2N_3X$ isomers. Nonetheless, the new data continue to emphasize that the role of the "orientating" A group depends critically on whether it is cis or trans to the leaving group. It is no longer generally true that, if A can bond to the metal ion by $L \xrightarrow{\pi} M$ donation, appreciable stereochemical change can be observed for *both cis-* and *trans*-$Co(en)_2AX^{n+}$. The systems $A = Cl^-$, Br^-, OH^-, OH_2, N_3^- follow this pattern, but $Co(en)_2(NCS)X$ and $Co(en)_2(SO_3)X$ (and $A = CH_3CO_2^-$ it seems, ref. 132) afford sharp exceptions. For $A = NCS^-$, we have confirmed the retention claimed (ref. 124) for *cis*-$Co(en)_2(NCS)X^+$ ($X = Cl^-$, Br^-), and obtained the same result for other X groups (DMSO, N_4O). Also, we find appreciable rearrangement for *trans*-$Co(en)_2(NCS)X^{n+}$ as did Tobe *et al.* (ref. 124). The sulfito systems are quite the reverse; the *trans*-$Co(en)_2SO_3X$ ions are highly retentive, while the cis isomers aquate with some (but not complete) rearrangement.

Recent work (ref. 133) on Λ-*cis-* and *trans*-$Co(en)_2(py)Cl^+$ reveal an additional example of retentive cis aquation (geometric and optical), but rearrangement for the trans isomer. Although once thought to be the norm, it is significant that the extensive review of the steric course of aquation of $Co(en)_2AX^{n+}$ ions has confirmed that some systems do indeed behave this way. However, the pyridine systems are unusual in that the Hg^{2+} induced aquation of *trans*-$Co(en)_2(py)Cl^+$ affords the first example of a pentaamine complex which is not topologically* retentive.

Perhaps the result of greatest significance revealed by the study of the trans ions is emergence of *several* examples which exhibit an X dependence for the stereochemical outcome. This is very clear for *trans*-$Co(en)_2(OH_2)(N_4O)^{3+}$ (from *trans*-$Co(en)_2(OH_2)N_3^{2+}$ + NO^+), in comparison with the data for other *trans*-$Co(en)_2(OH_2)X^{n+}$, but large differences are also apparent for *trans*-$Co(en)_2(NCS)X^{n+}$, *trans*-$Co(en)_2ClX^+$ and *trans*-$Co(en)_2BrX^+$. While it remains true that the steric course of hydrolysis of $Co(en)_2AX^{n+}$ is *essentially* independent of the leaving group, it has become very difficult to hold that this is *strictly* so. In retrospect, the anomaly in the cis series, Hg^{2+} induced aquation of *cis*-$Co(en)_2Br_2^+$ (ref. 36,40), may well reflect a genuine influence of the leaving group rather than the suggested special effect. This is supported, but not proven, by the results for $HgBr^+$ assisted aquation of Λ-*cis*-$Co(en)_2Br_2^+$. Both Hg^{2+} and $HgBr^+$ yield the

* The *p*-$Co(Metren)(NH_3)X^{n+}$ systems also aquate with rearrangement (ref. 58), but the *p* isomers do not give *t*-products, i.e. there is no overall topological change, but an apparent redistribution of the position of the CH_3 substituent on the ligand.

same stereochemical result, yet it seems more difficult for $HgBr^+$ to form the double bridge precursor (60) which was proposed (ref. 36) to accommodate the anomaly (A = BrHg in effect). Similarly, the explanation (ref. 30) for the *trans*-$Co(en)_2(OH_2)N_3^{2+}$ + NO^+ anomaly (ref. 112) could be challenged, and there now seems little point in testing the suggestion that *trans*-$Co(en)_2(ONO)N_3^+$ is the effective substrate here. There is no obvious correlation of the "anomalies" with specific leaving groups, although the old idea of a difference between spontaneous and induced aquation gets some support from the results. However, the lack of *consistent* differences is a major drawback.

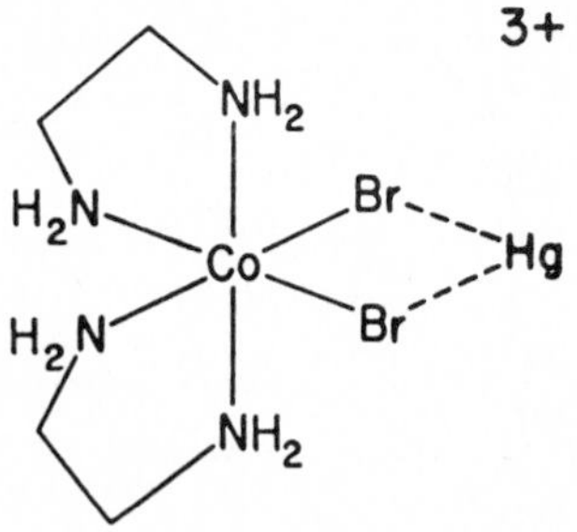

(60)

In conclusion, it seems certain that these aquation reactions are highly dissociative, because of the generally small influence of the leaving group, but it is harder to argue a case for a limiting dissociative reaction for those reactions which at least fulfil the stereochemical criterion. The steric course of aquation does not appear to be a sufficiently sensitive probe, but the stereochemistry of anion capture could be; no relevant results have yet been reported.

7.2 The role of the solvent

Water has been, by far, the solvent of choice for studies of substitution mechanisms. Nonetheless, there have been several significant investigations in non-aqueous media. The solvents of prominence are methanol, acetone, tetramethylenesulfone, dimethylacetamide, dimethylformamide, and dimethylsulfoxide. The last five belong to the group known as dipolar aprotic, studied extensively and systematically by Watts *et al.* (ref. 134). Work has been reported also for liquid ammonia and related media, and mixed solvents. Those including water as a component have been dubbed "troubled waters" (ref. 135), an apt label which may well be applicable to mixed solvents in general.

The emphasis of the research for non-aqueous solution has not been on the stereochemistry of substitution, but rather on various other facets of mechanism. These include the role of ion-pairs, the molecularity of solvolysis reactions with respect to the coordinating solvent component for solvent mixtures, and the role of the solvent lyate ion in both specific and general base catalyzed processes.

There has always been a need for a direct comparison between aqueous and non-aqueous media, but until recently the diversity in observed behaviour has obscured the clear identification of real analogies or contrasts. For example, *trans*-$Co(en)_2Cl_2^+$ hydrolyses to a mixture of *trans*-$Co(en)_2(OH_2)Cl^{2+}$ (74%) and *cis*-$Co(en)_2(OH_2)Cl^{2+}$ (26%) in water (ref. 110), whereas in dimethylsulfoxide, only *cis*-$Co(en)_2(OSMe_2)Cl^{2+}$ is the first observed product (ref. 136). In tetramethylenesulfone, *trans*-$Co(en)_2Cl_2^+$ isomerizes to *cis*-$Co(en)_2Cl_2^+$ without any apparent solvolysis (ref. 137). In methanol, the reverse cis to trans isomerization is observed (ref. 138). Part of the difficulty in the rationalization and possible unification of these observations is the uncertainty of the behaviour of the potential solvent containing intermediate complexes, $Co(en)_2Cl(solvent)^{2+}$. Several such species have been isolated and independently investigated (ref. 134), for all but the poorest coordinating solvents (acetone, tetramethylenesulfone). However only one isomeric form (cis) has been obtained.

The rate of solvolysis of Cl^- is greatly reduced in dipolar aprotic media compared to water, and the then relatively rapid trans to cis isomerization of $Co(en)_2(solvent)Cl^{2+}$, coupled with an enhanced stability of the cis isomer of the solvento complex, could accommodate the observation of *cis*-$Co(en)_2(solvent)Cl^{2+}$ as the exclusive product (ref. 139). Likewise, relatively rapid Cl^- anation of *cis*- or *trans*-$Co(en)_2(solvent)Cl^{2+}$, coupled with the instability of such ions in the more poorly coordinating media such as methanol and especially tetramethylenesulfone, could account for *cis-/trans*-$Co(en)_2Cl_2^+$ isomerization (ref. 134).

Recent work has uncovered successful synthetic strategies to the previously unknown *trans*-$Co(en)_2(solvent)A^{n+}$ ions.* (ref. 139,111,112). The *trans*-$Co(en)_2(solvent)A^{n+}$ ions are not elusive as once thought, provided they can be generated rapidly. Also problems associated with hydrolysis and subsequent trans to cis isomerization seem no more severe than for water.

Despite the solution of this major synthetic problem, there seems little point in a re-examination of the stereochemistry of the substitution of *cis*- and *trans*-$Co(en)_2X_2^+$ ($X = Cl^-, Br^-$) in dipolar aprotic solvents. It is clear that subsequent $Co(en)_2(sol)X^{2+}$ isomerization is too rapid (sol = CH_3CN excepted), and the steric course indeterminate. At best, it can be said that either

---

* Crystalline salts of $(NH_3)_5Co(solvent)^{3+}$ species containing ligands such as methanol, ethanol, alkyl phosphates, as well as ligands as poorly coordinating as acetone and iso-propanol have been obtained (ref. 56). The "non-coordinating" tetramethylenesulfone complexes have to date defied isolation, although the $(NH_3)_5Co$ complex has apparently been observed in solution (ref. 140).

isomeric product is consistent with the data. A move to better leaving groups is an obvious solution. Prior to the isolation of the *trans*-$Co(en)_2(sol)A^{2+}$ ions, Watts *et al.* had already observed some of these species in solvent interchange process of the kind,

$$\textit{cis-} \text{ or } \textit{trans-}Co(en)_2Cl(sol_1)^{2+} + sol_2 \longrightarrow \textit{cis-} + \textit{trans-}Co(en)_2Cl(sol_2)^{2+} + sol_1$$

$$\textit{trans-}Co(en)_2Cl(sol_2)^{2+} \longrightarrow \textit{cis-}Co(en)_2Cl(sol_2)^{2+} \qquad (61)$$

commencing with a range of *cis*-$Co(en)_2Cl(sol_1)^{2+}$ ions or with the only known trans-solvento species at that time, *trans*-$Co(en)_2Cl(OH_2)^{2+}$ ($sol_{1,2}$ = DMSO, DMA, DMF and DMSO) (ref. 141). The steric course of substitution could not be determined, but it was reported for the spontaneous ***hydrolysis*** of *cis*-$Co(en)_2Cl(sol)^{2+}$ * (ref. 120). The reason for these first observations of the trans solvento ions was clear. The neutral leaving groups were dissociated very much more readily than Cl^- or Br^-; these anions are poorly solvated in dipolar aprotic media and slow to leave in dissociative solvolysis.

By analogy with the aqueous chemistry, $CF_3SO_3^-$ and ClO_4^- as leaving groups are also candidates for stereochemical study (ref. 56,107), but have received little attention as yet. Also the principles behind the rapid induced aquation reactions such as

$$(NH_3)_5CoN_3^{2+} \xrightarrow[H_2O]{NO^+} (NH_3)_5CoOH_2^{3+} + N_2O + N_2$$

$$(NH_3)_5CoI^{2+} \xrightarrow[H_2O]{Hg^{2+}} (NH_3)_5CoOH_2^{3+} + HgI^+ \qquad (62)$$

could well be extended to non-aqueous media. Certainly they are useful synthetically (ref. 56,142); the putative *trans*-$Co(en)_2A(sol)^{2+}$ ions were generated in these ways (ref. 111,112,139), as have been the previously unknown optically resolved cis solvent complexes (from suitable chiral precursors, e.g.

---

* It is not widely recognized that this constituted the first report of non-stereoretentive spontaneous cis aquation, shown subsequently to be a general phenomenon.

Λ-*cis*-$Co(en)_2Cl(DMSO)^{2+}$ from Λ-*cis*-$Co(en)_2Cl_2^+$ in $DMSO/Hg^{2+}$, Λ-*cis*-$Co(en)_2N_3(DMSO)^{2+}$ from Λ-*cis*-$Co(en)_2(N_3)_2^+$ in Me_2SO/NO^+ or from Λ-*cis*-$Co(en)_2N_3X^+$ ($X = Cl^-$, Br^-) in $DMSO/Hg^{2+}$, ref. 121).

A moderately comprehensive study of the stereochemistry of substitution has now been completed for the spontaneous and induced solvolysis reactions of $Co(en)_2AX^{n+}$ species (ref. 121), in parallel with their hydrolysis reactions. The following subdivision of reactions is essentially for convenience rather than having intrinsic significance, and largely reflects the spectroscopic probes for the determination of the initial steric course.

(i) Solvent Exchange:

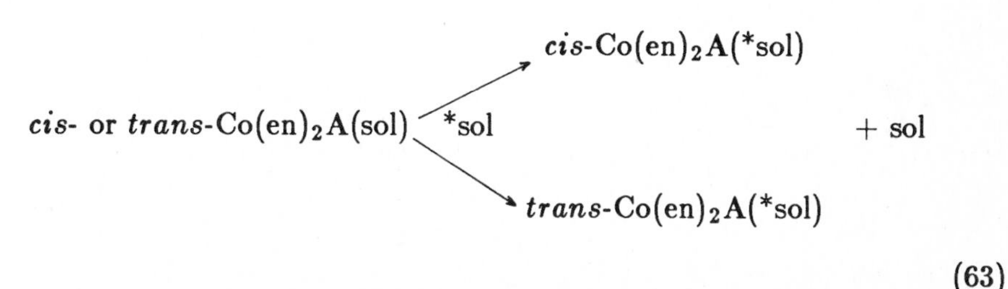

(63)

Nmr spectroscopy readily enables the determination of the rate of loss of coordinated solvent (ref. 143). Moreover, provided the opposite isomeric product does not exchange solvent too much more rapidly ($\leq$ 5-fold), the technique allows the determination of the intrinsic nature of the rearrangement—intramolecular (no exchange during isomerization), or dissociative* (intermolecular), i.e. one solvent molecule exchanged for each act of isomerization (ref. 139). To date the stereochemical studies have been confined to DMSO, but may be extended to readily available deuterated solvents such as DMF or MeCN, where there is already a body of data dealing purely with *rates* of solvent exchange.

For trans starting material, the solvent exchange rate (k_{tex}) gives the total rate of loss of solvent ($k_{tt} + k_{tc}$), while the rate of appearance of the cis product gives k_{tc}; the latter was also determined spectrophotometrically, in separate experiments. The steric course of substitution, expressed as % cis product, is then simply $10^2\ k_{tc}/(k_{tc} + k_{tt}) = 10^2\ k_{tc}/k_{tex}$. Note that the isomerization rate, when the process is intermolecular, cannot exceed the exchange rate, and this was always found to be the case.

With cis starting material, the % cis product for substitution is given by $10^2\ (k_{cex}\text{-}k_{ct})/k_{cex}$. The specific rate of cis to trans isomerization k_{ct} could not

---

* Dissociative, in the sense that the rearrangement process requires the complete dissociation of coordinated solvent. This is distinct from an intramolecular (non-dissociative) rearrangement, e.g. a twist mechanism.

be directly determined; the trans/cis equilibrium is totally (> 99%) towards the cis side. Nevertheless, if Λ-*cis*-$Co(en)_2A(sol)^{2+}$ racemize exclusively by isomerization to the achiral *trans*-$Co(en)_2A(sol)^{2+}$ ions (as do *all* the aqua analogs examined so far), then their racemization rates conveniently provide k_{ct}. The results for non-aqueous solution have been computed on this basis; further support for the assumption is given in (iv) ahead.

TABLE 13

Rate coefficients and stereochemistry of substitution for solvent exchange reactions of $Co(en)_2A(DMSO)^{2+}$ in DMSO

			10^5k, s^{-1}			% cis[d]
Reactant geometry	A	T,°C	k_{ex}	k_{tc}	k_{ct}[e]	% cis[d]
trans	Cl	35.0	700	62.7		85–90[a]
cis	Cl	35.0	7.0[b]		1.55	78
trans	N_3	35.7	448	361		81
cis	N_3	35.0	22.3	333	3.6	84
cis	DMSO	35.0	8.8[c]	–	–	–
trans	NCS	35.6	7.6	5.5		72
cis	NCS	35.6	1.4		< 0.1	> 93

[a] 89.5% cis calculated from data given; 85% minimum allows for possible underestimation of k_{ex} (see Table 14).
[b] Data from ref. 143.
[c] Data from ref. 143, corrected for statistical factor (see text).
[d] % cis product = $10^2\, k_{tc}/k_{ex}$ for trans and $10^2(k_{ex} - k_{ct})/k_{ex}$ for cis substrates.
[e] $k_{ct} = k_r$ assumed (see text and Table 16).

The rate and stereochemical data so far obtained are summarized in Table 13. In every case it has been possible to demonstrate that, where geometrical rearrangement occurs, labelled solvent is incorporated, i.e. cis and trans products arise exclusively from substitution; there is no (0 ± 5%) measurable contribution from an intramolecular (non-dissociative) process. This is analogous to the situation in water where, for the few complexes examined (*cis*- and *trans*-$Co(en)_2(OH_2)_2^{3+}$, for example), exactly one water molecule is exchanged for each act of isomerization. For completeness, in Table 13 we include the early data of Watts *et al.* (ref. 143); there is as yet no entry for the *trans*-$Co(en)_2(OSMe)_2^{3+}$ ion, although it has been synthesized, and we need note that the *observed* exchange rate for its cis isomer, as determined, refers to the secondary process,

$$cis\text{-Co(en)}_2(\text{OSMe}_2)(^*\text{OSMe}_2)^{3+} \xrightarrow{^*\text{Me}_2\text{SO}} cis\text{-Co(en)}_2(^*\text{OSMe}_2)_2^{3+} + \text{Me}_2\text{SO} \quad (64)$$

and the correct entry for the rate of the primary substitution process,

$$cis\text{-Co(en)}_2(\text{OSMe}_2)_2^{3+} \xrightarrow{^*\text{Me}_2\text{SO}} cis\text{-Co(en)}_2(^*\text{OSMe}_2)(\text{OSMe}_2)^{3+} \quad (65)$$

is therefore twice the reported value. This apparently trivial difficulty with statistical factors in exchange processes has not always been appreciated, as emphasized by the Mønsteds (ref. 126), and Merbach and Moore (ref. 127), and noted earlier.

There are too few data for the directly analogous aqueous systems to comment on the rate information in Table 13, other than to note that, *ceteris paribus*, the rate of substitution of coordinated water by solvent water is not greatly different to that for the analogous process in DMSO. The more relevant stereochemical information is discussed ahead.

(ii) Solvent Interchange.

The stereochemistry of the substitution process has been investigated for a range of solvent pairs and $\text{Co(en)}_2\text{AX}^{n+}$ complexes. Predictably, only where the replaced solvent is an appreciably poorer ligand than bulk solvent are the stereochemical data reliable; the processes are essentially dissociative (in the usual mechanistic sense), and thus the rate of subsequent *trans/cis*-$\text{Co(en)}_2(\text{sol})\text{A}^{2+}$ isomerization exceeds that for the primary solvent interchange reaction when, e.g. the leaving group is DMF and the solvent is the more weakly bound Me_2SO.

Most of the data of Table 14 were obtained spectrophotometrically. Generally, because rearrangement occurred, the spectra of reactant and products differed considerably despite the similarity of the spectra of analogous isomeric species (e.g. *cis*-$\text{Co(en)}_2\text{Cl(DMF)}^{2+}$ and *cis*-$\text{Co(en)}_2\text{Cl(DMSO)}^{2+}$), and this coupled with slower subsequent trans to cis isomerization of the products enabled a reasonably accurate determination of the stereochemistry of substitution.

(iii) Induced Solvolysis.

As in water, *cis*- and *trans*-$\text{Co(en)}_2\text{AX}^{n+}$ ($\text{X} = \text{Cl}^-, \text{Br}^-$) lose X^- in the presence of Hg^{2+} so rapidly that the initial *cis-/trans*-$\text{Co(en)}_2\text{A(sol)}^{2+}$ product distribution is readily defined. The results (Table 15) include the first determinations of the steric course of substitution for chiral cis complexes of this kind in non-aqueous solution. These data were obtained polarimetrically using the known chiro-optical properties for independently resolved *cis*-$\text{Co(en)}_2\text{A(sol)}^{2+}(\Lambda)$.

TABLE 14

Stereochemistry of solvent interchange processes for $Co(en)_2AX^{n+}$

Solvent	Reactant geometry	A	X	T,°C	% trans[a]
DMSO	trans	N_3	DMF	25	51[b]
	trans	N_3	DMF	35[c]	53[b]
	trans	Cl	DMF	25	15[b]
	cis	N_3	DMF	25	13
DMF	trans	N_3	DMSO	25	63
		Cl	DMSO	25	26

[a] ±(2 – 3)%, except where noted.
[b] ±10%.
[c] Solvent interchange rate coefficient = 5.9×10^{-4} sec^{-1} (spectrophotometric), 5.3×10^{-4} sec^{-1} (NMR). The shortfall in the NMR number suggests k_{ex} for *trans*-$Co(en)_2(DMSO)Cl^{2+}$. (Table 13) may be underestimated by several %.

Silver ion induced solvolysis is slower than that for Hg^{2+} and subsequent racemization, via the *trans*-$Co(en)_2A(sol)$ ion, and trans to cis isomerization were comparable in rate, inhibiting a quantitative study. It was clear nonetheless that, with cis reactants at least, some trans product was formed, and that the cis product was essentially optically pure *cis*-$Co(en)_2A(sol)^{2+}$.

TABLE 15

Stereochemistry of induced solvolysis in DMSO at 25°C

Reactant geometry	A	X	% trans	% cis	% Λ-cis
trans	N_3	BrHg	69	31	–
	N_3	ClHg	79	21	–
	Cl	ClHg	37	63	–
	NCS	N_4O	63	37	–
Λ-cis	N_3	BrHg	16	84	–
	N_3	ClHg	12	88	–
	Cl	ClHg	13	87	85.5
	NCS	N_4O	0	100	–
	NH_3	BrAg	0	100	100[a]

[a] Data from ref. 51.

(iv) Racemization of Λ-(+)-$Co(en)_2A(sol)^{2+}$.

These first studies on racemization of Werner type complexes in dipolar aprotic solvents yielded the results given in Table 16, along with analogous data for aqueous solution. This work is less demanding on compound purity than the steric course determinations since optically pure and optically impure materials alike give the same value for k_r. The aqueous data show $k_r = k_{ct}$. This result, together with the results of (iii) above, show that substitution with inversion ($\Lambda cis \rightarrow \Delta cis$) does not compete with substitution leading to either optical retention ($\Lambda cis \rightarrow \Lambda cis$) or geometric isomerization ($\Lambda cis \rightarrow trans$) in both aqueous and non-aqueous solution. The values of k_r for DMSO therefore define k_{ct}.

TABLE 16

Racemization rate data for Λ-*cis*-$Co(en)_2A(sol)^{n+}$

Solvent	A	T,°C	10^6 k_r,s^{-1}	10^6 k_{ct},s^{-1}	Ref.
DMSO	Cl	25	3.5	–	121
	Cl	35	15.5	–	121
	N_3	35	36	–	121
H_2O	Cl	25	20	20	36
	N_3	25	42	42	36
	OH_2	25	0.15	0.14	41
	OH	25	2180	2200	122
	Br	25	47	52	36

Direct measurements of the position of the equilibrium,

$$trans\text{-}Co(en)_2A(sol)^{2+} \overset{K}{\rightleftharpoons} cis\text{-}Co(en)_2A(sol)^{2+} \tag{66}$$

indicate > 99% cis isomer for all systems examined ($A = Cl^-$, N_3^-, NCS^-; sol = DMSO, DMF). This may be compared to results for water (ref. 36,110–112) where the equilibrium also favours the cis form (~ 80%), but the preference is much less pronounced. For the chloro and azido complexes in DMSO, K for (66) can be calculated ($k_{tc}/k_{ct} = k_{tc}/k_r = K$) and the results K = 100 ($A = Cl^-$, 35°C) and K = 400 (N_3^-) confirm that the solvent effect in terms of energy is quite significant. This has been explained by preferential solvation of the dipolar cis isomers by these solvents which have large dipole moments (ref. 139).

The clear solvent effect on the equilibria leads next to a consideration of its effect on the steric course of substitution. Results are summarized in Table 17. First, it is noted that the *trans*-$Co(en)_2AX^{n+}$ ions consistently form more cis product in DMSO and DMF compared to water. Second, the *cis*-$Co(en)_2AX^{n+}$ ions display a substitution stereochemistry not too different to water. Thus it seems the pronounced selective solvation of the cis form reflected in the equilibria is not felt to the same degree in the kinetic isomer distributions. The overall result is a shift *towards* common product distributions for a particular *cis*- and *trans*-$Co(en)_2AX^{n+}$ pair, but it remains true, as in water (section 7.1), that a cis substrate gives more cis product.

TABLE 17

Summary: stereochemistry of substitution of $Co(en)_2AX^{n+}$ in dipolar aprotic media

				% cis product		
				solvent		
Reactant geometry	T,°C	A	X	DMSO	DMF	H_2O [a]
trans	35	N_3	DMSO	81	37[c]	8
	25	N_3	ClHg	21	–	13
	25	N_3	BrHg	31	–	14
	25	N_3	DMF	49	–	8
	35	N_3	DMF	47	–	8
cis	35	N_3	DMSO	84	–	86
	25	N_3	ClHg	88	–	83
	25	N_3	BrHg	84	–	82
	25	N_3	DMF	87	–	85
trans	35	Cl	DMSO	85–90	74[c]	29.5
	25	Cl	ClHg	63	–	31
	25	Cl	DMF	85	–	32
	25	Cl	O_3SCF_3	77	73	40.5
cis	35	Cl	DMSO	79	–	75(85)[b]
	25	Cl	ClHg	87	–	76
trans	35.6	NCS	DMSO	72	–	44.5
	35.6	NCS	N_4O	37	–	25.5
cis	35.6	NCS	DMSO	> 93	–	100
	25	NCS	N_4O	100	–	100

[a] Data from Table 12; all 25°C.
[b] For induced aquation; number in parenthesis—spontaneous aquation.
[c] 25°.

The most interesting result to be gleaned from Table 17 is the obvious leaving group dependence of the steric course. This seems more pronounced than in water (Table 12, section 7.1) and is consistent with the proposal that five coordinate intermediates of appreciable lifetime do not exist in any of these simple systems (ref. 36). A reasonable view is that the leaving group has not completely escaped the solvation sheath of the complex at the transition state, and it remains to influence the steric course of solvent entry into an incipient $Co(en)_2A$ intermediate of a dissociative reaction. This "postdissociation" effect (ref. 28) should be more pronounced in dipolar aprotic solvents, especially for anionic leavings groups which are very poorly solvated, and possibly for neutral (solvent) and cationic (HgX^+) leaving groups as well, because of an enhanced cage effect. Solvent molecules are bulkier and the solvent structure immediately surrounding the complex is likely to resist the passage of the leaving group into bulk solution before the vacant coordination site is filled.

Although the leaving group affects the steric course, and to this extent it is coupled with the effect of A, some comment on the role of the A group can still be made. Qualitatively, it is much the same as found for water, and the variations on changing A are not offset by the solvent variation. For example, the solvolysis reactions of *cis*- and *trans*-$Co(en)_2(NO_2)X$ are retentive (ref. 144), as in water. This also is true for both the *cis*- and *trans*-$Co(en)_2(NH_3)X$ ions, in DMSO, DMF and $PO(OMe)_3$. Further, all four species *cis*- and *trans*-$Co(en)_2ClX$ and $Co(en)_2N_3X$ show steric change, as in water. In short, the gross effects found for water are mirrored in the non-aqueous chemistry. We have also seen that the cis product fully retains the parent chirality, even when there is an accompanying cis to trans rearrangement, and this also is consistent with the aqueous chemistry.

The steric course of anation for the *cis*- and *trans*-$Co(en)_2A(sol)^{n+}$ ions in dipolar aprotic media is an obvious follow-up area. There are some scattered data for *cis*- and *trans*-$Co(en)_2(NO_2)(sol)^{2+}$ (sol = DMSO, DMF, DMA); predictably, anation by Cl^-, NCS^-, Br^-, NO_2^-, N_3^-, and others is retentive, as are the solvent exchange/interchange processes (ref. 144,145). Watts *et al.* have recorded some data for Cl^- anation of *cis*-$Co(en)_2(sol)Cl^{2+}$ (ref. 136,146), but the attention has been focused more on the rate than the stereochemistry. The most interesting report has been that the steric course of Cl^- anation of *cis*-$Co(en)_2Cl(sol)^{2+}$ (sol = DMSO, DMF) depends upon the [anion] (ref. 147). This has been interpreted in terms of an ion-pair and ion-triplet which collapse to anated product with different cis/trans ratios.* This is an important result with a reasonable interpretation; the NCS^- work (ref. 148) could bear repeating in view of the finding (ref. 149,150) that $(NH_3)_5CoOH_2^{3+}/NCS^-$ in water and even $(NH_3)_5Co(DMSO)^{3+}/NCS^-$ in DMSO lead to both S- and N-bonded SCN^- products. Anation data are not confined to systems where the leaving group is the solvent, but the latter

* This observation suggests the prospect of an effect on the steric course of *aquation* by non-competing anions which ion-pair (e.g. SO_4^{2-}, $S_2O_6^{2-}$); it is yet to be tested (see section 7.1).

are easier to interpret. Long ago the racemization and Cl^- exchange reactions were studied for Λ-$Co(en)_2Cl_2^+$ in CH_3OH (ref. 138), and it was held that the results were not consistent with a *cis*-$Co(en)_2(MeOH)Cl^{2+}$ intermediate; however, the trans isomer has not been synthesized. Finally we should note that the anation reactions can be specific base catalyzed, and this can give rise to a steric course which apparently depends upon [anion]. Basic anions such as NO_2^- and N_3^- can function this way in unbuffered protic solvents such as H_2O and CH_3OH (ref. 151). Also, general base catalysis, in aprotic solvents, has been observed, e.g. N_3^-/DMSO (ref. 36). This is difficult to pinpoint and even more difficult to suppress. These points are important because the spontaneous processes can easily be obscured by a contribution from the base catalyzed pathway (see section 8.4).

There are scattered results for solvents other than dipolar aprotic media. In liquid NH_3, even containing NH_4^+ to suppress solvent autoprotolysis which yields strongly catalytic NH_2^-, the reactions are base catalyzed, and the catalysis has been shown to be due to NH_3 itself (ref. 67). The solvolysis of *trans*-$Co(en)_2(SO_3)(DMSO)^+$ appears to be the single exception (ref. 152). Balt has shown that in ND_3 the product *trans*-$Co(en)_2(SO_3)ND_3$ contains no D in the en ligands. This result is attributed to a combination of a reduction in the acidity of the amines,* all cis to SO_3^{2-}, and the unusual rapidity of the spontaneous process due to the well-known trans activation imposed by SO_3^{2-}. The steric course is the same as in water, 100% retention. It will be interesting to learn if the solvent dramatically affects the pronounced trans specificity of SO_3^{2-}. The *cis*-$Co(en)_2(SO_3)(sol)^+$ species are now accessible (ref. 117), and studies on isomerization, racemization and solvent exchange akin to those described earlier for other $Co(en)_2A(sol)^{2+}$ complexes can answer this question.

Triflic acid (CF_3SO_3H) has become an important reagent and solvent in both organic and inorganic chemistry. For metal complexes, it rapidly and irreversibly strips off ligands having a lone pair of e^-, e.g. Cl^-, NO_2^-, N_3^-, OH_2, and it is redox inactive. These are induced solvolysis reactions involving superb leaving groups, and they have rapidly emerging synthetic value (ref. 56,153,154), e.g.

* This seems also true of H_2O solvent; *trans*-$Co(en)_2(SO_3)OH$ at high pH undergoes anation at rates independent of $[OH^-]$, precluding base catalysis (ref. 155). Also, cis- and trans-isomers interconvert very slowly ($t_{1/2}$ ~ days), at $[OH^-]$ independent rates (ref. 117).

$$(NH_3)_5CrCl^{2+} \overset{CF_3SO_3H}{\rightleftharpoons} (NH_3)_5Cr(ClH)^{3+} \overset{CF_3SO_3^-}{\longrightarrow} (NH_3)_5CrO_3SCF_3^{2+} + HCl \quad (67)$$

But of relevance here was the conception that the CF_3SO_3H solvent system might induce steric change in otherwise retentive metal ion systems. This was suggested by early work on Co(III) complexes which indicated unusual stereochemistry, e.g. Λ-$Co(en)_2Cl_2^+$ in CF_3SO_3H very rapidly loses its activity, ultimately yielding *racemic* *cis*-$Co(en)_2(O_3SCF_3)_2^+$ (ref. 56). Subsequent observations have helped clarify the picture. At low temperature in CF_3SO_3H, Λ-$Co(en)_2OCO_2^+$ gives the Λ-$Co(en)_2(OH_2)(O_3SCF_3)^{2+}$ complex first,* without loss of activity, presumably by the usual C-O cleavage mechanism established for aqueous acid (ref. 156). Also, *cis*- and *trans*-$Co(en)_2(NH_3)(NO_2)^{2+}$ very rapidly yield the respective ammine triflato complexes in cold CF_3SO_3H, without geometric rearrangement, analogous to the behaviour of $Co(en)_2(NH_3)X^{n+}$ in other non-aqueous solvents as well as in H_2O (ref. 157). The pyridine complexes $Co(en)_2(py)(NO_2)^{2+}$ behave similarly, although *cis*-$Co(en)_2(py)(O_3SCF_3)^{2+}$ is obtained from trans reactant in warmer (25°C) CF_3SO_3H solutions over longer periods (30 min) (ref. 133). Finally, it has been observed that violet *cis*-$[Co(en)_2Cl_2]Cl$ dissolves in CF_3SO_3H at 0°C to give instantly a green solution from which *trans*-$[Co(en)_2Cl(O_3SCF_3)]CF_3SO_3$ (green) is readily crystallized in good yield (ref. 106). It is now known that the grass-green *trans*-$[Co(en)_2Cl_2]Cl$ complex very rapidly gives the same product, although visually there is no apparent reaction. All these observations may be rationalized without the need to propose a radically different substitution stereochemistry for the CF_3SO_3H solvent. The NH_3 and py species could be *shown* to be retentive because subsequent *cis*-/*trans*-$Co(en)_2(amine)O_3SCF_3^{2+}$ isomerization is relatively slow (although, in terms of absolute rates, it is very much more rapid than for the aqua complexes). However, for $Co(en)_2Cl(O_3SCF_3)^+$ derived from $Co(en)_2Cl_2^+$, the observed product distribution (largely trans) appears to be equilibrium controlled, even at 0°C. The aqueous substitution rates for $(NH_3)_5CoX^{2+}$ (X = OH_2, $CF_3SO_3^-$) indicate that $CF_3SO_3^-$ is dissociated ~ 10^4-fold more readily than H_2O (ref. 56,158). Thus, given the isomerization rate for *cis*-⇌ *trans*-$Co(en)_2(OH_2)Cl^{2+}$ in H_2O ($t_{1/2}$ 110 min, 25°C) (ref. 110), the isomerization rate for *trans*-⇌ *cis*-$Co(en)_2Cl(O_3SCF_3)^+$ may be estimated ($t_{1/2}$ ~ 1s, 25°C). This argument assumes a negligible solvent effect on the substitution

* At 25°C, subsequent racemization ensues rapidly, with the ultimate formation of *cis*-$Co(en)_2(O_3SCF_3)_2^+$ (ref. 56,156).

rates and steric course for the respective exchange processes. Notwithstanding this assumption, it is clear that $CF_3SO_3^-$ is such a superior leaving group that isomerization can proceed rapidly via a $CF_3SO_3^-$ exchange process in neat CF_3SO_3H, even for the amine (NH_3,py) complexes where most acts of exchange lead to retention. The qualitative order of subsequent isomerization rates $Cl^- \gg py \geq NH_3$ is consistent with this rationale.

A useful corollary is that isomerization for the inert metal ions may be induced under mild conditions for otherwise retentive metal ion systems such as Cr(III) or Ru(II). This seems to have synthetic value not yet realized, although the choice of non-displaced ligands is restricted to those not attacked by CF_3SO_3H, such as amines.

To summarize this section on solvent effects, it seems fair to comment that the nature of the solvent does not seem to profoundly effect the steric course of substitution, using the $Co(en)_2AX^{n+}$ complexes as models. Changes in the isomeric product ratios appear to be within a factor of five, corresponding to a relatively small energy difference in activation energies for alternate pathways, yet the solvents vary widely in acidity, dielectric constant, dipole moment, H-bonding capacity, steric bulk and geometry. The important inference is that the stereochemical outcome is controlled largely by the complex ion itself. At the moment it seems that the central metal ion is the key variable, since *measurable* steric change is largely if not entirely confined to cobalt(III). This persistent view that Co(III) is unique may change as stereochemical studies broaden to handle the labile metal ions (ref. 30), which in fact represent the bulk of the D-block transition metal ions.

8 BASE CATALYZED SUBSTITUTION

This topic has been reviewed in some depth up to about 1982 (ref. 28,30). We cover here selected aspects of subsequent findings, but the reader is referred to these reviews for detailed background.

8.1 Steric course of base hydrolysis of $Co(en)_2AX^{n+}$

Historically these reactions are important (ref. 44,159), although there are now a multitude of stereochemically more sophisticated systems which give the mechanism of the base hydrolysis reaction better definition.

Table 18 is a compilation of revised steric course data for the reactions,

$$\Lambda\text{-}cis\text{- or }trans\text{-}Co(en)_2AX + OH^- \longrightarrow \Lambda\text{-}cis\text{-} + \Delta\text{-}cis\text{-} + trans\text{-}Co(en)_2A(OH) + X \quad (68)$$

TABLE 18

Steric course of base hydrolysis of Λ-*cis*- and *trans*-$Co(en)_2AX^{n+}$ at 25°C

		cis reactant			trans reactant	
A	X	% cis	% Λ-*cis*	Ref.	% cis	Ref.
NH_3	Cl	77.5,78	57.5,61.5	51,52	64	51
	Br	77,78	57,56.5	51,52	–	
	ONO_2	77	61	52	63	51
	DMSO	77	67	51	–	
	$PO(OMe)_3$	77.5	70	51	–	
	SCN	–	–		64	161
	$NH_2(en)_2(NH_3)Co$[a]	74	–		–	160
N_3	Cl	48	–	53	24	53
	Br	47.5	–	53	26	53
	DMSO	39	–	53	30.5	53
	HCO_2	–	–		23	53
	N_3^-	–	–		26	53
	DMF	–	–		26	53
NCS	Cl	82	–	53	70	53
	Br	75	–	53	75	53
	DMSO	72	–	53	72	53
Cl	Cl,25°C	23	15.8	b	10.5	b
	0°C	18.5	13.6	b	8.5	b
OH	Cl,25°C	94.5	92.5	b	8.5	b
	0°C	95.5	94.0	b	7	b

[a] Steric course for bridge cleavage reaction of the complex ion ΛΛ; ΔΔ-*cis,cis*-$Co(en)_2(NH_3)CoNH_2Co(NH_3)(en)_2^{5+}$; the other reaction (10–25%, depending upon the conditions) involves the loss of a terminal ammonia, yielding $(en)_2(NH_3)CoNH_2Co(OH)(en)_2^{4+}$.

[b] Data transposed from Tables 1, 3, 4, 8, and 9.

The basis for the selection are results where the many and often gross errors in the early literature appear to have been removed. The variation in the nature of A and especially X are now sufficiently wide, and the steric course sufficiently well-defined (generally ± 2% or better), to make some general comments.

All these reactions follow the rate law $k(obsd) = k_{OH}[OH^-]$ and it is agreed (ref. 28,30) that the reactive species is the conjugate base $Co(en)(en\text{-}H)AX^{n+}$ which aquates to $Co(en)_2A(OH)$. Contributions from double substitution (sections 2,3) have been avoided, or corrected for where appropriate. For the several cases tested, the steric course is independent of $[OH^-]$ over a wide range (up to 1.0 M),* only very mildly temperature dependent, and independent of ionic strength (up to 1 M). The last point is important because both the rate and the raw law can depend strongly upon the medium (ref. 48,59), and this suggests that the step which determines the isomeric product distribution is not the rate determining step.

An outstanding problem in base hydrolysis is the *effective* site of deprotonation (ref. 28,30). This is not known for $Co(en)(en\text{-}H)AX^{n-1)+}$ and hence any variation in this site cannot be decoupled from the effects of a variation in A or X. Nonetheless the results which show the effect of a variation in the leaving group reveal that the steric course is *essentially independent of X*. However, the error is sufficiently small to say that this is not an absolute independence. The interpretation of this non-trivial point is discussed elsewhere (ref. 53), and it suffices to note here that this result implies a single site of deprotonation which is the same for each X.

As for aquation of $Co(en)_2AX^{n+}$ (section 7.1, Table 12), the steric course of aquation of $Co(en)(en\text{-}H)AX^{(n-1)+}$ shows a marked dependence on A. However, there appears to be no correction between the role of the A group in the two processes, and this could well be due to different (but only one) effective sites of deprotonation. Unlike normal aquation, base hydrolysis leads almost invariably to rearrangement ($A = CN^-$ a real exception? (ref. 130)), and there is a variable degree of (but always some) racemization in the cis product when derived from Λ-*cis*-$Co(en)_2AX^{n+}$ (ref. 23,51,52,159,162).

Recent data for Λ-*cis*- and the elusive *trans*-$Co(en)_2(py)Cl^{2+}$ ions, as well as Λ-*cis*- and *trans*-$Co(en)_2(OH)py^{2+}$, are not included in Table 18 but confirm the unusual behaviour of pyridine based complexes (ref. 133). For example, the *cis*-$Co(en)_2(py)OH^{2+}$ product from the Λ-*cis* substrate not only shows much more racemization than the aliphatic NH_3 analog (ref. 51,52), but more than any other chiral *cis*-$Co(en)_2AX^{n+}$ species (ref. 23,44,159,162). The $Co(en)_2(OH)py^{2+}$ ions behave normally because py is the leaving group. The steric course for $Co(en)_2(OH)X^{n+}$ is now seen to be constant for both anionic (1-) and neutral leaving groups, and the common result for *cis*-$Co(en)_2(OH)(O_2C_2O_2)$ adds a 2- leaving roup to the former list (ref. 163). Work in progress on the *cis*- and *trans*-$Co(en)_2(NO_2)X^{n+}$ systems where a wide range of neutral, 1- and 2- leaving groups have been used proposes to comment further on subtle leaving group effects (ref. 164).

Anions compete well with H_2O for the supposed coordinatively unsaturated $Co(en)(en\text{-}H)A^{n+}$ intermediate in base hydrolysis (ref. 51,52). The

* Though see ref. 59.

extent of capture is linear in [competing anion] and independent of [OH^-], consistent with the proposed mechanism (ref. 51,52). The steric course of NO_2^- and SCN^- scavenging has been reported (ref. 52), but this work predated the discovery that these ambidentate anions are captured at both ends (ref. 161,165–167), and the predominance of O- capture for NO_2^- (ref. 167) and S- capture for SCN^- (ref. 166) is not that assumed. The work will need revisiting. Azide ion has been more widely used, and the steric course results parallel those described for the capture of H_2O (ref. 51,52). The steric course is of course different, but not greatly different. The activities found for the cis products are very similar, but there is more *trans*-azido product for the reactions of both *cis*- and *trans*-$Co(en)_2(NH_3)X^{n+}$.* Results reported for the steric course of base hydrolysis of Λ-*cis*-$Co(en)_2X(NH_2CH_2CO_2)^+$ in N_3^- media, where additional competition arises from the pendant $-CO_2^-$ group, show similar trends, although here the [OH^-] is held to detectably affect the stereochemical outcome (ref. 59). Unfortunately there are yet no reliable data for competing anions other than N_3^-, nor data for complexes other than pentaamines which are probably the least stereochemically discriminating. Various accounts of the steric course of substitution have been given (ref. 30,44,51,52,59,162); none are compelling.

The issue of the existence and lifetime of the putative pentacoordinate species Co(en)(en-H)A remains active. However, fundamental studies other than purely stereochemical ones and using the simpler $(NH_3)_5CoX^{n+}$ species seem to be providing more useful information on this issue at the moment. These include work designed to find [anion] terms in the rate law and to examine their relationship to the product distribution, the use of ambidentate competitors such as NO_2^-, SCN^-, and $S_2O_3^{2-}$ which define an isomeric product ratio free from the problem of "inherited ion atmospheres", and mixed anion competition studies (ref. 48,62,166,167,169). The role of ion-pairs, which undoubtedly exist, is becoming clearer (ref. 48). No doubt these various approaches will be extended to the $Co(en)_2AX^{n+}$ systems where the additional handle of stereochemistry (ref. 51,52,59) should add to the armoury to tackle remaining problems.

8.2 Multidentate chelate systems

The proposition of an intermediate which is too reactive to be directly observed seems to require the demonstration of properties independent of its source. But it will be seen that another approach exists. Multidentate ligand systems permit increased isomer possibilities and subtle design fea-

---

* This is also true for the base hydrolysis of *trans*-$Co(NH_3)_4(^{15}NH_3)X^{2+}$ in N_3^- media (ref. 168).

tures to not only test the idea of an intermediate, but also probe its nature (lifetime, stereochemistry) (ref. 28,30). Stereochemical results for the elegant Co(Metren)(NH_3)X^{n+} systems, for example, have suggested that the common intermediate shown in (69) does not exist (ref. 58). Product distributions not only depend upon the starting geometry but even the leaving group, suggesting further that none of the reactions involve an intermediate. Results for *p*- and *t*-Co(tren)(NH_3)X^{n+} seem to reinforce this conclusion (ref. 170,171). However, for none of these reactions is the effective site of deprotonation known with certainty, despite proton exchange studies to assess it. It could change from complex to complex, especially if the pK_a/reactivity balance is subtle. Indeed, the balance could be decided by the leaving group. Similar comments apply to the many other elegant studies in this area, and it seems best to confine the discussion here to those few reactions where the site of deprotonation *is* known with certainty.

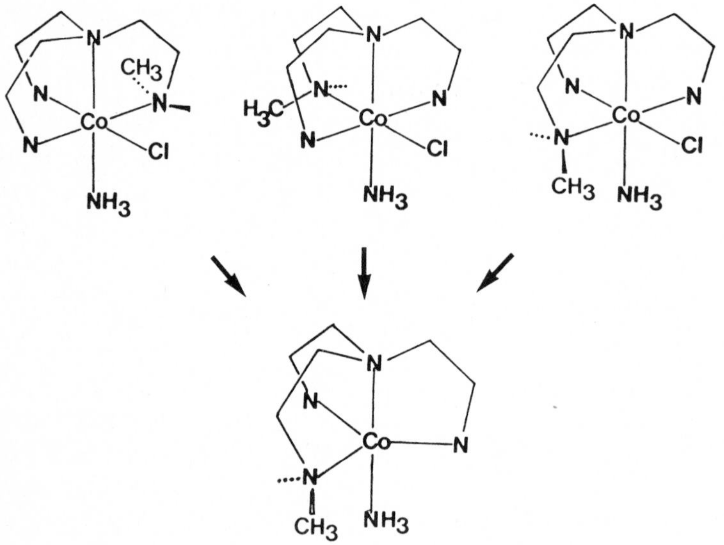

(69)

The work on the Co(tetraen)ONO_2^{2+} base hydrolysis reaction was a sophisticated attempt to test the π-bonding hypothesis, proposed originally by Basolo and Pearson to account for the enormous labilization afforded by aminate ions (ref. 172). Tobe *et al.* have recently reviewed the experimental data bearing on this issue (ref. 28,173), but we need to exhume the tetraen chemistry (ref. 30), first because the reaction (70) is the prototype for the others to be discussed, and second to correct a misinterpretation in respect of the π-bonding hypothesis.

The key feature of the reaction is the inversion at the chiral nitrogen centre adjacent to the leaving group. This has been shown to occur *during* the act of base catalyzed substitution, for both the H_2O and N_3^- entry paths,

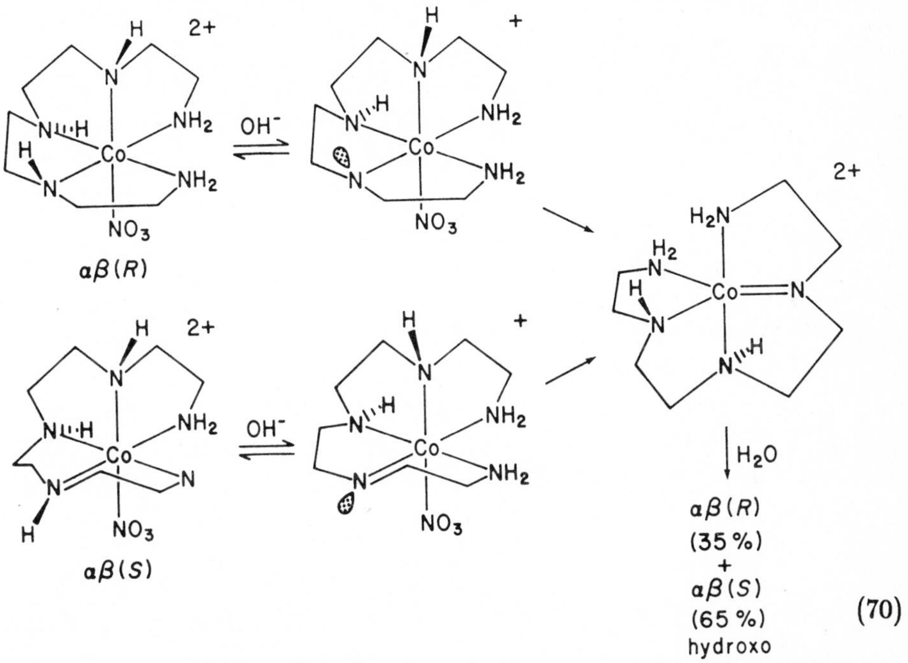

(70)

It is difficult to argue synchronous N-inversion, loss of NO_3^- and entry of H_2O or N_3^-. More reasonable is a sequence of events, and this must involve an intermediate *not* related to reactant or product simply by deprotonation.* N-inversion occurs after NO_3^- is lost but before a nucleophile is added to regenerate the octahedron. A five coordinate species is strongly implicated. Note that this conclusion follows without the need to demonstrate properties independent of its source, although this also is true. The effect of a variation in the leaving group was not studied. However, the diastereomeric substrates (optically resolved), which differ only in the chirality of the N-centre involved in the base hydrolysis reaction, give a common product distribution—an identical total % N_3^- capture for each isomeric reactant and the same (**R**-)/(**S**-)

---

* N-inversion cannot occur in the 6-coordinate aminato precursor complex on the time scale of hydrolysis. This would lead to mutarotation in one or both reactants ($\alpha\beta$(**R**),$\alpha\beta$(**S**)) since H-exchange is $\geq$ 100-fold faster than hydrolysis, and neither reactant nor product mutarotation is observed (ref. 172).

ratios for each of the hydroxo and azido products. There is no overall topological rearrangement, nor loss of chirality.

The results establish unequivocally the effective site of deprotonation. This site also appears to be the most acidic (ref. 30), i.e. the aminato complex generated by deprotonation at this site is the most abundant. It may not be the most reactive, but here this is immaterial since the *effective* centre is known with certainty. However, it has been repeatedly emphasized (ref. 28,173) that a less abundant but especially reactive aminato species can effect the base hydrolysis reaction, despite indications from H-exchange studies about the most abundant and hence the most *likely* effective centre. Ahead we consider an example *known* to be in this category. The alternative explanations for the observed stereochemistry of the base hydrolysis of s-Co(trenen)Cl^{2+}, based on different sites for deprotonation, emphasize the point (ref. 28,30). It is imperative that the site of effective deprotonation is known, and at the moment this can only be established with surety for reactions where N-inversion accompanies base catalyzed substitution.

Other informative features of the stereochemistry of substitution for the $\alpha\beta$-Co(tetraen)ONO_2^{2+} complexes are (i) non-equilibrium (**R**-)/(**S**-) isomer ratios, and (ii) *different* (**R**-)/(**S**-) ratios for the hydroxo (35:65) and azido (77:23) species. Previously (ref. 30) this was interpreted to mean that it was not possible to have two intermediates which differed only in the chirality of the crucial N centre and attendant chelate conformations. However, the different (**R**-)/(**S**-) ratios for N_3^- and H_2O capture can be accommodated provided N_3^- *vs* H_2O attack for each intermediate is different,

$$(\mathbf{R})\text{-Co(tetraen}')^{2+} \ (\rightleftharpoons) \ (\mathbf{S})\text{-Co(tetraen}')^{2+}$$

$$\begin{array}{cccc} N_3^- & H_2O & N_3^- & H_2O \\ \alpha\beta(\mathbf{R})\text{-}N_3^{2+} & \alpha\beta(\mathbf{R})\text{-}OH^{2+} & \alpha\beta(\mathbf{S})\text{-}N_3^{2+} & \alpha\beta(\mathbf{S})\text{-}OH^{2+} \\ x & 1-x & y & 1-y \end{array} \qquad (71)$$

This is reasonable since the intermediates are diastereomeric, not enantiomeric.

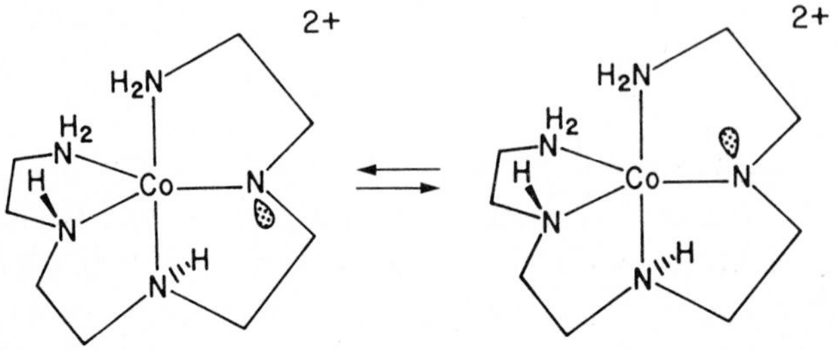

(72)

This explanation has its extreme in a single π-bonded species, shown in (70), which previously (ref. 30) was thought mandatory. However, the results cannot distinguish (70) from two rapidly* equilibrated intermediates (72). If one admits the possibility shown in (70) where the chirality at the fully π-bonded aminato N- centre is lost, an intermediate where π-bonding is almost complete must also be admitted. This automatically implies two intermediates (72) which differ in chirality at this almost flat N-centre.

In conclusion, the point is not so much the question of two rapidly equilibrating intermediates or a single π-bonded species, but the fact that N-inversion is apparently so much faster than the usual rate, $k \sim 10^4\ s^{-1}$. Significant π-bonding, which flattens the sec-N towards its fully π-bonded extreme, seems clear; the N-inversion rate is raised towards ∞ in the latter extreme! Thus whether it is fully π-bonded seems a less important question.

The second cases of N-inversion accompanying base hydrolysis are the Δ-*cis*-β(**R**,**R**)-$Co(trien)Cl(glyOEt)^{2+}$ systems (ref. 174). The effective site of deprotonation, the "flat" sec-NH centre, is clear. The stereochemical outcome is similar to that described for the $\alpha\beta$-$Co(tetraen)ONO_2^{2+}$ diastereoisomers. Different (**R**,**R**)-/(**S**,**R**)- ratios are found for H_2O ** and N_3^- competition. This involves no overall topological change, nor loss of chirality. Unfortunately insufficient of the *cis*-β_2-(**S**,**R**)-$Co(trien)N_3(glyO)^+$ diastereoisomer was obtained to convert (with retention) to the chloro complex. Access to this complex provides a check that the base hydrolysis product distribution is independent of the starting configuration at the crucial N-centre (Δ-*cis*-β_2-(**S**,**R**)- and Δ-*cis*-β_2-(**R**,**R**)).

An intermediate (73) (or two diastereomeric intermediates) similar to the tetraen analog (70) may be proposed. A partly to fully π-bonded sec-N amine in the trigonal plane accommodates the unusually rapid N-inversion process as before. Also, both (70) and (73) have 5-membered chelate arms to apical positions, argued (ref. 28,173) to favour π-bonding. Subsequent attack at position (i), regenerating the *cis*-β_2- configuration (or $\alpha\beta$, in the case of the tetraen analog, (70)), seems preferred to (ii) because of the ring strain in the

---

* It has to be unusually rapid ($k > 10^9$–$10^{10}\ s^{-1}$) since the intermediate is consumed by N_3^- and H_2O at these sort of rates, and the rate of equilibration is not competitive with the rate of consumption (ref. 172). We exclude the independent formation of each intermediate since one path requires synchronous loss of NO_3^- and N-inversion. Also the proportion of the two intermediates would have to be (coincidentally) the same for the $\alpha\beta$(**R**)- and $\alpha\beta$(**S**)-reactants, to accommodate the common product distributions.

** The pendant carboxylate or ester function also competes; this raises problems, considered in ref. 174, but which are not important here.

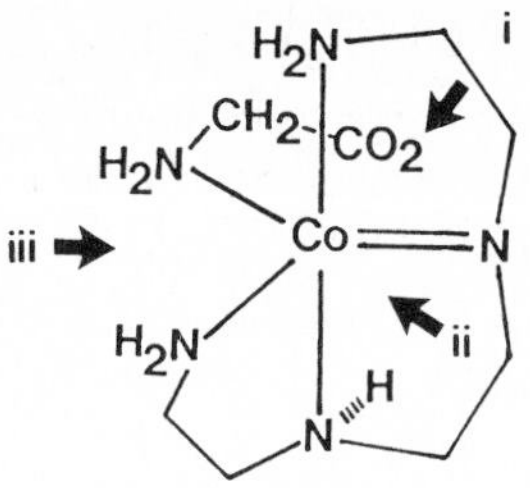

(73)

resultant *trans*- N_4 arrangement of the trien ligand, or trien portion of the ligand in the case of tetraen. Attack at position (iii) is blocked by the additional chelate for tetraen, but will yield the rearranged *cis*-β_1 configuration for the trien complex. Since *cis*-β_1 is not formed, it seems for this reason that the trigonal bipyramid (73) was not proposed originally (ref. 174). Instead a square pyramid was suggested, (where the possibility of L $\rightarrow$ M π-bonding by deprotonated sec-NH centre is lost). However, Comba (ref. 175) has pointed out that a slight distortion can accommodate both difficulties,

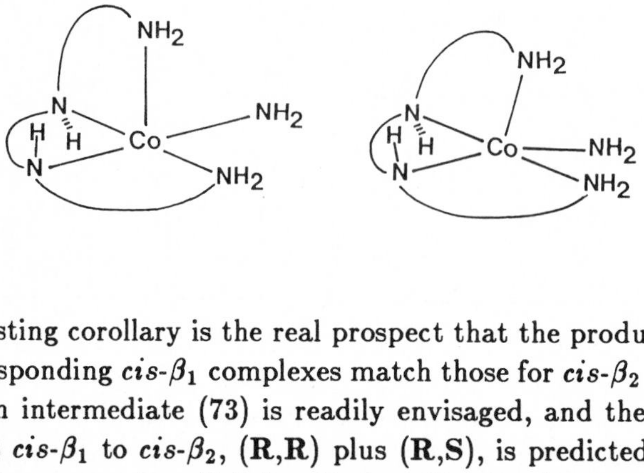

(74)

An interesting corollary is the real prospect that the product distributions for the corresponding *cis*-β_1 complexes match those for *cis*-β_2 reactants, (75). The common intermediate (73) is readily envisaged, and the topological rearrangement *cis*-β_1 to *cis*-β_2, (**R,R**) plus (**R,S**), is predicted. The outcome would seem to comment on Tobe's criterion of amine deprotonation cis to the leaving group (*cis*-β_2) for maximum effectiveness (ref. 28,173), as against the fact that the sec-NH centre in the *cis*-β_1 isomer is trans to Cl^-, known to enhance its acidity. The current debate over the relative importance of the most abundant *vs* the most reactive conjugate base is relevant (ref. 28,170), and it seems likely here that N-inversion during base hydrolysis could at least settle the question of the most *effective* deprotonated centre.

The final example of N-inversion accompanying base hydrolysis is possibly the most informative. Comba, Marty and Zipper (ref. 175–177) have shown that the base hydrolysis of κ-(and κ'-)Co(dien)(dapo)Cl^{2+} (76) is extraordinarily fast (k_{OH}1.3 $\times 10^5$ $M^{-1}s^{-1}$, μ = 1.0 M, 25°C), rivalled amongst

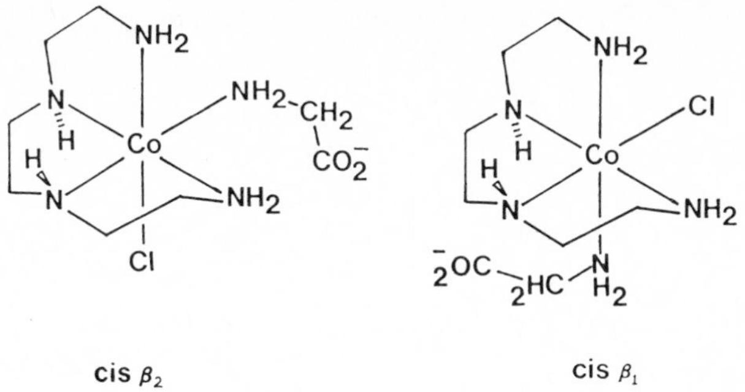

cis β_2 cis β_1 (75)

chloropentaaminecobalt(III) complexes only by the structurally very similar κ-Co(dien)(tn)Cl^{2+} ion (tn = $NH_2(CH_2)_3NH_2$). Interestingly, the stereochemical features fulfil all Tobe's criteria (ref. 28,173) for maximum π-bonding effectiveness of the deprotonated amine centre, reflected in both the rate and, discussed ahead, the stereochemical outcome.

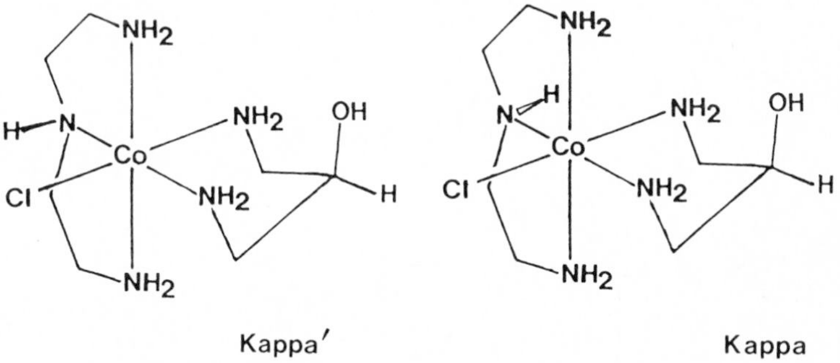

Kappa′ Kappa (76)

We do not take up the rate issue here—it is neither relevant nor straightforward, as emphasized by data for stereochemically related pyridine based complexes having comparable reactivity to the complexes shown in (76) but which have different π-bonding requirements and seemingly unrelated stereochemical outcomes (ref. 28,133,178).

An important difference between the dapo and tn species is that the former are chiral, although there are no formal chiral centres. This elegant conception leads to clear stereochemical distinctions between possible intermediates. For the logical intermediate,

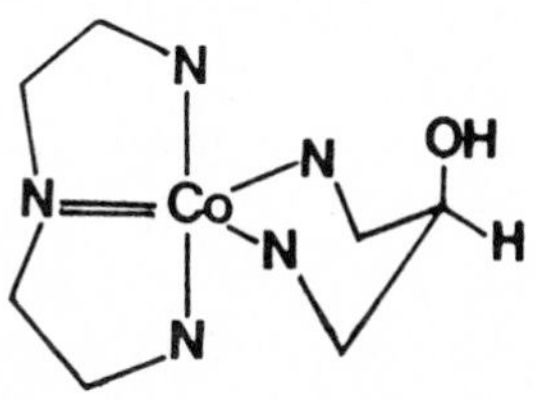

(77)

the best opportunities for π-bonding seem to exist. The structural features associated with the deprotonated sec-NH centre are very similar to those for the tetraen and trien complexes considered above. In addition, the six membered chelate ring in the plane of the idealized trigonal bipyramidal intermediate seems to confer the optimal bond angle ($\sim$ 120°). Finally, and the unique feature, *intermediate* (77), *and no other*, is achiral. As before, two rapidly equilibrated intermediates (78) are effectively equivalent to (77), and it is noted that these are enantiomers rather than diastereomers.

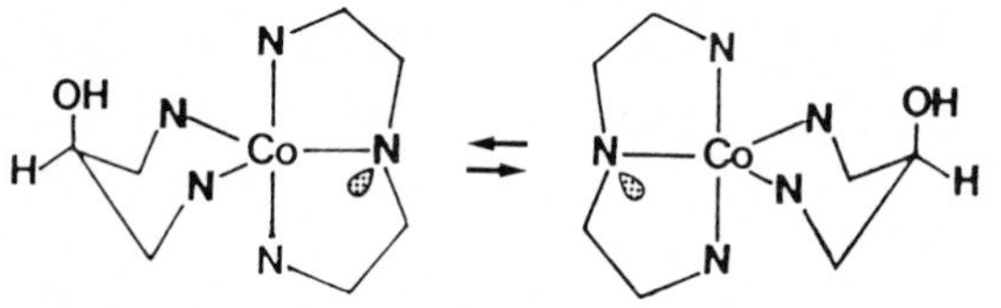

(78)

The first investigations of the Co(dien)(dapo)Cl^{2+} system were clouded by a difficulty associated with coordination of the pendant-OH group of the dien ligand (ref. 176). This problem has now been solved (ref. 175,177), justifying the early choice of the readily available dapo over an exotic, e.g. $NH_2CH_2CH(CH_3)CH_2NH_2$, which for the purpose is stereochemically equivalent. The use of the cheaper dapo ligand was demanded by the need for large scale preparations to extract the thermodynamically unstable κ- isomer, in minor abundance, from isomeric mixtures. Also there was a need to characterize *all* possible Co(dien)(dapo)Cl^{2+} isomers, demanding on ligand supply, and this was a major feat now essentially achieved (ref. 175).

The remarkable results for the steric course of base catalyzed substitution of κ-Co(dien)(dapo)Cl^{2+} have vindicated the effort (ref. 175–177). The observed hydroxo product is racemic, but this could be because κ- (and κ'-) Co(dien)(dapo)OH^{2+} subsequently racemize very rapidly (which they do; ref. 175). However, the products of N_3^- competition, which are also racemic, *must* arise as a *direct* result of substitution, since κ- and κ'-Co(dien)(dapo)N_3^{2+} are each optically stable under the conditions (ref. 177). Marty and Zipper have obtained the optically pure κ'-Co(dien)(dapo)N_3^{2+} isomer, by classical resolution methods, from κ/κ' mixtures. Furthermore, retentive conversion ($HCl/NaNO_2$) to κ'-Co(dien)(dapo)Cl^{2+} has enabled the demonstration that the same racemic κ/κ' (69:31)* azido product distribution obtains in base hydrolysis. Also, the extents of N_3^- capture are identical ($\sim$ 76% for 1.0 M N_3^-).

---

* This is a non-equilibrium distribution, found in separate experiments to be 81% κ, 19% κ'.

It transpires that the rates of base hydrolysis for κ and κ'-Co(dien)(dapo)Cl^{2+} are the same (although the sec-NH exchange rates are different) (ref. 177). This could suggest that the κ- and κ'- isomers are rapidly pre-equilibrated, (79), accommodating both the identical rates and the common product distribution. However, rapid reactant pre-mutarotation retains the chiral integrity of the κ and κ'- isomers, and cannot accommodate the observation of racemic κ- and κ'- azido competition products.

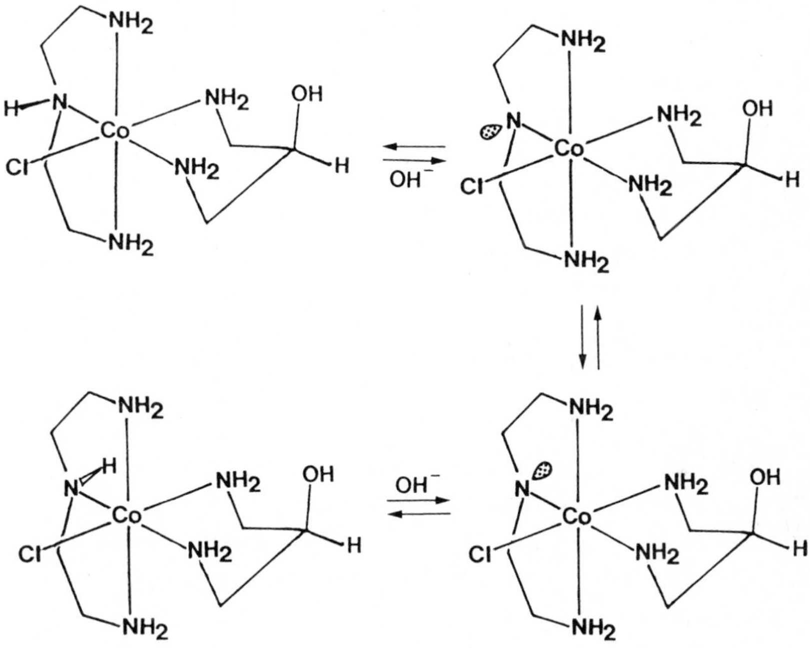

(79)

Comparable rates for the very similar κ- and κ'- isomers are not unexpected. However, for the base hydrolysis of κ- and κ'-Co(dien)(dapo)N_3^{2+}, the rates *are* measurably different, certainly precluding rapid $\kappa \rightleftharpoons \kappa'$ interconversion in the case of the azido complexes (ref. 177). Moreover, the key experiments for partially base hydrolyzed (racemic) κ- and κ'-Co(dien)(dapo)Cl^{2+} show that neither reactant is appreciably mutarotated, a result also excluding rapid $\kappa \rightleftharpoons \kappa'$ reactant preisomerization for the chloro species (ref. 179). The little mutarotation that is observed* is accommodated by Cl^- re-entry from the (achiral) intermediate, since (H_2O dissociated) $ZnCl_4^{2-}$ salts were used and the intermediate is highly discriminatory (i.e. for Cl^-—see ahead).

---

* The κ and κ'- isomers are separated readily by ion-exchange chromatography.

The confirmation of this inference awaits results for the active κ- and κ'-Co(dien)(dapo)Cl^{2+} species; any observed mutarotated product should be found racemic, and, depending upon the free $[Cl^-]$ attained, the recovered reactant could be found partly racemized.

The site of the effective deprotonation (sec-NH centre) in base hydrolysis is established by the observation of racemization at this centre. It is important to note that this is *not* the most acidic amine in κ- or κ'-Co(dien)(dapo)Cl^{2+}, but rather it is that trans to Cl^- (H-exchange studies).* Thus base hydrolysis proceeds via a less abundant but more reactive conjugate base, and this is one of the few cases for which this has been established.

In summary, there are at least three examples of base catalyzed substitution which involve N-inversion, and where the site of deprotonation is identified.** Moreover, here the case for penta coordinate intermediates seem strong. Appreciable π-bonding in the intermediate accommodates unusually rapid N-inversion, while the proposed geometry for the intermediate is consistent with the observed topological retention. More examples of N-inversion accompanying substitution are met in section 8.3, which reinforce these conclusions.

It has been suggested that rapid N-inversion could arise from e^- transfer, yielding a transient Co(II)-$\dot{N}\!\!\!\!\!\!\!\!\!\!\!\begin{smallmatrix}/\\ \backslash\end{smallmatrix}$ moiety for the intermediate (ref. 174). While a radical N-centre such as this may well be "flat", or sufficiently close to sp^2 hybridized N to accommodate very rapid N-inversion, its formation from Co(III)-$\ddot{N}\begin{smallmatrix}/\\ \backslash\end{smallmatrix}$ seems to be energetically prohibitive, as Comba (ref. 175) has cogently argued. We also note that the proposal of high spin Co(III) (ref. 46), which might accommodate the reactivity of the conjugate base, does not reasonably explain the enhanced rate of N-inversion. For high spin d^6 cobalt(III), the π-accepting orbitals on the metal are partly occupied, and less conducive to L $\rightarrow$ M π-bonding.

---

* These studies (ref. 175,177) show also that base hydrolysis has approximately the same rate as the fastest exchanging NH centre. However, it is not a case of rate limiting deprotonation for base hydrolysis, since the *effective* NH centre exchanges $\sim$ 10-fold more slowly.

** All these sites are cis to the leaving group. Trans-aminate ion labilization has been *demonstrated* for only one complex, *trans*-Co([14]-tetraene-N_4)$(NH_3)_2^{3+}$ (ref. 180). However, the popular view that cis deprotonation is more effective is not proven (ref. 28,162), although the evidence, all circumstantial, supports it. It cannot be *generally* true, since the X-ray structure of a Co(III)N_6 cage complex, deprotonated at one of the secondary amine centres, reveals a significant lengthening of the Co-N bond *trans* to this group (ref. 181).

8.3 Internal conjugate base reactions

The labilizing effect of deprotonated amine groups (aminate ions) coordinated to cobalt(III) and certain other metal ions is now well established. For complexes having more than one kind of amine proton the possibility exists for tautomerism within the conjugate base, and we saw in the last section the problem of ascertaining the balance of relative reactivities and relative abundances here.

The possibility exists also for hydroxo-amine complexes such as $Co(en)_2(OH)Cl^+$ to hydrolyze via the tautomeric $Co(en)(en\text{-}H)(OH_2)Cl^+$ complex. Superficially this is not an attractive proposition because the estimated relative abundances of the two tautomers require an extreme reactivity for the aminato form. However, where OH^- is the apparent leaving group in a substitution process, e.g.

$$(amine)_5CoOH^{2+} + N_3^- \xrightarrow{H_2O} (amine)_5CoN_3^{2+} + OH^- \qquad (80)$$

the credibility of an internal conjugate base pathway is boosted at least ~ 1000-fold, because H_2O is established to be a better leaving group than OH^- by this margin and the aminato tautomer provides H_2O as a leaving group.

The concept of an internal conjugate base substitution pathway dates at least to a paper by Kruse and Taube in 1961, where for $Co(en)_2(OH)_2^+$ the possibility was alluded to (ref. 119). Later Tobe *et al.* (ref. 182) suggested that *trans*-Co((**R**,S,S,**R**)-cyclam) $(OH)Cl^+$ could react via the *trans*-(cyclam′) $(OH_2)Cl^{+*}$ cation to explain why, in spontaneous hydrolysis, this was the only complex more reactive than its $Co(en)_2AX^{n+}$ counterpart. Also, the *cis*-(**R**,**R**,**R**,**R**)-Co(cyclam)$(OH)(OH_2)^{2+}$ to *trans*-(**R**,S,S,**R**)-Co(cyclam)$(OH)(OH_2)^{2+}$ rearrangement, which involves inversion at two of the four sec-NH centres, has been discussed in terms of an internal conjugate base (ref. 183). Similarly, the rapid ring closure or anation reactions of *cis*-β_2-Co(trien)(OH)$(NH_2CH_2CO_2R)$ (ref. 174) and *cis*-β_2-Co(trien)$(NH_3)OH^{2+}$ (ref. 184) could involve the rapid loss of OH^- via the β_2-Co(trien′)(amine) OH_2^{2+} conjugate base. Buckingham, Marzilli and Sargeson appear to be the first to have actually demonstrated the internal conjugate base reaction, although not referring to it as such, for the remarkable single step rearrangement of *trans*-(S,S)-Co(trien)$(OH_2)_2^{3+}$ to optically pure **Λ**-*cis*-β-(S,**R**)-Co(trien)$(OH_2)_2^{3+}$ in acid solution (ref. 185). This occurs at a rate

---

* cyclam = 1,4,8,11-tetraazacyclotetradecane; the prime denotes deprotonation at one of the four secondary amine centres.

inversely proportional to $[H^+]$, implying *trans*-(S,S)-Co(trien)(OH)$(OH_2)^{2+}$ as the precursor to the reactive species, and the inverted N-centre in the product testifies to the intervention of the internal conjugate base *trans*-(S,S)-Co(trien')$(OH_2)_2^{3+}$. The original account (ref. 185) has been revamped along these lines (ref. 30).

Stereochemistry obviously plays an important role in identifying such processes. A clear requirement is NH exchange during the act of substitution, possibly accompanied by N-inversion. Also, for simpler complexes such as $Co(en)_2(OH)X^{n+}$, the topological changes for normal (spontaneous) hydrolysis, as opposed to substitution via an internal conjugate base, are expected to be quite different (ref. 123). We have seen, for example, that the cis product is invariably fully active by the former route and partly racemized by the latter.

We believe internal conjugate base processes could be more prevalent in cobalt(III) and possibly Ru(III) chemistry than presently acknowledged. For this reason we introduce formal rate considerations which identify the kinetic criteria for their observation, and go on to further examples which appear to meet these criteria.

An aqua-amine metal ion complex has of course two kinds of proton acidity, albeit only the one is usually measurable. The following example is assumed representative of the general case,

$$H_3N-\overset{\overset{X}{|}}{\underset{|}{Co}}-OH_2^{3+} \overset{K_{a_1}}{\rightleftharpoons} H_3N-\overset{\overset{X}{|}}{\underset{|}{Co}}-OH^{2+} + H^+ \quad \text{(CB1)}$$

$$H_3N-\overset{\overset{X}{|}}{\underset{|}{Co}}-OH_2^{3+} \overset{K_{a_2}}{\rightleftharpoons} H_2N-\overset{\overset{X}{|}}{\underset{|}{Co}}-OH_2^{2+} + H^+ \quad \text{(CB2)}$$

$$(81)$$

Known pK_{a_1} ($\sim$ 6) (ref. 44) and estimated pKa_{a_2} ($\sim$ 15) values (ref. 43) lead to a value for the equilibrium constant governing the formation of the internal conjugate base,

$$K_3 = \frac{[H_2N\text{-}Co\text{-}OH_2^{2+}]}{[H_3N\text{-}Co\text{-}OH^{2+}]} = \frac{K_{a_2}}{K_{a_1}} \simeq 10^{-9} \qquad (82)$$

To some extent the metal ion, overall charge, amine substituents and the presence of other ligands, all of which affect individual pK_a values, are not critical variables for K_3 because these are common factors. The scheme below,

$$\begin{array}{ccc} & K_3 & \\ CB1 & \rightleftharpoons & CB2 \\ k_1 \searrow & & k_2 \swarrow \\ & \text{Products} & \end{array} \tag{83}$$

leads to the expression,

$$\%\text{reaction via CB2} = \frac{10^2(k_2/k_1)K_3}{1 + (k_2/k_1)K_3} \tag{84}$$

Hence, given $K_3 \simeq 10^{-9}$, appreciable reaction (say, > 50%) via the CB2 (in minuscule abundance) requires $k_2/k_1 > 10^9$. We contend that there are several systems where k_2/k_1 *is* greater than this. The Co(trien)(OH)OH_2^{2+} and Co((**R**,**S**,**S**,**R**)-cyclam)(OH)Cl^+ systems referred to above give an indication of the magnitude of k_2/k_1. Using an estimated pK_{a_2} for an amine proton of 15, relative rates of spontaneous aquation and base hydrolysis (k_{OH}/k_S) (ref. 28), and the relation $k_{OH} = k_2K_{a_2}$ for base hydrolysis (ref. 28,30), the enhancement in reactivity merely through amine deprotonation can be gauged. Using data for suitable Co(quadridentate amine)Cl^+ complexes as the basis for the calculation (ref. 28,186), k_2/k_1 comes to $\sim 10^{11} - 10^{13}$. Clearly the proposition has become very reasonable. We should add that k_1 is generally $\simeq 10^{-5}$ sec^{-1}, and thus the required magnitude of k_2 ($\geq 10^4$ sec^{-1}) does not approach the diffusion controlled limit ($\sim 10^{11}$ sec^{-1})* Also, recognizing that base hydrolysis of the complexes in question approach the limiting case of rate determining deprotonation (ref. 28), precisely because the aminato conjugate base is extraordinarily reactive, k_2 values estimated (incorrectly) from the non-limiting relation $k_{OH} = k_2K_{a_2}$ are in fact *underestimated.*

There are a number of multidentate ligand systems for which cobalt(III) complexes show a marked sensitivity towards base catalyzed reaction. The sensitivity appears to be maximized if the deprotonated amine centre is cis to the leaving group *and* can occupy the trigonal plane of a real or incipient trigonal bipyramidal intermediate with the right orientation of filled p-orbital on the N-centre in question. Tobe and Henderson have given a comprehensive discussion of the experimental data supporting these generalizations (ref. 173), and without lingering on the details, we note that here lie the best prospects for Co(tetraamine)(OH)X complexes hydrolyzing via the internal

---

* For other metal ion systems, much larger k_1 values could mean that k_2/k_1 is restricted by the upper diffusion controlled limited on k_2.

conjugate base with loss of X. The *trans*-(S,S)-Co(trien)(OH)OH_2^{2+} example given earlier fulfills all these criteria (ref. 30,185), but it remains the only established one. Since the required rate enhancements by amine proton abstraction are possible without fulfilling all of Tobe and Henderson's criteria (ref. 173), the stereochemical outcome of internal conjugate base reactions seems to represent another area for useful research activity.

The analysis above does not take account of the change from OH^- to H_2O which can affect the rate of loss of X. Also the site of deprotonation is not explicitly accommodated. The ratio k_2/k_1 was estimated just for the effect of deprotonation of an amine centre. There is evidence to suggest that OH^- is inherently* more labilizing than H_2O by a factor of $\sim 10^2$ whether cis or trans to the leaving group (ref. 187). Taking this into account, the estimated ratio of reactivities (k_2/k_1) of (amine')Co(OH_2)X and (amine)Co(OH)X need be reduced by 10^2, after calculating first the effect of amine deprotonation.

Hydroxo-pentaaminecobalt(III) or CoN_4(A)OH** systems offer additional advantages. As before, rate enhancements of at least 10^{11} can arise for amine deprotonation, but since OH^- is now the leaving group its inherent labilizing effect for CB1 over CB2 ($\sim 10^2$) is lost. This coupled to the inherent leaving group advantage of H_2O (CB2) over OH^- (CB1) has the effect of raising k_2/k_1 by 10^5.

There are several systems in this category where displacement is apparently faster than that for H_2O from an otherwise identical complex, and which suggest reaction via an internal conjugate base. However, some of these are isomerization reactions which may be *intramolecular*, i.e. they may not involve OH^- dissociation, and we need dispose of these first. The racemization of the complex (+)-Co(EN3A)OH_2*** follows the unusual rate law $k_r(\text{obsd}) = k/[H^+]$ in the pH region below the pK_a of the aqua ion (ref. 188). It appears (+)-Co(EN3A)OH^{2+} racemizes much more rapidly, and in the absence of ^{18}O exchange data an internal conjugate base substitution process may be proposed. There are several isomeric possibilities on route to racemate, but the intervention of the achiral intermediate shown in (85), along the line to products, accommodates the observation.

Further examples are $Co(en)_2A(OH)^{n+}$ where A = NH_3, NO_2^-, and CN^-, where cis/trans isomerization is rather slow, although faster than the corresponding process for the aqua complex (or of comparable rate; A = NO_2^-) (ref. 125,130,189). This is not to say that the aqua group is not more labile than OH^-. For *trans*-$Co(en)_2(NO_2)OH_2^{2+}$ for example, this is demonstrably not

* Other than by internal conjugate base formation!

** A is a non-displaced group other than an amine.

*** EN3A = $(^-O_2CCH_2)_2NCH_2CH_2NHCH_2CO_2^-$.

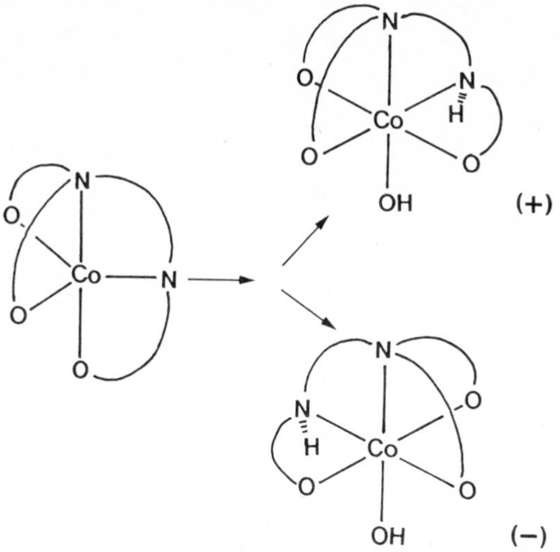

$$(85)$$

so; the H_2O group is exchanged or otherwise substituted in a matter of seconds, the OH^- group much more slowly (ref. 112). The key point is that the substitution reactions of complexes having $A = NH_3$, CN^- or NO_2^- are highly stereoretentive, and cis/trans rearrangements are thus very much slower. Indeed, the isomerization and racemization reactions of these complexes are amongst the slowest in $Co(en)_2AX^{n+}$ chemistry, and for this reason we believe the rates for the normal substitution processes are "undercut" by lower energy non-dissociative intramolecular pathways, not normally encountered in cobalt(III) chemistry. The isomerization of *trans*-$Co(en)_2(NH_3)OH^{2+}$ is certainly in this class, since in $H_2{}^{18}O$ the *cis*-$Co(en)_2(NH_3)OH^{2+}$ product contains no ^{18}O label (ref. 125). The corresponding isomerization reactions of the cyano and nitro complexes may be proven the same if high pH conditions can prevent ^{18}O incorporation into the reactant *trans*-$Co(en)_2(CN)(OH)^{2+}$ and *trans*-$Co(en)_2(NO_2)OH^{2+}$ complexes via cyano-aqua and especially the very labile *trans*-$Co(en)_2(NO_2)OH_2^{2+}$ ions, although in low abundance.

Intramolecular isomerization processes also seem to operate for $Co(en)_2(OH)_2^+$ (ref. 39,113). *Cis*-$Co(en)_2(OH)_2^+$ isomerizes to the trans isomer in $H_2{}^{18}O$ with less than one equivalent of ^{18}O incorporation. Also, Λ-*cis*-$Co(en)_2(OH)_2^+$ racemizes somewhat faster than cis to trans isomerization; this is not the normal stereochemical feature of a spontaneous substitution, but it is for base catalyzed substitution. This raises the possibility that internal conjugate base formation could accelerate intramolecular rearrangements as well as substitution processes. Certainly aminate ion does catalyze linkage isomerization reactions, e.g. of $(NH_3)_5CoSCN^{2+}$ and $(NH_3)_5CoONO^{2+}$, and these are known to be intramolecular (ref. 30). However, base also catalyzes the ***substitution*** of SCN^- or NO_2^-, which is lost irreversibly by this path.

Moreover, it seems generally true that catalysis of the substitution path is much greater than that for the intramolecular rearrangement (ref. 62). On this basis we could probably rule out intramolecular rearrangement of Λ-*cis*-$Co(en)_2(OH)_2^+$ via its internal conjugate base $Co(en)(en\text{-}H)(OH)OH_2^+$.

It is worth noting that, in principle, the acceleration of non-dissociative (intramolecular) paths by coordinated aminate ion *can* be experimentally tested. It requires the demonstration of no exchange for any potential leaving group *and* the demonstration of NH exchange during the act of rearrangement. However, since NH exchange is often a faster process (ref. 28), this will generally be difficult unless deprotonation occurs at a chiral nitrogen and it can be shown that N-inversion accompanies the rearrangement. An inherent acceleration of the intramolecular path by bound OH^- also merits consideration, but this seems very difficult to test experimentally. At present there seems to be no strong evidence in its favour, but the possibility need be borne in mind.

Returning to the $Co(en)_2(OH)_2^+$ reactions, we note that the normal pattern of reactivity obtains, *viz* the aqua group of $Co(en)_2(OH)OH_2^{2+}$ is faster to leave than OH^- from the corresponding hydroxo complex $Co(en)_2(OH)_2^+$. This is a valid comparison* since the inherently labilizing group OH^- is common to both species. Thus kinetic considerations lead to the conclusion that the labilization due to internal conjugate base formation, giving $Co(en)(en\text{-}H)(OH)OH_2^{2+}$ from $Co(en)_2(OH)_2^+$ and $Co(en)(en\text{-}H)(OH_2)_2^{2+}$ from $Co(en)_2(OH)OH_2^{2+}$, is insufficient to offset their very low abundance. This is predicted by calculations of k_2/k_1 (83), and supported experimentally by the stereochemical results for the spontaneous substitution reactions of $Co(en)_2(OH)X^{n+}$ (section 7.1)—the steric course is normal in the sense that, for loss of X, the cis product is fully active. Also the steric course is constant for each of Λ-*cis*- and *trans*-$Co(en)_2(OH)X^{n+}$, implying reaction via the usual CB1, (81).** In summary, the stereochemistry of *substitution* for all $Co(en)_2(OH)X^{n+}$ complexes is normal, but when the poor leaving group OH^- is involved the very slow *rearrangements* which occur can find alternate routes to a normal substitution mechanism, much like $Co(en)_2A(OH)^+$ ($A = NO_2^-$, NH_3, CN^-) (ref. 125,130,189). Since the effect of bound OH^- is not specific to its position (cis or trans to the leaving group, ref. 187), and also since coordinated OH^- promotes steric change in substitution (ref. 30), we

* The comparison $Co(en)_2(OH)OH_2^{2+}/Co(en)_2(OH_2)_2^{3+}$ leads to the reverse conclusion. However, the aqua group in $Co(en)_2(OH)(OH_2)^{2+}$ is activated by OH^-, and not vice versa, and hence the superior reactivity.

** *Trans*-$Co(en)_2(OH)Cl^+$ shows no evidence of NH exchange in D_2O at 0°C for 10 min. However, these conditions are insufficient to hydrolyze much of the complex (ref. 190).

add that $Co(en)_2(OH)_2^+$ should rearrange by ***substitution of OH***$^-$ more readily than $Co(en)_2(NO_2)OH^{2+}$ or $Co(en)_2(NH_3)OH^{2+}$. This appears to be the case, although there remains a non-dissociative intramolecular component in the Λ-*cis*-$Co(en)_2(OH)_2^+$ isomerization reactions (ref. 39,119).

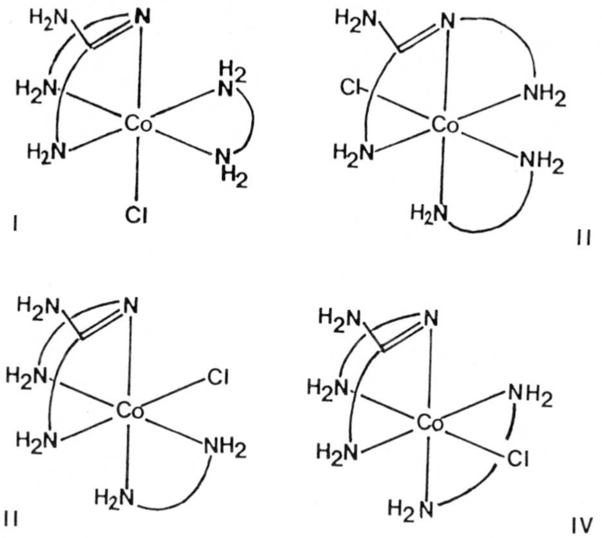

(86)

Buckingham *et al.* (ref. 191) have reported the synthesis, structure and base hydrolysis of the optically resolved isomers I,II, (86), two of the four possible geometric forms of $Co(en)(NH_2(CH_2)N{=}C(NH_2)CH_2NH_2)Cl^{2+}$, all of which are chiral. For two of these diastereomers, their enantiomers are generated by interchanging the two arms of the facially (I) or meridionally (II) coordinated unsymmetrical tridentate ligand. This system is interesting for three reasons. First, a limiting rate of base hydrolysis for I is observed at high $[OH^-]$, consistent with complete deprotonation of the exo-NH_2 moiety of the amidine group (pK_a ~13.0). This of course does not imply that base hydrolysis proceeds via this most abundant conjugate base. Several alternatives exist (the "proton anomaly", ref. 28). Second, base hydrolysis of II is stereoretentive while I leads to a net inversion of configuration about cobalt.* There is no racemization in either reaction, shown in (87).

Note that the "inversion" arises because of the way the chirality is designated. Certainly the chiro optical properties of I-OH and II-OH are inverted, but if one considers the following reaction of the structurally similar but achi-

* The steric course is not known with certainty since III- or IV-OH could be amongst first formed products and decay rapidly to II-OH. The properties of III- and IV-OH are unknown.

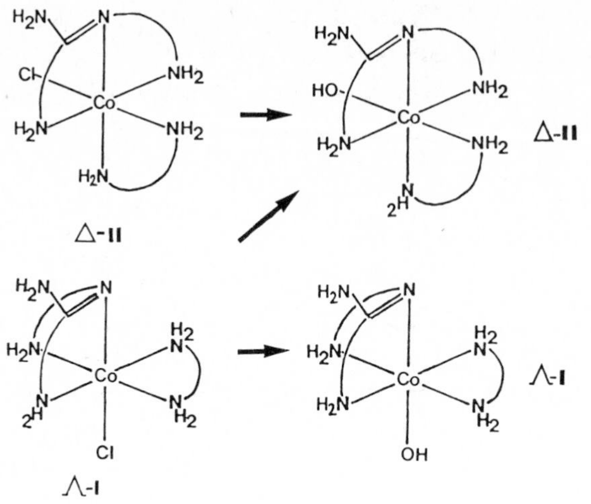

(87)

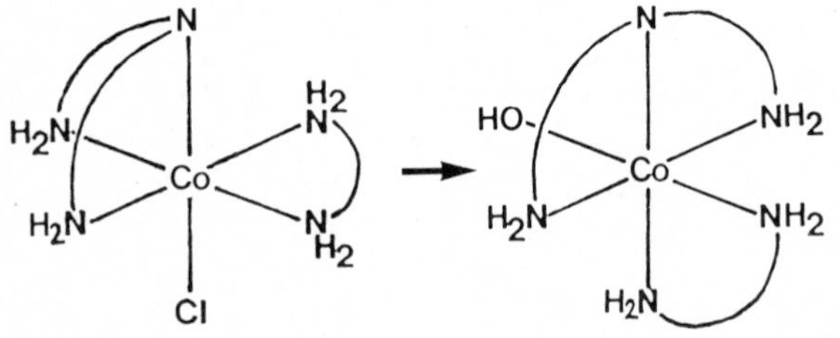

(88)

ral analog, (88), it is clear that this is not inversion but geometric isomerization which accompanies the hydrolysis. The label "inversion" for (87) arises merely because the tridentate ligand is unsymmetrical, and this example amplifies the point made in the Introduction on the distinction between inversion and geometric isomerization reactions. The third reason for the interest in reaction (87), especially relevant to this section, is the report that I-OH^{2+} rearranges to II-OH^{2+} much more rapidly than the corresponding aqua complex (ref. 191). In the absence of exchange data, a dissociative process via an internal conjugate base (deprotonated at an amine or amidine centre), or an intramolecular rearrangement, cannot be distinguished. The base hydrolysis rate constants for I-Cl^{2+} and II-Cl^{2+} are not abnormally large, a fact militating against the internal conjugate base route. The strain associated with the facially coordinated amidine ligand may be a driving force for an intramolecular rearrangement via usual mechanisms, and it is interesting that the ligand strain aspect does not show up in the k_{OH} values for the dissociative reactions of I-Cl^{2+} and II-Cl^{2+}—in fact, the more strained isomer I is the less reactive.

We move to the Co(dien)(dapo)X^{n+} systems which seem the more clear cut. Anwander and Marty (ref. 192) first noted the unusual lability of the κ isomer of Co(dien)(dapo)OH^{2+}. It racemizes with a half-life of ~ 18 sec (25°C), compared to hours for the corresponding aqua derivative. More significantly, it is anated by N_3^- on the second timescale, and unlike the racemization reaction, this is clearly a substitution process. Again, the aqua complex is much less reactive. This is now known to be true for a variety of nucleophiles Cl^-, NCS^-, OAc^-, and NO_2^- (ref. 193).

We need to recall relevant facts recorded in section 8.2. Base hydrolysis of optically resolved κ- and κ'-Co(dien)(dapo)Cl^{2+} in N_3^- media give both κ- and κ'- azido and hydroxo products; all products are racemic, and the κ/κ' product ratio for at least the azido products, as well as the % total azido product, is independent of starting material. Also, N_3^- competes very effectively with H_2O. The intermediate or equilibrated intermediates (89) accommodate these observations.

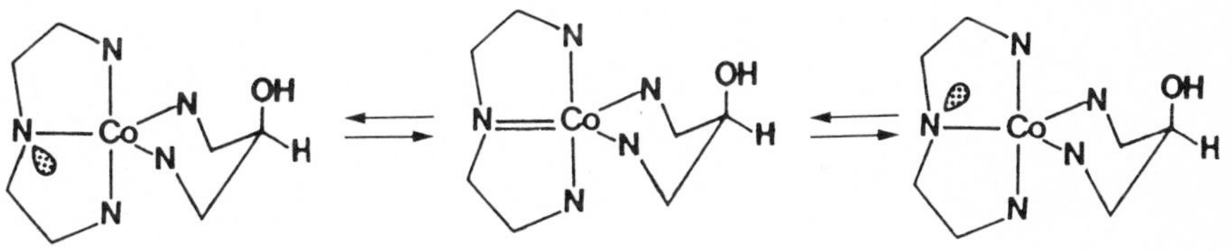

(89)

Reactions of κ- or κ'-Co(dien)(dapo)OH^{2+} which occur exclusively via the respective internal conjugate bases κ- or κ'-Co(dien-H)(dapo)OH_2^{2+} therefore are open to several checks, the collection of which can be viewed as a stringent test of the proposal of a common intermediate. First, N_3^- anation should give the same κ/κ' isomer ratio (kinetic) found for Co(dien)(dapo)Cl^{2+}, independent of starting configuration (κ or κ'). Second, if possible to determine, the extent of capture of N_3^- (*vs* H_2O) should match that for the chloro complexes, and again this should be independent of the κ and κ' reactant geometries. Finally, the limiting rates of N_3^- anation should match the respective rates of racemization of the κ- and κ'-OH^{2+} species. Implicit in this last prediction is curvature in the plot of the observed N_3^- anation rate constant *vs* $[N_3^-]$, because of the unusually high efficiency of N_3^- capture found for κ- and κ'-Co(dien)(dapo)Cl^{2+} (ref. 175–177).

The more important of the above predictions have been experimentally tested (ref. 193). First, azide ion anation of κ- and κ'-Co(dien)(dapo)OH^{2+} yields the common 69% κ-, 31% κ'-Co(dien)(dapo)N_3^{2+} product distribution found for Co(dien)(dapo)Cl^{2+} + OH^-. Second, the rate of N_3^- anation for each of κ- and κ'-Co(dien)(dapo)OH^{2+} shows a marked non-linear dependence on $[N_3^-]$; the rate "flattens" as $[N_3^-]$ is raised. Analysis of the data according to

$$k(\text{obsd}) = \frac{k_1 K[N_3^-]}{1 + K[N_3^-]} \tag{90}$$

yields $k_1 = 4 \times 10^{-2}$ sec^{-1} * and $K = 3.2$ M^{-1}. At least two chemically reasonable mechanisms can accommodate this rate law. For the ion-pair mechanism,

$$CoOH^{2+} + N_3^- \overset{K_{IP}}{\rightleftharpoons} CoOH^{2+},N_3^- \overset{k_1}{\longrightarrow} CoN_3^{2+} \tag{91}$$

the K of (90) must be identified with K_{IP} of (91). The observed value is far too large for a 2+/1− ion-pair interaction at μ =1.0 M.

$$CoOH \underset{k_{-1}}{\overset{k_1}{\rightleftharpoons}} I \underset{N_3^-}{\overset{k_2}{\longrightarrow}} CoN_3 \tag{92}$$

The second mechanism (92) is the internal conjugate base process, and here K is identified with k_2/k_{-1}, which is the conventional R factor for anion competition experiments,

$$CoCl^{2+} \xrightarrow{OH^-} I \begin{cases} \xrightarrow{k_2[N_3^-]} CoN_3^{2+} \\ \xrightarrow[k_{-1}]{H_2O} CoOH^{2+} \end{cases}$$

$$R = \frac{[CoN_3]}{[CoOH][N_3]} = \frac{k_2}{k_{-1}}. \tag{93}$$

The value of k_2/k_{-1} derived kinetically corresponds to ~ 80% N_3^- capture at 1 M N_3^- for both κ- and κ'- substrates, the same as that found (ref. 175–177) for the corresponding Co(dien)(dapo)Cl^{2+} isomers. This impressive agreement between R values, derived in one case kinetically and in the other by direct measurement, strongly supports the case for the common

* The κ- and κ'- isomers show, coincidentally, the same reactivity. The mechanism (92) *requires* a common K value.

intermediate (89) in the OH^- catalyzed hydrolysis reactions of κ- and κ'-$Co(dien)(dapo)Cl^{2+}$ and internal conjugate base anation reactions of κ- and κ'-$Co(dien)(dapo)OH^{2+}$. It remains to measure the rates of racemization (k_r) of both κ- *and* κ'-$Co(dien)(dapo)OH^{2+}$; the proposed mechanism (92) requires $k_1 = k_{rac}$ for each isomer, providing an acid test of the analysis. Also the racemic nature of the separated* κ- and κ'-$Co(dien)(dapo)N_3^{2+}$ products needs to be verified.

In summary, internal conjugate base processes for κ- and κ'-$Co(dien)(dapo)OH^{2+}$ seem secure. Furthermore, the case for the coordinatively unsaturated intermediate seems strong. In view of the general attack on the proposition of genuine reduced coordination number intermediates in cobalt(III) substitution reactions (ref. 29,58–60,171), the latter is an extremely important result. The extraordinarily high selectivity of the intermediate suggests value in extending competition studies to include nucleophiles additional to N_3^-, for comparison with selectivity scales for $[Co(CN)_4(SO_3)]^{3-}$, for example. Already imidazole has been found to be an effective nucleophile, and this is highly unusual pentaaminecobalt(III) chemistry which is being pursued (ref. 193).

8.4 Solvent effects

Little is yet known of the effect of solvent on the stereochemistry of base catalyzed substitution. Long ago Basolo and Pearson examined the anation of $Co(en)_2NO_2Cl^+$ in DMSO containing OH^- and NO_2^- (ref. 64). The product was shown to be *trans*-$Co(en)_2(NO_2)_2^+$ rather than $Co(en)_2(NO_2)OH^+$ (the latter ion is unreactive to anation by NO_2^- under the conditions). This and similar experiments for Rh(III) by Panunzi and Basolo (ref. 65) constituted a proof for the role of OH^- as a base rather than a nucleophile. For the purpose of examining the role of solvent on *stereochemical* outcome, the retentive Rh(III) and nitro-Co(III) reactions are far from optimal, yet little has been published since for viable alternatives.

Substitution reactions in liquid NH_3, with and without added NH_4^+ ion, proceed entirely by the base catalyzed pathway,** arising from NH_2^- or NH_3 catalysis depending upon $[NH_4^+]$ (ref. 67). For $Co(en)_2AX^{n+}$ ions, for example, the steric course however should be independent of the general base, since the reactive species is the common product determining conjugate base $Co(en)(en\text{-}H)AX^{n+}$. Unfortunately, it is premature to comment on the new

* This has been demonstrated for *each* of the azido products derived from $Co(dien)(dapo)Cl^{2+} + OH^- + N_3^-$.

** *trans*-$Co(en)_2(SO_3)(DMSO)^+$ excepted (section 7.2) (ref. 152).

results* of extensive investigations in process (ref. 70), conducted primarily to interpret the stereochemical data for double substitution in $NH_3(\ell)$ (section 4).

Balt has reported the solvent dependence for the steric course of base catalyzed substitution of *trans*-$Co(NH_3)_4(^{15}NH_3)Cl^{2+}$ (ref. 194). The results for water (44% cis product) agree with those of Buckingham, Olsen and Sargeson (ref. 168). Also the steric course is remarkably insensitive to a solvent variation, although a pentaammine system may not be as stereochemically discriminatory as, for example, $Co(en)_2AX$ ($A \neq$ amine), to properly gauge the effect, as remarked in section 8.1.

The tetraammine *trans*-$Co(NH_3)_4Cl_2^+$ solvolyzes in $^{15}NH_3(\ell)$ at low temperature to yield 100% *trans*-$Co(NH_3)_4(^{15}NH_3)Cl^{2+}$ (ref. 71). The stereochemistry for the analogous process in H_2O is not known, although the *trans*-$Co(en)_2Cl_2^+$ complex gives largely (93%, 0°C) trans product. Temperature differences will need to be accommodated when the data base for wider comparisons is broadened.

The results recorded in section 7.2 for the spontaneous substitution reactions of Co(III) complexes suggest the prospect of stronger solvent dependences than the few noted here for base catalyzed substitution, and no rationale can be offered until more data are obtained. The solvent dependence of the stereochemistry of base catalyzed *anation* also should not be overlooked. For water as solvent the contrast between the stereochemistry of substitution by H_2O and N_3^- in the base catalyzed process has proved informative (ref. 51,52,59,168), and opportunities exist for similar determinations for solvents such as DMSO, CH_3CN, $NH_3(\ell)$, and others.

9 CONCERTED COBALT(III) SUBSTITUTION PROCESSES

Most inorganic kineticists tend to think of cobalt(III) substitution processes as dissociative—S_N1 or S_N1(lim) in the original nomenclature system, or I_d or D in the new. Swaddle (ref. 29) has levelled criticism at this now widely used D, I_d, I_a, A Langford-Gray mechanistic classification scheme because of difficulties in defining reasonable "operational" distinctions between mechanisms. The proper description of the mechanism can depend upon how a reaction is probed. We also believe the classification scheme has deficiencies. For example, in section 8 there were reactions which could be labelled I_d according to the defined "operational" criteria (i.e. no intermediate) (ref. 195), but an intermediate (energy minimum) was clearly required by the stereochemical

* There exist conflicting reports in the literature on the steric course of ammoniolysis (see, e.g. ref. 52).

results. Fortunately the message in the chemistry is unchanged if mechanistic labels are avoided.

The stereochemical probes described in sections 7 and 8 were specifically aimed at testing the dissociative aspects of substitution. Any associative character can only be inferred by default, and the question of degree remains unanswered. In this section we take the opposite tack, using examples which permit direct probing of bond-making aspects of substitution processes. We do not intend to answer the difficult question of the balance of dissociative and associative components to a reaction, but wish to highlight the fact that cobalt(III) substitution can be shown to have *measurable* associative characteristics.

A number of rearrangements have been described recently which may be regarded as concerted. This we interpret to mean a single step substitution process where bond making by the incoming group is felt, especially in the rate but also in the stereochemical course of reaction.

Linkage isomerization processes such as

$$(NH_3)_5Co\text{-}O\text{-}S(S)O_2^+ \xrightarrow{H_2O} (NH_3)_5Co\text{-}S(S)O_3^+ \qquad (94)$$

are a class of intramolecular reactions of intense current interest (ref. 30,62). Most probes on the mechanism of these rearrangements, especially for the classic linkage isomerization (ref. 196),

$$(NH_3)_5Co\text{-}ONO^{2+} \xrightarrow{H_2O} (NH_3)_5CoNO_2^{2+} \qquad (95)$$

have led to the conclusion that they are "essentially dissociative" (ref. 30). Yet there are other lines of argument (ref. 62) which suggest an appreciable associative component. For example, the thiosulfate-**O** complex rearranges to the thiosulfate-**S** complex at least 500-fold faster than the isostructural and isoelectronic sulfato complex $(NH_3)_5Co\text{-}O(S)O_3^+$ aquates (ref. 62,197), suggesting some Co...S bonding activation in (94). Given the magnitude of a typical bond energy, it is not difficult to see how little bond making in the transition state can accommodate a rate enhancement of ~ 500 (~ 15 kJ mol^{-1}). The example illustrates that associative character can be detected in the classical way (ref. 195). The **a** component of the activation energy, for the spontaneous processes (94) and (95), is also apparent from the marked increase in the contribution from competitive hydrolysis to the total reaction, on going from spontaneous to base catalyzed linkage isomerization, and is discussed elsewhere (ref. 62).

The Hg^{2+} ion induced reaction of the Λ-*cis*-$Co(en)_2X(NH_2CH_2COY)^{2+}$ (Y = OH, NH_2, NRR′, OR) complexes involve exclusive capture of the pendant functional group (ref. 199),

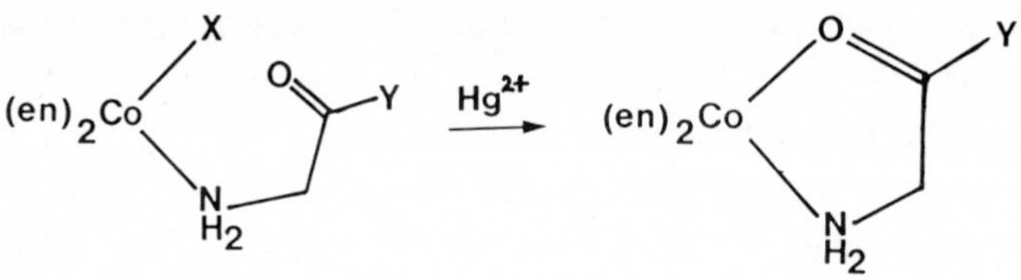

(96)

There is no detectable competition by H_2O (shown using ^{18}O-tracer studies where appropriate), nor by added nucleophiles.* Also the reaction occurs with complete retention of chirality. Note that O- is captured in preference to N- from the glycinamide and N-substituted derivatives $Co(en)_2Br(NH_2CH_2CO NRR')^{2+}$. This is not surprising since for *cis*-$Co(en)_2Cl(NH_2(CH_2)_2NH_2)^{2+}$ the dangling amine N-centre does not compete with solvent H_2O, and *cis*-$Co(en)_2(OH_2)(NH_2(CH_2)_2NH_2)^{3+}$ appears to be the only product (ref. 200). Furthermore, other studies establish that O- of an amide $RCONH_2$ is a better "nucleophile" than N- (ref. 201).

The data are accommodated by either of two extreme mechanisms. One is the conventional account (ref. 47)—the generation of a reduced coordination number intermediate in which solvent, added nucleophile, and the dangling end of the chelating ligand compete for the vacant coordination site. The other is a concerted process involving some bond making by the incoming group,

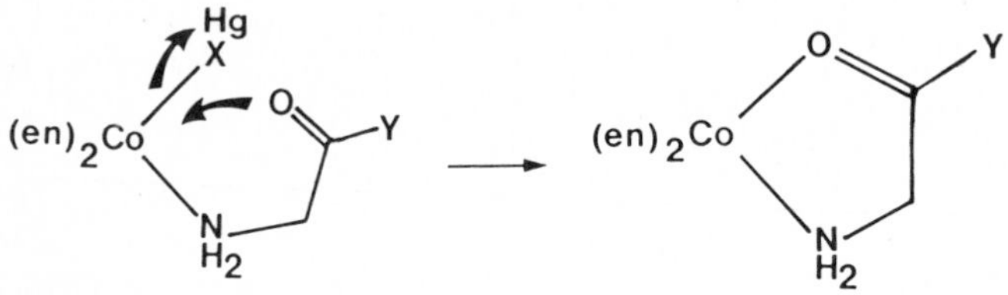

(97)

The latter mechanism seems preferable to an interchange process where bond making is minimal since it accommodates the marked preference for capture of $-CO.Y$. Now, if reaction (96) were concerted, it might be expected to be much more rapid than the corresponding reactions of, e.g. $Co(en)_2Br(NH_2(CH_2)_2NH_2)^{2+}$ or *cis*-$Co(en)_2(NH_3)Br^{2+}$, which lack the anchimeric assistance. However, in the absence of kinetic studies (though see ref.

---

* In the Hg^{2+} induced aquation of *cis*-β_2-$Co(trien)(glyOR)Cl^{n+}$, there is some (3%) competition by NO_3^- (1 M) (ref. 174). This can arise by direct injection from $HgNO_3^+$, and also this possibly has associative character (see ref. 198).

199), there can be no distinction between mechanisms on this basis. Also, the stereochemical result does not allow a decision, since pentaaminecobalt(III) complexes of this type invariably give geometric and optical retention. This consideration does, however, suggest value in examining analogous complexes of the kind Λ-*cis*-$Co(en)_2Br(OCO.Y)$ because competition with geometric rearrangement* is then possible through attack remote to the leaving group by H_2O or added competitors.

The exclusive capture by the dangling arm of the chelate in reaction (96) contrasts sharply with the less selective capture found for base hydrolysis of the same or closely related complexes (ref. 59,199). This fact alone suggests mechanism (97). Indeed, this is quite similar to the difference in behaviour found for spontaneous and base catalyzed linkage isomerization (ref. 30,62). These reactions have a common structural feature—an inbuilt potential nucleophile, and it remains to show unequivocally that the pendant group can function as a genuine nucleophile, i.e. that it can be captured by virtue of bond-making to cobalt(III), with a concomitant rate enhancement over competitive dissociative processes, rather than captured preferentially merely because of its enforced presence.

Buckingham *et al.* have recently described several novel rearrangements. *Trans*-$Co(en)_2(OH_2)(NH_2CH_2CONH_2)^{3+}$ isomerizes directly to the *cis*-$Co(en)_2(NH_2CH_2CONH_2\text{-}O)^{3+}$ chelate complex, without going through the *cis*-$Co(en)_2(OH_2)(NH_2CH_2CONH_2)^{3+}$ intermediate (ref. 199). Although the latter complex cyclizes approximately 100-fold faster and escapes direct detection, it yields largely (95%) the chelated glycinate species, *cis*-$Co(en)_2(NH_2CH_2CO_2)^+$ (ref. 200). Thus solvent H_2O does not compete with $-CONH_2$ in the direct substitution process yielding chelated cis product,

(98)

However, more surprising is the specific rate for this rearrangement, k =

* Based on results given in section 7.1 for complexes having a CoN_4OX chromophore.

$4 \times 10^{-6}\ s^{-1}$ (25°C), compared to a rate constant for exchange of $2 \times 10^{-6}\ s^{-1}$ (25°C), and of trans- to cis- isomerization of $\sim 2.5 \times 10^{-8}\ s^{-1}$ for the closely related *trans*-$Co(en)_2(NH_3)OH_2^{3+}$ ion (ref. 125). The rearrangement (98) is much faster than one can accommodate by the unassisted dissociation of the H_2O leaving group, and the process obviously has a component of bond making by the incoming $-CONH_2$ group.

A similar case can be constructed for the corresponding and much faster reaction of the cis isomer,

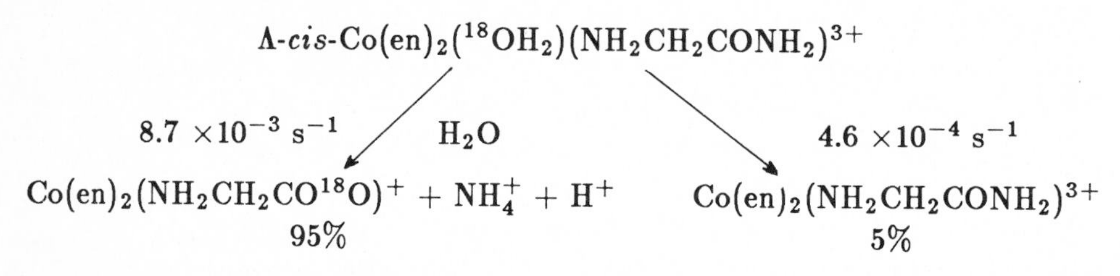

(99)

In (99) the specific rates for the individual steps are given, as is the fate of the ^{18}O label (ref. 202). The predominant process is not substitution at cobalt and does not concern us here. The labelling shows that the minor process involves displacement of bound water from the Co(III) centre by O-amide and this occurs ~ 400-fold faster than $H_2{}^{18}O$ exchange in the closely related *cis*-$Co(en)_2(NH_3)OH_2^{3+}$ ion (k_{ex} $1.1 \times 10^{-6}\ sec^{-1}$, 25°C, ref. 125). Again, anchimeric assistance is strongly inferred.

The other feature, uncharacteristic of a dissociative reaction, is the fact that *trans*-$Co(en)_2(OH)(NH_2CH_2CONH_2)^{2+}$ rearranges to *cis*-$Co(en)_2(NH_2CH_2CONH_2\text{-}O)^{3+}$ ~ 50-fold faster than the *trans*-aqua complex ($k = 2.8 \times 10^{-4}\ s^{-1}$ and $4 \times 10^{-6}\ s^{-1}$, respectively), yet OH^- is a poorer leaving group than OH_2 (ref. 199). Moreover, this again seems to be much faster than exchange of bound OH^- with $H_2{}^{18}O$.* Furthermore, the reaction cannot go via the *cis*-$Co(en)_2(OH)(NH_2CH_2CONH_2)^{2+}$ species ($k = 1.5 \times 10^{-4}\ s^{-1}$) on kinetic grounds,** and also because the latter yields largely (90%) the chelated glycinate complex rather than the (observed) chelated glycinamide (ref. 199,202). It is concluded that the unusual leaving group dependence of the rate of these trans to cis rearrangements ($OH^- > OH_2$) is inconsistent with a simple dissociative process.

---

* k_{ex} is not known, but it must be less than k_{tc} ($\sim 1 \times 10^{-5}\ s^{-1}$) since *trans*-$Co(en)_2(NH_3)(OH)^{2+}$ isomerizes to the cis form in $H_2{}^{18}O$ without ^{18}O incorporation (ref. 125).

** This intermediate ion would be detected since the specific rates k_1, k_2 for the consecutive reaction scheme $A \xrightarrow{k_1} B \xrightarrow{k_2} C$ do not differ sufficiently ($k_1 \sim k_2$ rather than $k_2 \geq k_1$).

Buckingham *et al.* have also presented strong evidence for the operation of concerted processes for the corresponding reactions of the glycinate complexes (ref. 203). They are not evident for the cis isomers $Co(en)_2(OH_n)(NH_2CO_2H_m)$ ($n = 1,2$; $m = 0,1$) because of especially rapid intramolecular cyclization reactions which yield *cis*-$Co(en)_2(NH_2CH_2COO)^{2+}$ (ref. 204). But the corresponding trans ions follow the pattern exposed for the glycinamide analogs. *Trans*-$Co(en)_2(OH_2)(NH_2CH_2CO_2H)^{3+}$ and *trans*-$Co(en)_2(OH_2)(NH_2CH_2CO_2)^{2+}$ (or its tautomer *trans*-$Co(en)_2(OH)(NH_2CH_2CO_2H)^{2+}$ isomerize directly to *cis*-$Co(en)_2(NH_2CH_2CO_2)^{2+}$ at indistinguishable rates ($k = 2 \times 10^{-6}$ s^{-1}), without the intervention of the *cis*-aqua or *cis*-hydroxo monodentate glycinato complex (^{18}O-tracer experiments). The hydroxo ion *trans*-$Co(en)_2(OH)(NH_2CH_2CO_2)^{+}$ behaves similarly, but again the process is more rapid ($k = 2.4 \times 10^{-4}$ s^{-1}, 25°C) than the reaction of the aqua complex. Indeed, it is even faster than the reaction of the cis complex which does not cyclize by a metal substitution process. Also, N_3^- does not compete for a coordination site in the trans- to cis- rearrangement of the $Co(en)_2(OH)(NH_2CH_2CO_2)^{+}$ ion (ref. 203), a fact also excluding a normal dissociative process.

In summary, the dangling arm of a metal bound but potential chelating ligand appears to be especially effective in promoting a substitution process. The rates for the rearrangement process far exceed actual or expected exchange rates for the leaving group, and this coupled with the inverted rate dependence for the effect of the leaving group ($OH^- \geq OH_2$) support a case for a component of associative activation. It might be interesting to compare the kinetic and stereochemical behaviour of complexes such as *trans*-$Co(en)_2Br(NH_2CH_2CONH_2)^{2+}$ or $Co(en)_2(O_3SCF_3)(NH_2CH_2CONH_2)^{2+}$, which do not contain OH_2 or its lyate ion OH^- as the leaving group, although precedent suggests first retentive aquation of X followed by the same reactions (depending upon pH) described for the aqua or hydroxo complex.

The complexes *cis*- and *trans*-$Co(en)_2(OH_n)(O_2CCO_2)^{(n-1)+}$ ($n = 1,2$) might be expected to behave similarly. Although not all experimental details are presently available (ref. 163), it seems concerted processes do not operate here. A reasonable explanation seems to be that cis/trans rearrangement via the usual dissociative H_2O exchange process, with ring closure of monodentate oxalate ion, is more readily accessible because of the higher exchange rates afforded by O-bonded oxalate.

In conclusion, it seems genuine nucleophilic reactions of Co(III) complexes can be promoted by having the appropriate centre held in proximity to the metal ion in the ground state. Clearly this must be contingent upon the geometry and the length of the appendage associated with the organic ligand bearing the nucleophile, and warrants further testing.

Acknowledgements

This work has been supported by the Australian Research Grants Scheme. We thank especially Professor S. Balt for permission to quote unpublished data and for stimulating discussion. Thanks are also due to Professor H. Brunner for directing us to relevant literature on organometallic inversion reactions, and to Professor R.D. Archer for communications. The author is grateful to his research students, especially Mrs. C. Begbie and Lt. L. McJames, for painstaking work on bis(ethylenediamine)cobalt(III) chemistry. We thank Professor W. Marty and his past and present students, Dr. Peter Comba and Mr. L. Zipper in particular, for helpful discussions, permission to cite their very elegant work on Co(dien)(dapo)X chemistry, and for the gracious hospitality at the University of Neuchâtel where this review took shape. Finally, Mrs. N. Moon is acknowledged for her excellent typing of the manuscript.

REFERENCES

1 P. Walden, *Ber.*, **26** (1893), 210; *ibid*, **29** (1896), 133.
2 R.S. Cahn, C.K. Ingold and V. Prelog, *Angew. Chem.*, **78** (1966), 413; *Angew. Chem. int. Ed. Engl.*, **5** (1966), 385.
3 J. March, *Advanced Organic Chemistry*, 2nd edn, McGraw-Hill, New York, 1977, p. 97.
4 B. Bosnich, W.G. Jackson and S.B. Wild, *J. Amer. Chem. Soc.*, **95** (1973), 8269.
5 A. Werner, *Ber.*, **47** (1914), 3087.
6 F.G. Mann, *J. Chem. Soc.*, 412 (1933).
7 R.D. Gillard and F.L. Wimmer, *J. Chem. Soc. Chem. Commun.*, 936 (1978).
8 R.H. Prince, *in* MTP International Review of Science, Series 1, Butterworths, London, ed. M.L. Tobe, Vol. 9, pp 353–393.
9 H. Brunner, *Adv. Organomet. Chem.*, **18** (1980), 152.
10 H. Brunner, *Top. Curr. Chem.*, **56** (1975), 67.
11 A. Werner and V.L. King, *Ber. Deut. Chem. Ges.*, **44** (1911), 1887.
12 J.C. Bailar, Jr. and R.W. Auten, *J. Amer. Chem. Soc.*, **56** (1934), 774.
13 F.P. Dwyer, A.M. Sargeson and I.K. Reid, *J. Amer. Chem. Soc.*, **85** (1963), 1215.
14 J.C. Bailar, Jr. and J.C. Peppard, *J. Amer. Chem. Soc.*, **62** (1940), 820.
15 J.C. Bailar, Jr., J.H. Haslem and E.M. Jones, *J. Amer. Chem. Soc.*, **58** (1936), 2226.
16 R.D. Archer, L.J. Suydam and D.D. Dollberg, *J. Amer. Chem. Soc.*, **93** (1971), 6837.

17 R.D. Archer, D.D. Dollberg and J.A. Chlastawa, Abstracts XVIth ICCC, Dublin, Ireland, 1974, 3.2.
18 R.D. Archer and C.J. Hardiman, *ACS Symp. Ser.,* **119** (1980), 357.
19 R.D. Gillard, *Coord. Chem. Rev.,* **16** (1975), 67.
20 N. Serpone, G. Ponterini, M.A. Jamieson, F. Bolletta and M. Maestri, *Coord. Chem. Rev.,* **50** (1983), 209.
21 J.C. Bailar, Jr., *Rev. Pure Appl. Chem.,* **16** (1966), 91.
22 R.D. Archer, *in* Coordination Chemistry, Plenum Press, New York, ed. S. Kirschner, 1969, pp 18–28.
23 W.G. Jackson and C.M. Begbie, *Inorg. Chem.,* **22** (1983), 1190.
24 E. Kyuno, L.J. Boucher and J.C. Bailar, Jr., *J. Amer. Chem. Soc.,* **87** (1965), 4458.
25 E. Kyuno and J.C. Bailar, Jr., *J. Amer. Chem. Soc.,* **88** (1966), 1120.
26 A.M. Sargeson and G.H. Searle, *Inorg. Chem.,* **6** (1967), 787.
27 J.A. Chambers, T.J. Goodwin, M.W. Mulqi, P.A. Williams and R.S. Vagg, *Inorg. Chim. Acta,* **75** (1983), 241; *ibid,* **81** (1984), 55.
28 M.L. Tobe, *in* Advances in Inorganic and Bioinorganic Mechanisms, Vol. 2, Academic Press, New York, ed. A.G. Sykes, 1983, pp 1–94.
29 T.W. Swaddle, *in* Advances in Inorganic and Bioinorganic Mechanisms, Vol. 2, Academic Press, New York, ed. A.G. Sykes, 1983, pp 95–138.
30 W.G. Jackson and A.M. Sargeson, *in* Rearrangement in Coordination Complexes, Vol. 2, Academic Press, New York, ed. P. de Mayo, 1980, pp 273–378.
31 L.J. Boucher, E. Kyuno and J.C. Bailar, Jr., *J. Amer. Chem. Soc.,* **86** (1964), 3656.
32 J.C. Bailar, Jr. and J.P. McReynolds, J. Amer. Chem. Soc., **61** (1939), 3199.
33 E.A. Dittmar and R.D. Archer, *J. Amer. Chem. Soc.,* **90** (1968), 1468.
34 W.S. Kwak and R.D. Archer, *Inorg. Chem.,* **15** (1976), 986.
35 S.C. Chan and M.L. Tobe, *J. Chem. Soc.,* 4531 (1962).
36 W.G. Jackson and A.M. Sargeson, *Inorg. Chem.,* **17** (1978), 1348.
37 M.E. Farago, B. Page and M.L. Tobe, *Inorg. Chem.,* **8** (1969), 388.
38 M.E. Baldwin, S.C. Chan and M.L. Tobe, *J. Chem. Soc.,* 4637 (1961).
39 M.E. Farago, B.A. Page and C.F.V. Mason, *Inorg. Chem.,* **8** (1969) 2270.
40 C.G. Barraclough, R.W. Boschen, W.W. Fee, W.G. Jackson and P.T. McTigue, *Inorg. Chem.,* **10** (1971), 1994.
41 L.G. Vanquickenborne and K. Pierloot, *Inorg. Chem.,* **23** (1984), 1471.
42 W.G. Jackson and W. Marty, results to be published.
43 D.A. Buckingham, L.G. Marzilli and A.M. Sargeson, *J. Amer. Chem. Soc.,* **91** (1969), 5227 and references therein.

44 F. Basolo and R.G. Pearson, *Mechanisms of Inorganic Reactions,* 2nd edn, Wiley, New York, 1967.
45 D.A. Buckingham, C.R. Clark and T.W. Lewis, *Inorg. Chem.,* **18** (1979), 2041.
46 R.D. Archer, *Adv. Chem. Ser.,* **62** (1967), 452.
47 A.M. Sargeson, *Pure Appl. Chem.,* **33** (1973), 527.
48 W.G. Jackson, C.N. Hookey, M.L. Randall, P. Comba and A.M. Sargeson, *Inorg. Chem.,* **23** (1984), 2473.
49 D.A. Buckingham, I.I. Olsen and A.M. Sargeson, *J. Amer. Chem., Soc.,* **88** (1966), 5443.
50 W.G. Jackson, C.M. Begbie and M.L. Randall, *Inorg. Chim. Acta,* **70** (1983), 7.
51 D.A. Buckingham, C.R. Clark and T.W. Lewis, *Inorg. Chem.,* **18** (1979), 1985.
52 D.A. Buckingham, I.I. Olsen and A.M. Sargeson, *J. Amer. Chem., Soc.,* **90** (1968), 6654.
53 W.G. Jackson and C.M. Begbie, *Inorg. Chem.,* **23** (1984), 659.
54 P.J. Staples and M.L. Tobe, *J. Chem. Soc.,* 4801 (1960).
55 W.G. Jackson and C.M. Begbie, unpublished data.
56 W.G. Jackson, G.A. Lawrance, N.E. Dixon and A.M. Sargeson, *Inorg. Chem.,* **20** (1981), 470.
57 W.G. Jackson, results to be published.
58 D.A. Buckingham, J.D. Edwards and G.M. McLaughlin, *Inorg. Chem.,* **21** (1982), 2770.
59 C.J. Boreham, D.A. Buckingham and C.R. Clark, *Inorg. Chem.,* **18** (1979), 1990.
60 W.L. Reynolds, S. Hafezi, A. Kessler and S. Holly, *Inorg. Chem.,* **18** (1979), 2860.
61 W.G. Jackson and C.M. Begbie, *Inorg. Chem.,* **20** (1981), 1654.
62 W.G. Jackson, D.P. Fairlie and M.L. Randall, *Inorg. Chim. Acta,* **70** (1983), 197.
63 R.D. Archer and J.C. Bailar, Jr., *J. Amer. Chem. Soc.,* **83** (1961), 812.
64 F. Basolo, R.G. Pearson and H.H. Schmidtke, *J. Amer. Chem. Soc.,* **82** (1960), 4434.
65 A. Panunzi and F. Basolo, *Inorg. Chim. Acta,* **1** (1967), 223.
66 H. Dutton, unpublished data.
67 S. Balt, W.E. Renkema and H. Ronde, *Inorg. Chim. Acta,* **86** (1984), 87, and references therein (in particular, ref. 71).
68 For sufficiently acidic amine centres (such as substituted ammonium ions), both H_2O and HO^- are effective bases (see ref. 185).
69 S. Balt, unpublished data.

70 S. Balt, personal communication.
71 S. Balt, H.J. Gamelkoorn, H.J.A.M. Kuipers and W.E. Renkema, *Inorg. Chem.,* **22** (1983) 3072.
72 A.J. Pöe and C.P. Vuik, *J. Chem. Soc. Dalton Trans.,* 661 (1976).
73 M.P. Hancock, B.T. Heaton and D.H. Vaughan, *J. Chem. Soc. Dalton Trans.,* 761 (1979).
74 H. Ogino and J.C. Bailar, Jr., *Inorg. Chem.,* **17** (1978), 1118.
75 G.W. Watt and J.F. Knifton, *Inorg. Chem.,* **7** (1968), 1159.
76 P. Gütlich, B.R. McGarvey and W. Kläui, *Inorg. Chem.,* **19** (1980), 3704.
77 H. Toftlund and S. Yde-Andersen, *Acta Chem. Scand.,* **A35** (1981), 575 and references therein.
78 M. Gerloch, B.M. Higson and E.D. McKenzie, *Chem. Commun.,* 1149 (1971).
79 I.C. Templeton, A. Zalkin, D.H. Templeton and T. Veki, *Inorg. Chem.,* **12** (1973), 1641.
80 S.F. Mason and B.J. Peart, *J. Chem. Soc. Dalton Trans.,* 949 (1973).
81 J. Burgess, *J. Chem. Soc. (A),* 1085, 1899 (1969) and references therein.
82 A. Yamagishi, *Inorg. Chem.,* **21** (1982) 1778.
83 G. Nord, *Acta Chem. Scand.,* **37** (1973), 743.
84 R.D. Gillard, C.T. Hughes, L.A.P. Kane-Maguire and P.A. Williams, *Trans. Met. Chem.,* **1** (1976), 114.
85 H. Brunner, B. Hammer, I. Bernal and M. Draux, *Organometallics,* **2** (1983), 1595.
86 S. Quinn, A. Shaver and V.W. Day, *J. Amer. Chem. Soc.,* **104** (1982), 1096.
87 F. Morandini, G. Consiglio, G. Ciani and A. Sironi, *Inorg. Chim. Acta,* **82** (1984), L27.
88 H. Brunner and E. Schmidt, *J. Organomet. Chem.,* **36** (1972), C18.
89 H. Brunner and E. Schmidt, *J. Organomet. Chem.,* **50** (1973), 219.
90 H. Brunner and M. Langer, *J. Organomet. Chem.,* **87** (1975), 223.
91 M. Lappus, Doktorarbeit, TU München, 1972.
92 H. Brunner and M. Lappus, *Angew. Chem.,* **84** (1972), 955; *Angew. Chem. int. Ed. Engl.,* **11** (1972), 923.
93 A. Wojcicki, *Adv. Organomet. Chem.,* **11** (1973), 87.
94 A. Davison and N. Martinez, *J. Organomet. Chem.,* **74** (1974), C17.
95 F. Calderazzo, *Angew. Chem.,* **89** (1977), 305; *Angew. Chem. int. Ed. Engl.,* **16** (1977), 299.
96 G.M. Whitesides and D.J. Boschetto, *J. Amer. Chem. Soc.,* **91** (1969), 4313.
97 G.M. Whitesides and D.J. Boschetto, *J. Amer. Chem. Soc.,* **93** (1971), 1529.

98 P.L. Bock, D.J. Boschetto, J.R. Rasmussen, J.P. Demers and G.M. Whitesides, *J. Amer. Chem. Soc.,* **96** (1974), 2814.

99 T.C. Flood and D.L. Miles, *J. Amer. Chem. Soc.,* **95** (1973), 6460.

100 S.L. Miles, D.L. Miles, R. Bau and T.C. Flood, *J. Amer. Chem., Soc.,* **100** (1978), 7278.

101 H. Brunner and H.-D. Schindler, *Chem. Ber.,* **104** (1971), 2467.

102 H. Brunner and H.-D. Schindler, *Z. Naturforsch.,* **26b** (1971), 1220.

103 H. Brunner and M. Langer, *J. Organomet. Chem.,* **87** (1975), 223.

104 H. Brunner, J. Aclasis, M. Langer and W. Steger, *Angew. Chem.,* **86** (1974), 864; *Angew. Chem. int. Ed. Engl.,* **13** (1974), 810.

105 P. Hofmann, *Angew. Chem.,* **39** (1977), 551; *Angew. Chem. int. Ed. Engl.,* **16** (1977), 536.

106 P. Comba, N.J. Curtis, W.G. Jackson and A.M. Sargeson, submitted to Inorg. Chem.

107 J.MacB. Harrowfield, A.M. Sargeson, B. Singh and J.C. Sullivan, *Inorg. Chem.,* **14** (1975), 2864.

108 J.L. Laird and R.B. Jordan, *Inorg. Chem.,* **21** (1982), 855.

109 W.G. Jackson, *Inorg. Chim. Acta,* **47** (1981), 159.

110 W.G. Jackson and C.M. Begbie, *Inorg. Chim. Acta,* **61** (1982), 167.

111 W.G. Jackson and C.M. Begbie, *Inorg. Chim. Acta,* **60** (1982), 115.

112 C.M. Begbie and W.G. Jackson, results to be published.

113 W.G. Jackson and A.M. Sargeson, unpublished data.

114 D.A. Buckingham, I.I. Olsen and A.M. Sargeson, *Inorg. Chem.,* **6** (1967), 1807.

115 W.G. Jackson and A.M. Sargeson, *Inorg. Chem.,* **15** (1976), 1986.

116 D.A. Loeliger and H. Taube, *Inorg. Chem.,* **4** (1965), 1032; *ibid,* **5** (1966), 1376.

117 W.G. Jackson, results to be published.

118 J.K. Yandell and L.A. Tomlins, *Aust. J. Chem.,* **31** (1978), 561.

119 W. Kruse and H. Taube, *J. Amer. Chem. Soc.,* **83** (1961), 1280.

120 I.R. Lantzke and D.W. Watts, *Aust. J. Chem.,* **19** (1966), 1821.

121 L.J.M. McJames, B.Sc. Honours thesis, University of New South Wales, 1981.

122 W.G. Jackson, submitted to Inorg. Chem.

123 W.G. Jackson, *Inorg. Chim. Acta,* **10** (1974), 51.

124 M.E. Baldwin and M.L. Tobe, *J. Chem. Soc.,* 4275 (1960).

125 D.F. Martin and M.L. Tobe, *J. Chem. Soc.,* 1388 (1962).

126 L. Mønsted and O. Mønsted, *Acta Chem. Scand.,* **A34** (1980), 259.

127 A.E. Merbach, P. Moore, O.W. Howarth and C.H. McAteer, *Inorg. Chim. Acta,* **39** (1980), 129.

128 J.MacB. Harrowfield, D.A. Buckingham and A.M. Sargeson, *J. Amer. Chem. Soc.,* **95** (1973), 7281.
129 J.P. Mathieu, *Bull. soc. chim.,* **4** (1937), 687.
130 S.C. Chan and M.L. Tobe, *J. Chem. Soc.,* 514 (1963).
131 M.L. Tobe and C.K. Williams, *Inorg. Chem.,* **15** (1976), 918.
132 T.P. Dasgupta, W. Fitzgerald and M.L. Tobe, *Inorg. Chem.,* **11** (1972), 2046.
133 W.G. Jackson and W. Marty, results to be published.
134 D.W. Watts, *Rec. Chem. Progr.,* **29** (1968), 131.
135 C.H. Langford, *J. Chem. Educ.,* **46** (1969), 557.
136 D.A. Palmer and D.W. Watts, *Aust. J. Chem.,* **21** (1968), 2895.
137 W.R. Fitzgerald and D.W. Watts, *J. Amer. Chem. Soc.,* **89** (1967), 821.
138 B. Bosnich, C.K. Ingold and M.L. Tobe, *J. Chem. Soc.,* 4074 (1965).
139 W.G. Jackson, *Aust. J. Chem.,* **34** (1981), 215.
140 L.M. Jackman, J.F. Dormish, R.M. Scott, R.H. Portman and R.D. Minarol, *Inorg. Chem.,* **18** (1979), 1503.
141 I.R. Lantzke and D.W. Watts, *J. Amer. Chem. Soc.,* **89** (1967), 815.
142 R.B. Jordan, A.M. Sargeson and H. Taube, *Inorg. Chem.,* **5** (1966), 1091.
143 I.R. Lantzke and D.W. Watts, *Aust. J. Chem.,* **20** (1967), 173.
144 W.G. Jackson, unpublished data.
145 R.D. Archer, *Coord. Chem. Rev.,* **4** (1969), 243.
146 M.L. Tobe and D.W. Watts, *J. Chem. Soc.,* 2991 (1964).
147 W.R. Fitzgerald and D.W. Watts, *J. Amer. Chem., Soc.,* **90** (1968), 1734.
148 I.R. Lantzke and D.W. Watts, *Aust. J. Chem.,* **19** (1966), 949.
149 W.G. Jackson, S.S. Jurisson and B.C. McGregor, *Inorg. Chem.,* **24** (1985), in press.
150 S.S. Jurisson and W.G. Jackson, unpublished results.
151 R.G. Pearson, P.M. Henry and F. Basolo, *J. Amer. Chem. Soc.,* **79** (1957), 5379, 5382.
152 S. Balt and H.J. Gamelkoorn, *Inorg. Chim. Acta,* **86** (1984), L61.
153 N.E. Dixon, G.A. Lawrance, P.A. Lay and A.M. Sargeson, *Inorg. Chem.,* **22** (1983), 846.
154 P.A. Lay, R.H. Magnuson, H. Taube and J. Sen, *J. Amer. Chem. Soc.,* **104** (1982), 7658.
155 D.R. Stranks and J. Yandell, *Inorg. Chem.,* **9** (1970), 751.
156 D.A. Buckingham, personal communication of unpublished results.
157 W.G. Jackson, unpublished data.
158 H.R. Hunt and H. Taube, *J. Amer. Chem. Soc.,* **80** (1958), 2642.
159 R.B. Jordan and A.M. Sargeson, *Inorg. Chem.,* **4** (1965), 433.
160 F.P. Rotzinger and W. Marty, *Inorg. Chem.,* **22** (1983), 3593.

161 D.A. Buckingham, I.I. Creaser, W. Marty and A.M. Sargeson, *Inorg. Chem.,* **11** (1972), 2738.
162 F.R. Nordmeyer, *Inorg. Chem.,* **8** (1969), 2780.
163 G.M. Miskelly, C.R. Clark, J. Simpson and D.A. Buckingham, *Inorg. Chem.,* **22** (1983), 3237.
164 C.M. Begbie, unpublished data.
165 D.A. Buckingham, I.I. Creaser and A.M. Sargeson, *Inorg. Chem.,* **9** (1970), 655.
166 W.G. Jackson and C.N. Hookey, *Inorg. Chem.,* **23** (1984), 668.
167 W.G. Jackson, M.L. Randall, A.M. Sargeson and W. Marty, *Inorg. Chem.,* **22** (1983), 1013.
168 D.A. Buckingham, I.I. Olsen and A.M. Sargeson, *J. Amer. Chem., Soc.,* **89** (1967), 5129; *ibid,* **90** (1968), 6539.
169 N.E. Dixon, W.G. Jackson, W. Marty and A.M. Sargeson, *Inorg. Chem.,* **21** (1982), 688.
170 D.A. Buckingham, P.J. Cresswell and A.M. Sargeson, *Inorg. Chem.,* **14** (1975), 1485.
171 D.A. Buckingham, C.R. Clark and W.S. Webley, *Aust. J. Chem.,* **33** (1980), 263.
172 D.A. Marzilli, B.M. Foxman and A.J. Herlt, unpublished data, quoted in ref. 30.
173 R.A. Henderson and M.L. Tobe, *Inorg. Chem.,* **16** (1977), 2576.
174 D.A. Buckingham, W. Marty and A.M. Sargeson, *Helv. Chim Acta,* **61** (1978), 2223.
175 P. Comba, Ph.D thesis, Université de Neuchâtel, 1981.
176 P. Comba and W. Marty, *Helv. Chim. Acta,* **63** (1980), 693.
177 L. Zipper, unpublished data.
178 U. Tinner and W. Marty, *Inorg. Chem.,* **20** (1981), 3750.
179 W.G. Jackson and W. Marty, results to be published.
180 D.P. Rillema, J.F. Endicott and J.R. Barber, *J. Amer. Chem. Soc.,* **95** (1973), 6987.
181 P. Comba and A.M. Sargeson, *J. Chem. Soc. Chem. Commun.,* 1985, in press.
182 C.K. Poon and M.L. Tobe, *Inorg. Chem.,* **7** (1968), 2398.
183 C.K. Poon and M.L. Tobe, *Inorg. Chem.,* **10** (1971), 225.
184 D.A. Buckingham, M. Dwyer and A.M. Sargeson, unpublished data quoted in ref. 174.
185 D.A. Buckingham, P.A. Marzilli and A.M. Sargeson, *Inorg. Chem.,* **6** (1967), 1032.
186 D.A. House, *Coord. Chem. Rev.,* **23** (1977), 223.
187 M.L. Tobe, *Acc. Chem. Res.,* **3** (1970), 377.

88 C.A. Bernhard, Ph.D thesis, Université de Neuchâtel, 1975.
89 M.N. Hughes, *J. Chem. Soc. A,* 1284 (1967).
90 G. Marangoni, M. Panayotou and M.L. Tobe, *J. Chem. Soc. Dalton Trans.,* 1989 (1973).
191 D.A. Buckingham, C.R. Clark, B.M. Foxman, G.J. Gainsford, A.M. Sargeson, M. Wein and A. Zanella, *Inorg. Chem.,* **21** (1982), 1986.
92 A. Anwander, travail de diplome, Université de Neuchâtel, 1981.
93 W.G. Jackson, L. Zipper and W. Marty, results to be published.
94 S. Balt, H. Gamelkoorn and W.E. Renkema, *J. Chem. Soc. Dalton Trans.,* 2415 (1983).
95 C.H. Langford and H.B. Gray, *Ligand Substitution Processes,* Benjamin, New York, 1966.
96 W.G. Jackson, G.A. Lawrance, P.A. Lay and A.M. Sargeson, *Inorg. Chem.,* **19** (1980), 904.
197 F. Monacelli, *Inorg. Chim. Acta,* **7** (1973), 65.
198 W.G. Jackson, G.A. Lawrance and A.M. Sargeson, *Inorg. Chem.,* **19** (1980), 1001.
199 C.J. Boreham, D.A. Buckingham and F.R. Keene, *Inorg. Chem.,* **18** (1979), 28.
200 M.D. Alexander and C.A. Spillert, *Inorg. Chem.,* **9** (1970), 2344.
201 D.P. Fairlie, Ph.D thesis, University of New South Wales, 1983.
202 C.J. Boreham, D.A. Buckingham and F.R. Keene, *J. Amer. Chem. Soc.,* **101** (1979), 1409.
203 C.J. Boreham and D.A. Buckingham, *Inorg. Chem.,* **20** (1981), 3112.
204 C.J. Boreham, D.A. Buckingham, D.J. Francis, A.M. Sargeson and L.G. Warner, *J. Amer. Chem. Soc.,* **103** (1981), 1975.

CHAPTER 5

STEREOCHEMISTRY OF ACETYLENES COORDINATED TO COBALT

GYULA PÁLYI, GYULA VÁRADI[a] and LÁSZLÓ MARKÓ

(Research Group for Petrochemistry of the Hungarian Academy of Sciences, H-8200 Veszprém, Schönherz Z.u.8. Hungary)

1 INTRODUCTION, INDUSTRIAL BACKGROUND

Acetylenes are high-energy, reactive molecules, however their thermal reactions with less-polar reactants (H_2, CO, acetylene, olefins, etc.) are characterized by high activation energy barriers. These reactions belong to the primeval examples of transition metal catalysis.

Due to their easy preparation (initially based on coal) acetylene and its derivatives dominated the organic chemical industry as primary starting materials in the first half of this century. The discovery and exploitation of rich oil fields after the IInd World War caused a rapid switch-over to an olefin-based organic (petrochemical) industry and acetylene has rapidly lost ist importance since the early 50's. As a result, first, the basis of acetylene production has been retooled from coal to natural gas (mostly partial oxidation of methane)[1,2] and second, the relative importance and absolute volume of acetylene(s) used suffered a considerable regression from the mid-sixties [3,4] on. It should be added that most of the current production is used for welding and similar non-chemical purposes.

Thus, industrial-scale developments in acetylene production were restricted to a few specialized companies (Union Carbide, GAF, some pharmaceutical works [5,6]) to cover demands of the steel industry and the preparation of some fine chemicals as 1,4-butynediol, ethynylsteroids, etc.

---

[a] Present address: Department of Molecular Biophysics and Biochemistry, Yale University, 260 Whitney Ave., New Haven, CT 06511, USA

An opportunity for a renaissance in acetylene chemistry [7] seemed to develop in the early 70's following a sudden increase of oil prices and as prophecies of an expected shortage of mineral oil and gas gained credit. Nuclear power based cheap energy was predicted for the mid-80's. This factor combined with the "oil--crisis" and the indisputably favorable chemical characteristics of acetylene(s) induced a new wave of research programs which aimed at modernizing acetylene production technologies [8-20] together with new efforts in acetylene-based organic syntheses.

A large fraction of these latter efforts was focused on the transition metal vs. acetylene chemistry (some characteristic examples: [21-30]) producing really spectracular results such as acetylenes attached to metal triangles [31] and the coupling reactions of acetylenes in the presence of low-valent cobalt compounds. Some of these results will certainly survive fluctuations of the large-scale industrial background, being of fundamental theoretical interest or being linked to the preparation (or production) of fine chemicals for which higher raw material costs can be endured. This review analyses the stereochemical aspects of acetylene coupling reactions in the presence of organocobalt complexes (mainly carbonyls). First, the organic products will be surveyed; and then, those organocobalt complexes will be discussed which contain coordinated acetylene and/or ligands derived from reactions of acetylene in the coordination sphere, and which can reasonably be believed to bear structurally coded information of the stereochemical course of the "organic" reactions.

2 ACETYLENE + ACETYLENE REACTIONS

2.1 Organic products

One of the most striking examples of the high activation barriers in reactions of acetylenes with apolar reagents is the reaction of two acetylene moieties. Thus, for example, the oligocyclization of acetylene to benzene or to cyclooctatetraene is highly exotermic (594 KJ/mol and 602 KJ/mol product resp. [32]) but the non-catalytic reaction needs a fairly high temperature (400-500°C), where several side-products are also formed [33]. The use of transition metal complex catalysts allows to diminish the reaction temperature by at least 200° and enables tayloring reaction paths and/or sequences which lead to products hardly or not accessible by any other known method.

2.1.1 Open-chain products

Several papers, dealing with the chemistry of acetylenes in the presence of metal carbonyls describe the formation of insoluble, "resin-like", most probably polymeric materials. However until quite recently (when the electric properties of polyacetylenes started to become interesting [34,35]) not much effort had been made to clear up the structure of these products. The results of the first systematic studies are appearing nowadays [36-38].

Since the geometry and the distribution of the substituents on the polyacetylene chains seem to have an important influence on the conductance properties [39] it can be predicted that the stereochemical aspects of the coordinative polymerization of acetylene(s) will receive increasing attention in the near future. Studies employing cobalt carbonyls as catalysts are under way [39].

2.1.2 Cyclic products

The formation of cyclic products is the classical field of acetylene + acetylene reactions catalyzed or assisted by transition metals, first of all by Co, Ni and Fe. The earliest findings were reported with Ni [40]. About a decade later it was discovered that cobalt carbonyls are also active in the cyclotrimerization of acetylenes.

A broad variety of cobalt carbonyls was tested as catalyst in the early stages of this research. $Co(CO)_3(NO)$ [41], $Co_2(CO)_8$ [42], $[Co(CO)_4]_2Hg$ [42], $(\mu_2\text{-}RC_2R')Co_2(CO)_6$ [43,44] and $(\mu_4\text{-}RC_2R')Co_4(CO)_{10}$ [45] were all found to be active; however, a more careful inspection of these reactions [42,43,46,47] showed that among these only the $(\mu_2\text{-}RC_2R')Co_2(CO)_6$ type complexes are involved effectively in the catalytic cycle and the others (if used as catalyst precursors) have first to be transformed into this compound type, thus requiring more drastic conditions and prolonged reaction times (induction period).

A large number of acetylenes can be trimerized by this method including C_2H_2 and its mono- and disubstituted derivatives. The substituents tested range from simple primary and secondary alkyl groups to various (substituted) phenyl and alkoxycarbonyl groups. The reactivity of these acetylenes is largely influenced by the substituent on the C_3 moiety. The following approximate order of reactivity has been found: COOR > Ph > alkyl $\simeq$ H [42,47].

Some acetylenes such as those bearing carboxylic, ortho-halophenyl or nitrophenyl groups, as well as mono- or di(t-butyl)- and bis(trimethylsilyl)acetylene , cannot be trimerized catalytically; however, several of these acetylenes react under stoichiometric conditions.

The reaction conditions thus depend on the catalyst (precursor) and the chemical nature of the acetylene to be trimerized: 20-100°C and reaction times of 1-8 h are required.

These trimerizations are characterized by high regioselectivity and represent a practical and useful preparative entry into the chemistry of some substituted benzene derivatives. This point, of course, is only of importance for cyclizations of asymmetrically substituted acetylenes as shown in Scheme 1.

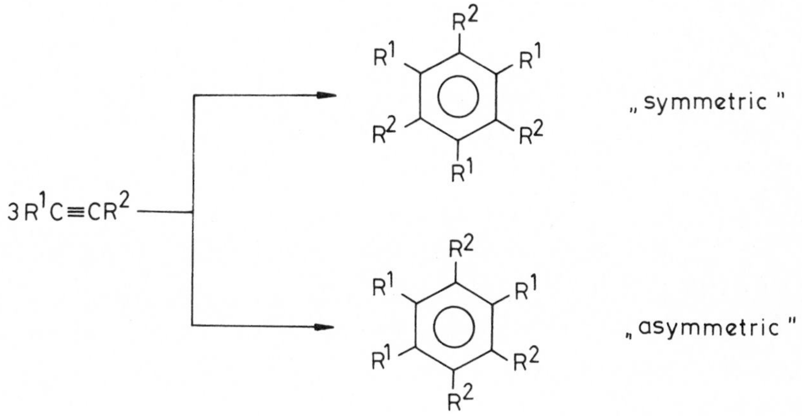

$3R^1C\equiv CR^2$

„symmetric"

„asymmetric"

Scheme 1

The "symmetric" 1,3,5-R^1-2,4,6-R^2-isomer is usually detected only in minor amounts, and the main product shows the "asymmetric" 1,2,4-R^1-3,5,6-R^2-distribution (some characteristic examples are collected in Table 1).

The most noteworthy product of these reactions was the first (however only stoichiometric) synthesis of such benzene derivatives which contain two t-butyl-groups in ortho-positions (1,2,4-tri(t--butyl)benzene [48-50]) and some hardly accessible hexasubstituted benzene derivatives e.g. hexa(i-propyl)- [51,52] or hexaphenylbenzene [42,47]. Some reactions leading to these stereochemically hindered benzenes are shown in Scheme 2.

TABLE 1

Cyclotrimerization of acetylenes with cobalt carbonyls

Item	Acetylene	Catalyst	Solvent	Tempe-rature (^{o}C)	Reaction time (h)	Product (yield %)
	Comparison of catalyst systems [42]					
1	PhC_2Ph	$Co_2(CO)_8$	-	280	1	Ph_6C_6 (60)
2	PhC_2Ph	$Hg[Co(CO)_4]_2$	-	270	1	Ph_6C_6 (70)
3	PhC_2Ph	$(PhC_2Ph)Co_2(CO)_6$	-	150	1	Ph_6C_6 (70)
4	PhC_2Ph	$Fe_3(CO)_{12}$	-	260-280	1	Ph_6C_6 (75)
5	PhC_2Ph	$Ni(CO)_4$	-	260	1	Ph_6C_6 (5)
6	PhC_2Ph	$Mn_2(CO)_{10}$	-	270	1	Ph_6C_6 (55)
7	PhC_2Ph	$Mo(CO)_6$	-	270	1	Ph_6C_6 (50)
8	PhC_2Ph	$Hg[Co(CO)_4]_2$	THF	65	1	Ph_6C_6 (55)
9	PhC_2Ph	$Hg[Co(CO)_4]_2$	$CH_3C_6H_{11}$	101	1	Ph_6C_6 (80)
10	PhC_2Ph	$Hg[Co(CO)_4]_2$	dioxane	101	1	Ph_6C_6 (95)
11	PhC_2Ph	$(PhC_2Ph)Co_2(CO)_6$	dioxane	101	1	Ph_6C_6 (95)
	Regioselectivity					
12	HC_2Pr^n	$Hg[Co(CO)_4]_2$	dioxane	101	2	1,2,4-(n-Pr)$_3C_6H_3$ (10) 1,3,5-(n-Pr)$_3C_6H_3$ (1) [42]
13	HC_2Ph	$Hg[Co(CO)_4]_2$	dioxane	101	1	1,2,4-$Ph_3C_6H_3$ (70) [42]
14	HC_2COOMe	$Co_2(CO)_8$	dioxane	101	2.5	1,2,4-(MeOOC)$_3C_6H_3$ (56) 1,3,5-(MeOOC)$_3C_6H_3$ (14) [47]

TABLE 1 (continued 1)

Item	Acetylene	Catalyst	Solvent	Temperature (^oC)	Reaction time (h)	Product (yield %)
15	HC_2CN	$Hg[Co(CO)_4]_2$	THF	65	2.5	1,2,4-$(NC)_3C_6H_3$ (38) [47]
16	HC_2CH_2OR	$(HC_2CH_2OR)Co_2(CO)_6$	dioxane	101	1	1,2,4-$(ROCH_2)_3C_6H_3$ (R = H,14; R = Me,17) [42]
17	$HC_2C(OH)Me_2$	$(NO)Co(CO)_3$	n-hexane	60	4	1,2,4-$(Me_2(HO)C)_3C_6H_3$ (95) [41]
18	MeC_2Ph	$Hg[Co(CO)_4]_2$	dioxane	101	1	1,2,4-Me_3-3,5,6-Ph_3C_6 (87) 1,3,5-Me_3-2,4,6-Ph_3C_6 (3) [42]
19	PhC_2COOMe	$Hg[Co(CO)_4]_2$	dioxane	101	1	1,2,4-Ph_3-3,5,6-$(MeOOC)_3R_6$ (55) [42]
20	PhC_2CN	$Co_2(CO)_8$	dioxane	101	4	1,2,4-Ph_3-3,5,6-$(NC)_3C_6$ (22) 1,3,5-Ph_3-2,4,6-$(NC)_3C_6$ (41) [47]
21	PhC_2Cl	$Hg[Co(CO)_4]_2$	dioxane	101	2	1,2,4-Ph_3-3,5,6-Cl_3C_6 (14) 1,3,5-Ph_3-2,4,6-Cl_3C_6 (2) [42]
22	MeC_2C_2Me	$Hg[Co(CO)_4]_2$	n-heptane	100	2	1,2,4-$(MeC_2)_3$-3,5,6-Me_3C_6 (4) 1,3,5-$(MeC_2)_3$-2,4,6-Me_3C_6 (8)[53]
23	$Me_3SiC_2C_2SiMe_3$	$Hg[Co(CO)_4]_2$	n-heptane	100	5	1,2,4-$(Me_3SiC_2)_3$-3,5,6-$(Me_3Si)_3C_6$ (25) 1,3,5-(Me_3SiC_2)-2,4,6-Ph_3C_6 (50) [53]
24	$PhC_2C_2SiMe_3$	$Hg[Co(CO)_4]_2$	n-heptane	100	2.5	1,3,5-(Me_3SiC_2)-2,4,6-Ph_3C_6 (50) [53]

TABLE 1 (continued 2)

Item	Acetylene	Catalyst	Solvent	Tempe- rature (^{o}C)	Reaction time (h)	Product (yield %)
	Cocyclization					
25	$PhC_2H + HC_2H$	$Hg[Co(CO)_4]_2$	dioxane	101	1	C_6H_6 (< 0.1); 1,4-$Ph_2C_6H_4$ (3.2); 1,2,4-Ph_3-C_6H_3 (3.4) [42]
26	$PhC_2H + PhC_2Ph_5$	$Hg[Co(CO)_4]_2$	dioxane	100	3	1,2,4-Ph_3-C_6H_3 (48); 1,2,3,5-Ph_4-C_6H_2 (7-12); Ph_6C_6 (6-34) [42]

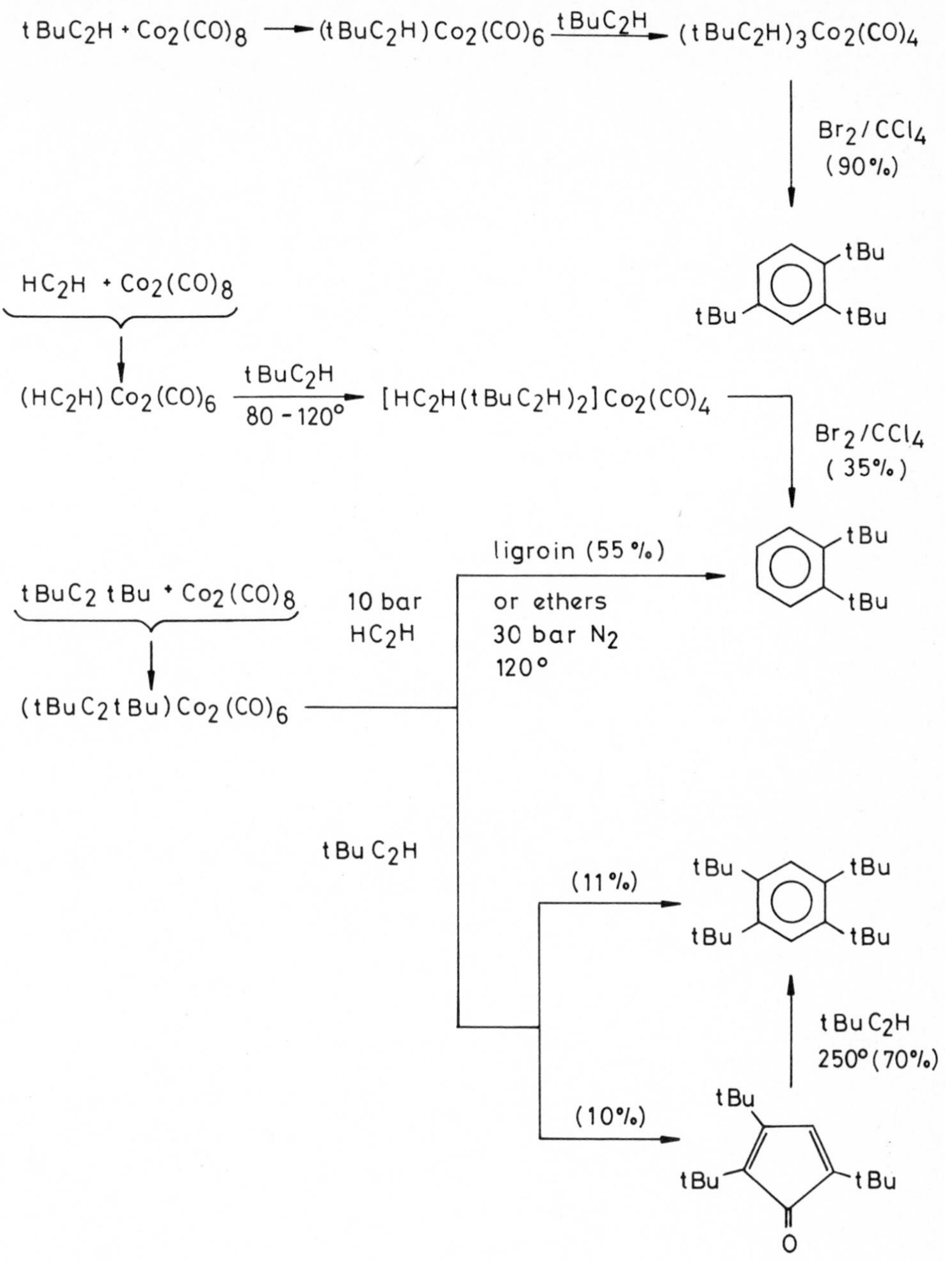

$tBuC_2H + Co_2(CO)_8 \longrightarrow (tBuC_2H)Co_2(CO)_6 \xrightarrow{tBuC_2H} (tBuC_2H)_3Co_2(CO)_4$

Br_2/CCl_4 (90%)

tBu, tBu, tBu

$HC_2H + Co_2(CO)_8$

$(HC_2H)Co_2(CO)_6 \xrightarrow[80-120°]{tBuC_2H} [HC_2H(tBuC_2H)_2]Co_2(CO)_4$

Br_2/CCl_4 (35%)

tBu, tBu

$tBuC_2tBu + Co_2(CO)_8$

$(tBuC_2tBu)Co_2(CO)_6$

10 bar HC_2H

ligroin (55%) or ethers 30 bar N_2 120°

$tBuC_2H$

(11%)

tBu, tBu, tBu, tBu

$tBuC_2H$ 250° (70%)

(10%)

tBu, tBu, tBu, O

Scheme 2 Reactions leading to ortho-di-tert.-Bu-benzenes [48-50, 54-56]

Acetylenes of more or less comparable reactivity give co-cyclization reactions under similar conditions as described above. Small differences in reactivity can be leveled off by applying an excess of the less reactive acetylene. Some selected examples of these syntheses are shown also in Table 1.

The fact that (η^5-C_5H_5)Co(CO)$_2$ reacts differently with alkylenes than other cobalt carbonyls was first discovered by Wender et al. [57] and studied later by R.B.King [58], who isolated cobaltorganic products (η^4-cyclobutadiene-derivatives). This mononuclear catalyst has been proved later (by the ingenious work of Vollhardt) to be very useful in synthetic organic chemistry.

The most outstanding results of Vollhardt's school include the formation of benzocyclobutenes [59-61], indans and tetralins [61-63] (some selected examples of these reactions are collected in Table 2) as well as the spectacular cobalt catalyzed synthesis of the steroid framework [63-65] (this and some related syntheses are collected in Table 3). The results initiated the highly stereo-selective synthesis of some other natural products [66-69]. The use of CpCo(CO)$_2$ as catalyst also enabled cyclizations which lead to benzene derivatives of high theoretical interest, especially in the field of strained fused ring systems [63,70,71].

The related triphenylphosphine complex, CpCo(PPh$_3$)$_2$ can be used with essentially similar results [73,74], however the presence of a non-volatile ligand (PPh$_3$ instead of CO) during the catalytic process makes this catalyst less active (vide infra).

2.2 Organometallic products, mechanism

Acetylenes and cobalt carbonyls react yielding a broad variety of cobaltorganic products. We shall discuss here mostly those which are relevant to the catalytic and stoichiometric processes described in Section 2.1.2. The reactions discussed here can be classified into two groups: dinuclear and mononuclear catalytic or template reactions. Since these show essential differences in the nature of the individual reaction steps the discussion will be divided according to this classification.

2.2.1 Dinuclear reactions

Reactions which involve organocobalt reactants non containing η^5-cyclopentadienyl ligand, like $Co_2(CO)_8$, (acetylene)$Co_2(CO)_6$,

TABLE 2

Examples of the application of acetylene cyclization for the synthesis of benzocyclobutenes, indans and tetralins [59-63]

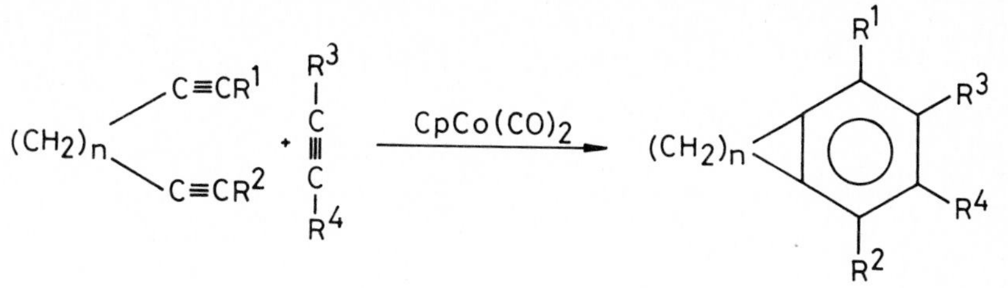

n	R^1	R^2	R^3	R^4	Yield /%/
2	H	H	nHex	H	13
	H	H	Ph	H	17
	H	H	CH_2OH	H	14
	H	H	Ph	Ph	48
	H	H	Me_3Si	Me_3Si	65
	H	H	COOMe	COOMe	14
	Me	Me	Ph	Ph	20
	Me	Me	COOMe	COOMe	28
	Me_3Si	H	Me_3Si	H	13
	Me_3Si	H	Me_3Si	COOMe	16
	Me_3Si	Me_3Si	Me_3Si	H	2
3	H	H	nHex	H	14
	H	H	Ph	H	26
	H	H	Ph	Ph	24
	H	H	COOMe	COOMe	20
	H	H	Me_3Si	Me_3Si	82
4	H	H	nHex	H	14
	H	H	Ph	H	18
	H	H	Ph	Ph	21
	H	H	COOMe	COOMe	26
	H	H	Me_3Si	Me	34
	H	H	Me_3Si	Me_3Si	85

TABLE 3 Cyclotrimerization of diynes [64,65,72]

Starting diyne component	Product	Yield (%)
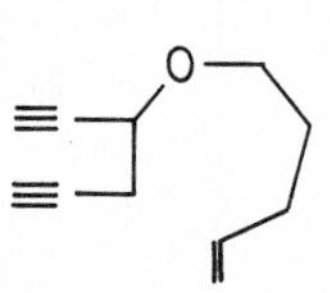 O	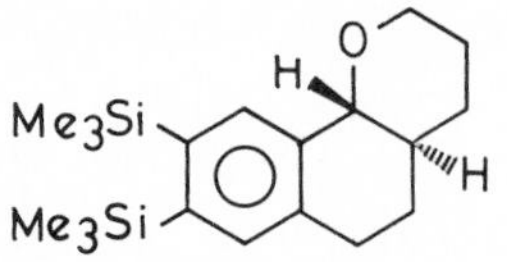 H, O, Me_3Si, Me_3Si, H	60
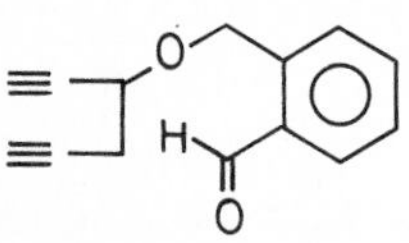 O, H, O	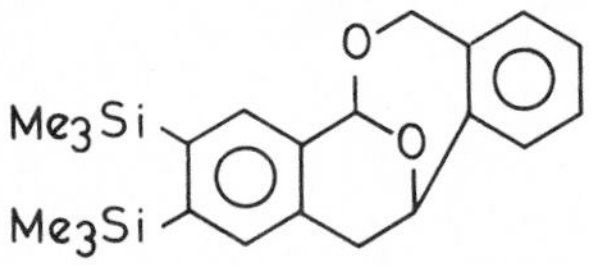 O, Me_3Si, O, Me_3Si	50
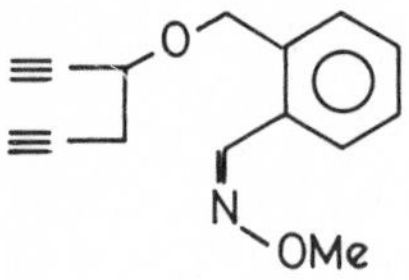 O, N, OMe	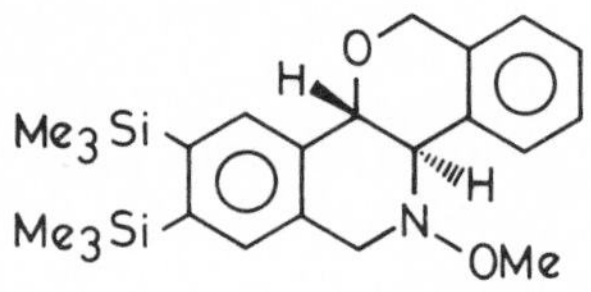 H, O, Me_3Si, Me_3Si, H, N, OMe	45
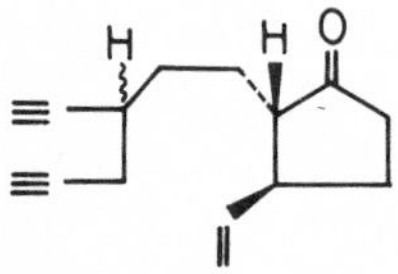 H, H, O	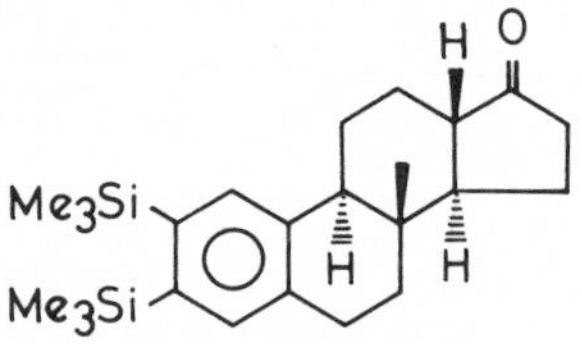 H, O, Me_3Si, Me_3Si, H, H	71
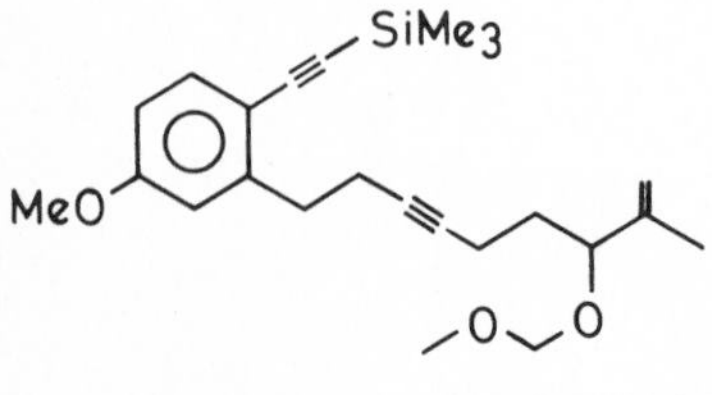 * $SiMe_3$, MeO, O, O	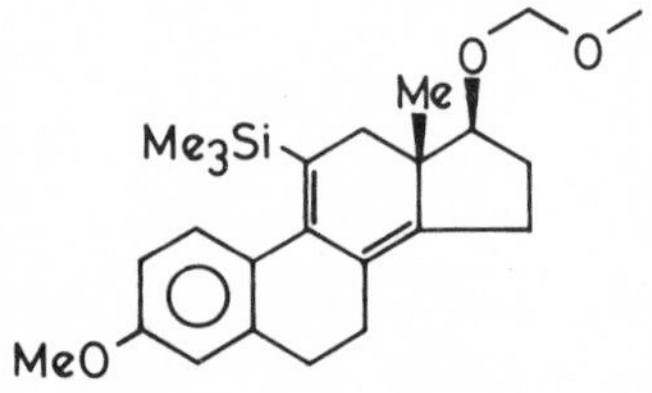 O, O, Me, Me_3Si, MeO	81 (24% overall yield from p-$MeOC_6H_4COCl$)

* Without $Me_3SiC_2SiMe_3$

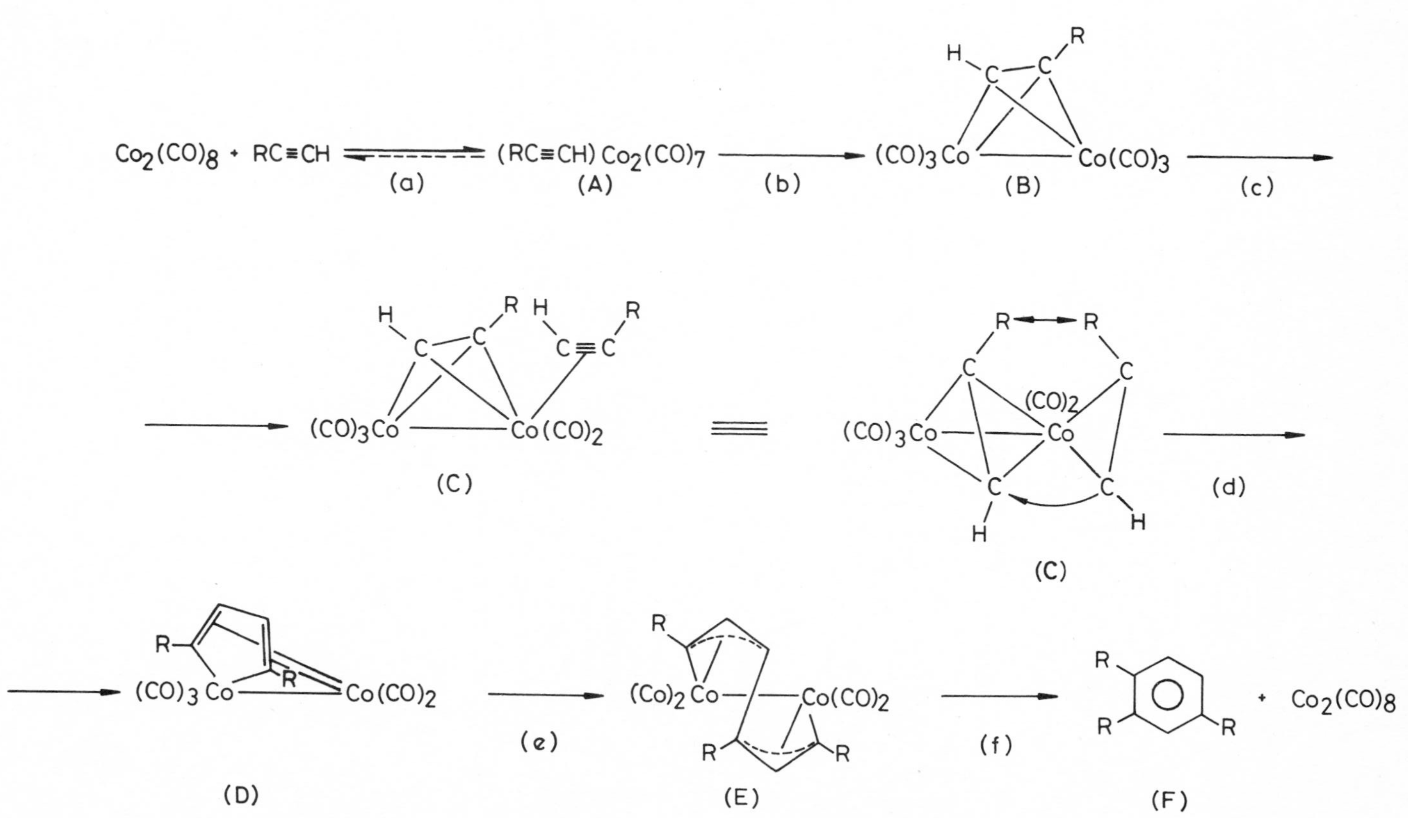

$Co_2(CO)_8$ + RC≡CH ⇌ (a) (RC≡CH)$Co_2(CO)_7$ (A) → (b) $(CO)_3Co$–[HC–CR]–$Co(CO)_3$ (B) → (c)

→ $(CO)_3Co$–[HC–CR]–$Co(CO)_2$ (HC≡CR) (C) ≡ $(CO)_3Co$–$Co(CO)_2$ (R↔R, H, H) (C) → (d)

→ $(CO)_3Co$–$Co(CO)_2$ (R, R) (D) → (e) $(Co)_2Co$–$Co(CO)_2$ (R, R, R) (E) → (f) 1,2,4-$R_3C_6H_3$ (F) + $Co_2(CO)_8$

Scheme 3 Mechanism of the cyclotrimerization of acetylenes with $Co_2(CO)_8$ exemplified by the formation of a 1,2,4-trisubstituted benzene derivative from a terminal alkyne

$Hg[Co(CO)_4]_2$, etc. belong to this group.

Product distribution studies, isolation and structure determination of the intermediates and kinetic studies concerning the cyclotrimerization of acetylenes via dinuclear cobalt complexes can be rationalized by a mechanism depicted in Scheme 3. The individual steps and intermediates are based on the following pieces of evidence:

(i) Equilibrium (a) can reasonably be supposed on the basis of kinetic studies [75,76]; however (A) could not be identified, either by preparation or by spectroscopic techniques.

(ii) The sum of steps (a) + (b) is well documented. Compounds (B) are known since three decades [77,78] and their formation from acetylenes and $Co_2(CO)_8$ constitutes a useful preparative method with high yields [78]. Complexes (B) were characterized by IR [79-87], NMR [88-9] and X-ray diffraction [91-94]. The bonding conditions were analyzed by EH [95,96] and CNDO/2 [97,98] MO calculations.

(iii) The reactivity of compounds (B) can be interpreted by considering the following facts: the acetylene in (B) may be regarded as being in a cis-excited (olefin-like) state [86,92]; the $\mu - \eta^2$-ligand in (B) can be further activated by twisting it towards the $\mu - \eta^1$-coordination [95,99-101]; the driving force of such a displacement can be the repulsion between the bulkier R substituents in (C) which, on the other hand, draws the less bulky CH ends of the two alkynes near each other and, consequently, the C,C bond between these carbon atoms becomes established.

(iv) Another important fact provides further support for this picture: both in the free and the complexed terminal acetylenes, the terminal (CH) carbon atoms seem to be the more shielded ones according to the ^{13}C-NMR spectra (c.f. [88-90] and vide infra), (this should be true for the terminal alkyne ligand in (C) too) which means that the orbitals on these carbon atoms are more extended; therefore a favourable extent of overlap between the two coordinated C(H)(tail-to-tail) atoms is more probable than between the C(H) and C(R) (head-to-tail) or the two C(R)(head-to-head) carbons. This, provides the reason why we prefer a tail-to-tail coupling in (D) over the possibility of head-to-tail or even head-to-head isomers of (D).

(v) Step (d) could be documented directly, although only in a rather particular case [102a]. Later, the formation of a compound which can be regarded as a PPh_3-derivative of (D) was reported but this complex was prepared by an indirect way [102b]. Analogous Rh [103] and Ir [104] complexes are known.

(vi) Steps (d) + (e) were observed directly and the structure of complexes (E) was proved by IR [45,105], NMR [105,106] and X-ray diffraction [107-109]. The structure of the co-trimerization product $[(HC_2H)(tBuC_2H)_2]Co_2(CO)_4$ [107] merits a remark. The two tBu groups are situated in this complex at the two ends of the new C_6-chain; that is, the farthest possible. The final result of trimerization is, however, the formation of 1,2-di-tBu-benzene in good yield. A case of Mother Nature having fun, or an excellent example of the really new reaction possibilities offered by using transition metal catalysts or templates! The role of complexes (E) in the synthesis of benzene derivatives with two bulky substituents in <u>ortho</u>-position can be seen in Scheme 2.

(vii) The fact that complexes (E) are chiral seems to be a challenging aspect of this chemistry [109] which, however, still has to be exploited. Recent results on diastereoselective cyclizations with the mononuclear cobalt complex $CpCo(CO)_2$ give a hint for possible fields of application.

(viii) Step (f) is again well documented [47,56,78] and the structure of the final product (F) is well known (c.f. Table 1 and refs. cited there).

2.2.2 Mononuclear reactions

A considerable amount of evidence concerning the mechanism of acetylene cyclizations with $CpCoL_2$ (L = CO or PPh_3) has been accumulated. The most important findings are summarized in Scheme 4. Some comments are, however, necessary.

(i) Reaction (a) could be carried out separately with acetylenes containing substituents with π-bonds such as $R^1 = R^2$ = Ph or R^1 = Ph, R^2 = COOMe and L = PPh_3 [110-111]. An alternative to the activation of the first acetylene through a zwitterionic monohapto complex $[CpLCo^+{-}C(R^1){=}C^-(R^2)]$, as suggested for more polar systems [113,114], needs not be considered in the present case.

(ii) Reaction (b) and intermediate (C) could not be observed directly. A recent kinetic and theoretical study [115] documented the intermediacy of (C), which cyclizes to the coordinatively unsaturated intermediate (D) in an oxidative coupling reaction (d). Of course, the unsaturated intermediate (D) could not be detected

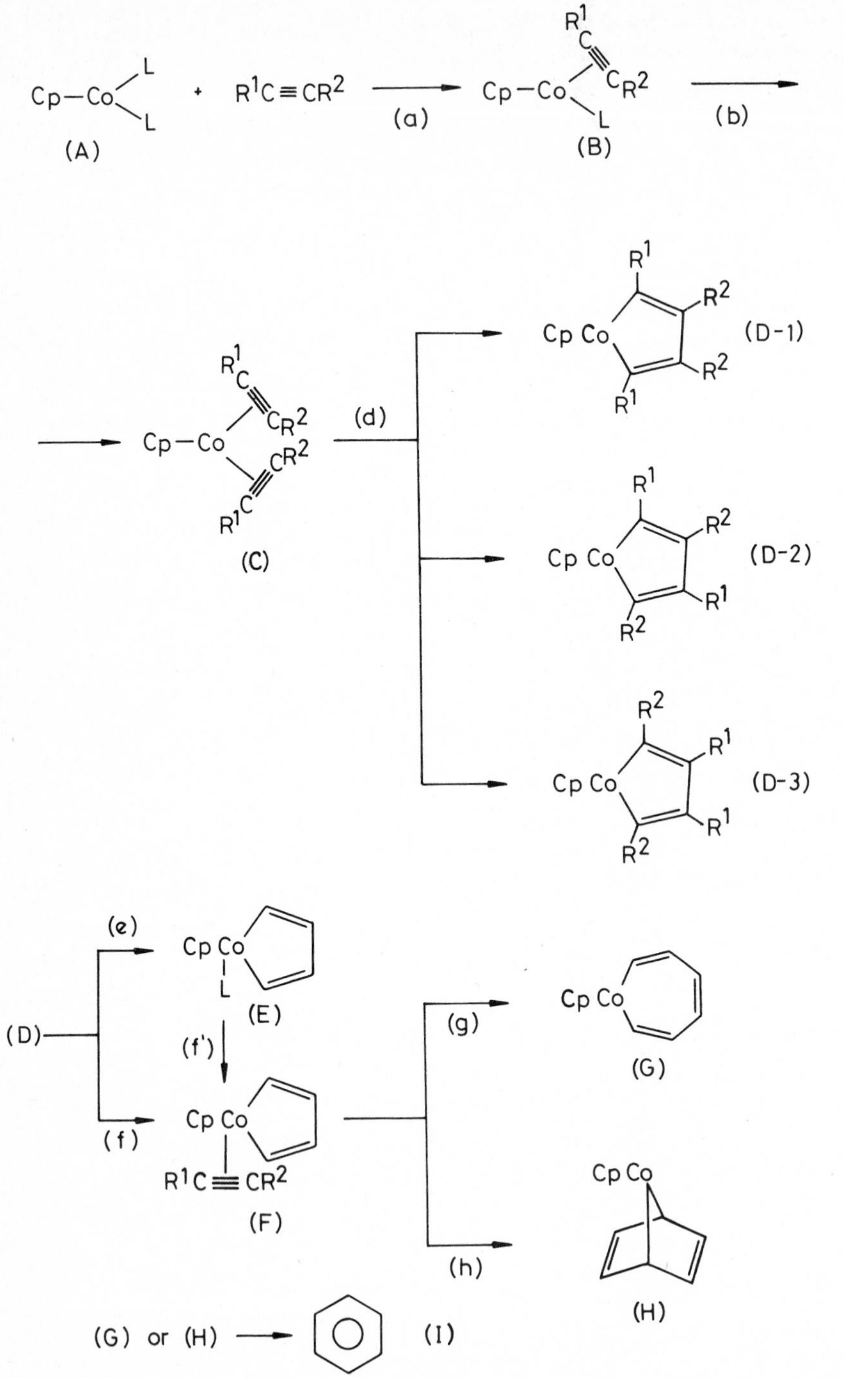

Cp—Co(L)(L) (A) + $R^1C{\equiv}CR^2$ —(a)→ Cp—Co(L)($R^1C{\equiv}CR^2$) (B) —(b)→

—→ Cp—Co($R^1C{\equiv}CR^2$)($R^1C{\equiv}CR^2$) (C) —(d)→

Cp Co (R^1, R^2, R^2, R^1) (D-1)

Cp Co (R^1, R^2, R^1, R^2) (D-2)

Cp Co (R^2, R^1, R^1, R^2) (D-3)

(D) —(e)→ Cp Co (L) (E)

(D) —(f)→ Cp Co ($R^1C{\equiv}CR^2$) (F)

(E) —(f')→ (F)

(F) —(g)→ Cp Co (G)

(F) —(h)→ Cp Co (H)

(G) or (H) —→ (I)

Scheme 4 Cyclotrimerization of acetylenes with mononuclear cobalt complexes $CpCoL_2$, L = CO or PPh_3

by direct methods but the phosphine derivative (E) (L = PPh_3) could be isolated in several cases [74,111,112,116,117]. These compounds were characterized by analytical, spectroscopic and (in a few cases) X-ray diffraction [118-120] methods. It should be noted that, generally, not all isomers of complexes (E) can be detected [112]. A series of interesting derivatives of complexes (D) were prepared [102,112,121-123]; these are important as intermediates in co-cyclizations of two acetylene moieties with a third non-acetylenic unsaturated molecule.

(iii) The coupling reaction (d) was found to depend more on steric than on electronic factors [115]. This seems to contrast with the rules governing the binuclear reactions (c.f. Sections 2.1.2., 2.2.1. and 3.2.3.2.). An explanation for this behaviour may be the difference in the chemical environment of the two <u>sp</u> carbons of an asymmetrically substituted acetylene which increases in the order free ligand < (dihapto terminal) < bridging, as reflected by the ^{13}C-NMR spectra. The acetylene ligand is therefore more flexibly coordinated in the terminal than in the bridging mode.

(iv) Intermediate (F) and (G) or (H) are, so far, assumed model systems only [63,73]. Transformation (E) → (I) was performed also in separate experiments, e.g. [112].

(v) The reaction course of the cyclizations involving α,ω-diynes is somewhat different. The most probable variants, depicted in Scheme 5, were chosen on the basis of suggestions from refs. [63,74]. It seems likely that the reaction path (c), (d), (e) or (f) is operative if more than two CH_2 groups link the two acetylenic units, while path (g), (h), (i) or (j) is more likely to operate in reactions leading to benzocyclobutadienes (I) [59,64,72,74].

3 ACETYLENE + CARBON MONOXIDE REACTIONS

A great number of acetylene carbonylations have been developed. Competent reviews [22,124,125] cover the whole spectrum of this field. We shall consider here only those reactions which are important from a stereochemical standpoint.

3.1 Open-chain products

The formation of open-chain products from acetylenes and CO generally requires the presence of a third reactive component H_2O, ROH, R_2NH, RSH, etc. usually with a mobile hydrogen atom. The most

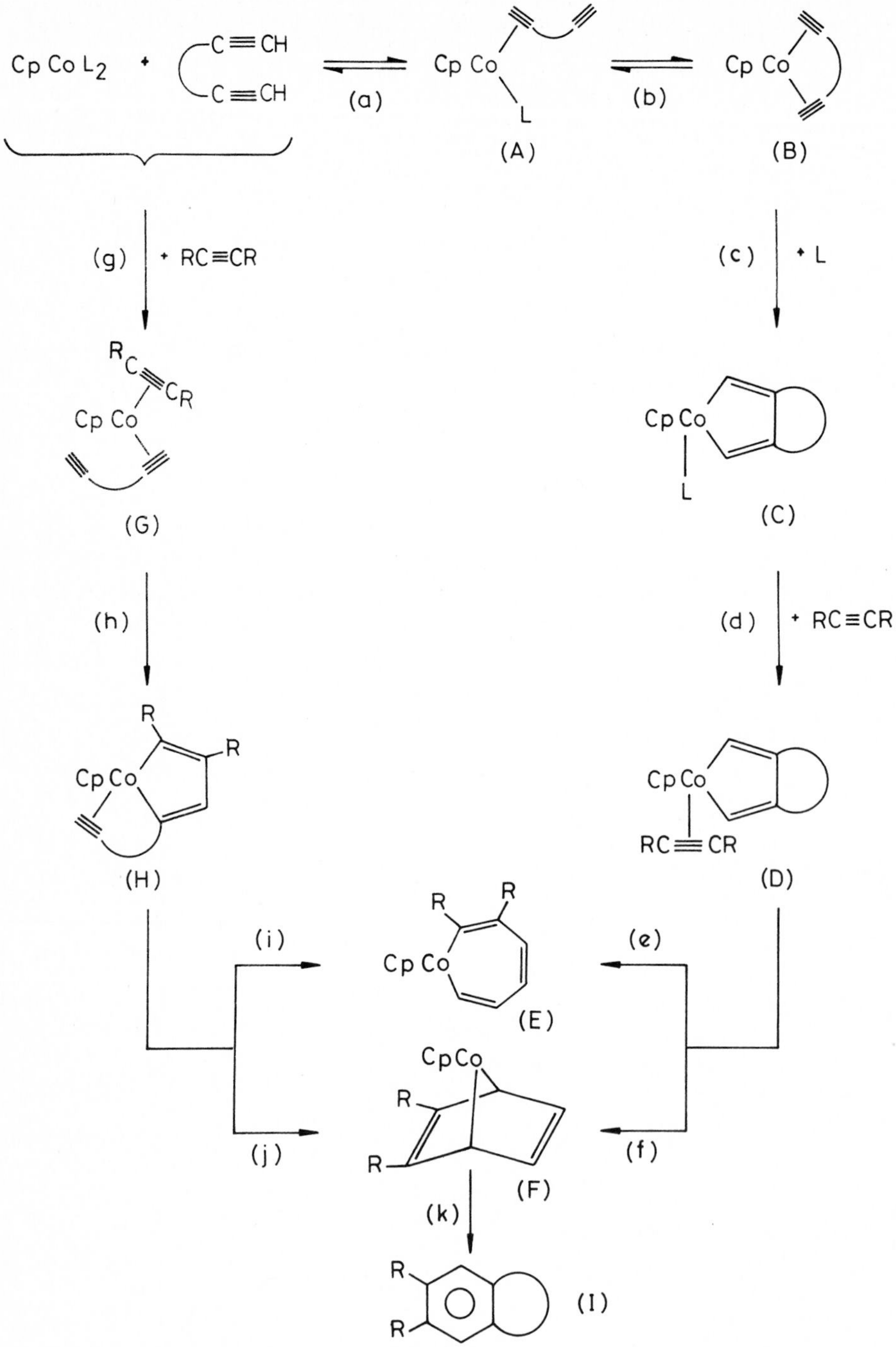

CpCoL2 + C≡CH C≡CH ⇌ (a) CpCo L (A) ⇌ (b) CpCo (B)

(g) + RC≡CR
(c) + L

RC C CR CpCo (G)
CpCo L (C)

(h)
(d) + RC≡CR

R R CpCo (H)
CpCo RC≡CR (D)

(i)
(e)
R R CpCo (E)
CpCo R R (F)
(j)
(f)
(k)
R R (I)

Scheme 5 Co-cyclo-trimerization of an α,ω-diyne and an acetylene, L = Co or PPh_3

important products obtained from this reaction are derivatives of acrylic and succinic acid which, unfortunately, do not provide much information about the stereochemical course of the reaction. Reactions of substituted acetylenes, however, can be more fruitfully evaluated from this aspect. The few relevant results are summarized in Table 4.

The carbonylation of acetylenes to acid derivatives using cobalt carbonyl catalysts is believed to proceed *via* the well-known (c.f. Section 2.2.1.)(μ_2-acetylene)$Co_2(CO)_6$ complex as intermediate. The bridge-coordinated acetylene can be attacked by two coordinated CO-s in a *cis*- or a *trans*-addition manner as depicted in Scheme 6.

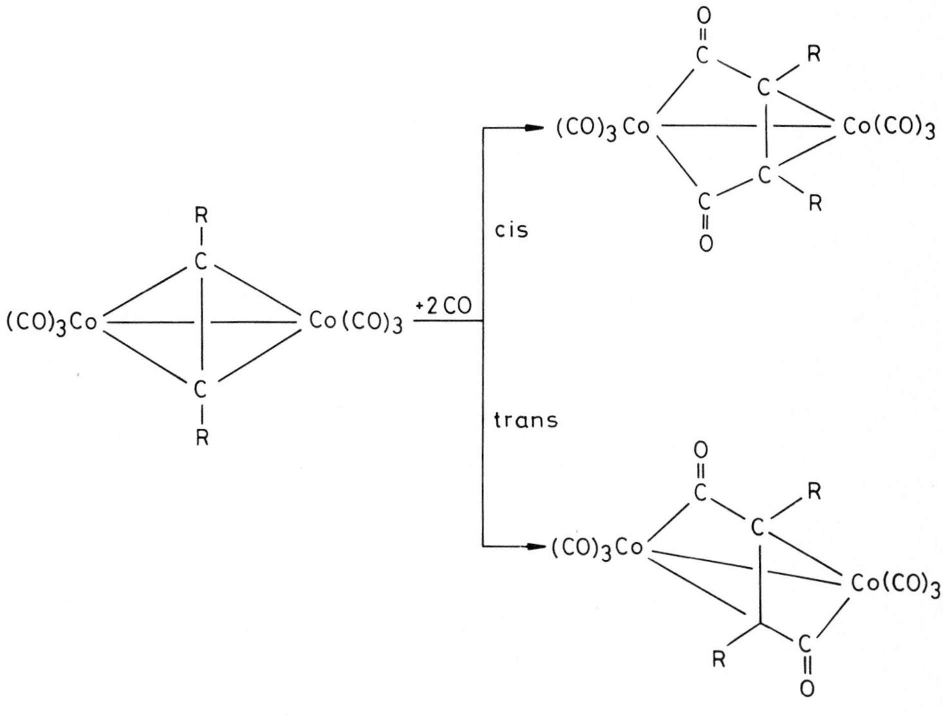

(CO)3Co, R, C, C, R, Co(CO)3, +2 CO, cis, trans

Scheme 6 Possibilities of addition of CO to bridge-coordinated acetylene

TABLE 4

Some open chain carbonylation products of acetylenes providing important stereochemical information
Catalyst: $Co_2(CO)_8$

Alkyne	Co-substr.	Temp. (°C)	Initial p^{CO} (bar)	Product	Yield (%) [Ref.]
HC_2H	MeOH	90-100	200-300	Dimethylfumarate	2.5
				Dimethylsuccinate	14 [124]
EtC_2H	H_2O	125	200	Ethylsuccinic acid n.r.	n.r. [22]
MeC_2Me	H_2O	140	200	meso-2,3-Dimethyl-succinic acid	n.r. [22]
PhC_2H	H_2O	140	200	racemic-Phenyl-succinic acid	44 [22, 126]
PhC_2Ph	H_2O	215	150	meso-Diphenyl-succinic acid (84%) racemic-Diphenyl-succinic acid (16%)	9 [22]
$R_2NCH_2C_2H$	-	100	600-1000	Citraconic acid	44 [127]*
(R = Me,Et)	-	110	280	bis(dialkylamide)	33 [128]

n.r. = not reported

* These authors reported the formation of an (isomeric) dihydro-amino-furamide which was later shown to be citraconic amide [128].

The results shown in Table 4 indicate that the cis-addition path seems to be the more favored one [22,129] under these conditions. This conclusion is further supported by the stereochemistry of the carbonylation of acetylenes without co-substrates (c.f. Section 3.2 formation of lactones).

3.2 Cyclic products

The reaction of acetylenes with CO in the absence of a co-substrate with a mobile H atom leads to cyclic products containing various ratios of the two unsaturated components. Thus cyclopentadienones (2:1), quinones (2:2) and five-membered lactones (1:2) may be formed. Since with cobalt catalysts quinones are merely formed in small amounts only the two other products will be discussed.

3.2.1 Cyclopentadienones

Cyclopentadienones were observed generally in stoichiometric reactions [45,57,130-133] or as by-products [134-136]. Only one case is known where (unsubstituted) cyclopentadienone is formed in higher yield (20-50%) [137].

No systematic studies were performed to establish rules of substituent distribution. From the few cases reported [45] one could conclude that the alkyl groups of terminal acetylenes are directed preferentially to positions near the carbonyl group. Considering that the attack of CO on a bridge-coordinated, terminal acetylene starts most probably at the CH-end of the molecule, (c.f. Section 3.2.2.) and that dimerization of acetylenes starts also with the interaction of these ends (c.f. Scheme 3) the reaction path (d), (e) on Scheme 7 seems to be the more probable [137].

An interesting cyclopentadienone derivative, symmetrical bis-cyclopentanone was recently obtained from the reaction of acetylene with CO under mild conditions [137]. This is the main ketonic product in benzene (46%). The formation of this compound may be immagined to proceed through a variety of pathways; however, the bis-carbonylation and bicyclization of the $(acetylene)_3Co_2(CO)_4$ "fly over" complex on Scheme 3 (intermediate (E))[137] seems to be an acceptable path, challenging further studies which are expected to provide a definite answer.

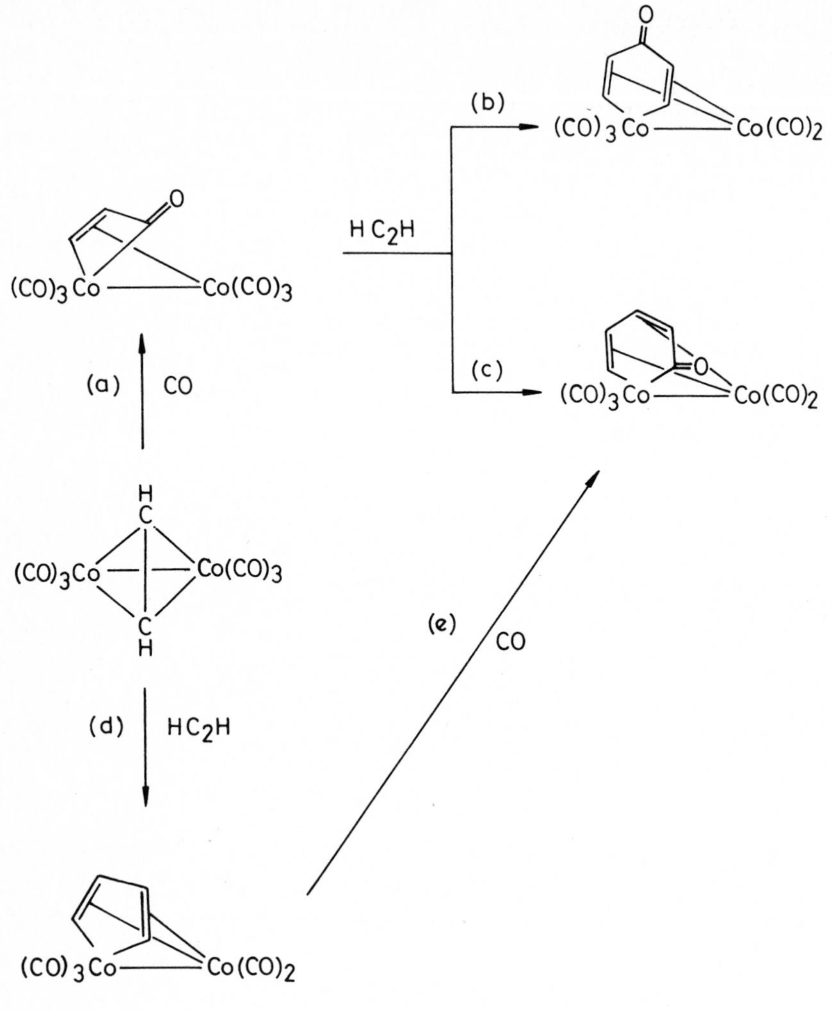

$(CO)_3Co$—$Co(CO)_2$ (b)

HC_2H

$(CO)_3Co$—$Co(CO)_3$

(c) $(CO)_3Co$—$Co(CO)_2$

(a) CO

$(CO)_3Co$—$Co(CO)_3$

(e) CO

(d) HC_2H

$(CO)_3Co$—$Co(CO)_2$

Scheme 7 Possible intermediates of the formation of cyclopentadienones

3.2.2 Monolactones

More than two decades ago, Heck [138,139] reported an interesting cyclization reaction (Scheme 8).

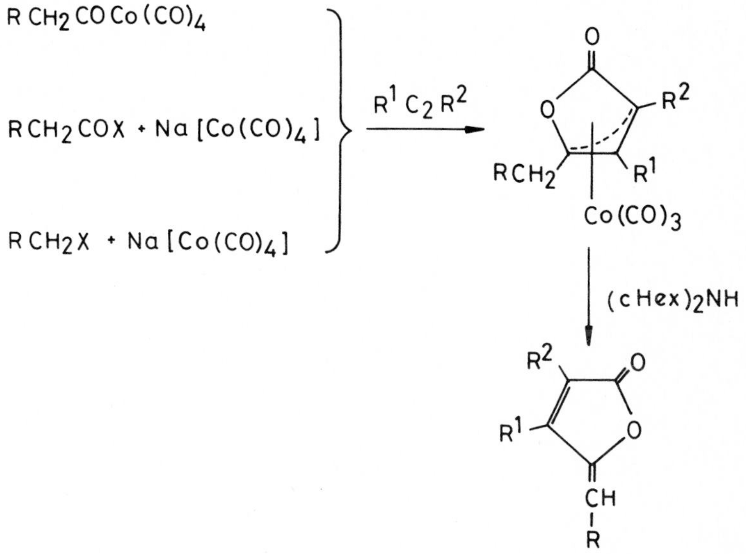

$RCH_2COCo(CO)_4$

$RCH_2COX + Na[Co(CO)_4]$ $\xrightarrow{R^1C_2R^2}$ (cobalt tricarbonyl lactone complex with R^2, R^1, RCH_2, $Co(CO)_3$) $\xrightarrow{(cHex)_2NH}$ (lactone with R^2, R^1, =CH–R)

$RCH_2X + Na[Co(CO)_4]$

Scheme 8

The intermediate cobalt tricarbonyls can be isolated and characterized. On the other hand, the reaction can be performed in one step without isolation of the intermediate. A broad range of R groups was tested: R = H, COOEt, CH=CHCOOMe, CN, CH_2CH_2OH, cHex, etc. R^1, R^2 = H, Et, tBu, $C(Me)_2OH$. However not very much is known about the distribution of the substituents in the product.

Strangely enough, this promising reaction has received almost no attention. Only one significant contribution has been reported. Alper et al. [140] developed a phase transfer-catalyzed variant of this synthesis:

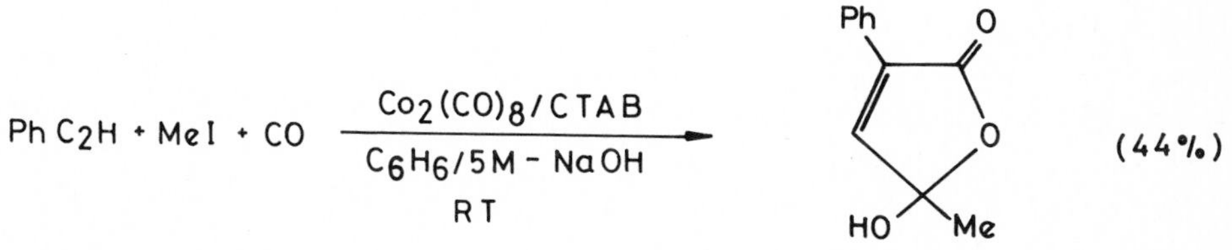

$$PhC_2H + MeI + CO \xrightarrow[C_6H_6/5M-NaOH,\ RT]{Co_2(CO)_8/CTAB} \text{(lactone: Ph, O, HO, Me)} \quad (44\%)$$

It would be early to speculate about the mechanism of this reaction but the importance of such an easy and mild lactone synthesis should be stressed.

3.2.3 Dilactones

The binary reaction of acetylene (and its derivatives) with CO provides an unexpected product. Almost exclusively 2,4,6-octatriene-4,5-diolides (bifurandiones) are formed with moderate to good yields, depending on the reaction conditions and the structure of the acetylenic substrate [141-150]:

$$2\,CH{\equiv}CH + 4\,CO \longrightarrow$$

$Co_2(CO)_8$ is used as catalyst and the yields can be improved by adding tertiary phosphorous, arsenic or antimony compounds [149,150]. Some characteristic preparative results are summarized in Table 5. These polyfunctional (conjugated triolefin and bilactonic) compounds seem to be versatile starting materials for various organic reactions [151,152].

Starting from an asymmetric acetylene, for example a terminal one, RC_2H, the formation of six isomers can be immagined: the 2,6- 2,7- and the 3,6-disubstituted bifurandiones and the (Z) and (E) isomers (with respect to the central double bond) of these.

It is generally akcnowledged that the 3,6-isomers are not formed in the carbonylation of acetylenes. There is, however, only one quantitative study reporting the relative amounts of the isomers: Albanesi [146] showed that the carbonylation of propyne provided the following products 37.5% 2,6-(Z)-, 54.3% 2,6-(E)- and 8.3% 2,7-(E)-dimethylbifurandiones. In accord with this observation, other studies report also the predominance of the 2,6-(E)- and -(Z)-isomers [143-145,149,153].

The structure of the unsubstituted (Z) and (E) isomers as well as of the (E)-2,7-dimethyl-derivative have been determined by X-ray crystallography [154].

Under milder conditions (90-110°, 40-60 bar total pressure) and higher C_2H_2/CO ratios ($\sim$1) the "ethynylogs" of the bifurandiones 2,4,6,8-decatetraene-4,7-diolides are formed in low ($<$ 10%) yields besides bifurandiones [147].

TABLE 5

Preparation of bifurandiones from acetylenes and CO in the presence of $Co_2(CO)_8$

Starting acetylene	ER_3	E/Co atomic ratio	ac/Co mole/atom ratio	Initial p CO (bar)	Temp. (^oC)	Reaction time (h)	Yield (%)	References and remarks
HC_2H	-	0	n.r.	200	90-100	2.5	14.0	[145] n.r. = not reported
HC_2H	-	0	n.r.	250	90-100	2.5	20-30	[144]
HC_2H	-	0	n.r	275-315	90	16	30	[143] solvent $(Me_2N)_2CO$ only <u>cis</u> isomer
HC_2H	-	0	n.r.	800-1000	90	17	69	[143]
HC_2H	$P(OMe)_3$	1:4	32	270	120	10	70.5	[149]
HC_2H	$P(nBu)_3$	1:4	32	270	120	10	72.5	[149]
$nPrC_2H$	-	0	16.4	270	120	10	5.0	[149]
$nPrC_2H$	$P(OMe)_3$	4:1	16.4	270	120	10	21.0	[149]
$nPrC_2H$	$P(OMe)_3$	1:4	16.4	270	120	10	28.4	[149]
$nPrC_2H$	$P(nBu)_3$	1:4	16.4	270	120	10	27.6	[149]
$nPrC_2H$	$AsPh_3$	1:4	16.4	270	120	10	18.7	[149]
$nBuC_2H$	-	0	n.r.	800-1000	100-110	15	58	[143]
$nBuC_2H$	-	0	16.4	270	120	10	9.0	[149]
$nBuC_2H$	$P(OMe)_3$	1:4	16.4	270	120	10	27.5	[149]
$nHexC_2H$	$P(OMe)_3$	1:4	16.4	270	120	10	25.6	[149]
PhC_2H	-	0	n.r.	200	110-120	2.5	4.0	[145]
PhC_2H	-	0	n.r.	800-1000	100-110	15	6-21	[143]
PhC_2H	-	0	16.4	270	120	10	4.0	[149]
PhC_2H	$P(OMe)_3$	1:4	16.4	270	120	10	21.2	[149]

Bifurandiones can also be made by conventional organic reactions [155,156]. These reactions are fairly laborious but provide moderate to good overall yields. The catalytic synthesis is more elegant; however, it should be mentioned that the "organic" techniques are also suitable for the preparation of 3,6-isomers which are inaccessible by the catalytic route.

3.2.3.1 Organometallic products, mechanism

The formation of the bilactonic products from acetylenes and CO attracted considerable attention for various reasons.

(i) The virtually one-step synthesis of such complicated oligoadducts proceeding under favourable conditions with high chemical selectivity presents a challenging problem for the chemist. It could (and can) be hoped that important general features of molecular catalysis may be deduced from the understanding of this reaction.

(ii) The absence of one of the possible isomeric products in the case of asymmetric acetylenes suggested that one of this rings is formed by quantitative regioselectivity while the other one is not. It seems interesting to learn how this happens and why?

(iii) At the very beginning it seemed possible, and as research in this field advanced, it became more and more probable that the two cobalt atoms of the catalyst ($Co_2(CO)_8$ can be regarded as the key compound in the catalytic cycle) remain linked together during the whole catalytic process. Thus, this reaction can be regarded as one of the very few examples of those cycles during which the catalyst is more than mononuclear throughout the whole process [157].

The accumulated knowledge relevant to the formation of bifurandiones has been rationalized in Scheme 9. The individual steps of the proposed mechanism are based on the following evidence and speculations.

(i) Intermediates (B) and (C) have been characterized as already described in the comments of Scheme 3. It has been proved that bifurandiones can also be made from intermediate (C) instead of starting from acetylene and dicobalt octacarbonyl.

(ii) Intermediate (D) could be isolated in one case (R^1 = OMe, R^2 = nBu) and characterized by analyses and IR spectra. It has been proved that high pressure carbonylation of (D) leads to the corresponding (F) complex [159].

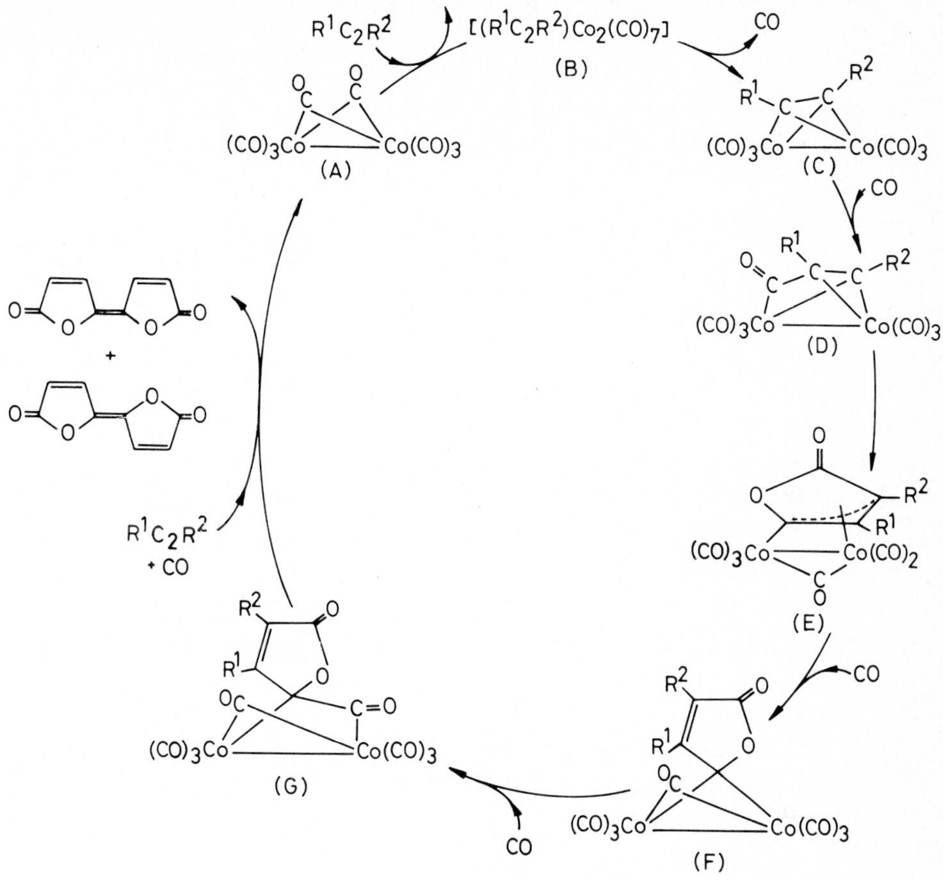

Scheme 9 Catalytic cycle of the formation of bifurandiones

(iii) Steps (A) to (F) or (C) to (F) can be performed separately by using saturated hydrocarbons (or in some cases benzene) as solvent. Several (F) complexes were prepared in this way [89, 160-162]. These were characterized by analyses, IR [79,89,161] and NMR [89,161,163] spectroscopy and (in one case, $R^1 = R^2 = H$) by X-ray diffraction [164].

(iv) The suggested structure of compound (E) is based on that of the analogous cobaltorganic intermediate of monolactone formation, (η^3-lactonyl)$Co(CO)_3$, which could be isolated and characterized [138] (Section 3.2.2).

(v) The structure of compounds (F) is interesting for two reasons. The overall structure shows a quantitative stereospecifity, i.e. the μ_2-lactone ring always turns its double-bond "edge" toward the bridging carbonyl group [164]. Only one isomer could be detected also with respect to the distribution of substituents R^1 and R^2 [89,162,163]. Thus, the carbonylation of an acetylene into the "first" lactone ring shows quantitative regio- and stereospecifity. This can be interpreted in terms of the ^{13}C-NMR spectra of the starting acetylenes and complexes (C) (Table 6) [89,163]. It can be deduced from these data that carbonylation starts only with those acetylenes in which the $\delta(^{13}C)$ value of at least one of the carbon atoms is lower than $\approx$ 80 ppm and, on the other hand, always that carbon atom is directed to the 3-position of the lactone ring which shows the lower $\delta(^{13}C)$ value (CR^1). These rules can be interpreted - as a crude approximation - by assuming that the more shielded (lower δ) carbon atoms would posses the more extended orbitals which, again, is favorable for a bond-forming overlap of an acetylene orbital with a coordinated CO orbital [167]. If so, it can be expected that carbonylation starts always at the same carbon atom and that the selective formation of one acyl group directs insertion of the second CO. We believe that the CO inserted first provides the "etheric" CO of the lactone ring. This is supported by the mechanism of monolactone formation (Section 3.2.2) where the acyl group furnishes this CO entity [138]. It can reasonably be supposed further that the difference in the fate of the "first" and the "second" acyl group is caused by an η^2-type coordination of the first one [168] in an intermediate between (D) and (E). If so, it cannot be excluded that the "second" carbonylation proceeds by "insertion" of CO into the acyl-O,Co bond [169]. Another alternative would be that the first acyl group pushes the other end

TABLE 6

^{13}C-NMR data of free acetylenes, $R^1C_2R^2$, and their $(\mu_2\text{-}R^1C_2R^2)Co_2(CO)_6$ derivatives [a] [δ] = ppm

Substituents		Free acetylene		Complexed acetylene		Differences		
R^1	R^2	1	2	1	2	1	2	n
H	nPr	66.7	81.8	73.2	98.5	6.5	16.7	23.2
H	nPent	67.4	82.9	74.0	99.0	6.6	16.1	22.7
H	Ph [c]	76.4	82.1	72.7	90.0	-3.7	7.9	11.6
H	COOMe	74.4	73.1	73.5	76.9	-0.9	3.8	4.7
H	tBu	66.9	91.2	73.4	112.0	6.5	20.8	27.3
Me	Me [b]	73.6	73.6	94.4	94.4	20.8	20.8	41.6
Me	nBu	74.2	77.6	94.0	100.1	19.8	22.5	42.3
Me	nPent	75.3	79.3	94.0	100.3	18.8	20.9	39.7
Me	Ph [e]	80.5	86.1	91.9	94.4	10.4	8.3	18.7
Et	Et	81.0	81.0	101.8	101.8	20.9	20.9	41.8
tBu	tBu	85.3	85.3	111.9	111.9	26.6	26.6	53.2
$SiMe_3$	$SiMe_3$	113.4	113.4	92.8	92.8	-20.6	-20.6	41.2
Ph	Ph [c]	88.9	88.9	91.0	91.0	2.1	2.1	4.2
COOMe	COOMe [c]	78.1	78.1	79.0	79.0	0.8	0.8	1.6
COOEt	Ph [e]	81.6	85.7	79.0	91.2	-2.6	5.5	3.1
nBu	OMe	35.7	87.8	6.2	90.3	29.5	2.5	31.0

[a] Generally CCl_4 solvent, TMS int. standard [89,163]

[b] From [165]

[c] From [88]

[d] From [166]

[e] Solvent C_6D_6

of the acetylene (CR^2) nearer to one of the cobalts and thus increases the extent of overlap between one of the orbitals of this carbon atom and one of the coordinated CO-s. This latter possibility, however, does not account for the formation of the C,O-bond in the lactone ring.

(vi) The formation of the second lactone ring may start in two different ways. One would be the coordination of acetylene and the construction of a second lactone ring followed by dimerization of the rings. Again this may happen

a/ on (two different) Co_2 entities or

b/ on the same Co_2 template where the first lactone ring was synthesized.

Possibility a/ can be ruled out since this would require the 2,7--isomer to be the only product from terminal acetylenes which it is not. Moreover, a thermal dimerization experiment provided a very low yield ($\sim$ 2%) of the corresponding 2,7-substituted bifurandione [162], showing that although this reaction may proceed it is, by no means, a principal reaction path. Alternative b/ cannot be ruled out but it seems more probable that the first step of the formation of the second ring is the insertion of a CO into the Co,C bond of the μ_2-lactonyl ligand. This is depicted in intermediate (G). Complexes of this type were not reported; however, the reaction of acetylenes with the μ_2-lactone on the Co_2 template was investigated in detail [170-172] (vide infra). The possibility that the acyl-CO may be formed on both sides of the lactone ring may account for the formation of the (Z) and (E) isomers with respect to the central double bond of the bifurandiones. This interpretation is supported by the two isomers of the acetylene adducts of complexes (F) [170-172].

(vii) No further speculation on the fate of (G) seems to be justified for the moment. It should be mentioned, however, that the lack of regioselectivity in the formation of the second ring may be due to the involvement of only one of the cobalt atoms in this process. This problem needs and merits further study.

It can reasonably supposed that the formation of the 3+4 adducts, the decatetraenediolides, may proceed according to essentially the same chemistry as described above. It is tempting to suppose that the crucial point of the formation of these bilactones is the insertion of that acetylene unit which links the two lactone rings. Such a possibility would also be interesting from the point of view

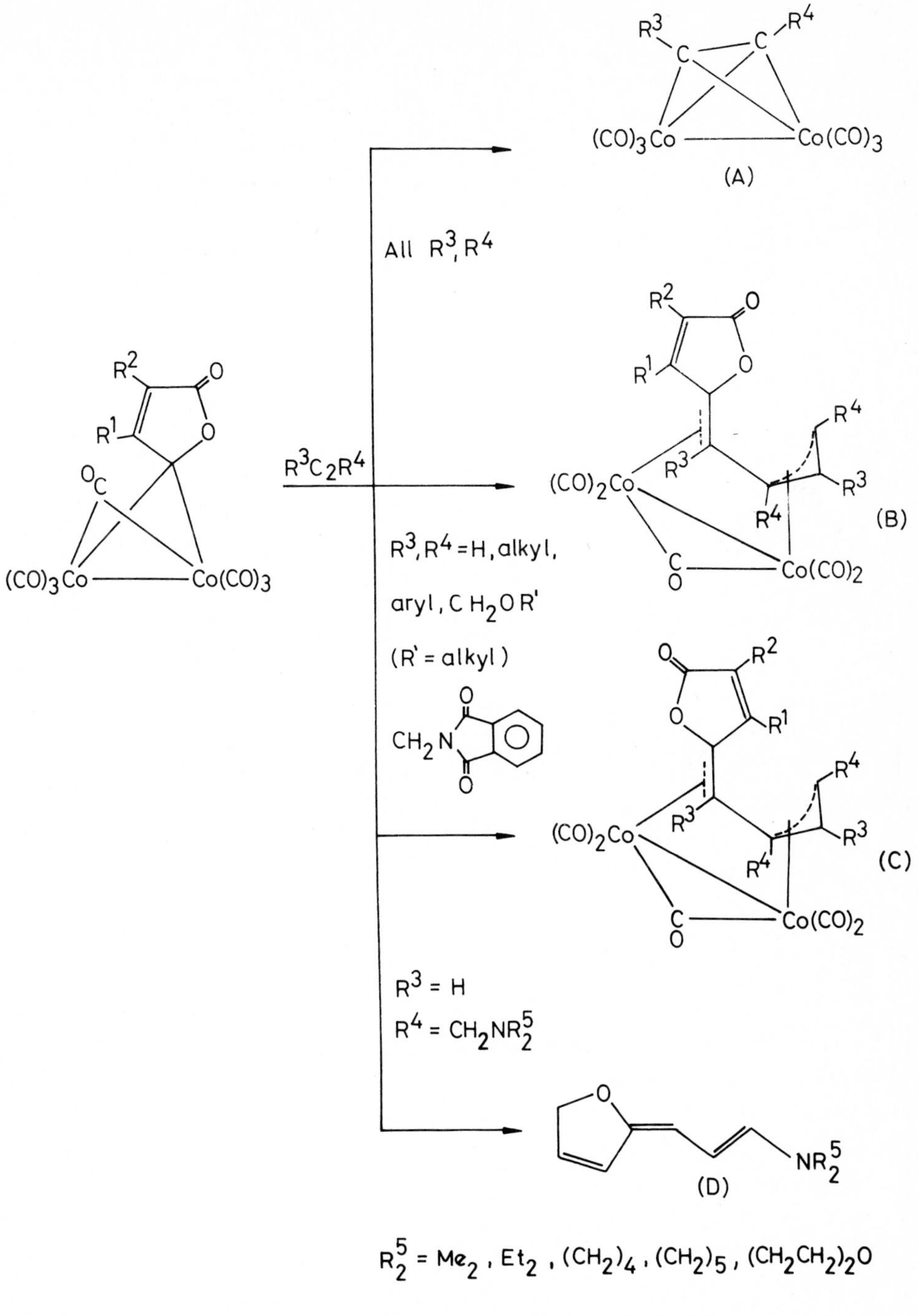

All R^3, R^4

$R^3C_2R^4$

R^3, R^4 = H, alkyl, aryl, CH_2OR' (R' = alkyl)

CH_2N (phthalimido)

R^3 = H
R^4 = $CH_2NR^5_2$

(A) (B) (C) (D)

R^5_2 = Me_2, Et_2, $(CH_2)_4$, $(CH_2)_5$, $(CH_2CH_2)_2O$

Scheme 10 Reaction of the (μ_2-butenolide)(μ_2-CO)-$Co_2(CO)_6$ complexes with acetylenes under mild conditions that, if the C_2H_2 unit linking the two lactones in the decatetraenediolide is really derived from an acetylene molecule, it would undergo considerable isomerization: the triple bond is transformed into a single bond.

The reaction of the acetylene with the μ_2-lactone ring could be modelled successfully as shown in Scheme 10. In the first series of experiments, bisadducts of the overall composition [(lactone)(acetylene)$_2$]$Co_2(CO)_5$ were found [170-172]. Complex (B) and (C) were characterized by analyses, IR, NMR and X-ray diffraction. The only difference between the structures of these complexes is the orientation of the lactone ring with respect to the rest of the molecule. This is in accord with the previous speculation that attack from one or the other side of the lactone ring may lead to stereoisomers.

Since complexes (B) and (C) contain two "new" acetylene moieties, they are not directly related to the octatrienediolides. Using acetylenes in which the double bond formed by isomerization becomes conjugated to a proper group (tertiary N), the monoadducts could also be prepared, shown as (D) in Scheme 10 [173]. If the conjugation capacity of the lone pair of the N atom is diminished (using N-propargylphtalimide as "new" acetylene component) the "usual" (B) type bis-adduct is formed.

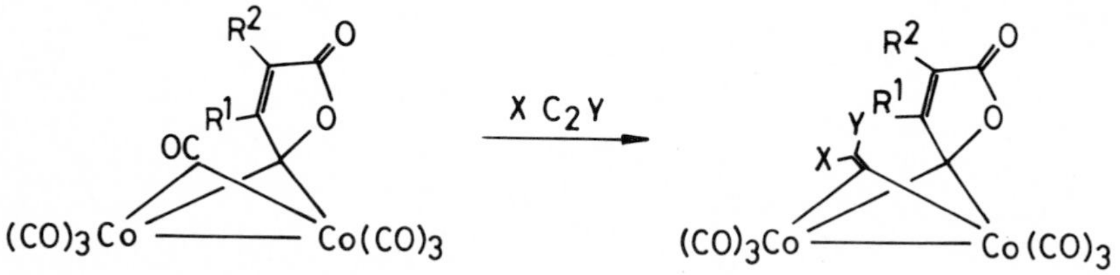

R^2 O R^1 O OC $(CO)_3Co$ $Co(CO)_3$ —— X C_2Y ——→ R^2 O Y R^1 O X $(CO)_3Co$ $Co(CO)_3$

X = alkyl, aryl, I

Y = Cl, Br, I

Scheme 11

As a relevant piece of information regarding the high degree of stereoselectivity in the formation of these dicobalt-complexes, the following observation should be mentioned: mono- and dihaloacetylenes react with the (μ_2-butenolide)(μ_2-CO)$Co_2(CO)_6$ complexes yielding the μ_2-vinylidene-complexes in Scheme 11 [174]. These complexes were characterized by analyses, IR, NMR- and X-ray diffraction. It is, again an interesting stereochemical aspect that in these complexes the vinylidene bridge is always in the same position with respect to the lactone ring as the (μ_2-CO) group in the starting complexes - that is "cis" to the double-bonded edge of the lactone ring.

4 ACETYLENE + OLEFIN + CARBON MONOXIDE REACTIONS

4.1 Organic products

These reactions lead to cyclopentenones and are therefore a logical extension of the formation of cyclopentadienones from two moles of acetylene and one molecule of CO. In spite of this, attention was drawn to this possibility relatively late when Pauson and Khand [175] reported their first observation in 1973. Research interest was then increased by its potential for the preparation of prostaglandine synthons [176] and other organic products of biological interest [177-181].

Pauson and Khand discovered this cyclization originally with strained (more reactive) olefins (as norbornene, nornbornadiene, diazonorbornadiene, etc.) and then extended it in an ingenious series of efforts to simpler and simpler olefinic partners achieving, finally, acceptable yields with ethylene itself. We summarized in Table 7 the most important types of these cyclizations (in the order of simple to more complicated olefins) together with some specific applications.

The most striking feature of the reaction [188] is the high degree of regio- and stereoselectivity, which makes it particularly attractive for the synthesis of structures requiring fine geometric selectivities. The following main rules can be formulated.

(i) The distribution of the substituents of the acetylene and the olefin (if not equal) shows always a preference for the bulkier substituent(s) to be directed into those positions of the new 5-atomic ring which are neighbouring the carbonyl. If the other isomer is needed, the steric effect of the substituent should be "masked": thus reacting methyl-(trimethylsilyl)-acetylene with an

TABLE 7 Co-cyclization of acetylenes, olefins and carbon monoxide

Acetylene	Olefin	Product	Yield (%)	Reference
HC_2Ph	C_2H_4	H; Ph; O	30	[182]
MeC_2nPent	C_2H_4	Me; nPent; O	17	[182]
$HC_2(CH_2)_6COOMe$	C_2H_4	COOMe; O	46	[183]
HC_2Ph	C_3H_6	(A): H; H; Me; H; H; Ph; O (B): Me; H; H; H; H; Ph; O	(A) 12 (B) 10.5	[182]
HC_2Ph	PhC_2H_3	(A): H; H; Ph; H; H; Ph; O (B): PhCH = CHCH = CHPh	(A) 12 (B) 39	[182]

TABLE 7 (continued)

Acetylene	Olefin	Product	Yield (%)	Reference
HC_2Ph		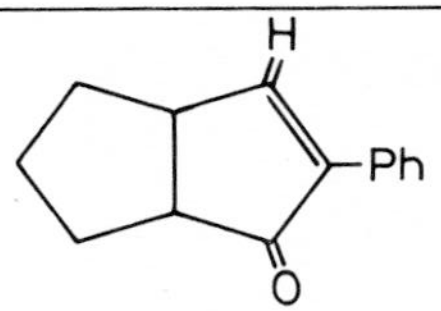 H, Ph, O	47	[182]
HC_2Me		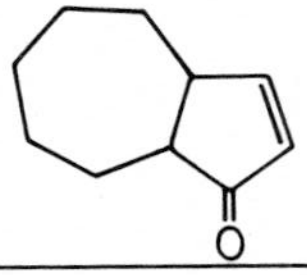 O	17	[182]
$HC_2(CH_2)_6COOMe$		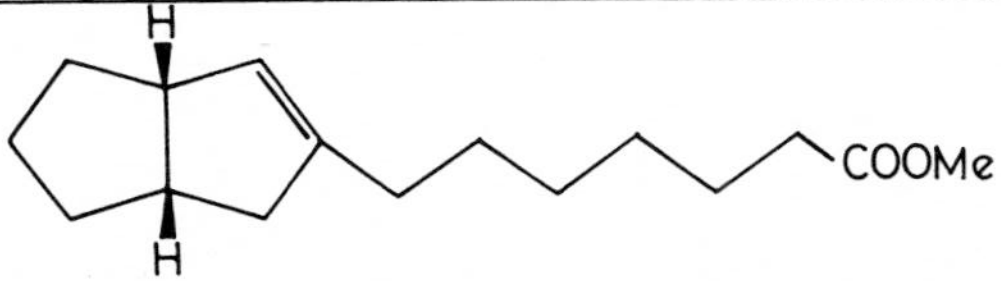 H, H, COOMe	33	[183]
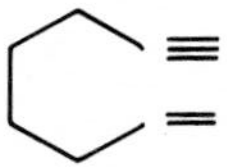		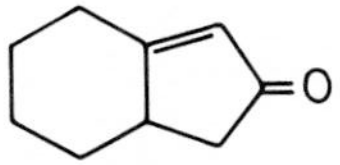 O	35-40	[184]
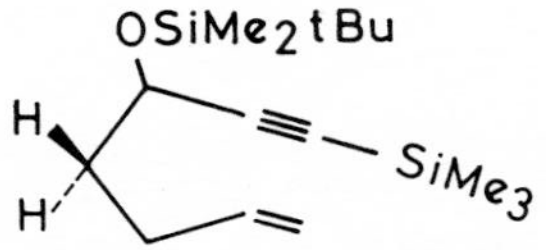 $OSiMe_2tBu$, H, H, $SiMe_3$		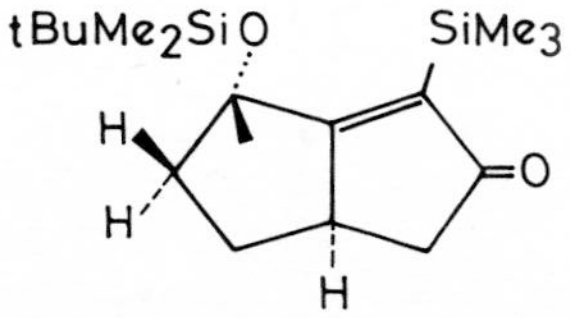 $tBuMe_2SiO$, $SiMe_3$, H, H, O, H	79	[185]

TABLE 7 (continued)

Acetylene	Olefin	Product	Yield (%)			Reference
			R^1	R^2		
$R^1C_2R^2$			H	H	52	[175]
			H	Me	33	
			H	Ph	59	
			Et	Et	23	
			Ph	Ph	35	
$R^1C_2R^2$			H	H	43	[175]
			H	Me	33	
			H	Ph	45	
			H	H	29	
			H	Me	17	
			H	Ph	13	
			H	H	4	
			H	Me	4	
			H	Ph	-	

TABLE 7 (continued)

Acetylene	Olefin	Product	Yield (%)	Reference
HC_2H	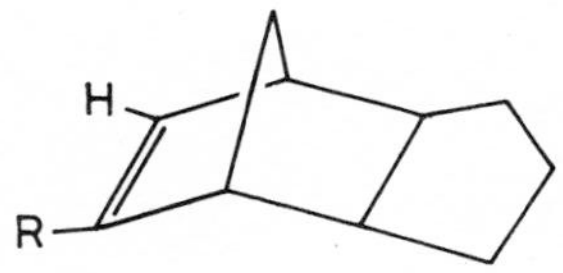 H, R	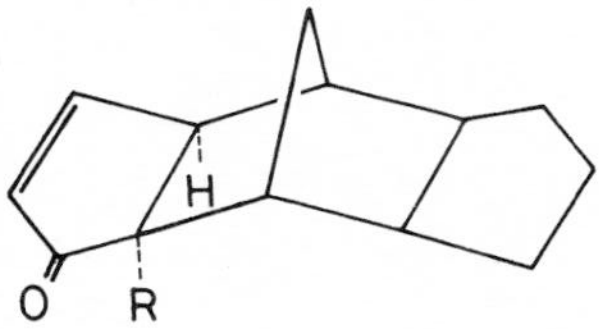 H, R, O	R Ph 60 Fc 65	[186]
HC_2R	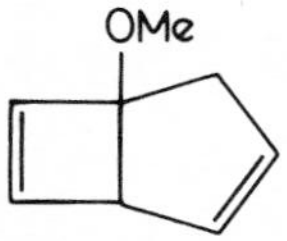 OMe	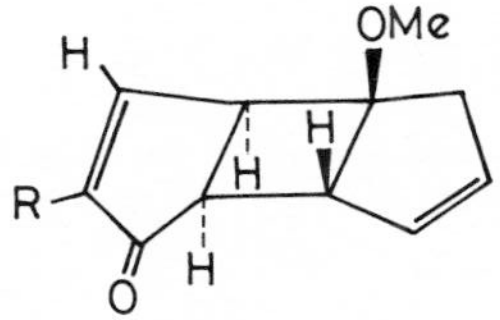 H, OMe, H, H, R, O, H	R H 46 Me 60 Ph 50	[187]

olefin followed by hydrolysis of the trimethylsilyl group provides the 3-methylcyclopentenone derivative.

(ii) The fused rings (if rigid) show the same stereochemistry in all cases, that is the annellation provides always an exo-configuration.

(iii) Even the geometry of remote rings influences markedly the configuration of the new fused ring systems. Norbornadiene (if reacted with both double bonds) gives a tetracyclic product in which both cyclopentenone rings are exo-oriented. It was proved in an elegant experiment that this is in fact the remote orientative effect of the "first" carbonyl gorup: cyclopentadiene dimer was used instead of norbornadiene and this provided an approximately 1:1 ratio of the corresponding cisoid and transoid isomers [186]. Another remote substituent effect was found in the reaction in which the olefinic component was 1-methoxybicyclo[3.2.0]-hepta--3,6-diene. Only one isomer was formed with the two five-membered rings trans with respect to the cyclobutane system and the methoxy group cis to the cyclopentenone ring [187].

These reactions of outstanding selectivity can be used fruitfully in the synthesis of complicated organic molecules, as shown in Table 7 and as reported in references [189-191]. It should be mentioned here that acetylene cyclotrimerization to benzene derivatives was combined in some cases with cyclizations involving double bonded functional groups [64,65,72] (c.f. Section 2.1.2.). Such cyclizations were found to be catalyzed also by a Ziegler/ /Natta type system generated from Co(II) salts and AlR_3 [192].

4.2 Mechanistic considerations

It is generally accepted that the regioselectivity of these reactions is controlled mostly by steric factors. In other words: the reaction starts with the substitution of a terminal carbonyl ligand of the (μ_2-acetylene)$Co_2(CO)_6$ complex by one olefin, then the less hindered ends of the two coordinated (organic) ligands couple and the ring is closed by reaction with CO at the free (bulkier) ends.

This overall picture is acceptable but it should be recalled that - at least for the acetylene - the less hindered end is also the more reactive one, as it could be deduced from the ^{13}C-NMR spectra. Most probably this is true for the coordinated olefin also; however, there are not enough NMR data on cobalt/olefin complexes to prove this hypothesis. It should also be rembered that the steric effect of the bulkier substituent(s) may be not only passive (geometric shielding of the adjacent sp or sp^2 carbons) but also active (pushing the less bulky ends of the unsaturated ligands nearer to each other).

The stereoselectivity of the ring-closure reaction can be understood if the coordination of both the acetylene and the olefin is considered to correspond to a cis-excited [86,92] state. Coupling is the consequence of the overlapping of the coordinatively generated, excited (π^*) orbitals of similar geometry which can be expected (due to the double coordination of the acetylene) to be of similar energy. It should be pointed out, however, that the coordination of the (terminal) olefin should be extremely sensitive to the geometry of the rest of the molecule. The clarification of this mechanism would merit a detailed preparative and structural study. The picture which can be deduced from the available data is sketched in Scheme 12.

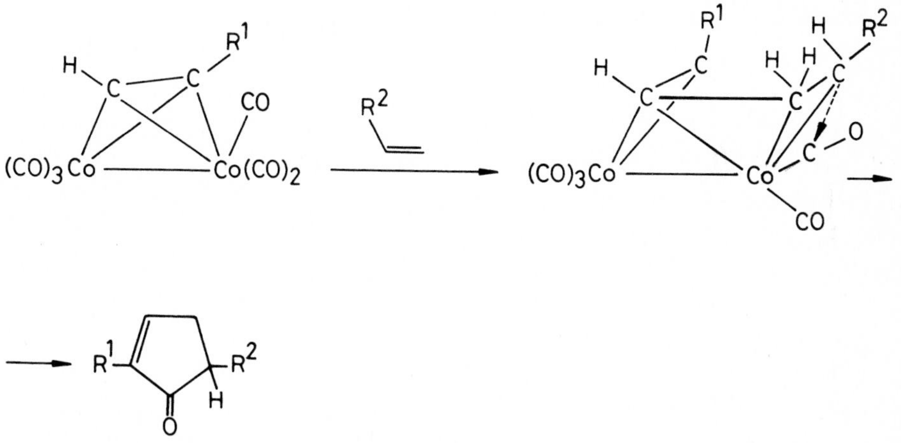

Scheme 12

4.3 Cyclization of acetylenes with various unsaturated ligands

Acetylenes either with or without CO react readily with a great variety of unsaturated molecules in the presence of low--valent cobalt compounds. The cocyclization of acetylene with nitriles (leading to pyridine derivatives) seems to be the most explored of these [23,28,193]. Recently even rules of orientation were established in terms of ^{13}C-NMR spectra [194].

The stereochemical aspects of some other reactions of this kind are only partly clear [102,112,121-123,195].

5 REACTIONS IN THE SIDE CHAIN OF ACETYLENES COORDINATED TO COBALT

The use of the coordinated $Co_2(CO)_6$ moiety as a protecting group for the C,C triple bonded entity emerged almost simultaneously in the early 70ies from two important US research centers. Seyferth reported [196] the Friedel-Crafts alkylation of arylacetylenes while Nicholas and Pettit [197] converted $Co_2(CO)_6$-protected enynes to the corresponding ynes or ynols by reduction (N_2H_4 or BH_3) or hydroboration. The protecting group can be effectively removed by oxidation with Fe(III) or Ce(IV) salts [196].

Nicholas later recognized that the latter reactions proceed through the formation of an α-carbenium ion center in the side chain of the coordinated alkyne [198]. He succeeded in isolating and characterizing salts of this carbenium ion [199] which is stabilized by the neighbouring organometallic group. Following these results, the Boston College group developed several interesting reactions of these complexed carbenium ions. These reactions, due to the bulky C_2Co_2 cluster adjacent to the reaction center, show outstanding regio- and stereoselectivity. Thus the $[\mu_2\text{-}R^1C_2C^+(R^1,R^2)]Co_2(CO)_6$ cations were used for the propargylation of aromatics [200], preparation of enynes (via coupling with allylsilanes) [201], alkylation (propargylation) of β-dicarbonyl compounds [202], selective α-alkylation (propargylation) of ketones [203] and in the synthesis of skipped 1,5-diynes [204]. Such carbonium ions are the intermediates of the regioselective acid-catalyzed reactions of 1,2-epoxyalkynes with nucleophiles [205]:

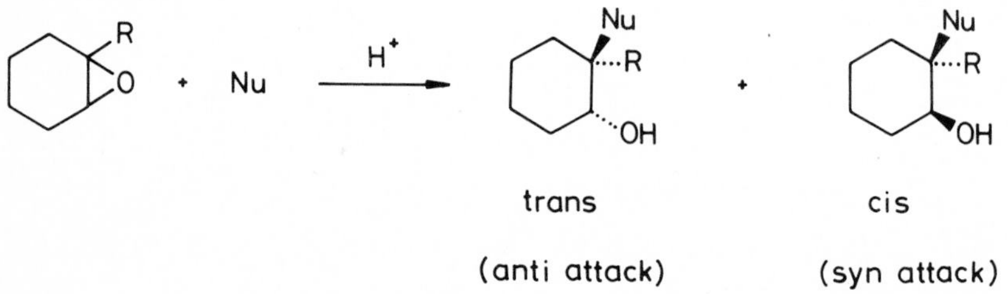

TABLE 8

Comparison of the reactivity of $(C_2H)Co_2(CO)_6$ substituted cyclohexene oxides with organic analogs [205] in acid catalyzed reactions

Epoxide R	Nucleophile	syn/anti
C ≡ CH	MeOH [a]	1:99
$(C\equiv CH)Co_2(CO)_6$	MeOH [b]	50:50
Me	H_2O [c]	< 0.2:100
C ≡ CH	H_2O [c]	2:98
Ph	H_2O [c]	63:37
$(C\equiv CH)Co_2(CO)_6$	H_2O [c]	59:41
Me	Cl_3CCOOH	6:94
C ≡ CH	Cl_3CCOOH	42:48
Ph	Cl_3CCOOH	100:0.2
$(C\equiv CH)Co_2(CO)_6$	Cl_3CCOOH	> 95:5

[a] TsOH catalyst

[b] HBF_4 catalyst

[c] H_2SO_4 catalyst

The $[C_2H(Co_2)CO]_6$ substituent strongly influences the stereoselectivity of the reactions by favoring the syn attack which leads to the cis product. Table 8 shows some examples of how dramatically the syn selectivity increases with respect to the "purely" organic reactions. The particular reactivity of these organometallic carbenium ions was successfully utilized in the synthesis of natural organic substances [206] or in further alkylation reactions [207].

It should be mentioned in this context that the stabilization of the α-carbenium ion can be attributed [208] to the electron releasing effect of the $C_2Co_2(CO)_6$ group, similarly as the stabilization of the $Co_3(CO)_9CCO^+$ acylium ions [209] can also be ascribed to the "electron-sink" represented by the $Co_3(CO)_9C$ tetrahedral cluster unit [210]. The stabilization of the ethynylcarbenium ions by complexation to the $Co_2(CO)_6$ moiety is not an isolated phenomenon, this "protecting" group can also be used as "stabilizing" group of otherwise unstable triple bonded organic [213] or inorganic [214-216] molecules.

6 CONCLUSIONS AND OUTLOOK

An attempt has been made to summarize those acetylene/ cobalt carbonyl compounds or reactions of acetylenes in the presence of cobalt carbonyls (and some similar low valent Co compounds) which are important from a stereochemical viewpoint. The review was not intended to be comprehensive, our aim was more to focus light on systems where coordination contributes to the development of reaction courses which result in virtually one-step syntheses yielding otherwise not, or hardly accessible substances.

It can be stated that many important features of the reactions of acetylenes with, or catalyzed by, cobalt carbonyls are already known. Further efforts seem, however, to be necessary because

- important steps in catalytic or assited reactions are presently only assumed on the basis of analogies and on the other hand
- reaction possibilities were explored in many cases only with simple models and the effective application of these reactions for more complex molecules has still to be developed.

Results culminating in the syntheses of complicated organic molecules in the last decade represent highlights of this chemistry and at the same time the first steps on the way to transform transition organometallic chemistry from a fascinating academic topic to a routine laboratory or even technical organic chemical method.

ACKNOWLEDGEMENTS

The authors acknowledge fruitful discussions to the following colleagues: Prof.M.Bán (Szeged), Prof.G.Bor (Zürich), Professa.A. Furlani (Roma), Dr.I.T.Horváth (Zürich), Prof.K.M.Nicholas (Boston, Mass.), Prof.P.L.Pauson (Glasgow), Prof.E.Sappa (Torino), Prof.D. Seyferth (Cambridge, Mass.) and Prof.P.L.Stanghellini (Torino). These discussions contributed significantly to the ideas put forward in this paper.

REFERENCES AND NOTES

1 E.Barthalome, Chem.-Ing.-Techn., 49, (6) (1977) 459.

2 D.A.Duncan, Kirk-Othmer Enzycl.Chem.Technol., 3rd Ed., 1, (1978) 211.

3 G.Kováts, Magyar Kém.Lapja, 34, (1979) 1, 65.

4 Chem.Engn.News. 57,(24)(1979) 32.

5 Chem.Engn.News, 57, (9)(1979) 9.

6 Chem.Engn.News, 58, (5)(1980) 11.

7 M.Blaiver and G.Lefebvre, Rev.Inst.Fr.Petrol., 31, (1) (1976) 149.

8 M.K.Sarkar and S.S.Gosh, Indian Chem.Engn., 18, (1976) 30.

9 R.Bonmati, M.Guttierrez and H.Tollet, Fr.Pat.Demande. 2,294,225 (1976).

10 I.M.Romanyuk and A.N.Andrushko, Khim.-prom.-st. (Moscow), (10) (1976) 733.

11 E.Yanagizawa, Japan Pat., 01,242 (1978).

12 J.Popowski, Ger.Pat., 2,748,893 (1979).

13 T.J.Manuccia, US Pat.Appl., 721,907 (1977).

14 I.Nemes, T.Vidróczy, L.Botár and D.Gál, Arch.Termodyn. Spalania, 8, (1977) 43.

15 G.Moegel, H.Oribold and H.Schlief, DDR Pat., 127,002 (1977).

16 Acad.Sci.Belorus SSR, Fr.Pat.Demande, 2,304,243 (1976).

17 J.Kulczyczka, Khim.Tverd.Topl. (Moscow), (3) (1978) 102.

18 D.Hebecker and H.Schmidt, DDR Pat., 123,084 (1976).

19 A.Szimanski, Khim.Tverd.Topl. (Moscow), (1) (1979) 84.

20 Chem.Engn.News. 53, (16) (1975) 37; 40.

21 R.S.Dickson and P.J.Fraser, Adv.Organomet.Chem., 12, (1974) 323.

22 P.Pino and G.Braca, Organic Syntheses via Metal Carbonyls (I.Wender and P.Pino Eds.). J.Wiley-Interscience, New York, 2, (1977) 419.

23 K.P.C.Vollhardt, Accounts Chem.Res., 10, (1977) 1.

24 K.P.C.Vollhardt, Nachr.Chem.Techn.Lab., 25, (10)(1977) 584.

25 P.L.Pauson and I.U.Khand, Ann.N.Y.Acad.Sci., 295, (1977) 2.

26 E.Müller, Synthesis, (1974) 761.

27 J.Hambrecht, H.Straub and E.Müller, Tetrahedron Lett., (1976) 1789.

28 H.Bönnemann and R.Brinkman, Synthesis, (1975) 600.

29 G.Pályi, G.Váradi and I.T.Horváth, J.Mol.Catalysis, 13, (1981) 61.

30 M.Saha and K.M.Nicholas, J.Org.Chem., 49, (1984) 417 and refs. cited therein.

31 An excellent review covering all aspects of this field appeared recently: E.Sappa, A.Tiripicchio and P.Braunstein, Chem.Revs., 83, (1983) 203.

32 S.W.Benson, Thermochemical Kinetics. J.Wiley, New York, 1968.

33 G.M.Badger, G.E.Lewis and I.M.Napier, J.Chem.Soc., (1960) 2825.

34 T.Ito, H.Shirakawa and S.Ikeda, J.Polym.Sci., Polym.Chem., 13, (1975) 1943.

35 C.K.Chiang, A.J.Heeger and A.G.McDiarmid, J.Phys.Chem., 83, (1979) 407.

36 A.Furlani, M.V.Russo, P.Carusi, S.Licoccia, E.Leoni and G.Valenti, Gazz.Chim.Ital., 113,(1983) 671.

37 J.Sen, Personal communication, 1983.

38 A.Furlani, S.Licoccia, M.V.Russo, A.Camus and N.Marsich, J.Polym.Sci., Polym.Chem., in the press.

39 A.Furlani, personal communication, 1984.

40 W.Reppe, O.Schlichting, K.Klager and T.Toepel, Justus Liebigs Ann.Chem., 560, (1948) 1.

41 SNAM Spa. Belg.Pat., 648.530 (1964).

42 W.Hübel and C.Hoogzand, Chem.Ber., 93, (1960) 103.

43 H.Greenfield, H.W.Sternberg, R.A.Friedel, J.H.Wotiz, R.Markby and I.Wender, J.Am.Chem.Soc., 78, (1956) 120.

44 M.R.Tirpak, C.A.Holingsworth and J.H.Wotiz, J.Org.Chem., 25, (1960) 687.

45 U.Krüerke and W.Hübel, Chem.Ber., 94, (1961) 2829.

46 W.Hieber and R.Breu, Chem.Ber., 90, (1957) 1259.

47 C.Hoogzand and W.Hübel, Organic Syntheses via Metal Carbonyls. (I.Wender and P.Pino, Eds.), Wiley-Interscience, New York, 1, (1968) 343.

48 U.Krüerke, C.Hoogzand and W.Hübel, Chem.Ber., 94, (1961) 2817.

49 C.Hoogzand and W.Hübel, Angew.Chem., 73, (1961) 680.

50 C.Hoogzand and W.Hübel, Tetrahedron Lett., (1961) 637.

51 E.M.Arnett and J.M.Bollinger, J.Am.Chem.Soc., 86, (1964) 4729.

52 H.Hopf and A.Gati, Helv.Chim.Acta, 48, (1965) 509.

53 W.Hübel and R.Merényi, Chem.Ber., 96, (1963) 930.

54 E.M.Arnett, M.E.Strem and R.A.Friedel, Tetrahedron Lett., (1961) 658.

55 E.M.Arnett and M.E.Strem, Chem.Ind.(London), (1961) 2008.

56 W.Hübel, Organic Syntheses via Metal Carbonyls. (I.Wender and P.Pino, Eds.), Wiley-Interscience, New York, 1, (1968) 273.

57 R.Markby, H.W.Sternberg and I.Wender, Chem.Ind.(London), (1959) 1381.

58 R.B.King and A.Efraty, J.Am.Chem.Soc., 94 (1972) 3021.

59 K.P.C.Vollhardt and R.G.Bergman, J.Am.Chem.Soc., 96, (1974) 4996.

60 W.G.L.Aalbersberg, A.J.Barkovich, R.L.Funk, R.L.Hillard III and K.P.C.Vollhardt, J.Am.Chem.Soc., 97, (1975) 5600.

61 R.L.Hillard III and K.P.C.Vollhardt, J.Am.Chem.Soc., 99, (1977) 4058.

62 R.L.Hillard III and K.P.C.Vollhardt, Angew.Chem., 87 (1975) 744.

63 K.P.C.Vollhardt, Accounts Chem.Res., 10, (1977) 1.

64 R.L.Funk and K.P.C.Vollhardt, J.Am.Chem.Soc., 99, (1977) 5483.

65 J.-C.Clinet, E.Dunach and K.P.C.Vollhardt, J.Am.Chem.Soc., 105, (1983) 6710.

66 E.D.Sternberg and K.P.C.Vollhardt, J.Am.Chem.Soc., 102 (1980) 4841.

67 T.R.Gadek and K.P.C.Vollhardt, Angew.Chem., 93 (1981) 801.

68 R.A.Earl and K.P.C.Vollhardt, J.Am.Chem.Soc., 105 (1983) 6991.

69 M.D'Alarcao and N.J.Leonard, J.Am.Chem.Soc., 105,(1983) 5958.

70 R.L.Funk and K.P.C.Vollhardt, Angew.Chem., Int.Ed.Engl., 15, (1976) 53.

71 C.J.Saward and K.P.C.Vollhardt, Tetrahedron Lett., (1975) 4539.

72 R.L.Funk and K.P.C.Vollhardt, J.Am.Chem.Soc., 98 (1976) 6755.

73 D.R.McAlister, J.E.Bercaw and R.G.Bergman, J.Am.Chem.Soc., 99 (1977) 1666.

74 L.P.McDonnell-Bushnell, E.R.Evitt and R.G.Bergman, J.Organomet.Chem., 157, (1978) 445.

75 F.Ungváry and L.Markó, Chem.Ber., 105 (1972) 2457.

76 P.C.Ellgen, Inorg.Chem., 11 (1972) 691.

77 H.W.Sternberg, R.A.Greenfield, J.H.Wotiz, R.A.Friedel, R.Markby and I.Wender, J.Am.Chem.Soc., 76 (1954) 1457.

78 R.S.Dickson and P.J.Fraser, Adv.Organomet.Chem., 12 (1974) 323.

79 G.Bor, Chem.Ber., 96, (1963) 2644.

80 G.Bor, J.Organomet.Chem., 94, (1975) 181.

81 G.Váradi, I.Vecsei, A.Vizi-Orosz, G.Pályi and A.G.Massey, J.Organomet.Chem., 114, (1976) 213.

82 G.Bor, P.L.Stanghellini and S.F.A.Kettle, Inorg.Chim.Acta, 18, (1976) L18.

83 R.S.Dickson and D.B.W.Yawney, Aust.J.Chem., 20, (1967) 77; 21, (1968) 97, 1077; 22, (1969) 1143.

84 Y.Iwashita, F.Tamura and A.Nakamura, Inorg.Chem., 8, (1968) 1179.

85 Y.Iwashita, Inorg.Chem., 9, (1970) 1178.

86 Y.Iwashita, A.Ishikawa and M.Kainosho, Spectrochim.Acta, 27A, (1971) 271.

87 A.Meyer and M.Bigorgne, Organometallics, 3 (1984) 1112.

88 S.Aime, L.Milone, R.Rosetti and P.L.Stanghellini, Inorg.Chim. Acta, 22, (1977) 135.

89 G.Váradi, I.Vecsei, I.Ötvös, G.Pályi and L.Markó, J.Organomet. Chem., 182, (1979) 415.

90 G.Pályi, G.Váradi and I.T.Horváth, J.Mol.Catalysis, 13, (1981) 61.

91 W.G.Sly, J.Am.Chem.Soc., 81, (1959) 18.

92 D.A.Brown, J.Chem.Phys., 33, (1960) 1037.

93 F.A.Cotton, J.D.Jamerson and B.R.Stults, J.Am.Chem.Soc., 98, (1976) 1774.

94 D.Gregson and J.A.K.Howard, Acta Cryst., 39c, (1983) 1024.

95 D.L.Thorn and R.Hoffmann, Inorg.Chem., 17, (1978) 126.

96 K.I.Goldberg, D.M.Hoffmann and R.Hoffmann, Inorg.Chem., 21, (1982) 3863.

97 M.I.Bán, M.Révész, I.Bálint, G.Váradi and G.Pályi, J.Mol. Struct. THEOCHEM, 88, (1982) 375.

98 M.I.Bán, G.Dömötör, A.Vizi-Orosz and G.Pályi, Atti Accad.Sci. Bologna, Cl.Sci.Fis., 271, (1982/3) [13/10] 217.

99 R.S.Dickson and G.N.Pain, J.Chem.Soc.Chem.Commun. (1979) 277.

100 Y.Iwashita, F.Tamura and H.Wakamatsu, Bull.Chem.Soc. Japan, 43,(1970) 1520.

101 N.M.Boag, M.Green, J.A.K.Howard, J.L.Spencer, R.F.D. Stansfield, F.G.A.Stone, M.D.O.Thomas, J.Vicente and P.Woodward, J.Chem.Soc.Chem.Commun., (1977) 930.

102 (a) M.A.Bennett and P.B.Donaldson, Inorg.Chem. 17,(1978) 1955. (b) H.Yamazaki, K.Yasufuku and Y.Wakatsuki, Organometallics, 2,(1983) 726.

103 B.L.Booth, R.N.Haszeldine and I.Perkins, J.Chem.Soc. Dalton Trans., (1981) 2593.

104 M.Angoletta, P.L.Bellon, F.Demartin and M.Manassero, J.Chem. Soc.Dalton Trans.,(1981) 150.

105 R.S.Dickson and P.J.Fraser, Aust.J.Chem., 23,(1970) 475.

106 R.S.Dickson and D.B.W.Yawney, Aust.J.Chem., 22,(1969) 533.

107 O.S.Mills and G.Robinson, Proc.Chem.Soc., (1964) 187.

108 F.S.Stephens, Acta Cryst., 21A,(1966) 154.

109 R.S.Dickson, P.J.Fraser and B.M.Gatehouse, J.Chem.Soc. Dalton, Trans., (1972) 2278.

110 H.Yamazaki and N.Hagihara, J.Organomet.Chem., 7, (1967) P22.

111 H.Yamazaki and N.Hagihara, J.Organomet.Chem., 21,(1970) 431.

112 H.Yamazaki and Y.Wakatsuki, J.Organomet.Chem., 139,(1977) 157.

113 Y.Wakatsuki and H.Yamazaki, J.Am.Chem.Soc., 95,(1973) 5781.

114 S.Otsuka and A.Nakamura, Adv.Organomet.Chem., 14,(1976) 245.

115 Y.Wakatsuki, O.Nomura, K.Kitaura, K.Morokuma and H.Yamazaki, J.Am.Chem.Soc., 105,(1983) 1907.

116 K.Yasufuku and H.Yamazaki, J.Organomet.Chem., 127,(1977) 197.

117 H.Yamazaki and N.Hagihara, Bull.Chem.Soc.Japan, 44,(1971) 2260.

118 Y.Wakatsuki, K.Aoki and H.Yamazaki, J.Am.Chem.Soc., 96, (1974) 5284.

119 R.G.Gastinger, M.D.Rausch, D.A.Sullivan and G.J.Palenik, J.Am.Chem.Soc., 98,(1976) 719.

120 H.Yamazaki and Y.Wakatsuki, J.Organomet.Chem., 272,(1984) 251.

121 H.Yamazaki and Y.Wakatsuki, Bull.Chem.Soc.Japan, 52, (1979) 1239.

122 Y.Wakatsuki, T.Kuramitsu and H.Yamazaki, Tetrahedron Lett., (1974) 4549.

123 Y.Wakatsuki, O.Nomura, H.Tone and H.Yamazaki, J.Chem.Soc. Perkin Trans.II, (1980) 1344.

124 J.Falbe: Carbon Monoxide in Organic Synthesis. Springer Berlin, 1970.

125 J.Falbe (Ed.): New Syntheses with Carbon Monoxide. Springer, Berlin 1980.

126 G.Natta and P.Pino, US Pat., 2,851,486 (1955).

127 (a) J.C.Sauer, B.W.Howk and R.T.Stiehl, J.Am.Chem.Soc., 81, (1959) 2339; (b) B.W.Howk and R.T.Stiehl, J.Am.Chem.Soc., 81,(1959) 2339; b B.W.Howk and J.C.Sauer, US Pat., 2,859,214 (1958).

128 I.T.Horváth, I.Pelczer, G.Szabó and G.Pályi, J.Mol. Catalysis, 20,(1983) 153.

129 J.A.Osborn, F.H.Jardine, J.F.Young and G.Wilkinson, J.Chem. Soc. A, (1966) 1711.

130 R.S.Dickson, C.Mok and G.Connor, Aust.J.Chem., 30,(1977) 2143.

131 M.Gerlock and R.Mason, Proc.Chem.Soc., (1963) 107; Proc. Royal Soc. A, 279,(1964) 170.

132 R.S.Dickson and G.Wilkinson, J.Chem.Soc., (1964) 2699.

133 M.D.Rausch and R.A.Genetti, J.Org.Chem., 35,(1970) 3888.

134 D.P.Tate, J.M.Augl, W.M.Ritchey, B.L.Loss and J.G.Grasseli, J.Am.Chem.Soc., 86,(1964) 3261.

135 J.M.J.Tetteroo: Dissertation, Univ.Aachen, 1965.

136 E.R.F.Gesing, J.P.Tane and K.P.C.Vollhardt, Angew.Chem. Int.Ed.Engl, 19,(1980) 1023.

137 N.E.Schore, B.E.LaBelle, M.J.Knudsen, H.Hope and X.J.Xu, J.Organomet.Chem., 272,(1984) 435.

138 R.F.Heck, J.Am.Chem.Soc., 81,(1964) 2819.

139 R.F.Heck, US Pat., 3,293,265 (1966).

140 H.Alper, J.K.Currie and H.des Abbayes, J.Chem.Soc.Chem. Commun., (1978) 311.

141 (a) J.C.Sauer, Ger.Pat., 1,054,086 (1955).
(b) J.C.Sauer, US Pat., 3,840,570 (1958)

142 W.Reppe and A.Magin, Ger.Pat., 1,071,077 (1955).

143 J.C.Sauer, R.D.Cramer, V.A.Engelhardt, T.A.Ford, H.E. Holmquist and B.W.Howk, J.Am.Chem.Soc., 81,(1959) 3677.

144 G.Albanesi and M.Tovaglieri, Chim.Ind.(Milano), 41,(1959) 189.

145 G.Albanesi and M.Tovaglieri, Swiss Pat., 376,900 (1964).

146 G.Albanesi, Chim.Ind.(Milano), 46,(1964) 1169.

147 G.Albanesi, R.Farina and A.Taccioli, Chim.Ind.(Milano), 48,(1966) 1151.

148 W.Reppe, N.V.Kutepow and A.Magin, Angew,Cnem.Int.Ed.Eng. 8,(1969) 727.

149 G.Váradi, I.T.Horváth, J.Palágyi, T.Bak and G.Pályi, J.Mol. Catalysis, 9,(1980) 457.

150 T.Bak, I.T.Horváth, L.Markó, J.Palágyi, G.Pályi and G.Váradi Hung.Pat., 180, 501 (1983); Swiss.Pat.Appl. 8208 (1980); Ger.Offen., 3,041,955 (1981).

151 (a) H.E.Holmquist, US Pat., 2,835,710 (1958);
(b) B.W.Howk and J.C.Sauer, US Pat., 2,840,548 (1958);
(c) H.E.Holmquist, US Pat., 2,849,457 (1958);
(d) J.C.Sauer, US Pat., 2,859,220 (1958);
(e) H.E.Holmquist, US Pat., 2,866,792 (1958);
(f) H.E.Holmquist, US Pat., 2,866,793 (1958);
(g) H.E.Holmquist, US Pat., 2,884,450 (1959);
(h) H.E.Holmquist, US Pat., 2,892,844 (1959);
(i) H.E.Holmquist and J.C.Sauer, 2,957,009 (1960).

152 (a) H.E.Holmquist, F.D.Marsh, J.C.Sauer and V.A.Engelhardt, J.Am.Chem.Soc., 81, (1959) 3681;
(b) H.E.Holmquist, J.C.Sauer, V.A.Engelhardt and S.W.Howk, J.Am.Chem.Soc., 81, (1959) 3686.

153 There is only one report about the selective formation of selective formation of one isomer: only (Z) isomer fo the unsubstituted bifurandione was found in $(Me_2N)_2CO$ [143]. This result would merit further study.

154 (a) G.Allegra, Atti Accad.Nazl.Lincei,Rend., Cl.Sci.Fis.Mat. Nat., 28,(1960) 197;
(b) A.Colombo and G.Allegra, Atti Accad.Nazl.Lincei., Rend., Cl.Sci.Fis.Mat.Nat., 36, (1964) (2) 187;
(c) A.Colombo and G.Allegra, Acta Cryst., 21, (1966) 124.

155 C.W.Bird and C.Y.Wong, Tetrahedron, 30,(1974) 2331.

156 C.W.Bird and D.Y.Wong, Tetrahedron, 31, (1975) 31.

157 It is one of the fundamental problems of "cluster catalysis" [158] whether the cluster remains intact during the whole process or the actual catalytic effect is due to dissociated, highly active, indetectable mononuclear species. There are only very few cases where this aspect is cleared beyond doubt.

158 Some leading references: E.L.Muetterties, Bull.Chem.Soc. Belge, 84, (1975) 959; 85,(1976) 451. E.Band and E.L. Muetterties, Chem.Rev., 78, (1978) 639. E.L.Muetterties and J.Stein, Chem.Rev., 79,(1979) 479. M.Tachikawa and E.L.Muetterties, Progr.Inorg.Chem., 28,(1981) 203. J.S. Bradley, Adv.Organomet.Chem., 22,(1983) 1. H.Vahrenkamp Adv.Organomet.Chem., 22,(1983) 169. R.D.Aadams and I.T. Horváth, Progr.Inorg.Chem., (1984).

159 L.Papp and G.Pályi, unpublished.

160 H.W.Sternberg, J.G.Shukys, C.Delle Donne, R.Markby, R.A.Friedel and I.Wender, J.Am.Chem.Soc., 81,(1959) 2339.

161 G.Pályi, G.Váradi, A.Vizi-Orosz and L.Markó, J.Organomet. Chem., 90,(1975) 85.

162 D.J.S.Guthrie, I.U.Khand, G.R.Knox, J.Kollmeyer, P.L.Pauson and W.E.Watts, J.Organomet.Chem., 90,(1975) 95.

163 G.Váradi, I.T.Horváth and G.Pályi, unpublished.

164 O.S.Mills and G.Robinson (a) Proc.Chem.Soc., (1964) 187; (b) Inorg.Chim.Acta, 1,(1967) 61.

165 R.A.Friedel and H.L.Rectofsky, J.Am.Chem.Soc., 85,(1963) 1300.

166 D.Rosenberg, J.W.de Haan and W.Drenth, Rec.Trav.Chim. Pay-Bas, 87,(1968) 1387.

167 A similar argumentation has been used to explain the acetylene/acetylene coupling in cyclotrimerization (c.f. Scheme 3 and comments).

168 E.J.Kuhlman and J.J.Alexander, Coord.Chem.Rev. 33,(1980) 185.

169 No systematic study is available on the relative reactivity of the Co,C and Co,O bonds towards CO, but comparison of the few relevant results suggest that the latter is more reactive: Co,C bond: [168], V.Galamb and G.Pályi, Coord. Chem.Revs. 59,(1984) 203 Co,O bond: M.Tasi and G.Pályi, Organometallics, accepted for publication.

170 P.A.Elder, D.J.S.Guthrie, J.A.D.Jeffreys, G.R.Knox, J.Kollmeier, P.L.Pauson, D.A.Symon and W.E.Watts, J.Organomet.Chem., 120,(1976) C13.

171 G.Váradi, I.T.Horváth, G.Pályi, L.Markó, Yu.L.Slovokhtov and Yu.T.Struchkov, J.Organomet.Chem., 206,(1981) 119.

172 J.A.D.Jeffreys, J.Chem.Soc.Dalton Trans., (1980) 435.

173 I.T.Horváth and G.Pályi, to be published

174 I.T.Horváth, G.Pályi, L.Markó and G.D;Andreetti, J.Chem. Soc.Commun., (1979) 1054; Inorg.Chem., 22,(1983) 1049.

175 I.U.Khand, G.R.Knox, P.L.Pauson, W.E.Watts and M.I.Foreman, J.Chem.Soc.Perkin Trans. 1., (1973) 977.

176 (a) R.E.Ireland and R.H.Mueller, J.Am.Chem.Soc., 94, 1972 5897; (b) E.J.Corey and G.Moinet, J.Am.Chem.Soc., 95, (1973) 6831; (c) G.Stork and S.Raucher, J.Am.Chem.Soc., 98, (1976) 1583.

177 R.A.Ellison, Synthesis, (1973) 397.

178 A.Smith, III., B.Toder, S.Branca and R.K.Dieter, J.Am.Chem. Soc., 103,(1981) 1996.

179 D.K.Klipa and H.Hart, J.Org.Chem., 46,(1981) 2815.

180 Y.Takahashi, K.Isobe, H.Hagiwara, H.Kosugi and H.Uda, J.Chem.Soc.Chem.Commun., (1981) 714.

181 P.F.Schuda, H.L.Ammon, M.R.Heimann and S.Bhattacharjee, J.Org.Chem., 47,(1982) 3434.

182 I.U.Khand and P.L.Pauson, J.Chem.Res., (S) (1977) 9; (M) (1977) 0168.

183 R.F.Newton, P.L.Pauson and R.G.Taylor, J.Chem.Res., (S) (1980) 277; (M) (1980) 3501.

184 N.E.Schore and N.C.Crondace, J.Org.Chem., 46,(1981) 5436.

185 C.Exon and P.Magnus, J.Am.Chem.Soc., 105 (1983) 2477.

186 I.U.Khand and P.L.Pauson, J.Chem.Soc., Perkin Trans. 1., (1976) 30.

187 P.Bladon, I.U.Khand and P.L.Pauson, J.Chem.Res., (S) (1977) 8; (M) (1977) 0153.

188 A review, covering the results of Pauson's group up to 1977: [25]

189 N.E.Schore, Synth.Commun., 9,(1979) 41.

190 D.C.Billington and P.L.Pauson, Organometallics, 1,(1982) 1560.

191 D.C.Billington, Tetrahedron Lett., (1983) 2905.

192 J.E.Lyons, H.K.Myers and A.Schneider, J.Chem.Soc.Chem. Commun., (1978) 636.

193 Y.Wakatsuki and H.Yamazaki, Synthesis, (1976) 26.

194 H.Bönnemann, W.Brijoux, R.Brinkmann, W.Meurers, R.Mynott, W. von Philipsborn and T.Eglof, J.Organomet.Chem., 272, (1984) 231.

195 Y.Wakatsuki and H.Yamazaki, J.Chem.Soc.Chem.Commun., (1973) 280.

196 D.Seyferth and A.Wehman, J.Am.Chem.Soc., 92,(1970) 5520.

197 K.M.Nicholas and R.Pettit, Tetrahedron Lett., (1971) 3475.

198 K.M.Nicholas and R.Pettit, J.Organomet.Chem., 44,(1972) C21.

199 R.E.Connor and K.M.Nicholas, J.Organomet.Chem., 125,(1977) C45.

200 R.F.Lockwood and K.M.Nicholas, Tetrahedron Lett., (1977) 4163.

201 J.E.O'Boyle and K.M.Nicholas, Tetrahedron Lett. (1980) 1595.

202 D.H.Hodes and K.M.Nicholas, Tetrahedron Lett. (1978) 4349.

203 K.M.Nicholas, M.Mulvaney and M.Bayer, J.Am.Chem.Soc., 102, (1980) 2508.

204 S.Padmanabhan and K.M.Nicholas, Tetrahedron Lett., (1983) 2239.

205 M.Saha and K.M.Nicholas, J.Org.Chem., 49, (1984) 417.

206 S.Padmanabhan and K.M.Nicholas, Synth.Commun., 10 (1980) 503.

207 S.Padmanabhan and K.M.Nicholas, J.Organomet.Chem., 212, (1981) 115.

208 C.Battistini, P.Crotti and F.Machia, J.Org.Chem., 46 (1981) 434.

209 D.Seyferth, J.E.Hallgren and C.S.Eshbach, J.Am.Chem.Soc., 96,(1974) 1730.

210 This was suggested on the basis of spectroscopic data [211] and was confirmed later by several independent sources [212].

211 G.Pályi, F.Piacenti and L.Markó, Inorg.Chim.Acta.Rev. 4, (1970) 109.

212 See for example: (a) P.T.Chesky and M.B.Hall, Inorg.Chem., 20,(1981) 4419; (b) G.Granozzi, E.Tondello, D.Ajo, M.Casarin, S.Aime and D.Osella, Inorg.Chem., 21,(1982) 1081; and refs. cited in these works.

213 A.Messeguer and F.Serratosa, Tetrahedron Lett., (1973) 2895.

214 A.S.Foust, M.S.Foster and L.F.Dahl, J.Am.Chem.Soc., 91,(1969) 5631.

215 A.Vizi-Orosz, G.Pályi and L.Markó, J.Organomet.Chem., 60, (1973) C25.

216 (a) D.Seyferth, J.S.Merola and R.S.Henderson, Organometallics, 1, (1982) 859; (b) I.C.T.R.Burkett-St.Laurent, P.T.Hitchcock, H.V.Croto and J.F.Nixon, J.Chem.Soc.Chem.Commun. (1981) 1141.

SUBJECT INDEX

(grouped by chapter)

Chapter 1
Stereochemistry of 1,3-Diene Complexes And the Steric Course of Their Reactions

Chromium Compounds

Cobalt Compounds

Coordination Modes for Dienes

Fluxionality in Metal Dienes

Geometries Around the Metals of Dienes

Hafnium Compounds

Iridium Compounds

Iron Compounds

Magnesium Compounds

Manganese Compounds

Molecular Structures

Molybdenum Compounds

Nickel Compounds

Niobium Compounds

Osmium Compounds

Palladium Compounds

Platinum Compounds

Rhodium Compounds

Ruthenium Compounds

Structural Correlations in Metal Dienes

Tantalum Compounds

Thermodynamic Information

Titanium Compounds

Tungsten Compounds

Vanadium Compounds

Zirconium Compounds

Chapter 2
Stereochemistry of the Phosphates of Divalent Metals

Abbreviations Used in Text

Metaphosphates by Formulae

Polymetaphosphates by Formulae

Phosphates with Given Names

Ultraphosphates

Ultraphosphates by Formulae

Chapter 3
Transition Metal Complexes With Carbon Disulfide Correlations Between Stereochemistry and Reactivity

Carbon Disulfide Bonding Modes

Chromium Compounds

Cobalt Compounds

Dipolar Character of CS_2 Ligands

Dithiocarbenes and Dithiolenes

Infrared Spectra

Iridium Compounds

Iron Compounds

Manganese Compounds

Nickel Compounds

Niobium Compounds

Osmium Compounds

Palladium Compounds

Platinum Compounds

Rhenium Compounds

Rhodium Compounds

Ruthenium Compounds

Structural Data

Tungsten Compounds

Vanadium Compounds

Chapter 4

Stereochemistry of the Bailar Inversion and Related Metal Ion Substitution Reactions

Asymmetric Centers

Bailar Inversion Reaction

Base Hydrolysis

Chromium Compounds

Cobalt Compounds

Configurational Features in Octahedral Compounds

Iridium Compounds

Iron Compounds

Coordination Compounds of Iron

Organometallic Compounds of Iron

Leaving Groups in Hydrolysis of cis- and trans-Co Compounds

Ligand Names and Their Abbreviations

Manganese Compounds(Organometallics)

Mercury Compounds in the Reactions of Co amines

Molybdenum Compounds(Organometallics)

Net Optical Inversion Path in

Optical Activity Effects in

Coordination Compounds

Organometallic Compounds

Rhodium Compounds

Solvents Other Than Water in Bailar Inversion and Double Substitution

Stereochemistry of the Aquation Reaction

Walden Inversion

Chapter 5

Stereochemistry of Acetylenes Coordinated to Cobalt